Crop Improvement

Siti Nor Akmar Abdullah • Ho Chai-Ling
Carol Wagstaff
Editors

Crop Improvement

Sustainability Through Leading-Edge Technology

Springer

Editors
Siti Nor Akmar Abdullah
Institute of Plantation Studies
Universiti Putra Malaysia
Serdang, Selangor, Malaysia

Ho Chai-Ling
Institute of Plantation Studies
Universiti Putra Malaysia
Serdang, Selangor, Malaysia

Carol Wagstaff
Food and Nutritional Sciences
University of Reading Food
and Nutritional Sciences
Reading, Berkshire, UK

ISBN 978-3-319-65078-4 ISBN 978-3-319-65079-1 (eBook)
DOI 10.1007/978-3-319-65079-1

Library of Congress Control Number: 2017953795

Printed on acid-free paper

This Springer imprint is published by Springer Nature
The registered company is Springer International Publishing AG
The registered company address is: Gewerbestrasse 11, 6330 Cham, Switzerland

Foreword

It is a great pleasure to introduce this important book: *Crop Improvement: Sustainability Through Leading-Edge Technology*. The book is published at a particularly timely moment in our history when new crop improvement technologies are being developed at the same time as Codon Genomics the world faces formidable challenges associated with population growth, climatic change, environmental deterioration and resource depletion. The impressive cast of editors and authors are well qualified to produce a book on such an important topic. Two editors, Professors Siti Nor Akmar Abdullah and Ho Chai-Ling, are from Universiti Putra Malaysia, which is an emerging powerhouse of crop and environmental research in Southeast Asia, while the third editor, Professor Carol Wagstaff, is from the University of Reading, one of the oldest centres of agricultural research in Europe.

The target audience of the book includes scientists and students around the world with interest in this crucial area that seeks to tackle the imperative of increasing crop quality and yield while also ensuring that such crops are grown and processed in the most sustainable manner possible. As demonstrated in the individual chapters of the book, sustainable crop improvement requires the collaboration of all relevant stakeholders including farmers, processers, academic researchers, governments, private sector agriculture-related companies and the constructive engagement of NGOs.

One of the most important advances of the twenty-first century has been the development of DNA-based technologies for the understanding and manipulation of key agronomic traits in crops. Such technologies include DNA marker assisted selection for crop breeding which is already having such a positive impact on some of the major commodity crops that provide the bulk of edible calories for global populations. The past decade has also witnessed a revolution in the application of genomics (including the related methods of transcriptomics, proteomics, metabolomics, phenomics, etc.) to an increasingly wide range of crops. The process that started with the sequencing of the small genome of the model plant, *Arabidopsis thaliana*, has now progressed to the sequencing of very large and complex genomes of such key crops as wheat, barley, maize and oil palm. Other biotechnologies are being applied to the surveillance and detection of new crop diseases that are increasingly emerging as climates change and international travel helps their spread to new regions.

One of the mantras of modern crop improvement is 'sustainable intensification', which involves the application of all forms of breeding technology, both old and new, to the betterment of crop (and especially food) production—but in a way that minimises adverse environmental impacts of agriculture. For example, genetic technologies are enabling us to breed crops that can synthesise their own biologically based defence compounds and thereby reduce or eliminate the need for chemical inputs such as fungicides, herbicides and pesticides. In other cases, better management of crops, including the use of more of their by-products, can improve both profitability and efficiency of production as well as reduce the enormous amount of wastage that continues to affect the food industry.

The most serious threats to global food security in the coming decades will be felt in developing countries, particularly in sub-Saharan Africa and Asia. As this book shows, scientists in these regions are increasingly taking the lead in the development and application of new methods and approaches to sustainable crop improvement. There are still many hurdles to climb but it is indeed encouraging to see how new generations of researchers are tackling these formidable challenges.

The University of South Wales, UK Denis J. Murphy

Preface

The chapters in this book are mainly from selected oral presentations at the International Conference on Crop Improvement held in Universiti Putra Malaysia in December, 2015. Additional chapters are from authors that were invited to contribute certain topics to strengthen the content of the book. The theme of the conference 'Sustainability Through Leading-Edge Technology' used as the book title is a clear reflection of the content. The book aims to enlighten the readers on various strategies utilising molecular-biotechnology tools and agricultural engineering innovations to overcome the challenges in crop production in a sustainable manner. Sustainability issues are addressed along the supply chain from crop production to preserving and processing crops after harvest.

Major impacts on crop production due to catastrophic diseases and global climate change need urgent and innovative solutions. One of the main issues is the limited availability of gene pool due to domestication that prevents further genetic improvement. Sourcing for new alleles from various untapped genetic resources for important traits like disease and abiotic stress resistance is critical and appropriately presented as one of the key strategies in current crop improvement efforts. Systems biology that relies on various 'omics platform and strong bioinformatics support is essential in understanding biological processes and the underlying mechanisms of importance for enhancing crops' performance and ensuring their survival under adverse conditions. The information generated forms the basis for subsequent genetic improvement efforts through molecular marker technology and genetic engineering. Some of the powerful genetic manipulation strategies as well as recent development that speed up the process of obtaining the desired trait with better public acceptance are elaborated.

The book also includes aspects of preserving crops after harvest and diversifying usage of by-products as these are key factors in promoting sustainable crop quality, addressing waste, and moving crops through the food and industrial supply chain. Agriculture engineering innovations in the form of biosensors for field application, bioreactors for culturing of plants for production of valuable secondary metabolites and crop processing technology to minimise loss as important development towards attaining sustainability are also included as food for thoughts for current and future generation.

We take this opportunity to thank all authors for accepting our invitation and their contribution in making this book a reality. Their continuous cooperation and commitment during the editorial process are much appreciated. We are also thankful to Prof. Denis Murphy for constructive comments and meaningful foreword for this book. We really appreciate the strong support of the current UPM's Vice Chancellor, Prof. Datin Paduka Dr. Aini Ideris, in promoting and encouraging publication of books as one of the best medium to reach out to the global community.

We are grateful to Springer International team for their support and valuable advices at every stage of the book production.

Selangor, Malaysia — Siti Nor Akmar Abdullah
Selangor, Malaysia — Ho Chai-Ling
Reading, Berkshire, UK — Carol Wagstaff

Contents

1 **RNA-seq Analysis in Plant–Fungus Interactions** 1
Ho Chai-Ling and Wai-Yan Yee

2 **Study of Oil Palm Photosynthesis Using Omics Technologies** 27
Bee Keat Neoh, Huey Fang Teh, Yick Ching Wong,
Tony Eng Keong Ooi, See Siang Cheah, and David R. Appleton

3 **Molecular Marker Technology for Genetic Improvement of Underutilised Crops** 47
Acga Cheng, Hui Hui Chai, Wai Kuan Ho,
Aliyu Siise Abdullah Bamba, Aryo Feldman, Presidor Kendabie,
Razlin Azman Halim, Alberto Tanzi, Sean Mayes,
and Festo Massawe

4 **Transcription Factors Associated with Abiotic Stress and Fruit Development in Oil Palm** 71
Siti Nor Akmar Abdullah, Azzreena Mohamad Azzeme,
Mortaza Ebrahimi, Engku Ahmad Khairi Engku Ariff,
and Farah Hanan Abu Hanifiah

5 **Sensing Crop Diseases** 101
Thangavel Lakshmipriya, Subash C.B. Gopinath, and Uda Hashim

6 **Development of Genomic Resources and Assessing Their Potential for Accelerated *Acacia* Breeding** 117
Wikneswari Ratnam, Chee Yen Choong, and Muhammad Asif Javed

7 **Engineering of Secondary Metabolites in Tissue and Cell Culture of Medicinal Plants: An Alternative to Produce Beneficial Compounds Using Bioreactor Technologies** 137
Mortaza Ebrahimi and Arash Mokhtari

8 **Bioreactor Technology for Sustainable Production of Valuable Plant Metabolites: Challenges and Advances** 169
Morteza Khanahmadi and Kee Yoep Paek

9 **Cisgenesis and Intragenesis as New Strategies for Crop Improvement** 191
Mahdi Moradpour and Siti Nor Akmar Abdullah

10 **Use of dsRNA in Crop Improvement** 217
Lau Su Ee, Purabi Mazumdar, Pooja Singh, and Jennifer Ann Harikrishna

11 **Alternative Strategy in Crop Protection: Protease Inhibitors from Turmeric** 253
Chan Seow-Neng, Norliza Abu Bakar, Maziah Mahmood, Ho Chai-Ling, and Noor Azmi Shaharuddin

12 **Sustainable Supply Chains: Utilisation of Whole Crops for New Product Development** 271
Rebecca Smith

13 **Food Losses in Rice Milling** 287
Ye Aung

14 **Elicitors and Their Roles in Plant Defence Against Pathogens Particularly Basidiomycetes** 305
Sathyapriya Hamid and Mui-Yun Wong

15 **Induction of Systemic Resistance for Disease Suppression** 335
Kalaivani K. Nadarajah

16 **Tools and Targeted Genes for Plant Disease Detection** 359
Umaiyal Munusamy and Siti Nor Akmar Abdullah

Index 381

Contributors

Siti Nor Akmar Abdullah Institute of Plantation Studies, Universiti Putra Malaysia, Serdang, Selangor, Malaysia

David R. Appleton Sime Darby Technology Centre Sdn Bhd, UPM-MTDC Technology Centre III, Universiti Putra Malaysia, Serdang, Selangor, Malaysia

Engku Ahmad Khairi Engku Ariff Department of Agriculture Technology, Faculty of Agriculture, Universiti Putra Malaysia, Serdang, Selangor, Malaysia

Ye Aung Buhler Asia Pacific Ltd, Yangon, Myanmar

Azzreena Mohamad Azzeme Institute of Plantation Studies, Universiti Putra Malaysia, Serdang, Selangor, Malaysia

Department of Biochemistry, Faculty of Biotechnology and Biomolecular Sciences, Universiti Putra Malaysia, Serdang, Selangor, Malaysia

Norliza Abu Bakar Biotechnology and Nanotechnology Research Center, Malaysian Agricultural Research and Development Institute (MARDI), Serdang, Malaysia

Aliyu Siise Abdullah Bamba Biotechnology Research Centre, School of Biosciences, The University of Nottingham Malaysia Campus, Semenyih, Selangor Darul Ehsan, Malaysia

Crops For the Future, Semenyih, Selangor Darul Ehsan, Malaysia

CSIR-Savannah Agricultural Research Institute, Nyankpala, Tolon District, N/R, Ghana

Ho Chai-Ling Institute of Plantation Studies, Universiti Putra Malaysia, Serdang, Selangor, Malaysia

Department of Cell and Molecular Biology, Faculty of Biotechnology and Biomolecular Sciences, Universiti Putra Malaysia, Serdang, Selangor, Malaysia

See Siang Cheah Plantation Research Advisory, Sime Darby Research, Pulau Carey, Selangor, Malaysia

Acga Cheng Biotechnology Research Centre, School of Biosciences, The University of Nottingham Malaysia Campus, Semenyih, Selangor Darul Ehsan, Malaysia

Institute of Biological Sciences, Faculty of Science, University of Malaya, Kuala Lumpur, Malaysia

Chee Yen Choong Faculty of Science and Technology, Universiti Kebangsaan Malaysia, Bangi, Selangor, Malaysia

Mortaza Ebrahimi Agricultural Biotechnology Research Institute of Iran-Isfahan Branch, Isfahan, Iran

Lau Su Ee Centre for Research in Biotechnology for Agriculture, University of Malaya, Kuala Lumpur, Malaysia

Aryo Feldman Crops For the Future, Semenyih, Selangor Darul Ehsan, Malaysia

Subash C.B. Gopinath Institute of Nano Electronic Engineering, Universiti Malaysia Perlis, Kangar, Perlis, Malaysia

School of Bioprocess Engineering, Universiti Malaysia Perlis, Arau, Perlis, Malaysia

Razlin Azman Halim Crops For the Future, Semenyih, Selangor Darul Ehsan, Malaysia

Sathyapriya Hamid Laboratory of Plantation Science and Technology, Institute of Plantation Studies, Universiti Putra Malaysia, Serdang, Selangor, Malaysia

Farah Hanan Abu Hanifiah Institute of Plantation Studies, Universiti Putra Malaysia, Serdang, Selangor, Malaysia

Jennifer Ann Harikrishna Centre for Research in Biotechnology for Agriculture, University of Malaya, Kuala Lumpur, Malaysia

Institute of Biological Sciences, Faculty of Science, University of Malaya, Kuala Lumpur, Malaysia

Uda Hashim Institute of Nano Electronic Engineering, Universiti Malaysia Perlis, Kangar, Perlis, Malaysia

School of Bioprocess Engineering, Universiti Malaysia Perlis, Arau, Perlis, Malaysia

Hui Hui Chai Biotechnology Research Centre, School of Biosciences, The University of Nottingham Malaysia Campus, Semenyih, Selangor Darul Ehsan, Malaysia

Muhammad Asif Javed Faculty of Forestry and Range Management, University of Agriculture Faisalabad, Faisalabad, Pakistan

Presidor Kendabie Plant and Crop Sciences, School of Biosciences, The University of Nottingham, Sutton Bonington Campus, Leicestershire, Loughborough, UK

Morteza Khanahmadi Agricultural Biotechnology Research Institute of Iran-Isfahan Branch, Agricultural Research, Education and Extension Organization (AREEO), Isfahan, Iran

Wai Kuan Ho Biotechnology Research Centre, School of Biosciences, The University of Nottingham Malaysia Campus, Semenyih, Selangor Darul Ehsan, Malaysia

Crops For the Future, Semenyih, Selangor Darul Ehsan, Malaysia

Thangavel Lakshmipriya Institute of Nano Electronic Engineering, Universiti Malaysia Perlis, Kangar, Perlis, Malaysia

School of Bioprocess Engineering, Universiti Malaysia Perlis, Arau, Perlis, Malaysia

Maziah Mahmood Faculty of Biotechnology and Biomolecular Sciences, Department of Biochemistry, Universiti Putra Malaysia (UPM), Serdang, Malaysia

Festo Massawe Biotechnology Research Centre, School of Biosciences, The University of Nottingham Malaysia Campus, Semenyih, Selangor Darul Ehsan, Malaysia

Crops For the Future, Semenyih, Selangor Darul Ehsan, Malaysia

Sean Mayes Crops For the Future, Semenyih, Selangor Darul Ehsan, Malaysia

Plant and Crop Sciences, School of Biosciences, The University of Nottingham, Sutton Bonington Campus, Leicestershire, Loughborough, UK

Purabi Mazumdar Centre for Research in Biotechnology for Agriculture, University of Malaya, Kuala Lumpur, Malaysia

Arash Mokhtari Agricultural Biotechnology Research Institute of Iran-Isfahan Branch, Agricultural Research, Education and Extension Organization (AREEO), Tehran, Iran

Mahdi Moradpour Institute of Plantation Studies, Universiti Putra Malaysia, Serdang, Selangor, Malaysia

Umaiyal Munusamy Institute of Plantation Studies, Universiti Putra Malaysia, Serdang, Selangor, Malaysia

Kalaivani K. Nadarajah Faculty of Science and Technology, School of Environmental and Natural Resource Sciences, National University of Malaysia, Bangi, Selangor, Malaysia

Bee Keat Neoh Sime Darby Technology Centre Sdn Bhd, UPM-MTDC Technology Centre III, Universiti Putra Malaysia, Serdang, Selangor, Malaysia

Tony Eng Keong Ooi Sime Darby Technology Centre Sdn Bhd, UPM-MTDC Technology Centre III, Universiti Putra Malaysia, Serdang, Selangor, Malaysia

Kee Yoep Paek Department of Horticultural Science, Chungbuk National University, Cheongju, South Korea

Wikneswari Ratnam Faculty of Science and Technology, Universiti Kebangsaan Malaysia, Bangi, Selangor, Malaysia

Chan Seow-Neng Faculty of Biotechnology and Biomolecular Sciences, Department of Biochemistry, Universiti Putra Malaysia (UPM), Serdang, Malaysia

Noor Azmi Shaharuddin Faculty of Biotechnology and Biomolecular Sciences, Department of Biochemistry, Universiti Putra Malaysia (UPM), Serdang, Malaysia

Pooja Singh Centre for Research in Biotechnology for Agriculture, University of Malaya, Kuala Lumpur, Malaysia

Rebecca Smith Department of Food and Nutritional Sciences, University of Reading, Reading, UK

Valley Produce, Springalls Farm, Swallowfield, Reading, UK

Alberto Tanzi Biotechnology Research Centre, School of Biosciences, The University of Nottingham Malaysia Campus, Semenyih, Selangor Darul Ehsan, Malaysia

Crops For the Future, Semenyih, Selangor Darul Ehsan, Malaysia

Huey Fang Teh Sime Darby Technology Centre Sdn Bhd, UPM-MTDC Technology Centre III, Universiti Putra Malaysia, Serdang, Selangor, Malaysia

Mui-Yun Wong Laboratory of Plantation Science and Technology, Institute of Plantation Studies, Universiti Putra Malaysia, Serdang, Selangor, Malaysia

Department of Plant Protection, Faculty of Agriculture, Universiti Putra Malaysia, Serdang, Selangor, Malaysia

Yick Ching Wong Sime Darby Technology Centre Sdn Bhd, UPM-MTDC Technology Centre III, Universiti Putra Malaysia, Serdang, Selangor, Malaysia

Wai-Yan Yee Codon Genomics S/B, Seri Kembangan, Selangor, Malaysia

About the Book

Technology and innovations go hand in hand with agricultural sustainability. New technologies provide rapid and accurate analytical data vital for monitoring crop status and environmental impact. Current emphasis on systems biology, on-site and real-time data acquisition enables targeted improvement of crops and site directed solutions that are effective, economical and labour saving. Continuous input of technology will sustain global crop contribution for food, feed and industrial materials. This book addresses inherent issues in crop improvement. Molecular-biotechnology tools together with innovations in agricultural engineering provide new insights on the biological system and solutions to the various problems.

About the Editors

Siti Nor Akmar Abdullah is currently the Director of the Institute of Plantation Studies and a Professor in Plant Molecular Biology at the Faculty of Agriculture, Universiti Putra Malaysia (UPM). Her current research focuses on functional genomics as well as genetic manipulation and biomarker technology for yield and oil quality improvement, enhancing nutrient uptake and disease tolerance in oil palm. She has led several research projects at the national level and secured research grants from oil palm industry members. She has more than 80 papers in citation-indexed journals and 5 granted patents on oil palm genetic manipulation tools and molecular disease diagnosis. Twenty-five Ph.D. students have graduated under her as the main or co-supervisor. She has delivered several talks as an invited speaker in various countries like Australia, China, India and Indonesia and also served as a reviewer for many international peer-reviewed journals. She has led national and international conferences and workshops on plant biotechnology and molecular biology in collaboration with Wageningen University, University of Reading and the University of Nottingham. She has established an internationally certified core research facility for functional genomics and genetic engineering in UPM.

Ho Chai-Ling is a Professor in Plant Molecular Biology at the Faculty of Biotechnology and Biomolecular Sciences, Universiti Putra Malaysia (UPM). She is also a research associate at the Institute of Plantation Studies, UPM. She obtained her Ph.D. from Chiba University, Japan, unravelling the molecular biology of the phosphorylated serine biosynthetic pathway. She was the recipient of several scholarships/fellowships including the MONBUSHO Scholarship and Fulbright Visiting Scholarship. Her current research interests include molecular interactions of oil palm and Ganoderma, transcriptional regulation of rice grain filling process and molecular characterisation of transcripts related to seaweed agar biosynthesis. Dr. Ho has published more than 100 papers in citation-indexed journals including *Trends in Plant Sciences*, *Plant Journal*, *Journal of Biological Chemistry* and *BMC Genomics*. She has supervised 6 Ph.D. and 20 M.Sc. students. Dr. Ho has served as assessor for many grant applications and reviewer for many international peer-reviewed journals. She also serves in the organising committee of several national and international conferences and workshops on plant biotechnology, phycology, functional genomics and structural biology.

Carol Wagstaff is presently Professor of Crop Quality for Health in the Department of Food and Nutritional Sciences at the University of Reading and is the Director of the AgriFood Training Partnership. Her research team works on a range of projects spanning the fresh produce supply chain with the overall aim of improving nutrition, flavour and shelf life of fruit and vegetable crops, thereby encouraging consumption of a healthy diet and reducing waste. The research group is best known for its expertise in combining metabolite profiling with a genetic approach to crop improvement, but also has projects based in agronomy, supply chain modelling, innovative processing design and molecular regulation. Funding is currently from BBSRC, Innovate UK, international government schemes and a significant portion is from industry partners who represent crop breeders and processors.

Chapter 1
RNA-seq Analysis in Plant–Fungus Interactions

Ho Chai-Ling and Wai-Yan Yee

1.1 Introduction

Global food supply is dependent on about 150 plant sources (International Development Research Centre, IDRC; https://www.idrc.ca/). Of these, rice, wheat, and maize provide more than 50% of the world's food energy while sorghum, millet, potatoes, sweet potatoes, soybean, and sugar (beet and cane) provide another 25% (Food and Agriculture Organization of United Nations, FAO; http://www.fao.org/).

Plant crops interact with some microorganisms in a mutually beneficial manner. There are also pathogenic microbes that colonize plant hosts for food and to complete their reproduction cycle, thus reducing both crop yield and quality, if not destroying the crops. Most interactions between plants and microbes occur in the rhizosphere, thus many microbes can also indirectly affect plant growth and crop production by modifying the soil conditions. Hence, understanding the complex nature of plant–microbe interactions can potentially offer new strategies to enhance plant productivity and also to enhance plant health by reducing plant diseases. The emerging area of multispecies transcriptomics is promising in providing a comprehensive understanding of the complex interactions of plants with their biotic and abiotic environments.

H. Chai-Ling (✉)
Institute of Plantation Studies, Universiti Putra Malaysia,
UPM, 43400 Serdang, Selangor, Malaysia

Department of Cell and Molecular Biology, Faculty of Biotechnology and Biomolecular Sciences, Universiti Putra Malaysia, UPM, 43400 Serdang, Selangor, Malaysia
e-mail: clho@upm.edu.my

W.-Y. Yee
Codon Genomics S/B, 26, Jalan Dutamas 7, Taman Dutamas Balakong,
43200 Seri Kembangan, Selangor, Malaysia

S.N.A. Abdullah et al. (eds.), *Crop Improvement*,
DOI 10.1007/978-3-319-65079-1_1

1.2 Plant–Fungus Interactions

Arbuscular mycorrhiza (AM) originated 400 million years ago, coincident with land colonization by plants (Pirozynski and Malloch 1975; Simon et al. 1993; Remy et al. 1994; Redecker et al. 2000; Heckman et al. 2001). AM symbiosis was proposed to be ancestral not only to other plant–fungus interactions but also plant–microbe interactions (Kistner and Parniske 2002). The symbiotic mechanisms between legumes, and eukaryotic AM fungi and prokaryotic *Rhizobi,* were reported to be well conserved (Endre et al. 2002; Stracke et al. 2002; Ané et al. 2004; Imaizumi-Anraku et al. 2004; Lévy et al. 2004). The same signal-transduction components shared by both bacterial and fungal root symbioses were also recruited by pathogenic root-knot nematodes (Weerasinghe et al. 2005).

1.2.1 Importance

Plant–fungus interactions can be broadly divided into beneficial, pathogenic, or neutral interactions (Mora et al. 2011). The potential outcome of these plant–fungus interactions was shown to be affected by environmental factors such as drought, temperature, salinity, soil acidity, and waterlogging (Berg 2009). Under multiple abiotic and biotic stresses, prioritization of physiological pathways in plants can determine the outcome of the plant–microbe interactions. In the last few decades, transcriptome analysis has fueled a better understanding of host–pathogen interactions and the underlying molecular mechanisms. The prioritization of physiological pathways in plants was shown to be hormonally regulated (Memelink 2009; Depuyd and Hardtke 2011). The interplay of jasmonic acid (JA), salicylic acid (SA), and ethylene was found to affect the resistance and susceptibility of plants to pathogens. SA can increase the resistance of plant host to hemibiotrophs but promotes the susceptibility to necrotrophs (Veronese et al. 2004, 2006) while JA and ethylene play a role in the resistance of plant hosts to necrotrophs (Glazebrook 2005; Laluk et al. 2011).

Plants undergoing abiotic stresses or treated with SA have increased production of reactive oxygen species (ROS) and often display higher resistance against biotrophs (Bechtold et al. 2005; Kunz et al. 2008). By contrast, JA signaling is associated with wounding and physical damage which are of the advantage of necrotrophs. In addition, JA is also involved in induced systemic resistance (ISR) which provides resistance against subsequent infections (Pieterse et al. 2009; Matilla et al. 2010).

1.2.1.1 Beneficial Plant–Fungus Interactions

Beneficial fungi promote plant growth and/or suppress plant diseases via a variety of mechanisms, which include improved nutrient acquisition, production of growth regulators, and biosynthesis of pathogen-inhibiting compounds. Biological control of pathogens can also happen through direct interaction between beneficial and pathogenic fungi through competition for living space and mycoparasitism as demonstrated by coiling of hyphae around pathogen by *Trichoderma* spp. (Benitez et al. 2004). Some beneficial fungi can trigger induced systemic resistance (ISR), a type of priming that provides resistance against subsequent infections and protect the host from diseases (Pieterse et al. 2009; Matilla et al. 2010). Basal defenses could be triggered in host plants in their interactions with beneficial fungi as with pathogenic fungi. Similar expression was reported for a set of rice genes in symbiotic and pathogenic interactions, respectively, thus suggesting a conserved response to fungal colonization which may play a role in compatibility (Güimil et al. 2005). In multispecies interactions that involve *Trichoderma-Arabidopsis-Pseudomonas syringae* and *Piriformospora indica*-barley-powdery mildew, the beneficial fungi (e.g., *Trichoderma* and *Piriformospora*) were shown to modulate host gene expression profiles only in the presence of pathogens (*P. syringae* or *Blumeria graminis* f. sp. *hordei*, respectively) (Molitor et al. 2011; Brotman et al. 2012).

1.2.1.2 Pathogenic Plant–Fungus Interactions

In pathogenic plant–fungus interactions, the fungi are detrimental to the hosts. Pathogenic fungi can cause diseases on plants which not only reduce both yield and quality, but also kill the crop. Generally pathogenic fungi can be divided into three classes based on their feeding modes: biotrophs, necrotrophs, and hemibiotrophs. Biotrophs live on intact living plant tissue while necrotrophs, by contrast, would kill and benefit from dead plant cells.

The infection processes, the nature of effector proteins, and the elicited host defense responses are different in biotrophs and necrotrophs (Laluk and Mengiste 2010). Biotrophs establish a long-term feeding relationship with their living hosts by growing between host cells and invade only few host cells. They may secrete a limited amount of cell wall degrading enzymes and generally do not produce phytotoxic compounds (Mendgen and Hahn 2002). The plant immune responses of biotrophs include a hypersensitive response which causes cell death at the site of infection, thus confining the spread of a pathogen by abolishing its nutrient supply. Conversely, necrotrophs induce cell death and necrosis by secreting phytotoxic compounds and cell wall degrading enzymes (Govrin and Levine 2000). Many necrotrophs are actually hemibiotrophs which start as biotrophs and turn necrotrophs later.

In gene-for-gene resistance, successful pathogen has the ability to suppress the host defense responses that it triggers in a compatible host–pathogen interaction. However, the presence of resistance (R) genes enables plants to reverse the suppression of basal defenses and induce potent and effective defenses. The genes that are involved in R gene-mediated defenses in compatible interactions were found to be similar or identical to those induced during the basal defenses in incompatible interactions (Wise et al. 2007).

1.2.1.3 Neutral Plant–Fungus Interactions

In neutral plant–fungus interactions, the fungi may not be beneficial or detrimental to the plants directly. Since they utilize plant-derived organic compounds as substrates for energy production, they may still play important roles in nutrient cycling and modifying plant environments. For instance, saprophytic fungi break down organic material and the minerals released are made available for plant uptake (Kubartová et al. 2009). Another example is the extrusion of protons and/or the production of organic acids that cause a drop in soil pH. This leads to solubilization of phosphates in precipitate form into the soil solution, which is readily available for uptake by roots (Scervino et al. 2010).

1.2.2 Limitations in Transcriptome Data of Plant–Fungus Interactions

In the past, the gene expression patterns of plants and fungi have been investigated in many plant–fungus interactions (Güimil et al. 2005; Rotter et al. 2009; Samolski et al. 2009; Chen et al. 2014). However, most of these studies focus on either plant or fungus, and only a few reports have profiled the transcriptomes of both partners simultaneously (Schenk et al. 2012). Studies on multispecies transcriptomes are important in revealing the tightly linked gene expression of plants and fungi involved in symbiotic beneficial or parasitic relationships.

In the early days, most transcriptome data of plant–fungus interactions were generated by microarray approach. One major limitation of microarray analysis is that the number and sequence of DNA probes available on the array determine the transcripts that can be analyzed. Although many plant and fungal genomes have been completely sequenced in recent years, the sequence data that are currently available are still limited for only a small number of plant and fungal species. Despite that cross-species DNA microarray hybridization was possible, the DNA probes available on microarray are unlikely to account for all the genetic diversity found in natural communities, and may need to be updated regularly. In addition, output inaccuracy due to cross-hybridization and low sensitivity is still a challenge for microarray analysis that requires higher specificity in oligonucleotide design.

1.3 RNA-seq Analysis in Plant–Fungus Interactions

In recent years, advances in RNA-seq which involves next-generation sequencing (NGS) of mRNA-derived cDNA have offered an alternative to DNA microarray for transcriptome profiling of plant–fungal interactions. In the RNA-seq approach, the availability of genome data is not a prerequisite. The quantitative measurement of gene expression is based on the number of reads obtained for each cDNA (including low-abundance transcripts). In addition, RNA-seq could be a better option for multispecies transcriptome profiling wherein the reads generated from a pool of mixed cDNA from multispecies can be accurately assigned to genes from individual species. Although this technology has its own share of limitations, mainly due to sequencing costs, the demand for high computational power, and challenges in data analysis, RNA-seq is able to provide a higher sequence depth, thus higher accuracy and sensitivity compared to DNA microarray approach.

1.3.1 Principles of RNA-seq

RNA-seq is a recently developed approach that utilizes the deep sequencing platforms provided by Next-Generation Sequencing (NGS) technologies to profile the transcriptome. In general, a population of RNA will be reverse transcribed into a library of cDNA fragments with adaptors attached to one and/or both ends (Wang et al. 2009). The cDNA fragments will be selected based on size, amplified and sequenced to obtain a high number of single-end or paired-end short sequencing reads. The sequencing read length (~30–400 bp) generated depends on the sequencing platforms used. The NGS sequencing platforms include Roche 454 (Thomas et al. 2006; Sugarbaker et al. 2008), Life Technologies SOLiD (Pandey et al. 2008; Su et al. 2014), Illumina (Bentley et al. 2008; Mortazavi et al. 2008), and Life Technologies Ion Torrent (Rothberg et al. 2011; Mangul et al. 2014). However, Roche 454 and Life Technologies SOLiD platforms have been discontinued from market and development.

Currently, the Illumina HiSeq sequencing platform is the most frequently used NGS platform in RNA-seq analysis. This technology is based on simultaneous/parallel multiple individual sequencing-by-synthesis reactions on a miniature slide. The Illumina platform is popular for gene expression studies given its low cost for the coverage provided. The latest platform which is commonly used is the HiSeq 3000/4000 which can generate up to 750 Gb throughput of data per flow cell corresponding to 2.5 billion single-end reads in approximately 3.5 days (http://www.illumina.com/content/dam/illumina-marketing/documents/products/datasheets/hiseq-3000-4000-specification-sheet-770-2014-057.pdf; Goodwin et al. 2016). This is equivalent to 50 whole transcriptome samples per flow cell in fewer than 3.5 days (with an assumption of 50 million reads per sample for HiSeq 3000). The read lengths for the latest HiSeq 3000/4000 model can reach a maximum of 2 × 150 bp (paired-end sequence of 150-bp reads each). The development of sequencing platforms with much higher coverage, coupled with replicate profiling and statistical analyses, holds the promise of a more accurate representation of expression profiles involved in plant–microbe interactions.

For a comprehensive understanding of plant–microbe interactions, it is valuable to analyze the associated gene expression changes in both the interacting organism simultaneously (Kawahara et al. 2012; Westermann et al. 2012). In principle, the

RNA-seq allows the host and microbe transcriptomes to be analyzed in parallel. There are numerous applications where RNA-seq can be utilized in answering a biological question, e.g., infected vs. noninfected systems and time-course studies. However, to have a general or optimal pipeline that can fit all purposes or suitable for all organisms is not possible (Conesa et al. 2016). In this paper, we do not aim to provide an exhaustive survey of the RNA-seq analysis, but rather to provide an introduction to the bioinformatics approach used in plant–fungus interaction, which enables the readers to gain familiarity with basic bioinformatics analysis steps.

The major steps for a typical RNA-seq study involve filtering for high quality sequencing reads, read alignment, quantifying gene and transcript expression, and identification of differential gene expression between samples (Conesa et al. 2016). Firstly, the quality of all sequencing raw reads generated from NGS platforms can be assessed in the initial pre-processing step involving determination of base quality score and GC content, and removal of ambiguous base "N" in sequences and over represented sequences (possible adapter contamination); using FastQC (http://www.bioinformatics.babraham.ac.uk/ projects/fastqc/). Low quality reads/sequences (based on base quality, i.e., Phred cutoff value, and minimum fragment length), adapter sequences and ambiguous "N" bases can be trimmed and/or filtered by FASTX-Toolkit (http://hannonlab.cshl.edu/fastx_toolkit/). After the trimming and filtering, orphan reads (single reads for paired-end data) are normally removed from subsequent analysis.

For plants or fungi that have been completely sequenced, sequencing reads can be mapped to the annotated genome using software such as TopHat, which used Bowtie (Trapnell et al. 2012) as its algorithmic core (Fig. 1.1). For plants or fungi that do not have a complete sequence, de novo assembly can be performed with packages such as Oases (Schulz et al. 2012), SOAPdenovo-Trans (Xie et al. 2014), Trans-ABySS (Grabherr et al. 2011), or Trinity (Haas et al. 2013) with an optimized k-mer size for each data set (Fig. 1.1). The assembled sequences are then clustered with TGICL (Pertea et al. 2003) and CAP3 (Huang and Madan 1999) to produce non-redundant sequences or unigenes. Sequencing reads can then be mapped to the unigenes. The number of reads that are mapped to each reference sequence/unigene can be quantified and used for differential expression analysis using Cufflinks, Cuffdiff (Trapnell et al. 2010), or other methods. The coverage of reads being mapped can be translated to expression value or fragments per kilobase of transcript per million mapped fragments (FPKM) (Trapnell et al. 2010). Differential expressed genes/transcripts between two samples are normally defined as transcripts that have at least fourfold difference in their expression levels ($\log_2$ fold change $\leq|2|$) with a false discovery rate (q-value) less than 5% in one sample compared to the other. To assess the global quality of the RNA-seq dataset, the reproducibility among replicates can be examined for possible batch effects. Reproducibility among technical replicates should be generally high (Spearman $R^2 > 0.9$) (Mortazavi et al. 2008). Further discussion on experimental design and appropriate RNA-seq study approach leading data to meaningful biological interpretation is available in Sects. 1.3.3.1 and 1.3.3.3.

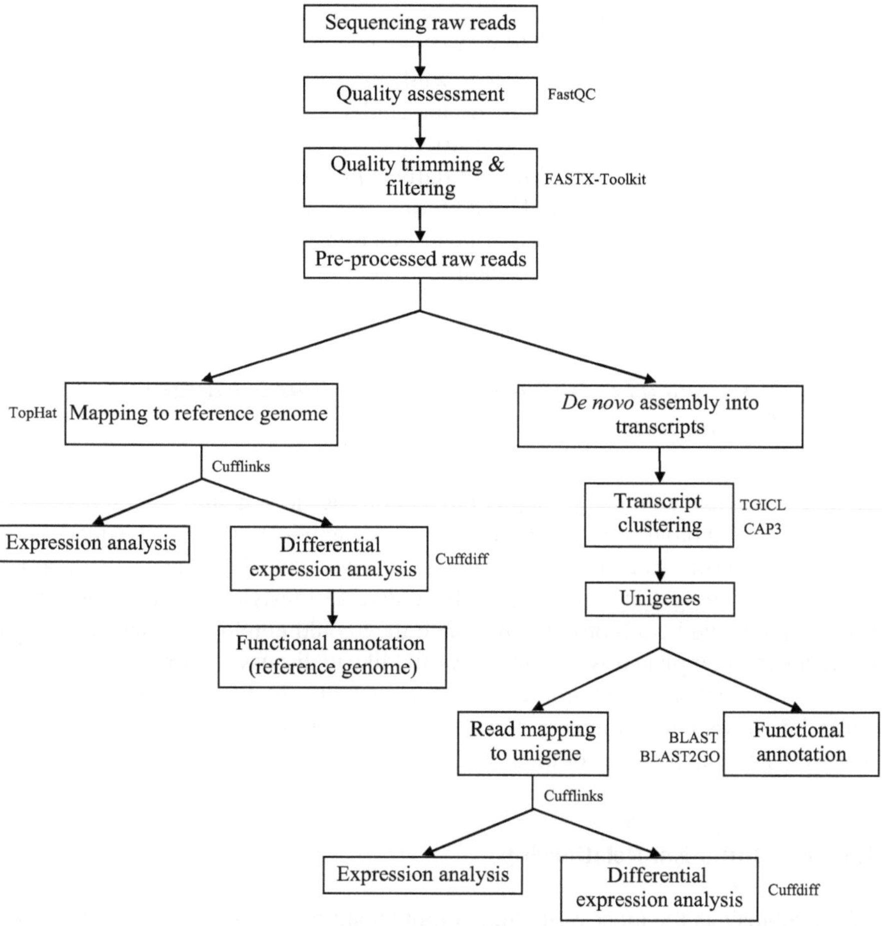

Fig 1.1 An overview of RNA-seq reference-based mapping and de novo assembly analysis flowchart

It is challenging to assign gene functions to short reads based on sequence homology, thus functional annotation is normally performed on unigenes upon assembly. To assign putative functions to unigenes, they are translated into six reading frames and compared against non-redundant (nr) database and/or SwissProt database at the National Center for Biotechnology Information (NCBI) with BLASTX (Altschul et al. 1997). In most studies, matches with E-values equal or less than 10^{-5} are treated as "significant matches" while sequences with no hits or matches with E-values more than 10^{-5} are classified as "non-significant matches." The protein-coding unigenes can also be functionally annotated by sequence similarity matches to conserved protein domain databases such as InterPro (Hunter et al. 2012) and Pfam (Finn et al. 2014). Assignment of Gene Ontology (GO) and mapping of unigenes into Kyoto Encyclopedia of Genes and Genomes (KEGG) metabolic pathways can be performed by tools such as Blast2GO (Conesa et al. 2005). The standard vocabularies used in GO allow exchangeability and classification of functional information across orthologs while the KEGG metabolic pathways facilitate a systematic understanding of the molecular interactions network among the genes and their biological functions. Enrichment analyses based on GO and KEGG terms can also be performed on differentially expressed genes to compare the abundance of specific GO terms and KEGG terms in the tested sample compared to the control sample. These results reveal gene functions and pathways that respond to the treatment applied to the tested sample.

1.3.2 RNA-seq as a Tool for Studies on Plant–Fungus Interactions

To date, knowledge on plant–fungus interactions at the genomic level remains elusive. NGS technologies is starting to gain wider applications in understanding the fungal mechanisms (infection, colonization, survival), and the molecular response of plants against fungal infection. Fungi have diverse lifestyles including biotrophic, necrotrophic, and hemibiotrophic in which they could employ distinct strategies to interact with their plant hosts. Here, we would like to discuss a few recent studies that involve the application of transcriptome sequencing in fungus–plant interaction. Table 1.1 shows a list of next-generation short-read transcriptomic sequencing data that were generated. A few of them will be discussed here for case studies.

1.3.2.1 Mutualism Relationship

AM symbiosis is the most well-studied mutualism relationship between plants and fungi. RNA-seq was employed to understand the transcriptional changes that occur in the leaves of mycorrhizal tomato plants (*Solanum lycopersicum*) (Cervantes-Gámez et al. 2016). Transcriptome profiling of leaves from mycorrhizal and non-mycorrhizal tomato plants demonstrated 742 differential expressed genes including those involved in signaling, transport, biotic and abiotic stresses, and hormone metabolism. Transcriptional changes also occurred in P, N, and sugar transporters and genes related to systemic defense priming. Similarly, Zouari et al. (2014) reported that the beneficial symbiosis established between tomato with *Funneliformis mosseae* (another AM fungus) have an impact on primary metabolisms, particularly on mineral nutrition.

In another study by Hiruma et al. (2016), RNA-seq data revealed that a destructive pathogen, *Colletotrichum tofieldiae* can also serve as an endemic endophyte in natural *Arabidopsis thaliana* populations. *C. tofieldiae* which established root colonization initially can spread systemically into shoots while transferring macronutrient phosphorus to shoots, thus promoting plant growth, and increasing fertility

Table 1.1 Selected RNA-seq datasets on plant–fungus interactions

Type of interactions/ experimental design	Interaction partners		Deposition of RNA-seq dataset(s)/accession number(s)	References
	Plant host	Fungus/fungi		
Mutualism relationship				
Two biological replicates, each containing the pooled RNA from three fruits	*Solanum lycopersicum* cv. Moneymaker tomato	*Funneliformis mosseae* (formerly *Glomus mosseae*)	Array express database E-MTAB-2276	Zouari et al. (2014)
Leaves of three biological replicates (individual plants) of mycorrhizal and non-mycorrhizal plants	*Solanum lycopersicum* var. Missouri (Tomato)	*Rhizophagus irregularis* (previously *G. intraradices*)	NCBI Sequence Read archive (Bioproject No. PRJNA263841)	Cervantes-Gámez et al. (2016)
Infected roots grown under Pi-sufficient and Pi-deficient conditions at four time points (6, 10, 16, and 24 dpi) and also *Colletotrichum tofieldiae* hyphae grown in vitro	*Arabidopsis thaliana* Col-0	*Colletotrichum tofieldiae*	Gene Expression Omnibus (GEO) database (GSE70094)	Hiruma et al. (2016)
Mutualism relationship versus pathogenic relationship				
Three biological replicates	*Vitis vinifera* Pinot Noir leaves (Grapevine)	*Trichoderma harzianum* T39 and pathogenic oomycete *Plasmopara viticola* (Berk. and Curt.) Berl. and de Toni	NCBI Sequence Read archive (Bioproject No. PRJNA168987)	Perazzolli et al. (2012)
Rice roots infected by *H. oryzae* at 2, 6, and 20 days after inoculation (DAI) and rice roots infected by *M. oryzae* at 2 and 6 DAI—Three biological replicates each, pooled(?) 2-ways interaction	*Oryza sativa* L. ssp. *japonica* cv. Nipponbare (blast-susceptible rice cultivar CO-39)	*Harpophora oryzae* strain R5-6-1 *Magnaporthe oryzae* strain Guy11	Not found in the article	Xu et al. (2015)

(continued)

Table 1.1 (continued)

Type of interactions/ experimental design	Interaction partners		Deposition of RNA-seq dataset(s)/accession number(s)	References
	Plant host	Fungus/fungi		
Pathogenic relationship				
Leaves at 24 hpi, pooled from 2 biological replicates 2-ways interaction	*Oryza sativa* L. ssp. *japonica* cv. Nipponbare carrying the rice blast resistance gene *Pia*	*Magnaporthe oryzae*-compatible Ina86-137 (race 001.0) (COM) and incompatible P91-15B (race 007.0, MAFF 101511) (INC) fungal strains	DDBJ sequence read archive (DRA) (DRA000542)—5 samples (3 controls: rice, 2 fungi; 2 mixed and infected samples of rice and fungi)	Kawahara et al. (2012)
1 and 6 days post infection—pooled samples for each time point	*Arabidopsis thaliana* (ecotype Col-0)	*Fusarium oxysporum* (strain Fo5176) (hemibiotroph)	GEO database (GSM845430-GSM845433) -only plant genes were analyzed	Zhu et al. (2012)
Three independent experiments at 12, 24, and 48 hpi consisting of 2 leaves from each of 4 replicates at each time point	*Lactuca sativa cv. Salinas* (lettuce)	*Botrytis cinerea* (strain B05.10) (necrotrophic)	NCBI Sequence Read archive (SRA059059)	De Cremer et al. (2013)
Seedlings at 2 dpi, 5 individual biological replicates—2-ways interaction	*Theobroma cacao* cv "Comum" (Forastero genotype)	*Moniliophthora perniciosa* isolate BP10 (hemibiotroph)	NCBI Sequence Read archive (SRA066232)	Teixeira et al. (2014)
Roots at 2dpi	Tomato cultivar Micro-Tom (*Verticillium* susceptible)	*Verticillium dahliae* isolate Vd080 (necrotrophic)	NCBI Sequence Read archive (SRX1022130)	Tan et al. (2015)

Type of interactions/ experimental design	Interaction partners		Deposition of RNA-seq dataset(s)/accession number(s)	References
	Plant host	Fungus/fungi		
Five leaves collected from five independent plants; two fungal culture media, i.e., Czapek-Dox broth and PDB; and five time points, i.e., 1, 4, 9, 14, 21 dpi of plant infection. –2-ways interaction	*Triticum aestivum* “Riband”	*Zymoseptoria tritici* isolate IPO323	Not found in the article	Rudd et al. (2015)
Four biological replicates of each noble rot stage (a total of 3) and control -2-ways interaction	*Vitis vinifera* cv. Sémillon		GEO: (GSE67932)	Blanco-Ulate et al. (2016)

under phosphorus-deficient conditions. Phosphate starvation in the host might have facilitated the transition of *C. tofieldiae* from a pathogen to a beneficial endophyte. The phosphate starvation response in the host could control root colonization by *C. tofieldiae* and is needed for plant growth promotion.

The abovementioned studies only analyzed the host plant component in the interactions without providing data on the fungi involved. However, there are also RNA-seq data on multispecies involved in mutualism relationships such as those involve in the ectomycorrhizal interaction between *Laccaria bicolor* and aspen (*Populus tremuloides*) roots (Larsen et al. 2011). The RNA-seq data suggested that *L. bicolor* may provide various metabolites (e.g., glycine, glutamate, and allantoin) for aspen, which in return provides sugars such as fructose and glucose to the fungus.

1.3.2.2 Mutualism Relationship Versus Pathogenic Relationship

Multispecies transcriptomics profiling with RNA-seq enables direct comparison of the interactions between rice and two closely related pathogenic (rice blast fungus *Magnaporthe oryzae*) and mutualistic (rice endophyte *Harpophora oryzae*) fungi species (Xu et al. 2015). In this study, differences in the metabolome and transcriptome of rice challenged by *H.* or *M. oryzae* were analyzed. The levels of metabolites in the shikimate and lignin biosynthesis pathways increased continuously in the *M. oryzae*-challenged rice roots. In the *H. oryzae*-challenged rice roots, these pathways were initially induced before they were suppressed. The expression of most genes encoding enzymes involved in glycolysis and TCA cycle was suppressed in the *H. oryzae*-challenged rice roots, but enhanced in the *M. oryzae*-challenged rice roots.

RNA-seq data also helped to elucidate molecular events associated with ISR induced by *Trichoderma harzianum* T39 in grapevine against downy mildew caused by *Plasmopara viticola* (Perazzolli et al. 2012). The findings demonstrated complex transcriptional reprogramming of grapevine leaves during resistance induction and pathogen inoculation. *T. harzianum* T39 was found to be able to partially inhibit some disease-related processes by modulating the genes related to the microbial recognition machinery directly. It also specifically activated defense responses after *P. viticola* inoculation by enhancing the expression of defense-related genes.

1.3.2.3 Pathogenic Relationship

Most of the RNA-seq data reported on plant–fungus interactions focused on their pathogenic relationship. Kawahara et al. (2012) studied the molecular mechanisms underlying both rice defense and fungal attack caused by *M. oryzae*. The gene expression profiles of both interacting organisms were monitored simultaneously in the same infected plant tissue by the RNA-seq approach. Detection of fungal gene expression in infected leaves has been difficult because very few numbers of fungal cells are present. However, the authors managed to analyze the mixed transcriptome of rice and blast fungus in infected leaves at 24 h postinoculation, corresponding to the penetration of the primary hyphae into leaf epidermal cells. The mixed transcriptome analysis was shown to be useful for simultaneous elucidation of the tactics of host plant defense and pathogen attack, whereby 240 fungal transcripts encoding putative secreted proteins were found to be upregulated. In addition, transcripts encoding glycosyl hydrolases, cutinases, and LysM domain-containing proteins in the *M. oryzae* were found to increase whereas pathogenesis-related and phytoalexin biosynthetic genes were upregulated in rice. The findings also demonstrated that incompatible interactions have more drastic changes in expression compared with the compatible interactions in both rice and blast fungus at this stage.

The dual RNA-seq approach was also employed by Teixeira et al. (2014) to simultaneously assess the transcriptomes of cacao and the hemibiotrophic fungus, *Moniliophthora perniciosa* which causes Witches' broom disease (WBD). The pathogen's transcriptome was analyzed in exceptional detail and thereby characterized the fungal nutritional and infection strategies during WBD and identified putative virulence effectors. The biotrophic *M. perniciosa* was found to increase the availability of soluble nutrients by changing plant metabolism in the diseased tissues before plant death. On the other hand, ineffective defense responses were found to be induced in the infected shoots which were characterized by carbon deprivation. The finding also suggested the establishment of a senescence process in the host, which marks the end of the WBD biotrophic stage. The results improve understanding of the development of (hemi) biotrophic plant–pathogen interactions.

Deep RNA sequencing coupling with metabolomics was also used by Rudd et al. (2015) to investigate the physiology of wheat and the hemibiotrophic fungus *Zymoseptoria tritici* throughout an asexual reproductive cycle of the fungus on wheat leaves. The study managed to profile more than 3000 pathogen genes, 7000 wheat genes, and 300 metabolites that were differentially regulated. The slow-growing extracellular and nutritionally limited *Z. tritici* were found to suppress the defense system of wheat initially. It was subsequently followed by a defense (hyper) activation in the host (increased expression of JA biosynthesis genes and large-scale activation of other plant defense responses) by manipulating specific plant carbohydrates during the switch of the pathogen to necrotrophic growth and reproduction.

Global expression profiling using RNA-seq has also been used to identify a complex gene network involved in the necrotrophic interaction between lettuce and *Botrytis cinerea* (De Cremer et al. 2013). The biological pathways in which the differentially expressed genes are implicated generate a holistic picture at the inoculation site. The phenylpropanoid pathway and terpenoid biosynthesis in the host plant were induced, whereas photosynthesis was downregulated at 48 h post inoculation. The study also enables comparison with data on the interaction of *B. cinerea* with *A. thaliana* which revealed both general and species-specific responses to infection with this pathogen. Additionally, the study also revealed that similar pathways were induced during compatible interactions of lettuce with necrotrophic pathogen and biotrophic pathogen (*Bremia lactucae)*. However, the fungal transcriptome was not profiled in this study. Other single-species RNA-seq analysis involving only the host transcriptome include time-course analysis of *A. thaliana* inoculated with *Fusarium oxysporum* (Zhu et al. 2012), and comparative analysis on tomato–*Verticillium dahlia* interactions in Verticillium wilt (Tan et al. 2015).

1.3.3 Challenges

1.3.3.1 Experimental Design and Replication

In comparative transcriptome, mRNA transcripts that are significantly increased or decreased in a pair of samples are identified. Pair-wise comparisons can be conducted on samples under particular sets of conditions or treatments. In plant–fungus interactions, a "treatment" could be inoculation of a pathogen genotype that perturbs the transcript profile of a host in reference to the control or untreated host. Pair-wise comparisons can also be conducted on treated samples collected from a time-course study (early infection stage versus late infection stage) or on pathogens in different mode of actions (biotrophic turns necrotrophic; mutualism turn pathogenic). Samples collected from a time course combined with alternate plant genotypes or pathogen isolates that elicit a response are useful in testing some hypotheses in host–pathogen interactions.

Careful design (blocking, randomization, and replication) and rigorous statistical analysis are prerequisites to planning expression profiling or other experiments in large-scale biology. Variations (including biological and technical variations) may derive from multiple and diverse sources in experiments. Due to high sequencing cost and difficulties in some experiments (scarcity of samples or difficulties in sample collection), many RNA-seq studies for transcriptional profiling have not been replicated at present, and therefore defy statistical analyses. However, the requirement and the number of biological replicates for RNA-seq depend on the heterogeneity of the experimental system. Samples derived from biological replicates treated under the same condition are expected to be more similar as demonstrated by Champigny et al. (2013). In this study, RNA-seq was applied to compare the gene expression profiles of an extremophile plant, *Eutrema salsugineum* grown in different fields and controlled environment with three biological replicates per condition. Samples in each group (with the same condition) were clustered together in a principal component plot (PCA). Rigorous statistical design, replication, and analysis should be considered whenever possible if expression data are used to draw broad conclusions about the biology of a system (Meyers et al. 2004). Technical replicates are not required in RNA-seq for comparative transcriptome purpose. However, biological replicates are essential to draw broad conclusions about plant–fungal interactions.

1.3.3.2 Sampling and RNA Preparation

In multispecies transcriptome profiling as in plant–fungus interaction, separation of fungal mycelia from infected plant tissues could be difficult or close to impossible, especially during the early stage of fungal infection whereby the fungal mycelia could be in negligible amount. Normally, a mixed sample with plant and fungus is collected for RNA-seq. Profiling of a mixed transcriptome requires further processing at data analysis which will be discussed in the subsequent section.

Harvesting of infected plant tissues at the early stage of fungal infection can be challenging due to scarcity of infected tissues and is always inseparable from healthy tissue. Sampling of diseased samples according to infection duration may not be always the best choice in a time-course experimental design as the infection may not be always synchronized. For root treatment that happens in soil or some other substrates, sample collection is further complicated by difficulties in collection of diseased samples based on disease symptoms especially for destructive sampling.

1.3.3.3 Data Analysis

One of the key differences between Sanger sequencing and NGS platform is the number of nucleotides obtained from each fragment being sequenced, i.e., Illumina platform (short reads technology) generally produces a read length ranging from 100 to 300 bases with a throughput of ~1500 Gb per run, as opposing to Sanger sequencing which produces a read length which is about ~800 bp with a throughput of ~77 kb per run (https://flxlexblog.wordpress.com/2016/07/08/developments-in-high-throughput-sequencing-july-2016-edition/; Goodwin et al. 2016). The new "Third Generation-Sequencing" platform based on single-molecule sequencing technology, e.g., Pacific Biosciences (PacBio) can generate sequences that are ~20,000 bp in length with a throughput of ~1 Gb per run. As the complexity of the data sets and the sizes for various NGS data increase and vary, high-throughput data storage, management, and computation for large sample sizes are becoming a limitation and pose many challenges.

RNA-seq is an analytical technique leveraging the capacity of NGS platforms to quantitatively sample a population of RNA molecules with large number of sequencing reads from parallel reactions on different samples (Li et al. 2014). Selection of a sequencing platform depends on the research goal. It is important to understand the benefits and limitations of the chosen platform to fully capture its advantages to answer the research problems. Currently, Illumina is the best market accepted sequencing platform for RNA-seq application. By offering a high throughput per sequencing run, it allows easier detection of low expressed transcripts. However, having longer reads may improve the performance in transcript assembly, quantification, and gene fusion detection (Shendure and Ji 2008; Ozsolak and Milos 2011). Long reads were shown to be able to provide more information on gene structures and splicing patterns compared to short reads (Tilgner et al. 2013).

Besides, the choice of sequencing platform has a direct impact towards the data analysis strategy. Different sequencing platform may have different requirements in library synthesis, sequence alignment, and data processing (Li et al. 2014). Generally, the two basic aims of RNA-seq are to determine the regions of the genomes that have been transcribed in a sample, and to quantify the expression of these transcripts. There are three principal approaches to infer the transcripts in the

RNA samples using RNA-seq. The simplest approach is to assume the transcripts in a sample are a subset of the transcripts listed in a curated database. A more ambitious strategy involves aligning reads to a reference genome and using the alignments to infer the transcript structures, while the most challenging approach is to assemble reads into putative transcripts de novo without the aid of a reference genome (Jänes et al. 2015).

In reference-based mapping approach, the paired-end sequencing reads are mapped to reference genome by exact match or scoring sequence similarity. There are two major categories of aligners: (1) unspliced read aligners, e.g., Bowtie2 (Langmead and Salzberg 2012) and Stampy (Lunter and Goodson 2011); (2) spliced aligners, e.g., TopHat2 (Kim et al. 2013) and GSNAP (Wu and Nacu 2010). Unspliced read aligners do not allow large gaps such as those that are potentially generated from reads spanning exon boundaries or splice junctions (Benjamin et al. 2014). The unspliced read aligners can map the reads to potentially novel exons but are limited by their inability to handle reads spanning splice junctions in the reference genome. Spliced aligners are able to map reads spanning splice junction by splitting them into smaller segments and determine the best match based on alignment scores and known di-nucleotide splice signals (Benjamin et al. 2014). The reference-based mapping transcriptome assembly has several advantages: the assembly can be performed with parallel computing or computer with only a few gigabytes of RAM; it has a lower risk of sequencing artifacts and contamination; and it is sufficiently sensitive in detecting and assembly of low abundant transcripts. In addition, missing nucleotides can be filled by the reference sequences, and it facilitates the discovery of novel transcripts/non-coding RNAs that have not been annotated in the reference genome (Martin and Wang 2011). There are some disadvantages and drawbacks to this approach: the success of reference-based mapping transcriptome assembly depends on the quality of reference genome being used; spliced reads spanning large introns may not be identified or mapped; alignment of reads to multiple regions, and removal of "multi-mapped reads" that cannot be mapped uniquely in the reference genome. However, the reference-based mapping approach is useful in identifying and differentiating transcript isoforms in large and complex transcriptomes, such as those from plants and fungi. In general, having a closely related, reliable, and well-annotated reference genome will aid in reference-based mapping studies.

De novo assembly leverages the redundancy of short-read sequencing to find overlaps between the sequencing reads, to assemble and merge the contigs into unigene as part of the process to reconstruct the transcripts. De novo assembler such as Trans-ABySS assembles the reads multiple times using the *de Bruijn* graph-based approach to reconstruct transcripts from a broad range of expression levels, before merging the contigs to remove redundancy. Conversely, other de novo assemblers, e.g., Oases (Schulz et al. 2012) and Trinity (Haas et al. 2013), directly traverse the *de Bruijn* graph to assemble the isoforms (Martin and Wang 2011). The advantages of de novo assembly are: no reference genome is required in the assembly; it can detect transcripts that are in reference genome or other unknown exogenous source; it does not depend on the alignment of reads to known splice sites; or affected by long introns that are present in the genome (thus is not limited by some unspliced read aligners which do not allow large gaps from read spanning exon boundaries or splice junctions); and is abled to assemble trans-spliced transcripts (Martin and Wang 2011). However, the process requires large amounts of computational resources (100 of gigabytes of RAM) and is time consuming (days to weeks). Nevertheless, the process could be distributed over a cluster of computational nodes to speed up the assembly process. In addition, it also requires a higher sequencing depth for full transcript assembly than reference-based mapping (Benjamin et al. 2014). De novo assembly of transcriptome is very sensitive to sequencing errors and chimeric molecules present in the data because it is difficult to distinguish real trans-spliced event from chimeric artifacts in the library preparation. De novo assembly also has difficulties in discriminating transcript variants generated from different alleles or paralogues (Grabherr et al. 2011; Martin and Wang 2011). Furthermore, unigenes generated from overlapping reads can be mis-assembled, fragmented, or incomplete. However, the assembled unigenes can be validated via BlastN to the publicly available database or confirmed by alignment with closely related homologues genes.

In parallel-gene expression analyses, the RNA-seq sequencing reads are mapped to either available genome assemblies or de novo assemblies of transcriptomes (if neither the genomes of plant nor fungus is sequenced). The latter may present some difficulties as plant and fungus share highly conserved sequences that are difficult to be assigned to either partner in plant–fungus interactions. The sequencing reads can be mapped to closely related fungal genomes that are available, to segregate the fungal sequences from the plant sequences. De novo assembly can then be performed on the sequencing reads upon removal of the probable fungal sequences to obtain the plant sequences. The sequences mapped to the fungal reference genome are probably fungal transcripts expressed during fungal infection of host plant (Ho et al. 2016). Further detailed examination on the identity and functions of the expressed genes in fungus is required.

In differentially gene expression analysis, the statistical cutoff is normally associated with a low False Discovery Rate (FDR) or q-value (Storey and Tibshirani 2003) in combination with a user-imposed fold change in gene expression which can vary from two- to fourfolds (Rudd et al. 2015; Hiruma et al. 2016; Ho et al. 2016). In RNA-seq analysis, the most significant differentially expression genes or the pathways that they are involved may be extracted using a stringent cutoff value. However, a less stringent cutoff value may also be used to extract some of the less statistically significant yet interesting genes or patterns of gene expression that have not been included in the initial data.

1.3.3.4 Verification of RNA-seq Data

In global gene expression profiling experiments (microarray and RNA-seq alike), variation may derive from multiple and diverse sources. It is essential or it has become a norm to verify the gene expression data that are used to deduce changes in cellular processes and pathways by using a different technique before drawing conclusions about the biology of plant–pathogen interactions (Meyers et al. 2004; Nettleton 2006). Previously, Wise et al. (2007) opined that confirmation of expression data generated by cDNA microarray analysis by using a different technique such as reverse transcription-quantitative polymerase chain reaction (RT-qPCR) or RNA blot analysis is not necessary if blocking, randomization, and replication are carefully addressed in experimental design, coupling with rigorous statistical analysis. This is especially true if the experiment is designed and analyzed to survey for candidate genes for functional analysis, by mutation, overexpression, or gene silencing because functional characterization of the candidate genes is the ultimate validation of the expression profiling results. Furthermore, performing RT-qPCR or RNA blot analysis on a limited number of genes derived from large data analysis ultimately only confirms data for the selected genes (Wise et al. 2007). Microarray data is generally considered to be reliable, and in line with results based on RT-qPCR or RNA gel blot analysis (Boddu et al. 2006; Stein et al. 2006). Costa et al. (2013) has referred RT-qPCR as the gold standard replacing RNA blotting methods for RNA quantification in clinical applications while the RNA-seq methods is a potential complementary technique to RT-qPCR for cancer laboratories. Currently, RNA-seq-related studies are also performed as preliminary screening for selection of targeted genes for further RT-qPCR analyses (Costa et al. 2013; Zhuang et al. 2015; Hu et al. 2016). In addition, database mining from existing expression resources, such as expressed sequence tag (EST) libraries, microarray databases, and high-throughput sequence data from independent, published experiments can also be used for corroborating differential expression (Caldo et al. 2004; Boddu et al. 2006; Gjetting et al. 2007).

1.4 Future Directions of RNA-seq Analysis in Plant–Fungus Interactions

Multispecies transcriptomics may lead to the discovery of key plant and microbial genes that characterize the interaction and may provide new strategies for disease resistance. To obtain a comprehensive view of the processes involved, a systems biology approach which is multidisciplinary would be required. Integration of genomics, transcriptomics, proteomics and metabolomics data using bioinformatics and statistical analysis will facilitate modeling of biological system and help to identify key biological processes, as well as make predictions. It is important to identify not only the functions of individual genes and transcripts but also proteins and metabolites during plant–microbe interactions. A better understanding of beneficial and detrimental interactions between plants and microbes may offer unprecedented opportunities to increase crop productivity. Given the worldwide importance of food security and other usages of plants, the knowledge generated would provide enormous environmental and financial benefits. With the progress in high-throughput

RNA-seq analyses of plant–microbe interactions, future research holds promise to fuel new insights that will be extremely valuable for the development of new technologies for sustainable food production.

References

Altschul SF, Madden TL, Schäffer AA, Zhang J, Zhang Z, Miller W, Lipman DJ (1997) Gapped BLAST and PSI-BLAST: a new generation of protein database search programs. Nucleic Acids Res 25:3389–3402

Ané JM, Kiss GB, Riely BK, Penmetsa RV, Oldroyd GE, Ayax C, Lévy J, Debellé F, Baek JM, Kalo P, Rosenberg C, Roe BA, Long SR, Dénarié J, Cook DR (2004) *Medicago truncatula* DMI1 required for bacterial and fungal symbioses in legumes. Science 303:1364–1367

Bechtold U, Karpinski S, Mullineaux PM (2005) The influence of the light environment and photosynthesis on oxidative signalling responses in plant–biotrophic pathogen interactions. Plant Cell Environ 28:1046–1055

Benitez T, Rincon AM, Limon MC, Codon AC (2004) Biocontrol mechanisms of *Trichoderma* strains. Int Microbiol 7:249–260

Benjamin AM, Nichols M, Burke TW, Ginsburg GS, Lucas JE (2014) Comparing reference-based RNA-Seq mapping methods for non-human primate data. BMC Genomics 15:570

Bentley DR, Balasubramanian S, Swerdlow HP, Smith GP, Milton J, Brown CG, Hall KP, Evers DJ, Barnes CL, Bignell HR, Boutell JM, Bryant J, Carter RJ, Keira Cheetham R, Cox AJ, Ellis DJ, Flatbush MR, Gormley NA, Humphray SJ, Irving LJ, Karbelashvili MS, Kirk SM, Li H, Liu X, Maisinger KS, Murray LJ, Obradovic B, Ost T, Parkinson ML, Pratt MR, Rasolonjatovo IM, Reed MT, Rigatti R, Rodighiero C, Ross MT, Sabot A, Sankar SV, Scally A, Schroth GP, Smith ME, Smith VP, Spiridou A, Torrance PE, Tzonev SS, Vermaas EH, Walter K, Wu X, Zhang L, Alam MD, Anastasi C, Aniebo IC, Bailey DM, Bancarz IR, Banerjee S, Barbour SG, Baybayan PA, Benoit VA, Benson KF, Bevis C, Black PJ, Boodhun A, Brennan JS, Bridgham JA, Brown RC, Brown AA, Buermann DH, Bundu AA, Burrows JC, Carter NP, Castillo N, Chiara E, Catenazzi M, Chang S, Neil Cooley R, Crake NR, Dada OO, Diakoumakos KD, Dominguez-Fernandez B, Earnshaw DJ, Egbujor UC, Elmore DW, Etchin SS, Ewan MR, Fedurco M, Fraser LJ, Fuentes Fajardo KV, Scott Furey W, George D, Gietzen KJ, Goddard CP, Golda GS, Granieri PA, Green DE, Gustafson DL, Hansen NF, Harnish K, Haudenschild CD, Heyer NI, Hims MM, Ho JT, Horgan AM, Hoschler K, Hurwitz S, Ivanov DV, Johnson MQ, James T, Huw Jones TA, Kang GD, Kerelska TH, Kersey AD, Khrebtukova I, Kindwall AP, Kingsbury Z, Kokko-Gonzales PI, Kumar A, Laurent MA, Lawley CT, Lee SE, Lee X, Liao AK, Loch JA, Lok M, Luo S, Mammen RM, Martin JW, McCauley PG, McNitt P, Mehta P, Moon KW, Mullens JW, Newington T, Ning Z, Ling Ng B, Novo SM, O'Neill MJ, Osborne MA, Osnowski A, Ostadan O, Paraschos LL, Pickering L, Pike AC, Pike AC, Chris Pinkard D, Pliskin DP, Podhasky J, Quijano VJ, Raczy C, Rae VH, Rawlings SR, Chiva Rodriguez A, Roe PM, Rogers J, Rogert Bacigalupo MC, Romanov N, Romieu A, Roth RK, Rourke NJ, Ruediger ST, Rusman E, Sanches-Kuiper RM, Schenker MR, Seoane JM, Shaw RJ, Shiver MK, Short SW, Sizto NL, Sluis JP, Smith MA, Ernest Sohna Sohna J, Spence EJ, Stevens K, Sutton N, Szajkowski L, Tregidgo CL, Turcatti G, Vandevondele S, Verhovsky Y, Virk SM, Wakelin S, Walcott GC, Wang J, Worsley GJ, Yan J, Yau L, Zuerlein M, Rogers J, Mullikin JC, Hurles ME, McCooke NJ, West JS, Oaks FL, Lundberg PL, Klenerman D, Durbin R, Smith AJ (2008) Accurate whole human genome sequencing using reversible terminator chemistry. Nature 456:53–59

Berg G (2009) Plant–microbe interactions promoting plant growth and health: perspectives for controlled use of microorganisms in agriculture. Appl Microbiol Biotechnol 84:11–18

Blanco-Ulate B, Amrine KC, Collins TS, Rivero RM, Vicente AR, Morales-Cruz A, Doyle CL, Ye Z, Allen G, Heymann H, Ebeler SE, Cantu D (2016) Developmental and metabolic plasticity of white-skinned grape berries in response to *Botrytis cinerea* during noble rot. Plant Physiol 169:2422–2443

Boddu J, Cho S, Kruger WM, Muehlbauer GJ (2006) Transcriptome analysis of the barley-*Fusarium graminearum* interaction. Mol Plant Microbe Interact 19:407–417

Brotman Y, Lisec J, Méret M, Chet I, Willmitzer L, Viterbo A (2012) Transcript and metabolite analysis of the *Trichoderma* induced systemic resistance response to *Pseudomonas syringae* in *Arabidopsis thaliana*. Microbiology 158:139–146

Caldo RA, Nettleton D, Wise RP (2004) Interaction-dependent gene expressionin Mla-specified response to barley powdery mildew. Plant Cell 16:2514–2528

Cervantes-Gámez RG, Bueno-Ibarra MA, Cruz-Mendívil A, Calderón-Vázquez CL, Ramírez-Douriet CM, Maldonado-Mendoza IE, Villalobos-López MA, Valdez-Ortíz A, López-Meyer M (2016) Arbuscular mycorrhizal symbiosis-induced expression changes in *Solanum lycopersicum* leaves revealed by RNA-seq Analysis. Plant Mol Biol Rep 34:89–102

Champigny MJ, Sung WW, Catana V, Salwan R, Summers PS, Dudley SA, Provart NJ, Cameron RK, Golding GB, Weretilnyk EA (2013) RNA-Seq effectively monitors gene expression in *Eutrema salsugineum* plants growing in an extreme natural habitat and in controlled growth cabinet conditions. BMC Genomics 14:578

Chen YC, Wong CL, Muzzi F, Vlaardingerbroek I, Kidd BN, Schenk PM (2014) Root defense analysis against *Fusarium oxysporum* reveals new regulators to confer resistance. Sci Rep 4:5584

Conesa A, Götz S, García-Gómez JM, Terol J, Talón M, Robles M (2005) Blast 2GO: a universal tool for annotation, visualization and analysis in functional genomics research. Bioinformatics 21:3674–3676

Conesa A, Madrigal P, Tarazona S, Gomez-Cabrero D, Cervera A, McPherson A, Szcześniak MW, Gaffney DJ, Elo LL, Zhang X, Mortazavi A (2016) A survey of best practices for RNA-seq data analysis. Genome Biol 17:13

Costa C, Giménez-Capitán A, Karachaliou N, Rosell R (2013) Comprehensive molecular screening: from the RT-PCR to the RNA-seq. Transl Lung Cancer Res 2:87–91

De Cremer K, Mathys J, Vos C, Froenicke L, Michelmore RW, Cammue BP, De Coninck B (2013) RNAseq-based transcriptome analysis of *Lactuca sativa* infected by the fungal necrotroph *Botrytis cinerea*. Plant Cell Environ 36:1992–2007

Depuyd S, Hardtke CS (2011) Hormone signalling crosstalk in plant growth regulation. Curr Biol 21:365–373

Endre G, Kereszt A, Kevei Z, Mihacea S, Kaló P, Kiss GB (2002) A receptor kinase gene regulating symbiotic nodule development. Nature 417:962–966

Finn RD, Bateman A, Clements J, Coggill P, Eberhardt RY, Eddy SR, Heger A, Hetherington K, Holm L, Mistry J, Sonnhammer EL, Tate J, Punta M (2014) Pfam: the protein families database. Nucleic Acids Res 42:D222–D230

Gjetting T, Hagedorn PH, Schweizer P, Thordal-Christensen H, Carver TLW, Lyngkjær MF (2007) Single-cell transcript profiling of barley attacked by the powdery mildew fungus. Mol Plant Microbe Interact 20:235–246

Glazebrook J (2005) Contrasting mechanisms of defence against biotrophic and necrotrophic pathogens. Annu Rev Phytopathol 43:205–227

Goodwin S, McPherson JD, McCombie WR (2016) Coming of age: ten years of next-generation sequencing technologies. Nat Rev Genet 17:333–351

Govrin EM, Levine A (2000) The hypersensitive response facilitates plant infection by the necrotrophic pathogen *Botrytis cinerea*. Curr Biol 10:751–757

Grabherr MG, Haas BJ, Yassour M, Levin JZ, Thompson DA, Amit I, Adiconis X, Fan L, Raychowdhury R, Zeng Q, Chen Z, Mauceli E, Hacohen N, Gnirke A, Rhind N, di Palma F, Birren BW, Nusbaum C, Lindblad-Toh K, Friedman N, Regev A (2011) Full-length transcriptome assembly from RNA-Seq data without a reference genome. Nat Biotechnol 29:644–652

Güimil S, Chang HS, Zhu T, Sesma A, Osbourn A, Roux C, Ioannidis V, Oakeley EJ, Docquier M, Descombes P, Briggs SP, Paszkowski U (2005) Comparative transcriptomics of rice reveals an ancient pattern of response to microbial colonization. Proc Natl Acad Sci U S A 102:8066–8070

Haas BJ, Papanicolaou A, Yassour M, Grabherr M, Blood PD, Bowden J, Couger MB, Eccles D, Li B, Lieber M, Macmanes MD, Ott M, Orvis J, Pochet N, Strozzi F, Weeks N, Westerman R, William T, Dewey CN, Henschel R, Leduc RD, Friedman N, Regev A (2013) *De novo* transcript sequence reconstruction from RNA-seq using the Trinity platform for reference generation and analysis. Nat Protoc 8:1494–1512

Heckman DS, Geiser DM, Eidell BR, Stauffer RL, Kardos NL, Hedges SB (2001) Molecular evidence for the early colonization of land by fungi and plants. Science 293:1129–1133

Hiruma K, Gerlach N, Sacristán S, Nakano RT, Hacquard S, Kracher B, Neumann U, Ramírez D, Bucher M, O'Connell RJ, Schulze-Lefert P (2016) Root endophyte *Colletotrichum tofieldiae* confers plant fitness benefits that are phosphate status dependent. Cell 165:464–474

Ho CL, Tan YC, Yeoh KA, Ghazali AK, Yee WY, Hoh CC (2016) *De novo* transcriptome analyses of host-fungal interactions in oil palm (*Elaeis guineensis* Jacq.) BMC Genomics 17:66

Hu Y, Xie S, Yao J (2016) Identification of novel reference genes suitable for qRT-PCR normalization with respect to the zebrafish developmental stage. PLoS One 11:e0149277

Huang X, Madan A (1999) CAP3: a DNA sequence assembly program. Genome Res 9:868–977

Hunter S, Jones P, Mitchell A, Apweiler R, Attwood TK, Bateman A, Bernard T, Binns D, Bork P, Burge S, de Castro E, Coggill P, Corbett M, Das U, Daugherty L, Duquenne L, Finn RD, Fraser M, Gough J, Haft D, Hulo N, Kahn D, Kelly E, Letunic I, Lonsdale D, Lopez R, Madera M, Maslen J, McAnulla C, McDowall J, McMenamin C, Mi H, Mutowo-Muellenet P, Mulder N, Natale D, Orengo C, Pesseat S, Punta M, Quinn AF, Rivoire C, Sangrador-Vegas A, Selengut JD, Sigrist CJ, Scheremetjew M, Tate J, Thimmajanarthanan M, Thomas PD, Wu CH, Yeats C, Yong SY (2012) InterPro in 2011: new developments in the family and domain prediction database. Nucleic Acids Res 40:D306–D312

Imaizumi-Anraku H, Takeda N, Charpentier M, Perry J, Miwa H, Umehara Y, Kouchi H, Murakami Y, Mulder L, Vickers K, Pike J, Downie JA, Wang T, Sato S, Asamizu E, Tabata S, Yoshikawa M, Murooka Y, Wu GJ, Kawaguchi M, Kawasaki S, Parniske M, Hayashi M (2004) Plastid proteins crucial for symbiotic fungal and bacterial entry into plant roots. Nature 433:527–531

Jänes J, Hu F, Lewin A, Turro E (2015) A comparative study of RNA-seq analysis strategies. Brief Bioinform 16:932–940

Kawahara Y, Oono Y, Kanamori H, Matsumoto T, Itoh T, Minami E (2012) Simultaneous RNA-seq analysis of a mixed transcriptome of rice and blast fungus interaction. PLoS One 7:e49423

Kim D, Pertea G, Trapnell C, Pimentel H, Kelley R, Salzberg SL (2013) TopHat2: accurate alignment of transcriptomes in the presence of insertions, deletions and gene fusions. Genome Biol 14:R36

Kistner C, Parniske M (2002) Evolution of signal transduction in intracellular symbiosis. Trends Plant Sci 7:511–518

Kubartová A, Ranger J, Berthelin J, Beguiristain T (2009) Diversity and decomposing ability of saprophytic fungi from temperate forest litter. Microb Ecol 58:98–107

Kunz BA, Dando PK, Grice DM, Mohr PG, Schenk PM, Cahill DM (2008) UV-induced DNA damage promotes resistance to the biotrophic pathogen *Hyaloperonospora parasitica* in *Arabidopsis*. Plant Physiol 148:1021–1031

Laluk K, Mengiste T (2010) Necrotroph attacks on plants: wanton destruction or covert extortion? Arabidopsis Book 8:e0136

Laluk K, Luo H, Chai M, Dhawan R, Lai Z, Mengiste T (2011) Biochemical and genetic requirements for function of the immune response regulator BOTRYTIS-INDUCED KINASE1 in plant growth, ethylene signaling, and PAMP-triggered immunity in *Arabidopsis*. Plant Cell 23:2831–2849

Langmead B, Salzberg SL (2012) Fast gapped-read alignment with Bowtie 2. Nat Methods 9:357–359

Larsen PE, Sreedasyam A, Trivedi G, Podila GK, Cseke LJ, Collart FR (2011) Using next generation transcriptome sequencing to predict an ectomycorrhizal metabolome. BMC Syst Biol 5:70

Lévy J, Bres C, Geurts R, Chalhoub B, Kulikova O, Duc G, Journet EP, Ané JM, Lauber E, Bisseling T, Dénarié J, Rosenberg C, Debellé F (2004) A putative Ca^{2+} and calmodulin-dependent protein kinase required for bacterial and fungal symbioses. Science 303:1361–1364

Li S, Tighe SW, Nicolet CM, Grove D, Levy S, Farmerie W, Viale A, Wright C, Schweitzer PA, Gao Y, Kim D, Boland J, Hicks B, Kim R, Chhangawala S, Jafari N, Raghavachari N, Gandara J, Garcia-Reyero N, Hendrickson C, Roberson D, Rosenfeld J, Smith T, Underwood JG, Wang M, Zumbo P, Baldwin DA, Grills GS, Mason CE (2014) Multi-platform assessment of transcriptome profiling using RNA-seq in the ABRF next-generation sequencing study. Nat Biotechnol 32:915–925

Lunter G, Goodson M (2011) Stampy: a statistical algorithm for sensitive and fast mapping of Illumina sequence reads. Genome Res 21:936–939

Mangul S, Caciula A, Al Seesi S, Brinza D, Măndoiu I, Zelikovsky A (2014) Transcriptome assembly and quantification from ion torrent RNA-Seq data. BMC Genomics 15(Suppl 5):S7

Martin JA, Wang Z (2011) Next-generation transcriptome assembly. Nat Rev Genet 12:671–682

Matilla MA, Ramos JL, Bakker PA, Doornbos R, Badri DV, Vivanco JM, Ramos-González MI (2010) *Pseudomonas putida* KT2440 causes induced systemic resistance and changes in *Arabidopsis* root exudation. Environ Microbiol Rep 2:381–388

Memelink J (2009) Regulation of gene expression by jasmonate hormones. Phytochemistry 70:1560–1570

Mendgen K, Hahn M (2002) Plant infection and the establishment of fungal biotrophy. Trends Plant Sci 7:352–356

Meyers BC, Galbraith DW, Nelson T, Agrawal V (2004) Methods for transcriptional profiling in plants. Be fruitful and replicate. Plant Physiol 135:637–652

Molitor A, Zajic D, Voll LM, Pons-K Hnemann J, Samans B, Kogel KH, Waller F (2011) Barley leaf transcriptome and metabolite analysis reveals new aspects of compatibility and *Piriformospora indica*-mediated systemic induced resistance to powdery mildew. Mol Plant Microbe Interact 24:1427–1439

Mora C, Tittensor DP, Adl S, Simpson AGB, Worm B (2011) How many species are there on earth and in the ocean? PLoS Biol 9:e1001127

Mortazavi A, Williams BA, McCue K, Schaeffer L, Wold B (2008) Mapping and quantifying mammalian transcriptomes by RNA-Seq. Nat Methods 5:621–628

Nettleton D (2006) A discussion of statistical methods for design and analysis of microarray experiments for plant scientists. Plant Cell 18:2112–2121

Ozsolak F, Milos PM (2011) RNA sequencing: advances, challenges and opportunities. Nat Rev Genet 12:87–98

Pandey V, Nutter RC, Prediger E (2008) Applied Biosystems SOLiDTM System: ligation-based sequencing. Wiley-VCH Verlag GmbH & Co. KGaA, Weinheim, pp 29–42

Perazzolli M, Moretto M, FontanaP FA, Velasco R, Moser C, Delledonne M, Pertot I (2012) Downy mildew resistance induced by *Trichoderma harzianum* T39 in susceptible grapevines partially mimics transcriptional changes of resistant genotypes. BMC Genomics 13:660

Pertea G, Huang X, Liang F, Antonescu V, Sultana R, Karamycheva S, Lee Y, White J, Cheung F, Parvizi B, Tsai J, Quackenbush J (2003) TIGR gene indices clustering tools (TGICL): a software system for fast clustering of large EST datasets. Bioinformatics 19:651–652

Pieterse CMJ, Leon-Reyes A, Van der Ent S, Van Wees SCM (2009) Networking by small-molecule hormones in plant immunity. Nat Chem Biol 5:308–316

Pirozynski KA, Malloch DW (1975) The origin of land plants: a matter of mycotrophism. Biosystems 6:153–164

Redecker D, Kodner R, Graham LE (2000) Glomalean fungi from the Ordovician. Science 289:1920–1921

Remy W, Taylor TN, Hass H, Kerp H (1994) Four hundred-million-year-old vesicular arbuscular mycorrhizae. Proc Natl Acad Sci U S A 91:11841–11843

Rothberg JM, Hinz W, Rearick TM, Schultz J, Mileski W, Davey M, Leamon JH, Johnson K, Milgrew MJ, Edwards M, Hoon J, Simons JF, Marran D, Myers JW, Davidson JF, Branting A, Nobile JR, Puc BP, Light D, Clark TA, Huber M, Branciforte JT, Stoner IB, Cawley SE, Lyons M, Fu Y, Homer N, Sedova M, Miao X, Reed B, Sabina J, Feierstein E, Schorn M, Alanjary M, Dimalanta E, Dressman D, Kasinskas R, Sokolsky T, Fidanza JA, Namsaraev E, McKernan KJ, Williams A, Roth GT, Bustillo J (2011) An integrated semiconductor device enabling non-optical genome sequencing. Nature 475:348–352

Rotter A, Camps C, Lohse M, Kappel C, Pilati S, Hren M, Stitt M, Coutos-Thévenot P, Moser C, Usadel B, Delrot S, Gruden K (2009) Gene expression profiling in susceptible interaction of grapevine with its fungal pathogen *Eutypa lata*: extending MapMan ontology for grapevine. BMC Plant Biol 9:104

Rudd JJ, Kanyuka K, Hassani-Pak K, Derbyshire M, Andongabo A, Devonshire J, Lysenko A, Saqi M, Desai NM, Powers SJ, Hooper J, Ambroso L, Bharti A, Farmer A, Hammond-Kosack KE, Dietrich RA, Courbot M (2015) Transcriptome and metabolite profiling of the infection cycle of *Zymoseptoria tritici* on wheat reveals a biphasic interaction with plant immunity involving differential pathogen chromosomal contributions and a variation on the hemibiotrophic lifestyle definition. Plant Physiol 167:1158–1185

Samolski I, de Luis A, Vizcaíno JA, Monte E, Suárez MB (2009) Gene expression analysis of the biocontrol fungus *Trichoderma harzianum* in the presence of tomato plants, chitin, or glucose using a high-density oligonucleotide microarray. BMC Microbiol 9:217

Scervino JM, Mesa MP, Della Mónica I, Recchi M, Moreno NS, Godeas A (2010) Soil fungal isolates produce different organic acid patterns involved in phosphate salts solubilization. Biol Fertil Soils 46:755–763

Schenk PM, Carvalhais LC, Kazan K (2012) Unraveling plant–microbe interactions: can multi-species transcriptomics help? Trends Biotechnol 30:177–184

Schulz MH, Zerbino DR, Vingron M, Birney E (2012) Oases: robust *de novo* RNA-seq assembly across the dynamic range of expression levels. Bioinformatics 28:1086–1092

Shendure J, Ji H (2008) Next-generation DNA sequencing. Nat Biotechnol 26:1135–1145

Simon L, Bousquet J, Lévesque RC, Lalonde M (1993) Origin and diversification of endomycorrhizal fungi and coincidence with vascular land plants. Nature 363:67–69

Stein M, Dittgen J, Sánchez-Rodríguez C, Hou BH, Molina A, Schulze-Lefert P, Lipka V, Somerville S (2006) *Arabidopsis* PEN3/PDR8, an ATP binding cassette transporter, contributes to nonhost resistance to inappropriate pathogens that enter by direct penetration. Plant Cell 18:731–746

Storey JD, Tibshirani R (2003) Statistical significance for genomewide studies. Proc Natl Acad Sci U S A 100:9440–9445

Stracke S, Kistner C, Yoshida S, Mulder L, Sato S, Kaneko T, Tabata S, Sandal N, Stougaard J, Szczyglowski K, Parniske M (2002) A plant receptor-like kinase required for both bacterial and fungal symbiosis. Nature 417:959–962

Su Z, Łabaj PP, Li S, Thierry-Mieg J, Thierry-Mieg D, Shi W, Wang C, Schroth GP, Setterquist RA, Thompson JF, Jones WD, Xiao W, Xu W, Jensen RV, Kelly R, Xu J, Conesa A, Furlanello C, Gao H, Hong H, Jafari N, Letovsky S, Liao Y, Lu F, Oakeley EJ, Peng Z, Praul CA, Santoyo-Lopez J, Scherer A, Shi T, Smyth GK, Staedtler F, Sykacek P, Tan XX, Thompson EA, Vandesompele J, Wang MD, Wang J, Wolfinger RD, Zavadil J, Auerbach SS, Bao W, Binder H, Blomquist T, Brilliant MH, Bushel PR, Cai W, Catalano JG, Chang CW, Chen T, Chen G, Chen R, Chierici M, Chu TM, Clevert DA, Deng Y, Derti A, Devanarayan V, Dong Z, Dopazo J, Du T, Fang H, Fang Y, Fasold M, Fernandez A, Fischer M, Furió-Tari P, Fuscoe JC, Caimet F, Gaj S, Gandara J, Gao H, Ge W, Gondo Y, Gong B, Gong M, Gong Z, Green B, Guo C, Guo L, Guo LW, Hadfield J, Hellemans J, Hochreiter S, Jia M, Jian M, Johnson CD, Kay S, Kleinjans J, Lababidi S, Levy S, Li QZ, Li L, Li L, Li P, Li Y, Li H, Li J, Li S, Lin SM, López FJ, Lu X, Luo H, Ma X, Meehan J, Megherbi DB, Mei N, Mu B, Ning B, Pandey A, Pérez-Florido J, Perkins

RG, Peters R, Phan JH, Pirooznia M, Qian F, Qing T, Rainbow L, Rocca-Serra P, Sambourg L, Sansone SA, Schwartz S, Shah R, Shen J, Smith TM, Stegle O, Stralis-Pavese N, Stupka E, Suzuki Y, Szkotnicki LT, Tinning M, Tu B, van Delft J, Vela-Boza A, Venturini E, Walker SJ, Wan L, Wang W, Wang J, Wang J, Wieben ED, Willey JC, Wu PY, Xuan J, Yang Y, Ye Z, Yin Y, Yu Y, Yuan YC, Zhang J, Zhang KK, Zhang W, Zhang W, Zhang Y, Zhao C, Zheng Y, Zhou Y, Zumbo P, Tong W, Kreil DP, Mason CE, Shi L (2014) A comprehensive assessment of RNA-seq accuracy, reproducibility and information content by the sequencing quality control consortium. Nat Biotechnol 32:903–914

Sugarbaker DJ, Richards WG, Gordon GJ, Dong L, De Rienzo A, Maulik G, Glickman JN, Chirieac LR, Hartman ML, Taillon BE, Du L, Bouffard P, Kingsmore SF, Miller NA, Farmer AD, Jensen RV, Gullans SR, Bueno R (2008) Transcriptome sequencing of malignant pleural mesothelioma tumors. Proc Natl Acad Sci U S A 105:3521–3526

Tan G, Liu K, Kang J, Xu K, Zhang Y, Hu L, Zhang J, Li C (2015) Transcriptome analysis of the compatible interaction of tomato with *Verticillium dahlia* using RNA-sequencing. Front Plant Sci 6:428

Teixeira PJ, Thomazella DP, Reis O, do Prado PF, do Rio MC, Fiorin GL, José J, Costa GG, Negri VA, Mondego JM, Mieczkowski P, Pereira GA (2014) High-resolution transcript profiling of the atypical biotrophic interaction between *Theobroma cacao* and the fungal pathogen *Moniliophthora perniciosa*. Plant Cell 26:4245–4269

Thomas RK, Nickerson E, Simons JF, Janne PA, Tengs T, Yuza Y, Garraway LA, Laframboise T, Lee JC, Shah K, O'Neill K, Sasaki H, Lindeman N, Wong KK, Borras AM, Gutmann EJ, Dragnev KH, Debiasi R, Chen TH, Glatt KA, Greulich H, Desany B, Lubeski CK, Brockman W, Alvarez P, Hutchison SK, Leamon JH, Ronan MT, Turenchalk GS, Egholm M, Sellers WR, Rothberg JM, Meyerson M (2006) Sensitive mutation detection in heterogeneous cancer specimens by massively parallel picoliter reactor sequencing. Nat Med 12:852–855

Tilgner H, Raha D, Habegger L, Mohiuddin M, Gerstein M, Snyder M (2013) Accurate identification and analysis of human mRNA isoforms using deep long read sequencing. G3 (Bethesda) 3:387–397

Trapnell C, Williams BA, Pertea G, Mortazavi A, Kwan G, van Baren MJ, Salzberg SL, Wold BJ, Pachter L (2010) Transcript assembly and quantification by RNA-Seq reveals unannotated transcripts and isoform switching during cell differentiation. Nat Biotechnol 28:511–515

Trapnell C, Roberts A, Goff L, Pertea G, Kim D, Kelley DR, Pimentel H, Salzberg SL, Rinn JL, Pachter L (2012) Differential gene and transcript expression analysis of RNA-seq experiments with TopHat and Cufflinks. Nat Protoc 7:562–578

Veronese P, Chen X, Bluhm B, Salmeron J, Dietrich R, Mengiste T (2004) The BOS loci of *Arabidopsis* are required for resistance to *Botrytis cinerea* infection. Plant J 40:558–574

Veronese P, Nakagami H, Bluhm B, Abuqamar S, Chen X, Salmeron J, Dietrich RA, Hirt H, Mengiste T (2006) The membrane-anchored botrytis-induced kinase1 plays distinct roles in *Arabidopsis* resistance to necrotrophic and biotrophic pathogens. Plant Cell 18:257–273

Wang Z, Gerstein M, Snyder M (2009) RNA-Seq: a revolutionary tool for transcriptomics. Nat Rev Genet 10:57–63

Weerasinghe RR, Bird DM, Allen NS (2005) Root-knot nematodes and bacterial Nod factors elicit common signal transduction events in *Lotus japonicus*. Proc Natl Acad Sci U S A 102:3147–3152

Westermann AJ, Gorski SA, Vogel J (2012) Dual RNA-seq of pathogen and host. Nat Rev Microbiol 10:618–630

Wise RP, Moscou MJ, Bogdanove AJ, Whitham SA (2007) Transcript profiling in host-pathogen interactions. Annu Rev Phytopathol 45:329–369

Wu TD, Nacu S (2010) Fast and SNP-tolerant detection of complex variants and splicing in short reads. Bioinformatics 26:873–881

Xie Y, Wu G, Tang J, Luo R, Patterson J, Liu S, Huang W, He G, Gu S, Li S, Zhou X, Lam TW, Li Y, Xu X, Wong GK, Wang J (2014) SOAPdenovo-Trans: *de novo* transcriptome assembly with short RNA-Seq reads. Bioinformatics 30:1660–1666

Xu XH, Wang C, Li SX, Su ZZ, Zhou HN, Mao LJ, Feng XX, Liu PP, Chen X, Snyder JH, Kubicek CP, Zhang CL, Lin FC (2015) Friend or foe: differential responses of rice to invasion by mutualistic or pathogenic fungi revealed by RNAseq and metabolite profiling. Sci Rep 5:13624

Zhu QH, Stephen S, Kazan K, Jin G, Fan L, Taylor J, Dennis ES, Helliwell CA, Wang MB (2012) Characterization of the defense transcriptome responsive to *Fusarium oxysporum*-infection in *Arabidopsis* using RNA-seq. Gene 512:259–266

Zhuang H, Fu Y, He W, Wang L, Wei Y (2015) Selection of appropriate reference genes for quantitative real-time PCR in *Oxytropis ochrocephala* Bunge using transcriptome datasets under abiotic stress treatments. Front Plant Sci 6:475

Zouari I, Salvioli A, Chialva M, Novero M, Miozzi L, Tenore GC, Bagnaresi P, Bonfante P (2014) From root to fruit: RNA-Seq analysis shows that arbuscular mycorrhizal symbiosis may affect tomato fruit metabolism. BMC Genomics 15:221

Chapter 2
Study of Oil Palm Photosynthesis Using Omics Technologies

Bee Keat Neoh, Huey Fang Teh, Yick Ching Wong, Tony Eng Keong Ooi, See Siang Cheah, and David R. Appleton

2.1 Introduction

Improving crop yield to meet the demands of increasing human population is a critical issue to be addressed. The increase in yield of crop plants can come from many avenues, such as better adaptation to environmental conditions, greater resistance to pests and diseases, improved agronomic practice, increased genetic yield potential and interaction of the above-mentioned factors (Gifford and Evans 1981).

According to Yamori (2013), improvement of photosynthesis can be done through the following efforts: (1) improving the *ribulose 1,5-biphosphate carboxylase/ hydrogenase (RuBisCO)* performance via quality control and/or quantity control, (2) increasing the thermo tolerance of *RuBisCO activase* to sustain *RuBisCO* activity under high temperatures, (3) enhancing CO_2 levels around *RuBisCO* to maximize catalytic rate and minimize photorespiration, (4) enhancing rate of chloroplast electron transport, (5) enhancing the capacity of metabolite transport processes and carbon utilization, (6) enhancing enzyme activity of Calvin cycle and (7) other innovations such as quantitative trait locus (QTL) analyses, phenomic screening, manipulation of mitochondrial respiration and improving photosynthesis under fluctuating light conditions. In an earlier review by Zelitch (1982), researchers have shown that improving photosynthetic efficiency through photosynthetic energy transduction and carbon dioxide (CO_2) assimilation in crops may help to increase crop yield significantly. He also mentioned in his review that several attempts were made to relate seasonal changes in canopy photosynthesis with yield by constructing

B.K. Neoh (✉) • H.F. Teh • Y.C. Wong • T.E.K. Ooi • D.R. Appleton
Sime Darby Technology Centre Sdn Bhd, UPM-MTDC Technology Centre III, Universiti Putra Malaysia, 1st Floor, Block B, 43400 Serdang, Selangor, Malaysia
e-mail: neoh.bee.keat@simedarby.com

S.S. Cheah
Plantation Research Advisory, Sime Darby Research,
Lot 2664, Jalan Pulau Carey, 42960 Pulau Carey, Selangor, Malaysia

S.N.A. Abdullah et al. (eds.), *Crop Improvement*,
DOI 10.1007/978-3-319-65079-1_2

a carbon budget in relation to dry matter accumulation in the field studies of soybean, barley, wheat, sorghum, maize and tobacco. These investigations showed a clear positive correlation between net photosynthesis and yield. A 33% increase in leaf photosynthesis may be translated into an 18% increase in biomass (Sinclair et al. 2004). On average, a doubling of the current CO_2 concentration ([CO_2]) in the field or laboratory chambers results in no increase in leaf area, but can cause a 23–58% increase in leaf photosynthetic rate (Drake et al. 1997) and an average of 35% increase in crop yield (Kimball 1983). The increase in [CO_2] has two effects on C_3 plants: an increase in leaf photosynthesis and a decrease in stomatal conductance to water vapour (*gs*) (Drake et al. 1997; Long et al. 2004). In addition, the elevated [CO_2] increases net leaf photosynthetic rate primarily by (1) competitive inhibition of the oxygenase activity of RuBisCO and therefore photorespiration and (2) acceleration of carboxylation because the CO_2 binding site is not saturated with CO_2 at the current [CO_2] (Long et al. 2006).

As oil palm appears to be generally a source-limited plant (Corley et al. 1971), information on the partitioning of nitrogen and carbon assimilated from photosynthetic (fully opened autotrophic) leaves to various sink organs, particularly developing mesocarp and non-photosynthetic (unopened spears/heterotrophic) leaves for reproductive and vegetative growth will be valuable to palm breeding and genomic studies. The correlation between the fat, sugars and protein nitrogen of olive leaves and fruits has been reported (Donaire et al. 1975), and supports that carbon and nitrogen are transported from leaves to fruit for lipid synthesis in the mesocarp. A study on oil palm leaf biomass, nitrogen, sugars and photosynthesis using leaf ranks −6 to 57 suggested that carbon isotope composition of heterotrophic spear leaves (<rank 0) was ^{13}C sugar-enriched due to remobilization of reserve carbon from source organs (Lamade et al. 2009). This indicated that in oil crops such as olive and oil palm, the carbohydrate source of fruit development and vegetative growth comes from photosynthesis.

In oil palm, bunch yield is in linear relationship to intercepted radiation per palm (Squire and Corley 1987), and the photosynthetic rate of individual leaves shows a curvilinear relationship with light intensity (Dufrene et al. 1990). At low light-intensity, light is the limiting factor, and the photosynthesis rate is directly related to light intensity. The "light compensation point" is the light intensity where photosynthesis and respiration are equal; below this light intensity, there is a net outflow of carbon dioxide from the leaf. At high light intensity, photosynthesis becomes light-saturated and the rate of carbon dioxide intake through stomata becomes the main limitation. Photosynthetic rate of leaves of young palms increases rapidly 2 months before the first bunch is harvested (Henson 1990). The rapid increase in photosynthesis rate of leaves of young palms coincides with the time where mesocarp commonly experiences intense growth and lipid accumulation (Corley et al. 1973; Corley and Tinker 2003; Badger 2009). It has been suggested that if the oil palm leaf had a longer period of high photosynthesis rate, that dry matter production and bunch yield would increase significantly (Suresh and Nagamani 2006).

The above-mentioned studies on oil palm photosynthesis suggested that the measurement of photosynthesis efficiency of seedlings and young palms may provide information on desired traits, such as yield, abiotic stress and disease tolerance.

However, direct photosynthesis measurement on individual adult palms may be difficult due to the large canopy and height, and therefore limit the number of measurements can be conducted per day. The problems arise when a large number of biological replicates are required for statistical significance. Therefore, Omics technologies may offer a solution to this problem by profiling the expression of photosynthesis-related genes, protein and metabolite concentrations to complement the physiological measurements related to photosynthesis.

2.2 Contribution of Omics Technologies to Photosynthesis Study

Various omics-based approaches are now being deployed to complement conventional breeding, and to improve the understanding and selection efficiency for specific physiological traits, including photosynthesis.

2.2.1 Genomics

Plant genomics is a study in genetics that applies DNA sequencing approaches combined with bioinformatics approaches to sequence, assemble, and analyse the function and structure of plant genomes. Chloroplasts are photosynthetic intracellular organelles, hence it is important to sequence the plant chloroplast genome. Our understanding of plant chloroplast genome evolution has greatly advanced through genomic sequencing and comparative analysis of chloroplast genomes from different species. For example, Uthaipaisanwong et al. (2012) characterized the oil palm chloroplast genome by using 454-pyrosequencing and the features of the characterized genome were then compared with those from date palm and rice. Consequently, they identified 32 RNA editing events in 19 chloroplast protein-coding genes of oil palm. Information from analysis of the chloroplast genome of natural population is useful for obtaining desirable agronomic traits, such as yield enhancement and resistance to pathogens (Daniell et al. 2016). This can be achieved by the discovery of useful alleles from genomic analysis as well as identification of regions of the genome in which diversity has been lost in domestic breeding. Desirable agronomic traits can then be achieved by breeding cultivated crops with their wild relatives (Brozynska et al. 2015).

2.2.2 Transcriptomics

Transcriptomics is a study of a complete set of RNA transcripts that are produced by the genome by using high-throughput methods, such as microarray analysis and RNA sequencing. To close the gap between the real and potential photosynthetic rates under field conditions, genome-wide identification and characterization of the nuclear genes for photosynthetic traits in a segregating population of poplar was performed using microarray and bulked segregant analysis (Wang et al. 2014a). A total of 515 differentially expressed genes were identified in the study revealing that photosynthesis regulation mainly involved genes in transport, metabolism and response to stimulus functions. Those differentially expressed genes were *ATP-binding cassette*, major facilitator superfamily protein (transport), *glucanase*, *beta-D-xylosidase*, *NADH dehydrogenase* (metabolism), *WRKY and chitinase* (stress response). Additionally, they suggested that stress responses could be a factor to be considered for photosynthetic improvement in the field.

With the reduction in cost of next generation sequencing (RNA-seq), many comparative transcriptomes for photosynthesis studies have recently been reported on different plant species such as maize (Chang et al. 2012; Ding et al. 2015), *Cleome* spp. (Bräutigam et al. 2011), *Staria* spp. (Xu et al. 2013; Rao et al. 2016), *Megathyrsus* spp. (Bräutigam et al. 2014), switchgrass (Serba et al. 2016), *Brassica* spp. (Mushtaq et al. 2016), foxtail, rice and sorghum (Ding et al. 2015). Many transcriptome comparative studies set out to investigate gene regulation differences between C_3 and C_4 plant species (Bräutigam et al. 2011, 2014; Xu et al. 2013; Ding et al. 2015; Rao et al. 2016). The leaves of C_4 crops such as maize and sorghum normally have better CO_2 fixation, along with better water and nitrogen use efficiencies compared to C_3 species such as rice. C_4 photosynthesis therefore affords higher efficiency of carbon conversion that enables plants to accumulate biomass at a faster rate compared to C_3 species. Understanding the C_4 photosynthesis pathway and mechanism may help to improve crop productivity; these pathway and mechanism can even be transferred to C_3 plants. Comparative transcriptome analysis provides a useful platform for candidate gene identification related to photosynthesis in C_4 plants, for examples, *mitochondrial decarboxylate carrier* and *phosphoenolpyruvate/phosphate translocator* (transport protein) (Bräutigam et al. 2011). Several recent transcriptome studies related to photosynthesis are summarized in Table 2.1.

2.2.3 *Proteomics*

Proteomics is the large-scale study of proteomes. A proteome is defined as the set of proteins found in a particular cell, tissue or organism. It is the aim of proteomics to elucidate all the proteins in a proteome, including protein abundances, protein modifications and protein–protein interactions. The proteomes of chloroplast, including its envelope, stroma and thylakoid membrane fractions, of several higher plants have been analysed (Friso et al. 2004; Pineda et al. 2010; de Luna-Valdez et al. 2015). In addition, the chloroplast proteomes of plant growing under various biotic and abiotic stresses have been described for rice, wheat, pea, as well as for the plant model *Arabidopsis thaliana* (Aro et al. 2005; Cui et al. 2005; Curto et al. 2006; Zhou et al. 2006; Wang et al. 2016). In their studies on the effect of 8 h darkness on *Arabidopsis* chloroplasts, Wang and colleagues identified 81 darkness-responsive

Table 2.1 Comparative transcriptome analysis summary

Reference	Plant species	Highlighted finding
Bräutigam et al. (2011)	*Cleome spinosa* and *Cleome gynandra*	Large quantities of Asp, Ala and pyruvate in C_4 leaves compared to C_3. The genes involved in transport and metabolism were also expressed higher in C_4 leaves including genes for *phosphate translocator, mitochondrial dicarboxylate carrier* and *adenylate kinase* family protein. The genes for transcription factors such as *AP2-EREBP, bZIP* and *zinc finger* also showed higher expression level in C_4 leaves
Chang et al. (2012)	*Zea mays*	Pathway analyses revealed differences between mesophyll and bundle sheath cells of maize in various functional categories. Mesophyll plays more important roles including light reaction, protein synthesis and folding. Bundle sheath cells play a role in transport, signalling, protein degradation etc. Potential candidate genes identified include those for *phosphoenolpyruvate carboxylase, NADP-MDH* and *dicarboxylate transporter*. The genes for many transcription factors were found to be differentially expressed, including *WRKY, NAC, MYB, bHLH* and *GRAS*
Xu et al. (2013)	*Setaria viridis*	Whole transcriptome analysis revealed 7056 simple sequence repeats from 60,751 transcripts assembled. Many genes related to C_4 photosynthesis were identified, including those for *PEPC, NADP-ME, MDH* and *PCK* from the study
Wang et al. (2014a)	*Populus*	The microarray study identified 163 up-regulated and 352 down-regulated genes. Gene enrichment identified 48 significant GO terms in biological processes and cell components. The candidate genes identified were mainly involved in transport, metabolism and response to stimulus function including *ATP-binding cassette*, major facilitator superfamily protein, *oligopeptide transport, S-type anion channel, NADPH oxidases, NADH dehydrogenase, WRKY* and *MYB*
Bräutigam et al. (2014)	*Megathyrsus maximus, Dichanthelium clandestinum*	The study indicated that the core C_4 cycles are similar between two species with only a few exceptions including subcellular location of acid production and up-regulation of genes in C_4 enzymes. Sucrose and starch synthesis are critical component in C_4 metabolism. *M. maximus* appears to have simpler C_4 metabolism compared to other plants with less adjustment involved within intercellular transport capacity and electron transfer

(continued)

Table 2.1 (continued)

Reference	Plant species	Highlighted finding
Ding et al. (2015)	*Zea mays, Sorghum bicolor, Oryza sativa, Setaria viridis*	From gene co-expression and differentially co-expression network analysis, 128 genes were found to be specific to C_4. Many genes which are associated with light reaction, starch and sucrose metabolism, transportation and transcription regulation were identified to be involved in C_4 photosynthesis. They include *PEPC, PPDK, NADP-MDH, NADP-ME, PPDK-RP, aspartate aminotransferase, triophosphate phosphate translocator, FBA, FBP, phosphoglucan phosphatase* and *sucrose transporter 1/2*
Serba et al. (2016)	Switchgrass	The comparison of lowland and upland genotypes revealed that transcripts related to photosynthesis efficiency and development, and photosystem reaction centre subunits were up-regulated in lowland genotype. These transcripts for *fatty acid desaturase, cupin superfamily protein*, elongation factor, *glycine-rich protein*, membrane protein and *zinc finger* transcription factor. However, catalase isozymes, *late embryogenesis abundant group I, photosulphokinases, S-adenosyl methionine* and *HLH* transcription factor were up-regulated in upland genotype. The SNPs markers detected in the study will be useful for trait mapping for breeding improvement
Mushtaq et al. (2016)	*Brassica* spp.	From the comparison of leaf transcriptome analysis on Brassica, almost all late biosynthetic genes of anthocyanin such as *dihydroflavonol four-reductase, anthocyanin synthase* and *transparent Testa 19* were up-regulated in all purple leaves. Three genes related to degradation of photo-damaged protein in *photosystem II* and light respiration were down-regulated including *FTSH protease 8, glycolate oxidase 1* and *glutamine synthetase 1;4*. The accumulation of anthocyanin might have potential physiological function related to photosynthesis

proteins (Wang et al. 2016). Most of the identified proteins are nucleus-encoded, thus suggesting that darkness response in chloroplast is closely controlled by the nucleus. Incidentally, the abundance of 17 ribosomal proteins that are known to be involved in protein synthesis decreased after the dark treatment. The findings revealed that darkness triggers the inhibition of protein translation in the chloroplast, inhibition of *Photosystem II (PSII)* resulting in preferential cyclic electron flow around *Photosystem I (PSI)* and starch degradation, as well as enhancement of redox capacity. The study has improved our knowledge of molecular regulatory mechanisms in chloroplasts under light-limiting conditions.

Comparative proteomics of tea leaves of normal and abnormal colours revealed differential protein abundance for proteins related to photosynthesis machinery, including *PSI, PSII, cytochrome b6/f complex*, photosynthetic electron transport, light-harvesting complex and *F-type ATPase* (Ma et al. 2016). The decrease of photosynthetic protein abundance was suggested to be associated with the colour changes in tea leaves. In another comparative study, the proteome of chloroplast envelope of pea (C_3 plant) was compared with that of maize (C_4 plant) (Bräutigam et al. 2008) revealing specific adaptations of the plastid envelope to C_4 photosynthesis and the proteins that were possibly required for maintaining C_4 metabolite fluxes. The data could be useful for improving the efficiency of photosynthesis in future C_4 crops. The information obtained from proteomics studies would help in gaining insights into roles and functions of proteins associated with photosynthesis. The knowledge may prove vital for crop modification, particularly in engineering better crops to cope with various stresses in the future.

2.2.4 Metabolomics

Metabolomics is a large-scale study of chemical processes involving metabolites by using various analytical platforms such as liquid chromatography–mass spectrometry, capillary electrophoresis–mass spectrometry, and gas chromatography–mass spectrometry (Sumner et al. 2003). Comparative analyses of C_4 and C_3 photosynthesis in developing leaves of maize and rice were previously conducted by using both integrated metabolomics and transcriptomics approaches (Wang et al. 2014b). Since C_4 and C_3 photosynthesis exhibit different efficiency in water and nitrogen consumption, the study was conducted to investigate whether C_3 plants having C_4 photosynthesis could have improved carbon fixation in a hot and arid situation. Metabolomics analysis of photosynthesis and the related primary metabolites were also conducted in transgenic rice plants with both overexpressed and knockout leaf *RuBisCO* using capillary electrophoresis–time-of-flight mass spectrometry (Suzuki et al. 2012). In rice seedlings with reduced *RuBisCO* content, Suzuki et al. (2012) found that there was a decrease in photosynthesis, starch and carbohydrate levels with a substantial accumulation of ribulose bisphosphate, diphosphate and triphosphate compounds of other nucleosides, ATP, ADP and amino acids. In addition, starch and carbohydrate levels decreased in these plants. Thus, genetic manipulation of *RuBisCO* widely affected C and N metabolism in rice. A lipidomics approach has been deployed for comprehensive analysis of lipid composition in crude palm oil and showed that galactolipids such as monogalactosyldiacylglycerol, digalactosyldiacylglycerol and sulphoguinovosyl diacylyglycerol are crucial for maintaining the function of the photosynthesis machinery and involved in photosynthesis and metabolic regulation under various stress conditions (Cheong et al. 2014).

2.2.5 Phenomics

Plant phenomics is the study of plant growth, architecture, performance, and composition using high-throughput methods of imaging, data acquisition and analysis (Araus et al. 2015; Walter et al. 2015). The phenotyping approach in studying photosynthesis involves the application of fluorescence imager in high-throughput estimation of leaf chlorophyll content and photochemical yield of *PS II*(Baker and Rosenqvist 2004). Leaf chlorophyll content is a valuable indicator of plant health since its content is affected by both biotic and abiotic stresses. According to Gitelson et al. (2003), the F735/F700 ratio of chlorophyll fluorescence at 735 nm (F735) to that in the range of 700 nm and 710 nm (F700) shows a good linear relationship with the chlorophyll content ($R^2 > 0.95$). Therefore, this ratio seems to be a suitable measurement for the estimation of chlorophyll content of leaves.

Phenomic data is essential to plant breeding programmes for yield enhancement. However, the process of obtaining phenotypic data is the most time consuming, costly and labour intensive step of many biological experiments (Flood et al. 2016). In addition, the precision of phenotypic data is compromised due to lack of uniformity and reproducibility of the measurements. Moreover, the complex multidimensional nature of phenotypes especially in the field environment poses additional challenges. Hence, high-throughput and precise phenotyping is needed to complement other omics-based approaches. High precision screening of plants under nursery or field conditions can be aided by using innovative agricultural technologies such as drone or fieldcopter with measuring system and image capture function (van der Wal et al. 2013). The information from transcriptomics, proteomics, metabolomics and other omics technologies will be precise only if proper phenotyping is done. The integration of all omics data will help to further enhance our understanding of plant photosynthesis.

2.3 Investigation of Photosynthesis in Oil Palm Using Omics Technologies

In this chapter, we review studies conducted on oil palm photosynthesis, focusing on the potential role of omic technologies in oil palm yield enhancement.

2.3.1 Yield

Overall palm oil yield can be attained by the combination of several positive bunch traits: bunch number per year, bunch weight, fruit per bunch, ratio of mesocarp thickness to fruit (M/F) and oil content in dry mesocarp (O/DM) (Appleton et al. 2014a, 2014b). Of these, two key contributors of yield that are highly heritable are ratios of M/F and O/DM.

The metabolite profiles of oil palm leaves have contributed to a distinct clustering of photosynthetic and non-photosynthetic leaves in principal component analysis (Neoh et al. 2013). It was found that the differences between these two leaf types were caused by differential abundance of several groups of metabolites such as sugars, amino acids and nucleosides. In the non-photosynthetic leaves, the relative abundance of sugar content was 21.3% lower than the photosynthetic leaves, presumably to support leaf growth during the fast elongation stage (Lamade et al. 2009; Kusano et al. 2011). Concordant with the lower amount of extract as a percentage of dry tissue weight obtained from the non-photosynthetic leaves, higher wet biomass was assumed compared to the photosynthetic leaves. The rapid increase in spear leaf biomass requires a carbon allocation presumably from both trunk apex and photosynthetic leaves. Total sugar content produced from photosynthesis may vary and is believed to be related to leaf rank (non-photosynthetic to photosynthetic) (Kusano et al. 2011; Ibrahim and Jaafar 2012). Adenine and many amino acids such as arginine, asparagine, proline, valine and tyrosine are more concentrated in non-photosynthetic compared to the photosynthetic leaves. In leaves, the biosynthesis of amino acids is developmentally regulated, usually being most active in young leaves (Galili 1995; Zhu-Shimoni 1997; Teija Ruuhola et al. 2003). This trend is similar to that observed in developing mesocarp tissue before lipid biosynthesis commences. It has been reported that nucleosides in young leaves were converted to salvage products (nucleotides and nucleic acids), degradation products and purine alkaloids, such as caffeine (Koyama et al. 2003). In coffee, almost 75% reduction of purine alkaloids is observed during leaf development (Deng and Ashihara 2010). The concentration of nucleosides is also found to be markedly lower in photosynthetic leaves (Neoh et al. 2013). The regulation of nucleosides also appears to play an important role in fruit development and lipid production in mesocarp (Neoh et al. 2013; Teh et al. 2013).

From our unpublished data, we observed that the photosynthetic-related genes were differentially expressed between high- and low-yielding palms with different M/F and O/DM traits. The differentially expressed genes included those encoding *chlorophyll a-b binding protein (CAB), PSI* and *ferredoxin*. All these genes encoded for the main functional proteins in the upstream of photosynthesis mechanism. They exhibited higher expression in the high-yielding population as compared to the low-yielding population (Fig. 2.1a).

The differential expression of these genes indicates that high-yielding palms may have high abundance of these proteins, suggesting they are better in light energy harvesting and electron transfer during photosynthesis, which then lead to higher lipid production in this population. It has been reported that the Single Nucleotide Polymorphisms (SNPs) in gene encoding *CAB* in barley were associated with at least one of the six agronomic traits including plant height, spike length, a number of grains per spike, 1000 grain weight, flag leaf area and leaf colour (Xia et al. 2012).

A higher concentration of G6P and sucrose was found in higher yielder palms analysed using CE-MS (G6P) and GC-MS (Sucrose) metabolomics platform (Neoh et al. 2013). In the same experiment, several metabolites in photosynthetic leaves were found to be associated to O/DM. In photosynthetic leaves, both glucose 6-phosphate (G6P) and sucrose, two of the products of photosynthesis, were found to show significant difference ($pV < 0.05$) between high- and low-yielding palms (Fig. 2.1b). Both G6P and sucrose were found to be more concentrated in high O/DM palms. Several studies reported the presence of a close correlation between the G6P concentration in the cytosol and photosynthetic sucrose production in wheat

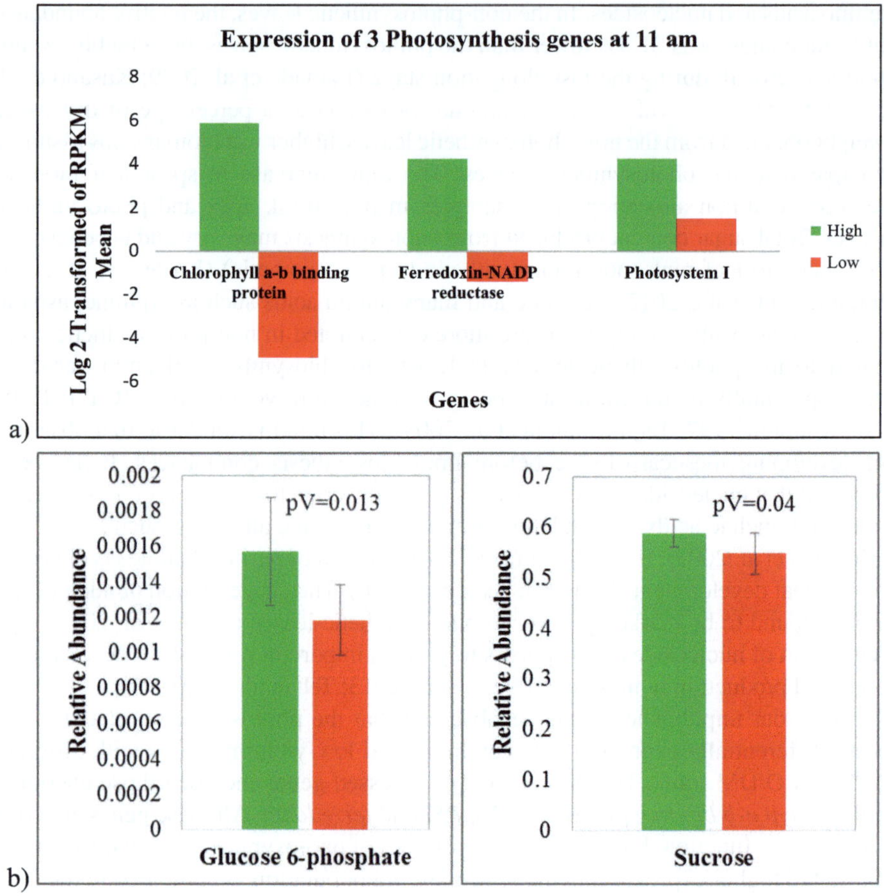

Fig. 2.1 Expression of photosynthesis-related genes and metabolites in oil palm leaf. (**a**) Differential expression of three photosynthesis-related genes in high and low yielder palms by RNAseq. Significant $Q > 0.85$ is explained by the probability of equivalent expression in Noiseq (Zheng and Moriyama 2013); (**b**) Abundance of G6P and sucrose in high and low yielder palms. pV indicated the *p*-value between high- and low-yielding palms, pV < 0.05 is considered significantly difference

protoplasts (Stitt et al. 1983). The increase in G6P and sucrose is attributed to increasing light intensities and CO_2 concentration, where both are key sources for photosynthesis. In oil palm, the elevated CO_2 concentration and interception radiation have led to the increase of biomass and bunch dry matter production (Squire and Corley 1987; Ibrahim et al. 2010).

2.3.2 Diurnal Cycle

Plants are exposed to changing environmental conditions, and one of the most common is the daily alternation between light and darkness. In an effort by Rees (1961) in monitoring midday closure of stomata in oil palm in Nigeria, he found that diurnal stomatal opening differed during the wet and the dry seasons. In the wet season, stomata opened early in the morning and remained wide open throughout the day before closing in the early evening. While in the dry season, partial closure of stomata occurred during the middle of the day (Rees 1961; Carr 2011). In Malaysia, Henson (Henson and Chang 1990; Henson 1991) found that stomatal conductance peaked in mid-morning and then progressively declined in the afternoon. The same diurnal pattern was obtained under clear sky conditions regardless of the age of the palms, although actual conductance increased with age.

Metabolic processes in plants and most other organisms operate in concert with day/night cycles. This coordination is accomplished by diurnal oscillations in transcriptional and posttranscriptional activities integrated with light, temperature, carbon status and circadian signalling (Usadel et al. 2008). Through this arrangement, metabolic pathways in the cell can sense and anticipate environmental cues (Mangelsen et al. 2010).

In our study, sampling of 20 adult palms was carried out at five time points (7 am, 11 am, 3 pm, 7 pm and 7 am the following day) to monitor the metabolites and expression changes in relation to the diurnal cycle. In the diurnal expression study, we observed the abundance of transcript encoding chlorophyll a-b binding protein reached its peak at 11 am and decreased slowly throughout diurnal cycle. The expression of the transcript encoding ferredoxin reached its peak, slightly later at 3 pm (supplementary data) and remained constant for several hours before eventually decreasing after sunset. As there was no measurement taken between 11 am and 3 pm, actual peak expression of these genes may occur during midday time. *CAB* which forms a complex structure with *PSI* and *II* captures solar energy for the excitation of the photosystems. The released electron from photosystems will be used to reduce *NADP*$^+$ to *NADPH* via *ferredoxin* and *ferredoxin-NADP reductase (FNR)*. The energy generated (ATP) is then transferred into the Calvin cycle, which is required by ribulose-biphosphate to capture carbon dioxide via *RuBisCO*. From our diurnal expression study, the expression of the gene encoding *CAB* reached the peak early into the sunlight hours, suggesting its importance in the capture of solar energy for photosynthesis under light-limiting conditions. The expression then appeared to decrease once light intensity was not a limiting factor. The expression continued to decrease until the beginning of the following diurnal cycle. *PSI* and *ferredoxin* were expressed at lower levels during the early stages of the diurnal cycle and increased gradually once light intensity increased and reached the peak expression at light saturation when photosynthesis reached its maximum (11 am–3 pm). When light intensity increases and more electrons are being transferred away from the photosystems, it is suggested that more *ferredoxin* and *FNR* are needed to reduce *NADP*$^+$ to *NADPH* and enter into Calvin cycle together with energy produced through *ATP synthase* complex. Study by Chang et al. (2017) also concluded that overexpression of *ferredoxin*-like protein improves production of rice by enhancing the capacity of photosynthetic carbon assimilation through higher efficiency of electron transport and gas exchange rate.

Metabolites such as sugar and sugar phosphates displayed a diurnal trend that was low during limited light (7–11 am), and peaked at midday (peak photosynthesis) at 11 am to 3 pm, and decreased as light reduced (3–7 pm) (Fig. 2.2). This is in agreement with the findings of Bläsing et al. (2005) whereby the photosynthetic carbon fixation in leaves supports the synthesis and export of sucrose to the remainder of the plant to support metabolism, storage, and growth during the day. At night, the plant becomes a net consumer of carbon. The supply of sugars from photosynthesis can increase the rates of nitrate and ammonium uptake and assimilation, the synthesis of organic acid acceptors, and the synthesis of amino acids (Stitt and Krapp 1999).

We also observed the accumulation of amino acids with the increase of sugars early in the day. The diurnal trend for Gln was similar to sugar where it peaked at 3 pm, whereas the concentration for other amino acids decreased after 11 am but increased again after 3 pm till the next morning (Fig. 2.2). In agreement with previous reports by Kovtun and Daie (1995) and Lam et al. (1995), we found the concentration of several amino acids (Fig. 2.2) were higher at night (except for Gln) compared to during the day time when photosynthesis was active. These amino acids are presumably important for nitrogen storage and transport for the allocation of nitrogen resources between source and sink organs, especially at the beginning of the night (Kovtun and Daie 1995; Lam et al. 1995). In oil palm, amino acids were found to be highly concentrated in high-yielding palms during mesocarp (sink) development (Teh et al. 2013), where 35% larger mesocarp was recorded in high-yielding palms compared to a low-yielding population. Teh et al. (2013) highlighted that the accumulation of amino acids at this stage could possibly support the production of proteins necessary for lipid biosynthesis and cell division required for mesocarp development.

2.3.3 *Abiotic Stresses*

Abiotic stresses in oil palm range from nutrient deficiency, flooding, climate change, drought and high salinity. In the oil palm industry, the El Nino weather effect is a serious concern, and prolonged drought can attribute to an 8% reduction in palm oil production in Malaysia (Shean 2016). The changes in weather pattern and low rainfall lead to soil water deficit and reduce bunch yield (Bakoumé et al. 2013; Gawankar et al. 2003). Reduced leaf production due to water stress by as much as 30% during the early growth phase and by 12.5% in the later growth phase has been observed (Gawankar et al. 2003). In addition, early closure of stomata due to moisture stress may reduce net photosynthesis, and ultimately fresh fruit bunch (FFB) production (Smith 1989). Other than that, photosynthesis-related parameters, such as content of

a)
b)

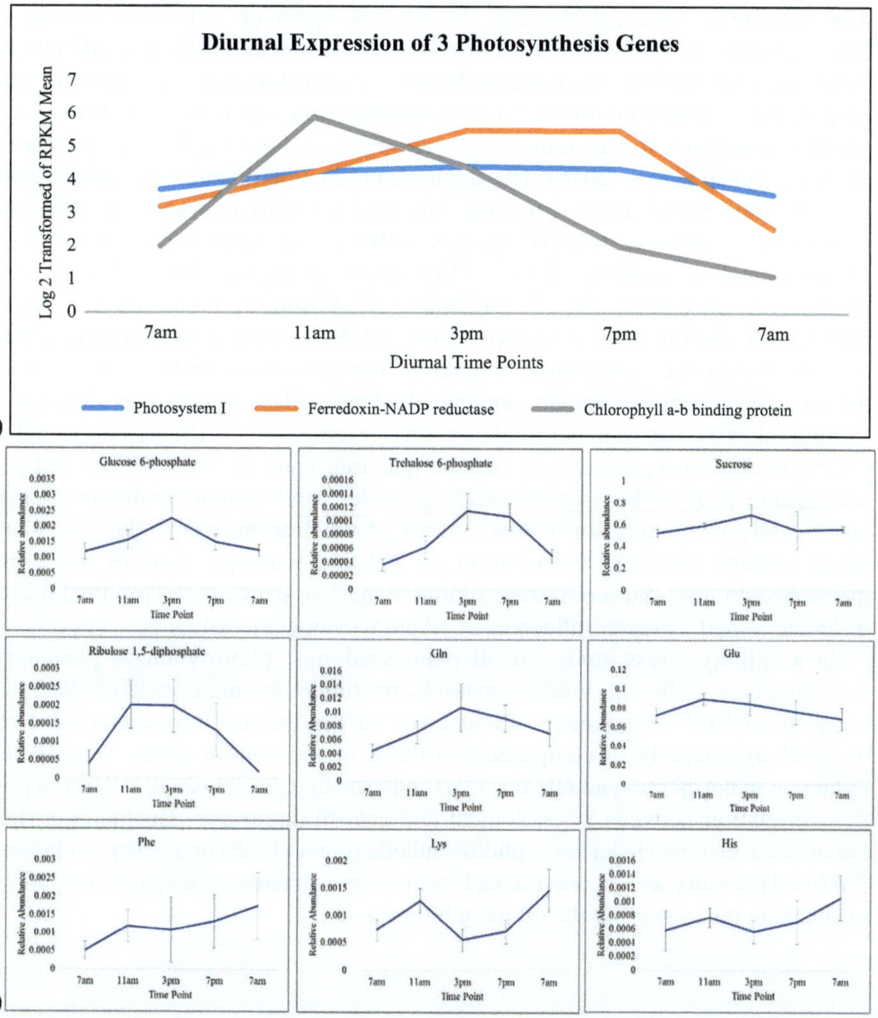

Fig. 2.2 Diurnal changes of photosynthesis-related genes and metabolites. (**a**) Expression of *Photosystem I, chlorophyll a-b binding protein* and *ferredoxin-NADP reductase* throughout diurnal cycle; (**b**) different trends of metabolites changes monitored throughout diurnal cycle from limited light to peak and back to limited light

chlorophyll a (chla), chlorophyll b (chlb) and total chlorophyll (chl), along with density, length and width of stomatal aperture, stomatal conductance and transpiration rate, photosynthetic rate and photosynthetic activity per plant have been reported to be used for selecting drought tolerant material for oil palm breeding (Kallarackal 1996; Suresh et al. 2010, 2012; Chaum et al. 2012; Méndez et al. 2012; Cha-um et al. 2013; Jazayeri et al. 2015; Putra and Purwanto 2015; Azzeme et al. 2016). According to Méndez et al. (2012), the selection of potential genotype for drought tolerant breeding material should encompass the following criteria: (1) highest photosynthetic rate, (2) the lowest respiratory rate, (3) highest efficiency to move its assimilates mainly towards the roots and (4) the ability to adjust its water potential (active accumulation of sugars) during stress. The reduction of chl content and the ratio of chla to chlb (chla:chlb) were significant in drought-stressed oil palm seedings (Azzeme et al. 2016). The significant reduction of chla was closely related to *PSII* deficiency. Genes encoding ethylene responsive binding protein, *late embryogenesis abundant (LEA)*, *dehydrin (DHN)*, cold-induced, *heat shock protein 70* and *metallothionein type 2* were differentially up-regulated in the leaves, while in the roots, only genes encoding *LEA* and *DHN* were up-regulated. The proline content increased in both vegetative tissues, while the total soluble protein content was affected by increasing drought severity. The activity of catalase was highest in the roots during severe drought, while guaicol peroxidase activity was shown to be the highest in the leaves under mild drought (Azzeme et al. 2016). In a polyethylene glycol induced drought stress, maximum quantum yield of *PSII* (Fv/Fm) and photon yield of *PSII* in the oil palm seedlings under water deficit conditions dropped significantly in comparison to the control group, leading to a reduction in net-photosynthetic rate (*Pn*) (Cha-Um et al. 2010b). Although some of the above-mentioned drought symptoms were similar to nutrient stress, it was reported that the water stress had a greater influence on oil palm (Sun et al. 2011).

In a salinity stress study on oil palm seedlings, photosynthesis phenotypes including chla, chlb, chl, total carotenoids, maximum quantum yield of *PSII*, photon yield of *PSII* and quantum efficiency of *PSII* in the seedlings under salt stress dropped significantly in comparison to those of the control group, leading to a reduction in net-photosynthetic rate (*Pn*) and growth (Cha-Um et al. 2010a). A positive correlation between physiological and growth parameters (including sodium ion, relative electrolyte leakage, photosynthetic pigments, *Pn* and water oxidation in *PSII*) and plant dry weight was found, in which, the salt stress affected photosynthesis mechanism thus lowering the oil palm biomass.

*2.3.4 Diseases (***Ganoderma** *and* **Fusarium***)*

Using a proteomics approach, Daim et al. (2015) found oil palm leaves infected with *Ganoderma boninense* revealed changes in proteins involved in photosynthesis. The abundance of photosynthesis-related enzymes such as *RuBisCO*, *RuBisCO activase*, *CAB* several different subunits of *ATP synthase* and *oxygen evolving enhancer protein 1 (OEE1)* of the light-harvesting complex were reduced in infected palms. Previous plant–fungi interaction studies have shown that reduction in photosynthesis rate during the infection process is attributed to a sugar-mediated repression of photosynthetic genes (Biemelt and Sonnewald 2006). However, the protein abundance of *FNR* from *PSI* was found to be increased two-fold by Daim et al. (2015). *Ferredoxin* may participate in other reactions in the chloroplast, including nitrogen and sulphur assimilation, amino acid, and fatty acid synthesis and also redox regulation (Bilgin et al. 2010). The increase of *ferredoxin* may reflect its participation in pathogen defence. De novo transcriptomics study of host–fungal interactions in oil palm was carried out by comparing the root transcriptomes of untreated oil palm seedlings with those inoculated with *G. boninense* and *T. harzianum*, respectively (Ho et al. 2016). Ho et al. (2016) found that among the down-regulated genes in *Ganoderma*-treated oil palm roots were those related to the generation of precursor metabolites and energy, photosynthesis, electron transport chain, ribosome biogenesis, *xyloglucan:xyloglucosyl transferase* activity and others. Both findings showed that the photosynthesis-related genes were down-regulated in *Ganoderma* infected palm.

Unlike *Ganoderma* which is prevalent in Malaysia coastal areas, *Fusarium* wilt is the most destructive disease of oil palm in Africa and causes severe losses in Côte d' Ivoire, Nigeria, Ghana, Cameroon and Congo (Flood 2006). To date, the photosynthesis of oil palms infected with *Fusarium* wilt has not been thoroughly studied. According to Mepsted et al. (1995), the stomata conductance and internal CO_2 decreased along with an increase of photosynthesis in *Fusarium* infected oil palm seedlings. According to the author, the result might be due to non-uniform stomata closure, as reflected in the negative correlation of stomata conductivity to photosynthesis rate. Therefore, further study is needed to determine if the cause of photosynthesis rate changes in *Fusarium* wilt is entirely due to by stomatal closure or the other environmental effects.

2.4 Conclusion

Photosynthesis is a mechanism that fuels plant growth and development through the conversion of light energy, carbon dioxide and water into sugars and carbohydrates. The effort to increase photosynthesis efficiency in oil palm is an important initiative to improve oil yield potential through the selection of genotypes with higher expression of photosynthesis genes such as *CAB, FNR* and *PSI*. Omics platforms can contribute to the selection of breeding materials that can perform better under light-limiting conditions, such as shading from neighbouring palms and cloud cover, and that have overall increases in photosynthesis early and late in the day. In fact, photosynthesis is also a useful tool to phenotype and determines the condition of palms under abiotic or biotic stresses. Omics studies can complement marker assisted breeding by providing more accurate or physiologically linked traits. Identification of selection markers associated with these traits can then increase the accuracy of drought and *Ganoderma* tolerant breeding material identification and selection, which is critical to the oil palm industry in Malaysia.

Acknowledgement The authors would like to thank Prof R. Hereward V. Corley for his contribution and proof reading of this book chapter.

References

Appleton DR, Teh HF, Neoh BK, Ooi TEK, Wong YC, Harikrisna K (2014a) Omics reveals subtle changes in carbon flux that lead to increased oil biosynthesis in oil palm. American Chemical Oil Society. http://www.aocs.org/Membership/informArticleDetail.cfm?itemnumber=40431

Appleton DR, Teh HF, Neoh BK, Ooi TEK, Wong YC, Kwong QB, Yusof HM, Chew FT, Harikrisna K (2014b) Omics: Mesocarp biochemistry provides insight into increased oil palm yield. Planters 90:241–254

Araus JL, Elazab A, Vergara O, Cabrera-Bosquet L, Serret MD, Zaman-Allah M, Cairns JE (2015) New Technologies for Phenotyping. In: Fritsche-Neto R, Borém A (eds) Phenomics: how next-generation phenotyping is revolutionizing plant breeding. Springer, New York, pp 1–14

Aro E, Suorsa M, Rokka A, Allahverdiyeva Y, Paakkarinen V, Saleem A, Battchikova N, Rintamäki E (2005) Dynamics of photosystem II: a proteomic approach to thylakoid protein complexes. J Exp Bot 56:347–356

Azzeme AM, Abdullah SNA, Aziz MA, Wahab PEM (2016) Oil palm leaves and roots differ in physiological response, antioxidant enzyme activities and expression of stress-responsive genes upon exposure to drought stress. Acta Physiol Plant 38:1–12

Badger MR (2009) Role of plant leaf development in optimising photosynthetic efficiency, capacity, growth and yield. Applying photosynthesis research to improvement of food crops. pp 20–26

Baker NR, Rosenqvist E (2004) Applications of chlorophyll fluorescence can improve crop production strategies: an examination of future possibilities. J Exp Bot 55:1607–1621

Bakoumé C, Norhazela S, Shahrakbah Y, Cheah SS, Mohamad Nazeeb AT (2013) Improved method for estimating soil moisture deficit in oil palm (*Elaeis guineensis* Jacq.) areas with limited climatic data. J Agric Sci 5:57–65

Biemelt S, Sonnewald U (2006) Plant-microbe interactions to probe regulation of plant carbon metabolism. J Plant Physiol 163:307–318

Bilgin DD, Zavala JA, Zhu J, Clough SJ, Ort DR, DeLUCIA E (2010) Biotic stress globally downregulates photosynthesis genes. Plant Cell Environ 33:1597–1613

Bläsing OE, Gibon Y, Günther M, Höhne M, Morcuende R, Osuna D, Thimm O, Usadel B, Scheible WR, Stitt M (2005) Sugars and circadian regulation make major contributions to the global regulation of diurnal gene expression in Arabidopsis. Plant Cell 17:3257–3281

Bräutigam A, Hoffmann-Benning S, Weber AP (2008) Comparative proteomics of chloroplast envelopes from C3 and C4 plants reveals specific adaptations of the plastid envelope to C4 photosynthesis and candidate proteins required for maintaining C4 metabolite fluxes. Plant Physiol 148:568–579

Bräutigam A, Kajala K, Wullenweber J, Sommer M, Gagneul D, Weber KL, Carr KM, Gowik U, Maß J, Lercher MJ, Westhoff P, Hibberd JM, Weber APM (2011) An mRNA blueprint for C4 photosynthesis derived from comparative transcriptomics of closely related C3 and C4 species. Plant Physiol 155:142–156

Bräutigam A, Schliesky S, Külahoglu C, Osborne CP, Weber APM (2014) Towards an integrative model of C(4) photosynthetic subtypes: insights from comparative transcriptome analysis of NAD-ME, NADP-ME, and PEP-CK C(4) species. J Exp Bot 65:3579–3593

Brozynska M, Furtado A, Henry RJ (2015) Genomics of crop wild relatives: expanding the gene pool for crop improvement. Plant Biotechnol J 14:1070–1085

Carr M (2011) The water relations and irrigation requirements of oil palm (*Elaeis guineensis*): a review. Exp Agric 47:629–652

Chang YM, Liu WY, Shih AC, Shen MN, Lu CH, Lu MY, Yang HW, Wang TY, Chen SC, Chen SM, Li WH, Ku MS (2012) Characterizing regulatory and functional differentiation between maize mesophyll and bundle sheath cells by transcriptomic analysis. Plant Physiol 160:165–177

Chang H, Huang HE, Cheng CF, Ho MH, Ger MJ (2017) Constitutive expression of a plant ferredoxin-like protein (pflp) enhances capacity of photosynthetic carbon assimilation in rice (*Oryza sativa*). Transgenic Res 26:279–289

Chaum S, Takabe T, Kirdmanee C (2012) Physio-biochemical responses of oil palm (*Elaeis guineensis* Jacq.) seedlings to mannitol-and polyethylene glycol-induced iso-osmotic stresses. Plant Production Sci 15:65–72

Cha-Um S, Takabe T, Kirdmanee C (2010a) Ion contents, relative electrolyte leakage, proline accumulation, photosynthetic abilities and growth characters of oil palm seedlings in response to salt stress. Pak J Bot 42:2191–2020

Cha-Um S, Takabe T, Kirdmanee C (2010b) Osmotic potential, photosynthetic abilities and growth characters of oil palm (*Elaeis guineensis* Jacq.) seedlings in responses to polyethylene glycol-induced water deficit. Afr J Biotechnol 9:6509–6516

Cha-Um S, Yamada N, Takabe T, Kirdmanee C (2013) Physiological features and growth characters of oil palm (*Elaeis guineensis* Jacq.) in response to reduced water-deficit and rewatering. Aust J Crop Sci 7:432

Cheong WF, Wenk MR, Shui G (2014) Comprehensive analysis of lipid composition in crude palm oil using multiple lipidomic approaches. J Genet Genomics 41:293–304

Corley R, Tinker P (2003) Growth, flowering and yield. In: The oil palm, 4th edn. Blackwell Science, Oxford, pp 89–131

Corley R, Hardon J, Tan G (1971) Analysis of growth of the oil palm (*Elaeis guineensis* Jacq.) I. Estimation of growth parameters and application in breeding. Euphytica 20:307–315

Corley R, Hardon J, Ooi S (1973) Some evidence for genetically controlled variation in photosynthetic rate of oil palm seedlings. Euphytica 22:48–55

Cui S, Huang F, Wang J, Ma X, Cheng Y, Liu J (2005) A proteomic analysis of cold stress responses in rice seedlings. Proteomics 5:3162–3172

Curto M, Camafeita E, Lopez JA, Maldonado AM, Rubiales D, Jorrín JV (2006) A proteomic approach to study pea (*Pisum sativum*) responses to powdery mildew (*Erysiphe pisi*). Proteomics 6:S163–S174

Daim LDJ, Ooi TEK, Ithnin N, Mohd Yusof H, Kulaveerasingam H, Abdul Majid N, Karsani SA (2015) Comparative proteomic analysis of oil palm leaves infected with *Ganoderma boninense* revealed changes in proteins involved in photosynthesis, carbohydrate metabolism, and immunity and defense. Electrophoresis 36:1699–1710

Daniell H, Lin CS, Yu M, Chang WJ (2016) Chloroplast genomes: diversity, evolution, and applications in genetic engineering. Genome Biol 17:134

Deng WW, Ashihara H (2010) Profiles of purine metabolism in leaves and roots of *Camellia sinensis* seedlings. Plant Cell Physiol 51:2105–2118

Ding Z, Weissmann S, Wang M, Du B, Huang L, Wang L, Tu X, Zhong S, Myers C, Brutnell TP, Sun Q, Li P (2015) Identification of photosynthesis-associated C4 candidate genes through comparative leaf gradient transcriptome in multiple lineages of C3 and C4 species. PLoS One 10:e0140629

Donaire JP, Sánchez AJ, López-Gorgé J, Recalde L (1975) Metabolic changes in fruit and leaf during ripening in the olive. Phytochemistry 14:1167–1169

Drake BG, Gonzàlez-Meler MA, Long SP (1997) More efficient plants: a consequence of rising atmospheric CO2? Annu Rev Plant Physiol Plant Mol Biol 48:609–639

Dufrene E, Ochs R, Saugier B (1990) Oil palm photosynthesis and productivity linked to climatic factors. Oleagineux 45:345–355

Flood J (2006) A review of Fusarium wilt of oil palm caused by *Fusarium oxysporum* f. sp. *elaeidis*. Phytopathology 96:660–662

Flood PJ, Kruijer W, Schnabel SK, van der Schoor R, Jalink H, Snel JFH, Harbinson J, Aarts MGM (2016) Phenomics for photosynthesis, growth and reflectance in Arabidopsis Thaliana reveals circadian and long-term fluctuations in heritability. Plant Methods 12:1–14

Friso G, Giacomelli L, Ytterberg AJ, Peltier J-B, Rudella A, Sun Q, van Wijk KJ (2004) In-depth analysis of the thylakoid membrane proteome of *Arabidopsis thaliana* chloroplasts: new proteins, new functions, and a plastid proteome database. Plant Cell 16:478–499

Galili G (1995) Regulation of lysine and threonine synthesis. Plant Cell 7:899–906

Gawankar M, Devmore J, Jamadagni B, Sagvekar V, Khan HH (2003) Effect of water stress on growth and yield of Tenera oil palm. J Appl Hortic 5:39–40

Gifford R, Evans L (1981) Photosynthesis, carbon partitioning, and yield. Annu Rev Plant Physiol 32:485–509

Gitelson AA, Gritz Y, Merzlyak MN (2003) Relationships between leaf chlorophyll content and spectral reflectance and algorithms for non-destructive chlorophyll assessment in higher plant leaves. J Plant Physiol 160:271–282

Henson I (1990) Photosynthesis and source-sink relationships in oil palm (*Elaeis guineensis*). Malaysian Soc Plant Physiol 1:165–171

Henson I (1991) Age-related changes in stomatal and photosynthetic characteristics of leaves of oil palm (*Elaeis guineensis Jacq*). Elaeis 3:336–348

Henson IE, Chang KC (1990) Evidence for water as a factor limiting performance of field palms in West Malaysia. Proceedings of 1989 PORIM international palm oil development conference-agriculture. Palm Oil Research Institute of Malaysia, Kuala Lumpur. pp 487–498

Ho CL, Tan YC, Yeoh KA, Ghazali AK, Yee WY, Hoh CC (2016) De novo transcriptome analyses of host-fungal interactions in oil palm (*Elaeis guineensis* Jacq.) BMC Genomics 17:66

Ibrahim MH, Jaafar HZ (2012) Impact of elevated carbon dioxide on primary, secondary metabolites and antioxidant responses of *Eleais guineensis* Jacq. (oil palm) seedlings. Molecules 17:5195–5211

Ibrahim MH, Jaafar HZE, Harun MH, Yusop MR (2010) Changes in growth and photosynthetic patterns of oil palm (*Elaeis guineensis* Jacq.) seedlings exposed to short-term CO_2 enrichment in a closed top chamber. Acta Physiol Plant 32:305–313

Jazayeri SM, Rivera YD, Camperos-Reyes JE, Romero HM (2015) Physiological effects of water deficit on two oil palm (*Elaeis guineensis* Jacq.) genotypes. Agronomía Colomb 33:164

Kallarackal J (1996) Water relations and photosynthesis of the oil palm in Peninsular India. KFRI Research Report

Kimball BA (1983) Carbon dioxide and agricultural yield: an assemblage and analysis of 430 prior observations. Agron J 75:779–788

Kovtun Y, Daie J (1995) End-product control of carbon metabolism in culture-grown sugar beet plants (molecular and physiological evidence on accelerated leaf development and enhanced gene expression). Plant Physiol 108:1647–1656. doi:10.1104/pp.108.4.1647

Koyama Y, Tomoda Y, Kato M, Ashihara H (2003) Metabolism of purine bases, nucleosides and alkaloids in theobromine-forming *Theobroma cacao* leaves. Plant Physiol Biochem 41:977–984

Kusano M, Fukushima A, Redestig H, Saito K (2011) Metabolomic approaches toward understanding nitrogen metabolism in plants. J Exp Bot 62:1439–1453

Lam HM, Coschigano K, Schultz C, Melo-Oliveira R, Tjaden G, Oliveira I, Ngai N, Hsieh MH, Coruzzi G (1995) Use of Arabidopsis mutants and genes to study amide amino acid biosynthesis. Plant Cell 7:887–898

Lamade E, Setiyo IE, Girard S, Ghashghaie J (2009) Changes in 13C/12C of oil palm leaves to understand carbon use during their passage from heterotrophy to autotrophy. Rapid Commun Mass Spectrom 23:2586–2596

Long SP, Ainsworth EA, Rogers A, Ort DR (2004) Rising atmospheric carbon dioxide: plants FACE the future. Annu Rev Plant Biol 55:591–628

Long SP, Zhu X-G, Naidu SL, Ort DR (2006) Can improvement in photosynthesis increase crop yields? Plant Cell Environ 29:315–330

de Luna-Valdez LA, León-Mejía P, Encarnación-Guevara S, Guevara-García AA (2015) Chloroplast omics. In: Barh D, Khan SM, Davies E (eds) Plant omics: the omics of plant science. Springer India, New Delhi, pp 533–558

Ma C, Cao J, Li J, Zhou B, Tang J, Miao A (2016) Phenotypic, histological and proteomic analyses reveal multiple differences associated with chloroplast development in yellow and variegated variants from *Camellia sinensis*. Sci Rep 6:33369

Mangelsen E, Wanke D, Kilian J, Sundberg E, Harter K, Jansson C (2010) Significance of light, sugar, and amino acid supply for diurnal gene regulation in developing barley caryopses. Plant Physiol 153:14–33

Méndez YDR, Chacón LM, Bayona CJ, Romero HM (2012) Physiological response of oil palm interspecific hybrids (*Elaeis oleifera* HBK Cortes versus *Elaeis guineensis* Jacq.) to water deficit. Braz J Plant Physiol 24:273–280

Mepsted R, Flood J, Cooper R (1995) Fusarium wilt of oil palm II. Stunting as a mechanism to reduce water stress. Physiol Mol Plant Pathol 46:373–387

Mushtaq MA, Pan Q, Chen D, Zhang Q, Ge X, Li Z (2016) Comparative leaves transcriptome analysis emphasizing on accumulation of anthocyanins in brassica: molecular regulation and potential interaction with photosynthesis. Front Plant Sci 7:311

Neoh BK, Teh HF, Ng TLM, Tiong SH, Thang YM, Mohd Amiron E, Mohaimi M, Chew FT, Harikrishna K, Appleton DR (2013) Profiling of metabolites in oil palm mesocarp at different stages of oil biosynthesis. J Agric Food Chem 61:1920–1927

Pineda M, Sajnani C, Barón M (2010) Changes induced by the pepper mild mottle tobamovirus on the chloroplast proteome of *Nicotiana benthamiana*. Photosynth Res 103:31–45

Putra E, Purwanto B (2015) Physiological responses of oil palm seedlings to the drought stress using boron and silicon applications. J Agron 14:49

Rao X, Lu N, Li G, Nakashima J, Tang Y, Dixon RA (2016) Comparative cell-specific transcriptomics reveals differentiation of C4 photosynthesis pathways in switchgrass and other C4 lineages. J Exp Bot 67:1649–1662

Rees A (1961) Midday closure of stomata in the oil palm *Elaeis guineensis*. J Exp Bot 12:129–146

Serba DD, Uppalapati SR, Krom N, Mukherjee S, Tang Y, Mysore KS, Saha MC (2016) Transcriptome analysis in switchgrass discloses ecotype difference in photosynthetic efficiency. BMC Genomics 17:1040

Shean M (2016) MALAYSIA: El Nino takes a bite out of 2015/16 palm oil production. Commodity Intelligence Report, Foreign Agriculture Service, United States Department of Agriculture

Sinclair TR, Purcell LC, Sneller CH (2004) Crop transformation and the challenge to increase yield potential. Trends Plant Sci 9:70–75

Smith B (1989) The effects of soil water and atmospheric vapour pressure deficit on stomatal behaviour and photosynthesis in the oil palm. J Exp Bot 40:647–651

Squire G, Corley R (1987) Oil palm. Tree Crop Physiol 18:141

Stitt M, Krapp A (1999) The interaction between elevated carbon dioxide and nitrogen nutrition: the physiological and molecular background. Plant Cell Environ 22:583–621

Stitt M, Wirtz W, Heldt HW (1983) Regulation of sucrose synthesis by cytoplasmic fructosebisphosphatase and sucrose phosphate synthase during photosynthesis in varying light and carbon dioxide. Plant Physiol 72:767–774

Sumner LW, Mendes P, Dixon RA (2003) Plant metabolomics: large-scale phytochemistry in the functional genomics era. Phytochemistry 62:817–836

Sun C, Cao H, Shao H, Lei X, Xiao Y (2011) Growth and physiological responses to water and nutrient stress in oil palm. Afr J Biotechnol 10:10465–10471

Suresh K, Nagamani C (2006) Variations in photosynthetic rate and associated parameters with age of oil palm leaves under irrigation. Photosynthetica 44:309–311

Suresh K, Nagamani C, Ramachandrudu K, Mathur R (2010) Gas-exchange characteristics, leaf water potential and chlorophyll a fluorescence in oil palm (*Elaeis guineensis* Jacq.) seedlings under water stress and recovery. Photosynthetica 48:430–436

Suresh K, Nagamani C, Kantha D, Kumar M (2012) Changes in photosynthetic activity in five common hybrids of oil palm (*Elaeis guineensis* Jacq.) seedlings under water deficit. Photosynthetica 50:549–556

Suzuki Y, Fujimori T, Kanno K, Sasaki A, Ohashi Y, Makino A (2012) Metabolome analysis of photosynthesis and the related primary metabolites in the leaves of transgenic rice plants with increased or decreased Rubisco content. Plant Cell Environ 35:1369–1379

Teh HF, Neoh BK, MPL H, JYS L, TLM N, Nalisha I, Thang YM, Mohaimi M, Chew FT, Harikrishna K, Appleton DR (2013) Differential metabolite profiles during fruit development in high yielding oil palm mesocarp. PLoS One 8:e61344

Teija Ruuhola VO, Lempa K, Haukioja E (2003) Amino acids during development of mountain birch leaves. Chemoecology 13:95–101

Usadel B, Bläsing OE, Gibon Y, Retzlaff K, Höhne M, Günther M, Stitt M (2008) Global transcript levels respond to small changes of the carbon status during progressive exhaustion of carbohydrates in Arabidopsis rosettes. Plant Physiol 146:1834–1861

Uthaipaisanwong P, Chanprasert J, Shearman JR, Sangsrakru D, Yoocha T, Jomchai N, Jantasuriyarat C, Tragoonrung S, Tangphatsornruang S (2012) Characterization of the chloroplast genome sequence of oil palm (Elaeis guineensis Jacq.) Gene 500:172–180

van der Wal T, Abma B, Viguria A, Prévinaire E, Zarco-Tejada PJ, Serruys P, van Valkengoed E, van der Voet P (2013) Fieldcopter: unmanned aerial systems for crop monitoring services. In: Stafford JV (ed) Precision agriculture '13. Wageningen Academic Publishers, Wageningen, pp 169–175

Walter A, Liebisch F, Hund A (2015) Plant phenotyping: from bean weighing to image analysis. Plant Methods 11:1–11

Wang B, Du Q, Yang X, Zhang D (2014a) Identification and characterization of nuclear genes involved in photosynthesis in Populus. BMC Plant Biol 14:1

Wang L, Czedik-Eysenberg A, Mertz RA, Si Y, Tohge T, Nunes-Nesi A, Arrivault S, Dedow LK, Bryant DW, Zhou W, Xu J, Weissmann S, Studer A, Li P, Zhang C, LaRue T, Shao Y, Ding Z, Sun Q, Patel RV, Turgeon R, Zhu X, Provart NJ, Mockler TC, Fernie AR, Stitt M, Liu P, Brutnell TP (2014b) Comparative analyses of C4 and C3 photosynthesis in developing leaves of maize and rice. Nat Biotechnol 32:1158–1165

Wang J, Yu Q, Xiong H, Wang J, Chen S, Yang Z, Dai S (2016) Proteomic insight into the response of Arabidopsis chloroplasts to darkness. PLoS One 11:e0154235

Xia Y, Ning Z, Bai G, Li R, Yan G, Siddique KHM, Baum M, Guo P (2012) Allelic variations of a light harvesting chlorophyll A/B-binding protein gene associated with agronomic traits in barley. PLoS One 7:e37573

Xu J, Li Y, Ma X, Ding J, Wang K, Wang S, Tian Y, Zhang H, Zhu XG (2013) Whole transcriptome analysis using next-generation sequencing of model species Setaria viridis to support C4 photosynthesis research. Plant Mol Biol 83:77–87

Yamori W (2013) Improving photosynthesis to increase food and fuel production by biotechnological strategies in crops. J Plant Biochem Physiol 1:113

Zelitch I (1982) The close relationship between net photosynthesis and crop yield. Bioscience 32:796–802

Zheng X, Moriyama EN (2013) Comparative studies of differential gene calling using RNA-Seq data. BMC Bioinformatics 14:S7

Zhou W, Eudes F, Laroche A (2006) Identification of differentially regulated proteins in response to a compatible interaction between the pathogen Fusarium graminearum and its host, *Triticum aestivum*. Proteomics 6:4599–4609

Zhu-Shimoni JX, Lev-Yadum S, Matthews B, Galili G (1997) Expression of an aspartate kinase homoserine dehydrokinase gene is subject to specific spatial and temporal regulation in vegetative tissues, flowers, and developing seed. Plant Physiol 13:695–706

Chapter 3
Molecular Marker Technology for Genetic Improvement of Underutilised Crops

Acga Cheng, Hui Hui Chai, Wai Kuan Ho, Aliyu Siise Abdullah Bamba, Aryo Feldman, Presidor Kendabie, Razlin Azman Halim, Alberto Tanzi, Sean Mayes, and Festo Massawe

3.1 Introduction

With the average global temperatures predicted to increase above 3.0 °C by mid-century, agricultural productivity is likely to significantly decline in many countries, particularly those currently in the hottest, most arid regions of the world (Abraham et al. 2014). There is growing evidence that rising temperatures are stifling the production of many important staples, notably maize (*Zea mays*), wheat (*Triticum aestivum*), and rice (*Oryza sativa*), due to their sensitivity to water shortage and heat stress (Khoury et al. 2014). For example, maize yields in Africa are modelled to

---

A. Cheng
Biotechnology Research Centre, School of Biosciences, The University of Nottingham Malaysia Campus, Jalan Broga, 43500 Semenyih, Selangor Darul Ehsan, Malaysia

Institute of Biological Sciences, Faculty of Science, University of Malaya, 50603 Kuala Lumpur, Malaysia

H.H. Chai
Biotechnology Research Centre, School of Biosciences, The University of Nottingham Malaysia Campus, Jalan Broga, 43500 Semenyih, Selangor Darul Ehsan, Malaysia

W.K. Ho • A. Tanzi • F. Massawe (✉)
Biotechnology Research Centre, School of Biosciences, The University of Nottingham Malaysia Campus, Jalan Broga, 43500 Semenyih, Selangor Darul Ehsan, Malaysia

Crops For the Future, Jalan Broga, 43500 Semenyih, Selangor Darul Ehsan, Malaysia
e-mail: festo.massawe@nottingham.edu.my

A.S.A. Bamba
Biotechnology Research Centre, School of Biosciences, The University of Nottingham Malaysia Campus, Jalan Broga, 43500 Semenyih, Selangor Darul Ehsan, Malaysia

Crops For the Future, Jalan Broga, 43500 Semenyih, Selangor Darul Ehsan, Malaysia

CSIR-Savannah Agricultural Research Institute, Off Tamale-Tolon Road, Nyankpala, Tolon District, N/R, Ghana

S.N.A. Abdullah et al. (eds.), *Crop Improvement*,
DOI 10.1007/978-3-319-65079-1_3

decrease by between 22 and 35% by 2030 because of the increased variability of rainfall patterns and changes in local temperatures (Ngwira et al. 2012; Shi and Tao 2014). Climate change, coupled with increasing global demand for food as the world population and its buying power grows, could also trigger a wave of price hikes for basic food commodities (Masters 2010).

While mainstream research has often focused on how to develop improved varieties of the world's mere 20 major food crops, thousands of indigenous and underutilised crops have been overlooked. Some of the underutilised (also called neglected, orphan or minor) crops are likely to be more resilient to harsh conditions than the major cereal crops as a result of the low-input farming systems in marginal environments in which they have been selected (Mayes et al. 2011). Until recently, unlocking the potential value of underutilised crops has not been seen as a mainstream approach to addressing the effects of climate change on agriculture. Although some of these crops, like quinoa (*Chenopodium quinoa*) and bambara groundnut (*Vigna subterranea* (L.) Verdc.), have recently received more research attention (although not necessarily more research funding), existing knowledge on the genetic potential of many other promising underutilised species, such as the high-protein legume winged bean (*Psophocarpus tetragonolobus*), remains limited (Massawe et al. 2016). There is, therefore, an urgent need for more innovative research to boost the productivity of potential future crops and molecular marker (or genetic marker) technology is considered to be a critical component in this endeavour. This far-reaching technology has indeed revolutionised the way plant genetics research is conducted (Semagn et al. 2006; Govindaraj et al. 2015) (Fig. 3.1).

Molecular markers are capable of detecting allelic variation or modification (e.g. methylation) in DNA sequence, and have been applied over a number of years to answer many complex biological questions in plant and animal science. These range from simple genetic fingerprinting for quality control in breeding programmes, to complex questions in population ecology and germplasm structure (Joshi et al. 1999; Schlotterer 2004). The last two decades have seen substantial investments in large-scale genetic and genomic resource development on major crops, including the development of simple sequence repeats (SSRs), expressed sequence tags (ESTs) and single nucleotide polymorphism (SNPs) (Schlotterer 2004; Kesawat and Kumar 2009). This is particularly the case for the major cereals, strengthening

A. Feldman • R.A. Halim
Crops For the Future, Jalan Broga, 43500 Semenyih, Selangor Darul Ehsan, Malaysia

P. Kendabie
Plant and Crop Sciences, School of Biosciences, The University of Nottingham, Sutton Bonington Campus, Leicestershire, LE12 5RD Loughborough, UK

S. Mayes
Crops For the Future, Jalan Broga, 43500 Semenyih, Selangor Darul Ehsan, Malaysia

Plant and Crop Sciences, School of Biosciences, The University of Nottingham, Sutton Bonington Campus, Leicestershire, LE12 5RD Loughborough, UK

(a) **Bambara groundnut** seeds (*Vigna subterranea*)

(b) **Winged bean** pods (*Psophocarpus tetragonolobus*)

Fig. 3.1 Examples of underutilised crops *Source:* http://www.cropsforthefuture.org/

the applications of marker-assisted selection (MAS) (Gupta et al. 2010; Miah et al. 2013; Huang et al. 2015). Compared to conventional breeding, MAS:

1. Does not require the plant to express the desired phenotypic trait to be able to select for plants carrying the trait.
2. Is not affected by the environment.
3. Is able to recognise heterozygous individuals, for example, hybrids can be identified by co-dominant markers.
4. Makes gene pyramiding possible by introgressing multiple quantitative trait loci (QTL) into an individual.
5. Is more time- and resource-efficient.

Advances in molecular biology, principally the invention of polymerase chain reaction (PCR) in the mid-1980s, have resulted in a wealth of molecular marker systems with potentially widespread utility for detecting genetic variation in plants (Semagn et al. 2006; Kumar et al. 2009). Although most research to date have been dedicated to major crops, the use of molecular marker technology to investigate the genetic basis of important traits for a range of underutilised species is increasing (Moe et al. 2012). The first section of this chapter will highlight the types of marker systems commonly used in crop improvement, followed by a thorough discussion on the past, current, and potential utilisation of molecular marker technology for the selection of the underutilised crop examples in the subsequent sections.

3.2 Commonly Used Molecular Marker Systems for Crop Improvement: Principles and Progress

Molecular marker systems can reveal genetic variation (or polymorphism) in DNA sequence, often arising from mutations that occur during DNA replication or repair. Plant molecular genetics has progressed dramatically since the establishment of the first molecular markers, allozymes (or isozymes), in the 1960s. Allozymes work on the principle of the electrophoretic mobility of protein variants in enzymes modified by amino acid substitution, when proteins are run on gel matrixes. Specific enzymatic stains can reveal the position of the protein bands on a gel, with amino acid changes effecting the migration of the band (Schlotterer 2004). In the past, these biochemical markers were widely used in plant population genetic studies, including measurements of population structure and divergence (Troconis-Torres et al. 2002). The popularity of allozymes decreased with the advent of DNA-based markers, which have higher discrimination level at the individual level and for which there are far larger number of markers available. These include restriction fragment length polymorphisms (RFLPs) (Tanksley et al. 1989), and a suite of PCR-based markers, such as SSRs (Varshney et al. 2005) and SNPs (Ganal et al. 2009). This section presents information on the most commonly used DNA-based marker systems in plants (Table 3.1).

Table 3.1 Comparison of advantages and disadvantages offered in different DNA-based marker systems

Marker	Advantages	Disadvantages
RFLPs	• High reproducibility	• Requires large amounts of high quality DNA
	• No sequence information required	• Difficult to automate
		• Involve radioactive methods
		• Cloning and characterisation of probe are required
RAPDs	• Inexpensive	• Low reproducibility
	• No sequence information required	• Mainly dominant
	• Produces a large number of bands, which can be further characterised individually	• Difficult to analyse
		• Difficult to automate
		• Cross-study comparisons are difficult
AFLPs	• High reproducibility	• Mainly dominant
	• A large number of loci can be analysed simultaneously	• Difficult to automate
SSRs	• Highly informative	• High mutation rate
	• High heterozygosity	• Complex mutation behaviour
	• High reproducibility	• Low to moderate abundance
	• Low ascertainment bias	• Special preparation required for cross-study comparison
	• Easy to isolate	• Sequence information required
SNPs	• Low mutation rate	• Substantial rate heterogeneity among sites
	• Easy to automate • High abundance	• Can be expensive to isolate
		• Ascertainment bias
		• Low information content of a single SNP

Adopted from Schlotterer 2004, Kumar et al. 2009

3.2.1 Restriction Fragment Length Polymorphism (RFLP)

The discovery of restriction enzymes (or restriction endonucleases) in the 1960s has led to the first DNA-based marker system known as RFLP (Schlotterer 2004). The marker system exploits variation in homologous DNA sequences, such as base pair changes within the endonuclease recognition sequence. The resulting fragments are separated by size using gel electrophoresis, followed by Southern blotting and hybridisation in solution of the filter to a labelled and denatured probe that identifies the locus of interest (Tanksley et al. 1989). Restriction enzymes are generally classified into four types; Type I, II, III, and IV, on the basis of cleavage position, sequence specificity, subunit composition, and cofactor requirements (Tomar 2010). A list of known restriction enzymes, along with their recognition sequences and cleavage sites, is provided in REBASE, a fully curated database available at http://rebase.neb.com/rebase/rebase.html (Roberts et al. 2003).

RFLPs were one of the most commonly used technologies to construct DNA fingerprints (Weising et al. 1994) until the emergence of PCR-based marker systems. RFLPs were one of the most commonly used technologies to construct DNA fingerprints (Weising et al. 1994) until the emergence of PCR-based marker systems. RFLPs were predominantly used to measure genetic diversity and construct genetic linkage maps for most major crop plants, notably maize, wheat and rice (Tanksley et al. 1989; Schlotterer 2004). Nonetheless, RFLPs are still very much in use today, mainly because of their high reproducibility, co-dominant nature and relatively low cost per data point. They have also been one of the major marker systems to confirm the presence and copy number of transgenes (Bisen 2014).

3.2.2 Random Amplified Polymorphic DNA (RAPD)

The discovery of PCR has led to the development of a handful of novel genetic assays characterised by the selective amplification of defined DNA fragments. In 1990, the first PCR-based marker system, RAPD, was described based on the amplification of random DNA fragments using short synthetic primers of arbitrary nucleotide sequence (Williams et al. 1990). These primers, usually 10 bp in length, occur at such high frequency in the genome that two random binding sites orientated to allow PCR are likely to occur a number of times in the genome of most species, leading to multiple band profiles. The amplified fragments are separated by gel electrophoresis, stained with ethidium bromide, and visualised under ultraviolet light. RAPD polymorphisms are indicated by the presence or absence of bands of particular sizes in comparison to other individuals, and hence treated as dominant markers (Williams et al. 1990). Inherited in a Mendelian fashion, their polymorphisms can arise as the result of base substitutions at the primer annealing sites or INDELs in the regions between the sites.

The RAPD technology represented a useful tool for plants with no existing markers, such as the endangered seven-son flower (*Heptacodium miconioides* Rehd.) (Liu et al. 2007a). From its inception until now, RAPDs have been largely used for genetic diversity and phylogenetic studies at the genus or species level, with the most recent example being *Pistacia* (Iranjo et al. 2016). RAPDs possess almost all the advantages of a PCR-based marker, offering rapid results, technical simplicity and no need for species-specific development. Beyond these advantages, RAPD analysis uses only small amounts of DNA (10–100 ng per reaction), and requires no prior knowledge of the sequences of the target genome (Arif et al. 2010). One inherent drawback of RAPDs is their low reproducibility, and thus highly standardised experimental procedures are required to achieve reproducible profiles. The reproducibility testing of RAPD in plants by a number of European laboratories revealed that the technique) is stable within labs, but not between labs (Jones et al. 1997). Another major drawback of RAPDs is their dominant nature of inheritance, which limits their use for genetic mapping and marker-assisted selection (MAS) (Bardakci 2000).

3.2.3 Amplified Fragment Length Polymorphism (AFLP)

AFLP (Vos et al. 1995) is another marker technology that has been widely used to assess the genetic diversity and phylogenetic studies in plants (Mba and Tohme 2005). This dominant marker system can be used in any species, including non-model organisms, as no prior sequence information is required. Unlike RAPDs, AFLPs are highly reproducible because they use a combination of restriction digestion and PCR amplification (Arif et al. 2010). However, the AFLP method is more technically challenging than RAPD, requiring highly purified genomic DNA. The standard AFLP protocol begins with the digestion of genomic DNA with two restriction enzymes, generally one enzyme with an average cutting frequency (e.g. 6-base cutter *Eco*RI), and the other with a higher cutting frequency (e.g. 4-base cutter *Mse*I). The resulting DNA fragments are then ligated to specific adapters, and a subset of these fragments are amplified using primer pairs that consist of common sequences of the adapter and arbitrary nucleotides extending into the unique sequence adjacent to the restriction enzyme recognition site. The amplified fragments, usually within 60–500 bp size range, are separated by the polyacrylamide gel electrophoresis (PAGE) or on a capillary sequencer. The former is rather cumbersome, time-consuming and labour intensive, the latter relatively expensive (Lin et al. 1996; Cheng et al. 2012).

The standard AFLP protocol has been modified by Ranamukhaarachchi et al. (2000) to develop a rapid plant genetic characterisation analysis with improved intensity and specificity, which is convenient to perform and reliable. The modified AFLP protocol involves the use of standard agarose as the medium of electrophoresis, and is carried out with only a single restriction enzyme and one type of adapter. However, as the gel resolution decreases, the chance of mis-calling bands as identical when they are in fact different in size increases. For high-throughput genotyping, specific AFLPs) can sometimes be converted into co-dominant and sequence-specific markers, such as sequence tagged sites (STSs) and cleaved amplified polymorphic sites (CAPSs) (Meksem et al. 2001).

3.2.4 Simple Sequence Repeat (SSR)

Plant genomes typically contain a large number of SSRs; DNA sequence motifs of up to 6 bp units repeated in tandem and widely scattered throughout the genome. Most SSR repeat motifs are composed of di-, tri-, or tetra-nucleotide sequences, with $(GA)_n$ and $(AT)_n$ being the most frequent repeat types observed (Powell et al. 1996). Highly reproducible, locus-specific, simple detection and abundance are all characteristics of SSRs, which have made them one of the most preferred PCR-based markers for assessing genetic diversity, population structure and evolution analysis in plants (Fu et al. 2013). Due to their co-dominant nature, high polymorphism and random distribution within the genome, SSRs are extensively used for genotyping and genetic map construction. For instance, a genetic linkage map consisting of 15 linkage groups with 406 SSRs covering 1143.5 cM of the genome was developed using a pseudo-testcross population derived from an intraspecific cross of two tea plants (*Camellia sinensis* varieties). The QTL analysis indicated that nine QTLs associated with catechins content were clustered into four linkage groups (LG03, LG11, LG12 and LG15), providing information for functional studies to improve tea plants (Ma et al. 2014).

Nonetheless, despite their many advantages, the de novo development of SSRs can be costly and time-consuming (Yu and Li 2007). Unlike sequence-arbitrary methods, such as AFLP, the sequence of each SSR locus needs to be known to allow the development of locus-specific, co-dominant, primers in the unique flanking regions of each SSR repeat. Classically, SSR analysis involves the construction of genomic DNA libraries. The preparation of these libraries is a tedious process, and the specific DNA sequencing required is expensive (Edwards et al. 1996). Nevertheless, the development of advanced molecular tools over the past two decades has enabled establishment of cost-effective, high-throughput methods for SSR analysis. These include the mining of sequences from DNA databases (e.g. those deposited at NCBI), and the sequencing method where the whole genome or parts of the genome (e.g. exome or transcriptome) are sequenced (Saeed et al. 2016). This has facilitated the development of Expressed Sequence Tag (EST) or gene-based SSRs (Zhao et al. 2012), which have higher recovery rates than genomic SSRs (Ellis and Burke 2007; Zhao et al. 2012).

Numerous studies have found that EST-SSRs are generally more transferable across taxonomic boundaries than the genomic SSRs (Ellis and Burke 2007). Despite their relatively low polymorphism (Yu and Li 2007), EST-SSRs have higher transferability rates, are more practical, more informative, and having a lower frequency of null alleles. These markers have been developed and utilised in many crop species, including several primary oil crops such as soybean (*Glycine max*) and rapeseed (*Brassica napus*) (Ellis and Burke 2007; Lai et al. 2013).

The SSRs have often been the marker chosen in studies of crops lacking of resources and species-specific information. One recent example is a study conducted on common vetch (*Vicia sativa*). The study reported the transferability of SSRs from *Vicia sativa* subsp. sativa across 22 *Vicia* species, which ranged between 32 and 82% (Raveendar et al. 2015). There have been increasing numbers of SSR being identified through the utilisation of the next-generation sequencing (NGS) technology particularly from transcriptomic data as the genic SSR are in the conserved coding regions, which, in turn, are highly transferable to related taxa and useful in association or comparative mapping. For example, a total of 3011 genic SSR have been identified as potential markers from 26 M reads of transcriptome data derived from root, stem and leaf tissues of rice bean (*Vigna umbellata* L.), with mono-nucleotide repeat motifs being the most abundant (47.3%) (Chen et al. 2016). Nevertheless, SSR motifs within coding sequence are constrained, as changes in repeat number will effect protein translation and can lead to a frameshift mutation. For this reason, trinucleotides repeat until are most common in coding sequence and most likely to be polymorphic, while dinucleotide motifs can be found in the upstream and downstream transcribed, but not translated regions. The biggest problem with such dinucleotide motifs is that the flanking sequence region is often highly constrained for primer design (Lin and Kussell 2012; Gonthier et al. 2015).

3.2.5 Single Nucleotide Polymorphism (SNP)

From the late 1990s to the early 2000s, the SSRs had indeed pervaded modern plant breeding and genetics research. The dominance of SSRs has been, however, replaced by SNPs during the last decade. Despite having lower polymorphism rates than the SSRs because of their biallelic nature (Kumpatla et al. 2012), SNPs have emerged as the current markers of choice because they are more stable, efficient, and amenable to high- and ultra-high-throughput automation. SNPs represent variation in a DNA sequence that affects only a single nucleotide; adenine (A), cytosine (C), guanine (G) or thymine (T) (Kumpatla et al. 2012). They are the most abundant class of polymorphisms in both plant and animal genomes, making them particularly suited for high-resolution mapping. Inherited as co-dominant markers, SNPs are able to discriminate efficiently between homozygous and heterozygous alleles (Arif et al. 2010). Nonetheless, the SNP system is not without limitations. The discovery of SNPs can be a difficult and complex process, depending on the level of genome complexity and the availability of reference genome sequences.

Although numerous protocols for SNP discovery have been described, the typical SNP genotyping protocol consists of three major components: target amplification, allelic discrimination and product detection (Chen and Sullivan 2003). The last half-decade has seen tremendous progress in SNP genotyping technologies, such as the TaqMan assay (allele specific hybridisation), SNP arrays (e.g. GoldenGate assay, BeadArray, microarray technology, KASP array), genotype-by-sequencing (GbS) and resequencing for variant calling. Each of these technologies renders a unique combination of throughput, accuracy, scale, and cost (Kumar et al. 2012; Huang et al. 2015). With the reduction in the cost of NGS technology, both de novo and reference-based SNP discovery and genotyping are now relatively feasible for various plant species.

Reference-based SNP identification using resequencing is more applicable to those model and major crops that have a complete or draft genome. For studies of underutilised crops, the GbS approach would be the choice of interest given ex ante sequence information is not required. NGS-derived SNPs have been recently reported in leguminous species such as alfalfa (*Medicago sativa*) and common bean (*Phaseolus vulgaris*) (Huang et al. 2015). More recently, restriction site associated DNA (RAD) sequencing has been adopted to produce 1894 high quality SNP markers and together with 68 SSRs, they were mapped onto seven chromosomes in barley (Ren et al. 2016). The genetic map allowed the identification of a new recessive dwarfing gene *btwd1* at 0.7 cM and 0.9 cM on chromosome 7H, giving a new insight into the use of SNPs markers for positional cloning of the gene and MAS in barley (Ren et al. 2016). While automation is often the most crucial factor in selecting a genotyping system for a large-scale SNP profiling), other factors such as cost, throughput, and data quality obtained are also of considerable importance (Huang et al. 2015).

3.2.6 Diversity Arrays Technology (DArT)

Generally, in the GbS approach, the genome complexity of a species is reduced using restriction enzymes prior to sequencing. For example, with the appropriate choice of methylation-sensitive restriction enzymes in a combination with a frequent cutting enzyme, repetitive fractions of the genome can be avoided targeting informative low copy sequences (Sánchez-Sevilla et al. 2015). There have been more than 30,000 wheat germplasm accessions at the International Maize and Wheat Improvement Centre (CIMMYT) genotyped using the GbS technique, specifically the Diversity Arrays Technology (DArT)-seq pipeline, under its Seeds of Discovery (SeeD) initiative (http://seedsofdiscovery.org/about/genotyping-platform/). Developed in the early 2000s, DArT offers a high-throughput genotyping platform for detecting sequence polymorphisms in a single experiment without species-specific sequence information.

Over the past few years, DArT marker system has been widely used for genetic diversity analysis, evaluation of population structure of a collection of germplasm, construction of genetic linkage maps, and alignment of genetic maps of different segregating populations (Phung et al. 2014; Sánchez-Sevilla et al. 2015). There are two types of platforms for DArT markers development, namely microarray-based DArT and DArTseq (Cruz and Kilian 2013). Practically, the DArT is a complexity reduction method that involves the use of methylation-sensitive restriction enzymes such as *Pst*I/*Taq*I (Semagn et al. 2006) or *Pst*I/*Bst*NI (Cruz and Kilian 2013). Genomic 'representations' are selected through amplification of primers complementary to the adaptors after genomic DNA is isolated and fragmented using enzymes, followed by the ligation of restricted fragments with adaptors. In effect, the technique amplifies the remaining *Pst*I-*Pst*I fragments. A 'Diversity Panel' is then generated when selected genomic 'representations' are purified and spotted onto a microarray in the DArT Arrays method (Jaccoud et al. 2001). When individual DNA samples are hybridised onto this 'Diversity Panel', polymorphic DArT markers are identified based on hybridisation signal intensities and assembled into a 'genotyping array' for routine genotyping consisting of presence and absence scores (Jaccoud et al. 2001; Semagn et al. 2006).

The DArTseq technique is suitable for detailed exploitation of genetic diversity and population structure, gene discovery for molecular breeding, high-resolution genetic mapping and MAS as no sequence information is required (Cruz and Kilian 2013). DArTseq genotyping was reported to produce 2835 polymorphic markers in *Eucalyptus* as compared to 1088 microarray-based DArT markers (Sansaloni et al. 2011). Due to the capability of producing a large quantity of polymorphic markers at a similar cost, DArTseq genotyping is considered to be more cost effective than microarray-based DArT.

3.3 Utilisation of Molecular Markers in the Studies of Underutilised Crops

It is apparent that there has been continuous progress and evolution of molecular marker systems used to meet the needs of modern plant breeding programmes. The selection of preferred molecular markers depends immensely on the crop species of interest, the intended research outcomes and the financial resources available (Kesawat and Kumar 2009). In the present era, DNA-based marker systems, notably SSRs and SNPs, have become increasingly important to underutilised crops by way of improving access and utilisation of germplasm resources, genetic analysis of breeding populations, parental selection and predicting progeny performance, marker-assisted selection, marker-enriched backcross breeding, comparative mapping, and gene isolation, function and manipulation. These systems have found their niche applications in various fields, which are discussed in detail in Sects. 3.2 and 3.3.

3.3.1 Underutilised Crops: What Are They and Where Do They Come From?

Underutilised crops are, by and large, indigenous crop species which are still used by communities at a local or regional level, but are largely overlooked by researchers, producers and consumers at a global level (Mayes et al. 2011). These underutilised

Table 3.2 List of recently available genomes or draft genomes on some underutilised crops

Crop example	Crop type	Centre(s) of diversity	Genome size (Mbp)	Year	References
Foxtail millet (*Setaria italica*)	Cereal	Asia, Mediterranean/ Southwest Asia	~490	2012	Bennetzen et al. (2012)
Muskmelon (*Cucumis melo*)	Vegetable	South Mexico, Central America	~375	2012	Garcia-Mas et al. (2012)
Peach (*Prunus persica*)	Fruit	Asia (Chinese Centre)	~265	2012	Verde et al. (2013)
Pigeon pea (*Cajanus cajan*)	Legume	Asia (Indian Centre)	~835	2012	Varshney et al. (2012)
Tausch's goatgrass (*Aegilops tauschii*)	Cereal	Middle East	~4400	2013	Jia et al. (2013)
Quinoa (*Chenopodium quinoa*)	Cereal	Latin America	~1450	2017	Jarvis et al. (2017)

crops are often bestowed with greater stability to cope with harsher and more dynamic environmental conditions (Thies 2000). Some of these crops, especially for underutilised cereals such as quinoa and teff (*Eragrostis tef*), are well adapted to the socio-economic conditions in their native habitats, and are preferred by both local farmers and consumers (Thies 2000; Cheng et al. 2015). Table 3.2 shows some examples of underutilised crop species and their known centre(s) of diversity.

The term 'underutilised', or other common synonymous terms such as 'orphan', 'neglected' and 'minor', does not necessarily reflect the crops' geographical distribution, nor their social or economic implications. For instance, many crop species are consumed as staple by millions of people within their centres of origin, but their poor marketability and limited accessibility make them largely underutilised in both social and economic terms (Padulosi and Hoeschle-Zeledon 2004). One of the key reasons behind their underutilisation is that they are neglected by mainstream research, mainly because they are not considered global crops and are mostly produced and consumed subsistently (Naylor et al. 2004; Massawe et al. 2016).

Nevertheless, in recent years, a small number of underutilised crops, bambara groundnut, winged bean, amaranth, and proso millet (*Panicum millaceum*), to name a few, has attracted worldwide research attention. This is mainly due to their ability to thrive in extreme environments with the increasing concern on global warming and other negative future climate changes, as well as their potential to contribute to nutritional security (Ray et al. 2013). The genome of some of these crops have also been fully sequenced (Table 3.2). Molecular genetics and genomics research on underutilised crops have gained momentum in the present decade, spurred largely by the advent of powerful yet affordable molecular marker technology. A handful of marker systems are being used extensively to investigate the genetic basis of agronomically important traits for a range of underutilised species, as discussed below.

3.3.2 *Summary of Recently Developed DNA-Based Markers in Underutilised Crops*

Recent years have been marked by significant progress in the development of DNA-based marker systems in underutilised crop species. Specifically, in the past 5 years, numerous markers representing various marker types, such as RAPDs, AFLPs, SSRs, and SNPs, have been used for genetic analysis in many potential underutilised crops (summarised in Table 3.3). Protein-coding sequences, functionally characterised genes and EST sequences have been made available to build up new generation markers like genic SSRs and SNPs. Given that SSRs and SNPs have rapidly emerged as the genetic markers of choice in view of their abundance, high transferability across related taxa, and amenability to automated analysis (Kumar et al.; as summarised in Table 3.1), this section focuses chiefly on the development of these marker systems.

SSR marker system has been extensively used in genetic mapping and molecular breeding in underutilised crops, in particular underutilised cereals/pseudo-cereals and legumes (Zeid et al. 2012; Bohra et al. 2014). Major progress has been seen in the development of EST-SSRs in underutilised species in the past decade, leading by a couple of highly potential underutilised species, including quinoa and finger millet (*Eleucine coracana*) (Dawson et al. 2007). EST-SSRs have also been developed for genetic analysis, for example, in adzuki bean (*Vigna angularis*) (Chen et al. 2015). SNP marker system has become more popular than SSRs in light of their higher abundance and suitability for automatic allele calling (Arias et al. 2012). The advent of NGS has allowed the identification of large collections of SNPs for some underutilised species, including Tausch's goatgrass (*Aegilops tauschii*) (Mochida and Shinozaki 2013). At present, SNP discovery is being carried out in many other underutilised species such as teff (Cheng et al. 2015), and roselle (*Hibiscus sabdariffa*) (Melo et al. 2016).

3.3.3 *Development of Choice of Marker Systems for the Development of Some Underutilised Crops*

3.3.3.1 Underutilised Cereal and Pseudo-cereal

Cereals are a major source of food in the world, and of the approximately 50,000 known edible plant species, more than 10,000 are cereals (Ji et al. 2013). Being staples that feed almost the entire human population, cereals are grown in far greater quantities than any other type of crop. Maize, wheat, and rice alone account for nearly two-thirds of the world's plant-derived food energy (Ji et al. 2013). Although the annual production of major cereals has seen a steady increase since the dawn of the Green Revolution, rapid changes in climate pose a huge threat to the sustainability of the current productions. This is, perhaps, the chief reason why there has been a marked

Table 3.3 DNA-based markers developed and genetic analysis of selected underutilised crops

Crop example	Marker type(s)	Research description	References
Bambara groundnut (*Vigna subterranean*)	SSR	SSR-based analysis of genetic diversity of Ghanaian bambara groundnut landraces	Abdullah Bamba and Massawe (2013)
	SSR, DArT	SSR-based analysis of genetic diversity and population structure in bambara groundnut landraces	Molosiwa et al. (2015)
	SSR, DArT	Construction of linkage map and QTL analysis of phenotypic traits in bambara groundnut	Ahmad et al. (2016)
Cucurbits (*Cucurbitaceae*)	AFLP, SSR, SNP	Construction of genetic map of and QTL analysis of fruit quality in melon	Harel-Beja et al. (2010)
	SNP	Construction of ultrahigh-density linkage map for cultivated cucumber using SNP genotyping array	Rubinstein et al. (2015)
Finger millet (*Eleusine coracana*)	RAPD SSR	RAPD-based analysis of genetic diversity in finger millet	Kumari and Pande (2010)
	Genic-SSR	QTL analysis of blast disease resistance in finger millet	Babu et al. (2014)
Grain amaranth (*Amaranthus* spp.)	SNP	Development and analysis of SNPs in grain amaranth; characterisation of the first linkage map in the genus	Maughan et al. (2011)
	SSR	SSR-based analysis of genetic diversity and population structure in *Amaranthus* species	Suresh et al. (2014)
Pigeon pea (*Cajanus cajan*)	RAPD	Identification of RAPDs linked to plant type gene in pigeon pea	Dhanasekar et al. (2010)
Proso millet (*Panicum miliaceum*)	SSR	SSR-based analysis of genetic diversity and phylogeography of proso millet across Eurasia	Hunt et al. (2011)
Teff (*Eragrostis tef*)	SSR	Linkage-map construction and QTL analysis of yield and lodging resistance in an enhanced SSR-based map	Zeid et al. (2011)
	SSR	SSR-based analysis of genetic diversity and relationships within and among *E. tef*, *E. pilosa* and *E. curvula*	Zeid et al. (2012)
Winged bean (*Psophocarpus tetragonolobus*)	SSR and SNP	Development of SSRs and SNPs in two Sri Lankan winged bean accessions	Vatanparast et al. (2016)
	Genic-SSR	Development of gene-based SSRs in winged bean for diversity assessment	Wong et al. (2017)

shift in research attention from a handful of major cereals to a wider range of cereals in the past decade (Pingali 2012). Two Andean cereal grains; quinoa and grain amaranth, have recently taken the spotlight as 'superfoods', due largely to their gluten-free property and nutrient composition (Massawe et al. 2016). They are often referred to as pseudo-cereals, as they do not belong to the grass family (Gramineae) but have similar uses as true cereals. Quinoa has had a slightly higher popularity over the past half-decade, receiving numerous titles and honours including 'Queen of Superfoods' and 'International Year of Quinoa' (http://www.fao.org/quinoa-2013/en/).

Quinoa has been an underutilised priority species since the early 2000s (Williams and Haq 2002). During the early years, its molecular studies were performed using AFLP, RAPD and SSR marker systems (Maughan et al. 2004). The first genetic linkage map of quinoa, covered an estimated 60% of its genome, was developed in 2004 based on AFLPs, RAPDs and SSRs (Maughan et al. 2004). The subsequent molecular studies were based primarily on SSRs, whereby more than 400 SSRs were developed and characterised (Jarvis et al. 2008). Christensen et al. (2007) used the developed SSRs to assess the genetic diversity among 121 accessions of quinoa within the United States Department of Agriculture (USDA) collection, reporting that the accessions can be clustered into two major groups: one comprising accessions from the lowlands of Chile (Coastal ecotype), and the other comprising accessions from the Andean Highlands (Altiplano ecotype). Tártara et al. (2012) conducted a more recent SSR-based genetic diversity analysis on quinoa using 35 germplasm accessions from Northwest Argentina.

The first EST libraries (424 ESTs) for quinoa were described in 2005 (Coles et al. 2005) using developing seed and floral tissue. The study yielded 38 SNPs and 13 INDELs based on 20 EST sequences derived from five quinoa accessions. Maughan et al. (2012) identified 14,178 putative SNPs for quinoa using pyrosequencing from five mapping populations with 511 were developed into KASPar-based SNP array to genotype 113 accessions of C. quinoa and eight from Chenopodium taxa. Subsequently, an integrated SNP-based linkage map was generated using two recombinant inbred line (RIL) populations. The map consisted of 29 linkage groups with an average distance of 3.1 cM between SNPs. The same research group also sequenced amaranth from four parental lines using the same technique, yielding 27,658 putative SNPs (Maughan et al. 2009). The developed SNPs were then used to generate the first complete linkage map of the *Amaranthus* genus (Maughan et al. 2011). In this map, a total of 411 SNPs were assigned to 16 linkage groups, spanning 1288 cM with an average distance of 3.1 cM between SNPs (Maughan et al. 2011). In 2014, Illumina Hi-Seq RNA sequencing (RNA-seq) technology has been used to perform the first large transcriptome analysis of drought-induced stress in quinoa (Raney et al. 2014).

3.3.3.2 Underutilised Legumes

Legumes, members of the Leguminosae (or Fabaceae) family, are an important component for low-input agriculture. Their desirability rests on their ability to fix atmospheric nitrogen, which consequently increases soil fertility for the production of the neighbouring or subsequent non-legume crop cycle such as rice (Rahman et al. 2014). While legumes are one of the most nutritious, versatile and relatively more affordable foods available all over the world, the utilisation of most leguminous species is relatively low with groundnut and soybean being exceptions due to their high oil and protein contents. Nevertheless, the importance of several leguminous species has been recently highlighted, as in the case for bambara groundnut (Massawe et al. 2016).

The groundwork for the genetic improvement in bambara groundnut began in the late 1990s, focusing primarily on increasing the understanding of its genetic diversity within and between landraces (Massawe et al. 2005). High levels of polymorphism among African bambara groundnut landraces were observed using different DNA-based marker systems, such as AFLP (Massawe et al. 2002) and RAPD (Massawe et al. 2003). These findings are further supported by a study conducted by Abdullah Bamba and Massawe (2013), which demonstrated the high level of SSR polymorphism among a collection of Ghanaian bambara groundnut landraces.

The first attempt in the development of SSR markers for bambara groundnut, which yielded ten SSRs, were reported by Basu et al. (2007). More recently, a total of 75 SSRs have been developed for this legume through various sequencing approaches for the analysis of its genetic diversity and population structure, as well as for the selection of pure lines from landraces (Molosiwa et al. 2015). SSR marker system has also been utilised to assess the breeding system and varietal purity of bambara groundnut (Ho et al. 2016). The first linkage map of bambara groundnut was developed from a F_3 population derived from a cross between two single genotypes derived from Tiga Nicuru (a landrace from Mali) and DipC (a landrace from Botswana) using 238 SSR and DArT array markers. In this map, a total of 36 QTLs were detected at eight linkage groups, associating with 19 phenotypic traits (Ahmad et al. 2016). These detected QTLs could be useful for future breeding programmes in bambara groundnut.

3.3.3.3 Other Types of Underutilised Crops

An abundance of vegetables, fruits, roots, and tubers has been important source of food and medicine for millennia, and yet, many of them are still widely underutilised and underexploited (Ebert 2014). Nonetheless, in the case of vegetables, numerous conservation and development initiatives have already been launched and championed by the World Vegetable Centre (AVRDC) (Ojiewo et al. 2010). Today, AVRDC maintains the largest vegetable germplasm collection of approximately 61,000 accessions, including about 12,000 accessions of indigenous species.

The Cucurbitaceae, also known as cucurbit or vine crop family, provides a wide variety of vegetables such as squash, melon, and several *Cucurbita* species with edible fruits. One example of potential underutilised cucurbits in which genetics have been fairly studied is the egusi (*Citrullus colocynthis*), also called bitter apple, bitter cucumber or desert gourd. Closely related to cultivated watermelon (*Citrullus lanatus*), egusi is an important medicinal plant that strives in arid environments

(Dane et al. 2006). Several marker systems have been used to estimate the genetic diversity among egusi and cultivated watermelon accessions, including RAPD (Levi et al. 2001), AFLP and SSR (Nimmakayala et al. 2010; Hwang et al. 2011). Two major clusters were observed using 965 AFLP and EST-SSR markers, separating 27 watermelon cultigens from four wild-type species. Si et al. (2009) found dynamic gene expression changes in response to drought stress in egusi root tissues using cDNA-AFLP. Time point leaf transcriptomic changes at seedling stage was further elaborated by Wang et al. (2014) utilising NGS technology. This drought tolerant species has been used as a model to elucidate genes responsible for stress tolerance in other cucurbit crops through genetic manipulation (Wang et al. 2014).

Yam (*Dioscorea* spp.), among a handful of tropical root and tuber crops, is an important staple food for millions of people in Africa. The white (*D. rotundata*) and yellow Guinea (*D. cayenensis*) yams are the two most important yam species in Africa, while the water yam (*D. alata*) is the most widely distributed species in the world (Mignouna and Abang Asiedu 2007). The initial effort in yam genetics began with the development of its polymorphic DNA-based markers in the late 1990s. The markers of choice in early studies of yam genetic variation were AFLP (Mignouna et al. 1998) and RAPD (Dansi et al. 2000). In 2003, a set of 36 SSRs was developed and the efficiency of these SSRs was assessed together with the previously developed AFLPs and RAPDs in analysing the genetic diversity among 45 accessions of *D. alata* (Mignouna et al. 2003). Each of the three marker systems showed polymorphism among the accessions, with AFLP revealing the highest level of polymorphism (Mignouna et al. 2003).

The first two comprehensive linkage maps of yams, one for *D. alata* (Mignouna et al. 2002a) and the other for *D. rotundata* (Mignouna et al. 2002b), were developed concurrently based on AFLPs. Since then, considerable progress has been made in the development of molecular tools for mapping and QTL analysis in yams (Petro et al. 2011). Narina et al. (2011) developed the first extensive public *Dioscorea* EST collection, yielding a total of 44,757 EST sequences. More recently, genomic resources for *D. alata* have been generated by Saski et al. (2015), including 288,505 high quality SNPs and more than 30,000 genomic SSRs. The availability of these resources will cater tools for wider compatibility across different *Dioscorea* species and for MAS in their breeding improvement programmes (Saski et al. 2015).

3.4 Potential Use of Molecular Marker Technology in Genetic Improvement of Underutilised Crops

The evolution of molecular marker technology constantly offers advanced tools for wider compatibility across different crop species, genetic analysis of complex traits and for MAS in crop breeding programmes (De La Fuente et al. 2013; Saski et al. 2015). In the case of major crops, the new generation marker systems such as SSR and SNP have been routinely used in the MAS breeding programmes worldwide to develop improved or new varieties (Gupta et al. 2010). The success of MAS depends largely on several important factors, including the genetic distance of the target gene(s) and the marker(s), and the number of target genes to be selected (Francia et al. 2005). The efficiency of MAS can be enhanced if the markers used are more closely linked to major genes or target loci (Liu et al. 2007b). It should be noted that, in any event, identification and validation of markers tightly linked to gene(s) controlling the target trait must be done prior to MAS.

It is evident that large-scale MAS has been successful in many major cereals, notably wheat, maize, and rice (Liu et al. 2007b; Gupta et al. 2010; Miah et al. 2013). Unfortunately, this is not the case for most of the underutilised crops (Kumar et al. 2011). Although the pace of development of markers and genomic resources has been hastened in some potential underutilised crops as discussed in Sect. 3.3, the overall progress in the use of MAS as part of their breeding programmes has been limited. To this end, researchers are urged to start employing the developed resources for improving underutilised crops, and making MAS an integral part of the current and future breeding programmes (Kumar et al. 2012). As for the less developed underutilised crops that have enormous potential, the winged bean being a notable example (Wong et al. 2016), the development of adequate genetics and genomic resources is a crucial first step towards their genetic improvement (Abdullah Bamba et al. 2015a, b).

On a positive note, the potential benefits and application of genomics resources developed in model and/or major plant species has been explored in some underutilised species. One of the approaches for cross-species study is through cross-hybridisation of heterologous nucleic acids from target species, i.e. underutilised crops, onto the Affymetrix oligonucleotide based microarray of a reference species, i.e. major crop (http://affy.arabidopsis.info/xspecies/). Several studies have reported the use of cross-species microarray, for instance, the development of more than 1000 single feature polymorphisms (SFPs) in cowpea through detection and validation using soybean genome array (Das et al. 2008). In general, the application of knowledge from model and/or major crops to underutilised crops may be effective in the following areas (Liu et al. 2007b):

1. Analyses of crop genetic diversity
2. Identification of useful alleles or perfect molecular markers
3. Development of new or improved varieties with specific allele integration using MAS schemes
4. Cloning and transfer of desirable alleles or traits among taxa

We are certain that the ever-decreasing cost of marker technologies, and availability of fully automated high-throughput approaches in this post-genomics era will contribute immensely to the research for underutilised crop improvement.

3.5 Conclusions and Future Prospects

Biodiversity loss, population growth and climate change are all interconnected and interdependent challenges facing global agriculture today. There is, in fact, no single recipe to ensure global food security in light of these challenges. From an agricultural perspective, diversification of global food sources and systems with underutilised crops is one of the most sensible means to deal with hunger and malnutrition in the climate change era. The slow progress made in the utilisation of molecular technology in a number of underutilised crops with high potential has been attributed to the limited efforts and finite resources dedicated to developing their genomic resources. Greater emphasis should be placed on generating a sophisticated molecular markers and genomic resources for each of these crops, which will lay a critical foundation for their future molecular breeding efforts.

References

Abdullah Bamba SA, Massawe F (2013) Microsatellites based marker molecular analysis of Ghanaian bambara groundnut (*Vigna subterranean* (L.) Verdc.) landraces alongside morphological characterization. Genet Resour Crop Evol 60:777–787

Abdullah Bamba SA, Massawe F, Mayes S (2015a) Beyond landraces: developing improved germplasm resources for underutilized species—a case for Bambara groundnut. Biotechnol Genet Eng Rev 30:37–41. doi:10.1080/02648725.2014.992625

Abdullah Bamba SA, Massawe F, Mayes S (2015b) Beyond landraces: developing improved germplasm resources for underutilized species—a case for Bambara groundnut. Biotechnol Genet Eng Rev 1-2:37–41

Abraham B, Araya H, Berhe T, Edwards S, Gujja B, Khadka RB, Sen D, Koma Y, Sharif A, Styger E, Uphoff N, Verma A (2014) The system of crop intensification: reports from the field on improving agricultural production, food security, and resilience to climate change for multiple crops. Agric Food Sci 3:1–12

Ahmad SA, Redjeki ES, Ho WK, Aliyu S, Mayes K, Massawe F, Kilian A, Mayes S (2016) Construction of a genetic linkage map and QTL analysis in bambara groundnut. Gen 59:459–472. doi:10.1139/gen-2015-0153

Arias RS, Borrone JW, Tondo CL, Kuhn DN, Irish BM, Schnell RJ (2012) Genomics of tropical fruit tree crops. In: Schnell RJ, Priyadarshan PM (eds) Genomics of tree crops. Springer Science, New York, pp 209–239

Arif IA, Bakir MA, Khan HA, Al Farhan AH, Al Homaidan AA, Bahkali AH et al (2010) A brief review of molecular techniques to assess plant diversity. Int J Mol Sci 11(5):2079–2096

Babu BK, Dinesh P, Agrawal PK, Sood S, Chandrashekara C, Bhatt JC, Kumar A (2014) Comparative genomics and association mapping approaches for blast resistant genes in finger millet using SSRs. PLoS One 9(6):e99182

Bardakci F (2000) Random amplified polymorphic DNA (RAPD) markers. Turk J Biol 25:185–196

Basu S, Roberts JA, Azam-Ali SN, Mayes S (2007) Development of microsatellite markers for bambara groundnut (*Vigna subterranea* L. Verdc.)—an underutilized African legume crop species. Mol Ecol Notes 7:1326–1328

Bennetzen JL, Schmutz J, Wang H, Percifield R, Hawkins J, Pontaroli AC, Estep P, Feng L, Vaughn JN et al (2012) Reference genome sequence of the model plant Setaria. Nat Biotechnol 30(6):555–561

Bisen PS (2014) Laboratory protocols in applied life sciences. CRC Press, Boca Raton

Bohra A, Pandey MK, Jha UC, Singh B, Singh IP, Datta D, Chaturvedi SK, Nadarajan N, Varshney RK (2014) Genomics-assisted breeding in four major pulse crops of developing countries: present status and prospects. Theor Appl Genet 127(6):1263–1291

Chen X, Sullivan PF (2003) Single nucleotide polymorphism genotyping: biochemistry, protocol, cost and throughput. Pharmacogenomics J 3:77–96

Chen H, Chen X, Tian J, Yang Y, Liu Z, Hao X et al (2015) Development of gene-based SSR markers in rice bean (*Vigna umbellata* L.) based on transcriptome data. PLoS One 11(3):e0151040

Chen HL, Chen X, Tian J, Yang Y, Liu Z, Hao X et al (2016) Development of gene-based SSR markers in rice bean (*Vigna umbellata* L.) based on transcriptome data. PLoS One 11(3):e0151040

Cheng A, Ismanizan I, Mohamad O, Habibuddin H (2012) Simple and rapid molecular techniques for identification of amylose levels in rice varieties. Int J Mol Sci 13:6156–6166

Cheng A, Mayes S, Dalle G, Demissew S, Massawe F (2015) Diversifying crops for food and nutrition security—a case of teff. Biol Rev 92:188–198

Christensen SA, Pratt DB, Pratt C, Nelson PT, Stevens MR, Jellen EN et al (2007) Assessment of genetic diversity in the USDA and CIP-FAO international nursery collections of quinoa (*Chenopodium quinoa* Willd.) using microsatellite markers. Plant Gen Res 5:82–95

Coles ND, Coleman CE, Christensen SA, Jellen EN, Stevens MR, Bonifacio A, Rojas-Beltran JA, Fairbanks DJ, Maughan PJ (2005) Development and use of an expressed sequenced tag library in quinoa (*Chenopodium quinoa* Willd.) for the discovery of single nucleotide polymorphisms. Plant Sci 168(2):439–447

Cruz VMV, Kilian A, Dierig DA (2013) Development of DArT marker platforms and genetic diversity assessment of the U.S. collection of the new oilseed crop Lesquerella and related species. PLoS One 8:e64062

Dane F, Liu J, Zhang C (2006) Phylogeography of the bitter apple, *Citrullus colocynthis*. Genet Resour Crop Evol 54:327–336

Dansi A, Mignouna HD, Zoudjihekpon J, Sangare A, Asiedu R, Ahoussou N (2000) Identification of some Benin Republic's Guinea yam cultivars using randomly amplified polymorphic DNA. Genet Resourc Crop Evol 47:619–662

Das AK, Anjaneluyu ASR, Gadekar YP, Singh RP, Pragati H (2008) Effect of full-fat soy paste and textured soy granules on quality and shelf-life of goat meat nuggets in frozen storage. Meat Sci 80:607–614

Dawson IK, Guarino L, Jaenicke H (2007) Underutilised plant species: impacts of promotion on biodiversity. Position Paper No. 2. ICUC, Colombo, Sri Lanka

De La Fuente GN, Frei UK, Lubberstedt T (2013) Accelerating plant breeding. Trends Plant Sci 18:667–672

Dhanasekar P, Dhumal KN, Reddy KS (2010) Identification of RAPD markers linked to plant type gene in pigeonpea. Indian J Biotechnol 9:58–63

Ebert AW (2014) Potential of underutilized traditional vegetables and legume crops to contribute to food and nutritional security, income and more sustainable production systems. Sustainability 6:319–335

Edwards KJ, Barker JHA, Daly A, Jones C, Karp A (1996) Microsatellite libraries enriched for several microsatellite sequences in plants. Biotechniques 20:758–760

Ellis JR, Burke JM (2007) EST-SSRs as a resource for population genetic analyses. Heredity 99:125–132

Francia E, Tacconi G, Crosatti C, Barabaschi D, Bulgarelli D, Dall'Aglio E, Valè G (2005) Marker assisted selection in crop plants. Plant Cell Tissue Org 82:317–342

Fu D, Ma B, Mason A, An Z (2013) MicroRNA-based molecular markers: a novel PCR-based genotyping technique in brassica species. Plant Breed 132(4):75–81

Ganal MW, Altmann T, Röder MS (2009) SNP identification in crop plants. Curr Opin Plant Biol 12:211–217

Garcia-Mas J, Benjak A, Sanseverino W, Bourgeois M, Mir G et al (2012) The genome of melon (*Cucumis melo* L.) Proc Natl Acad Sci U S A 109(29):11872–11877

Gonthier P, Sillo F, Lagostina E, Roccotelli A, Cacciola OS, Stenlid J, Garbelotto M (2015) Selection processes in simple sequence repeats suggest a correlation with their genomic location: insights from a fungal model system. BMC Genomics 16:1107

Govindaraj M, Vetriventhan M, Srinivasan M (2015) Importance of genetic diversity assessment in crop plants and its recent advances: an overview of its analytical perspectives. Genet Res Int 2015:431–487

Gupta PK, Kumar J, Mir RR, Kumar A (2010) Marker assisted selection as a component of conventional plant breeding. Plant Breed Rev 33:145–217

Harel-Beja R, Tzuri G, Portnoy V, Lotan-Pompan M, Lev S et al (2010) A genetic map of melon highly enriched with fruit quality QTLs and EST markers, including sugar and carotenoid metabolism genes. Theor Appl Genet 121(3):511–533

Ho WK, Muchugi A, Muthemba S, Kariba R, Mavenkeni BO, Hendre P, Song B, Deynze V, Massawe F, Mayes S (2016) Use of microsatellite markers for the assessment of bambara groundnut breeding system and varietal purity before genome sequencing. Genome 59(6):427–431

Huang CW, Lin YT, Ding ST, Lo LL, Wang PH, Lin EC, Liu FW, Lu YW (2015) Efficient SNP discovery by combining microarray and lab-on-a-chip data for animal breeding and selection. Microarrays 4:570–595

http://rebase.neb.com/rebase/rebase.html

http://seedsofdiscovery.org/about/genotyping-platform/

http://www.cropsforthefuture.org/

Hunt HV, Campana MG, Lawes MC, Park Y-J, Bower MA, Howe CJ et al (2011) Genetic diversity and phylogeography of broomcorn millet (*Panicum miliaceum* L.) across Eurasia. Mol Ecol 20:4756–4771

Hwang J, Ang JK, Son B, Kim K, Park Y (2011) Genetic diversity in watermelon cultivars and related species based on AFLPs and EST-SSRs. Notulae Botanicae Horti Agrobotanici Cluj-Napoca 39(2):285

Iranjo P, Nabati Ahmadi D, Sorkheh K, Memeari HR, Ercisli S (2016) Genetic diversity and phylogenetic relationships between and within wild *Pistacia* species populations and implications for its conservation. J Forestry Res 27(3):685–697

Jaccoud DK, Peng D, Feinstein D, Kilian A (2001) Diversity arrays: a solid state technology for sequence information independent genotyping. Nucleic Acids Res 29(4):e25

Jarvis DE, Kopp OR, Jellen EN, Mallory MA, Pattee J, Bonifacio A, Coleman CE, Stevens MR, Fairbanks DJ, Maughan PJ (2008) Simple sequence repeat marker development and genetic mapping in quinoa (*Chenopodium quinoa* Willd.) J Genet 87:39–51

Jarvis DE, Ho YS, Lightfoot DJ, Schmockel SM, Li B et al (2017) The genome of *Chenopodium quinoa*. Nature 542(7641):307–312

Ji Q, Xu X, Wang K (2013) Genetic transformation of major cereal crops. Int J Dev Biol 57:495–508

Jia J, Zhou S, Kong X et al (2013) *Aegilops tauschiidraft* genome sequence reveals a gene repertoire for wheat adaptation. Nature 496(7443):91–95

Jones CJ, Edwards KJ, Castaglione S, Winfield MO, Sala F et al (1997) Reproducibility testing of RAPD, AFLP and SSR markers in plants by a network of European laboratories. Mol Breed 3:381–390

Joshi SP, Ranjekar PK, Gupta VS (1999) Molecular markers in plant genome analysis. Crop Sci 16:102–105

Kesawat MS, Kumar BD (2009) Molecular markers: it's application in crop improvement. J Crop Sci Biotechnol 12:169–181

Khoury CK, Bjorkman AD, Dempewolf H, Ramirez-villegas J, Guarino L, Jarvis A, Rieseberg LH, Struik PC (2014) Increasing homogeneity in global food supplies and the implications for food security. Proc Natl Acad Sci U S A 111:4001–4006

Kumar P, Gupta VK, Misra AK, Modi DR, Pandey BK (2009) Potential of molecular markers in plant biotechnology. Plant Omics J 2(4):141–162

Kumar J, Choudhary AK, Solanki RK, Pratap A (2011) Towards marker-assisted selection in pulses: a review. Plant Breed 130:297–313

Kumar S, Banks TW, Cloutier S (2012) SNP discovery through next-generation sequencing and its applications. Int J Plant Genomics 2012:831460. doi:10.1155/2012/831460

Kumari K, Pande A (2010) Study of genetic diversity in finger millet (*Eleusine coracana* L. Gaertn) using RAPD markers. Afr J Biotechnol 9(29):4542–4549

Kumpatla SP, Buyyarapu R, Abdurakhmonov IY, Mammadov JA (2012) Genomics-assisted plant breeding in the 21st century: technological advances and progress. Plant Breed:131–184

Lai PS, Ho WS, Pang SL (2013) Development, characterization and cross-species transferability of expressed sequence tag-simple sequence repeat (EST-SSR) markers derived from kelampayan tree transcriptome. Biotechnology 12:225–235

Levi A, Thomas CE, Keinath AP, Wehner TC (2001) Genetic diversity among watermelon (*Citrullus lanatus* and *Citrullus colocynthis*) accessions. Gen Res Crop Evol 48:559

Lin WH, Kussell E (2012) Evolutionary pressures on simple sequence repeats in prokaryotic coding regions. Nucleic Acids Res 40(6):2399–2413

Lin J, Kuo J, Ma J (1996) A PCR-based DNA fingerprinting technique: AFLP for molecular typing of bacteria. Nucleic Acids Res 24(18):3649–3650

Liu P, Yang YS, Hao CY, Guo WD (2007a) Ecological risk assessment using RAPD and distribution pattern of a rare and endangered species. Chemosphere 68:1497–1505

Liu S, Banik M, Yu K, Park SJ, Poysa V, Guan Y (2007b) Marker-assisted election (MAS) in major cereal and legume crop breeding: current progress and future directions. Int J Plant Breed 1:74–88

Ma JQ, Yao MZ, Ma CL, Wang XC, Jin JQ, Wang XM, Chen L (2014) Construction of a SSR-based genetic map and identification of QTLs for Catechins content in tea plant (*Camellia sinensis*). PLoS One 9(3):e93131

Massawe FJ, Dickinson M, Roberts JA, Azam-Ali SN (2002) Genetic diversity in bambara groundnut landraces (*Vigna subterranea* (L.) Verdc) revealed by AFLP markers. Gen 45:1175–1180

Massawe FJ, Roberts JA, Azam-Ali SN, Davey MR (2003) Genetic diversity in bambara groundnut (*Vigna subterranea* (L.) Verdc) landraces assessed by random amplified polymorphic DNA (RAPD) markers. Genet Resour Crop Evol 50:737–741

Massawe F, Mwale SS, Azam-Ali SN, Roberts JA (2005) Breeding in bambara groundnut (*Vigna subterranea* (L.) Verdc.): strategic considerations. Afr J Biotechnol 4(6):463–471

Massawe F, Mayes S, Cheng A (2016) Crop diversity: an unexploited treasure trove for food security. Trends Plant Sci 21(5):365–368

Masters WA (2010) Africa's turnaround: from crisis to opportunity in African agriculture. In: Food and financial crises: impacts on Sub-Saharan Africa. CAB International, Wallingford

Maughan PJ, Bonifacio A, Jellen EN, Stevens MR, Coleman CE, Ricks M et al (2004) A genetic linkage map of quinoa (Chenopodium Quinoa) based on AFLP, RAPD, and SSR markers. Theor Appl Genet 109(6):1188–1195

Maughan PJ, Yourstone SM, Jellen EN, Udall JA (2009) SNP discovery via genomic reduction, barcoding and 454-pyro-sequencing in amaranth. Plant Gen 2:260–270

Maughan PJ, Smith SM, Jellen EN, Fairbanks D (2011) Development, characterization and linkage mapping of single nucleotide polymorphisms in the grain amaranths (*Amaranthus* sp.) Plant Gen 4:1–10

Maughan PJ, Smith SM, Rojas-Beltran JA, Elzinga D, Raney JA, Jellen EN, Bonifacio A, Udall JA, Fairbanks DJ (2012) Single nucleotide polymorphism identification, characterization, and linkage mapping in quinoa. Plant Genome 5:114–125

Mayes S, Massawe F, Anderson PG, Roberts JA, Azam-Ali SN, Hermann M (2011) The potential of underutilized crops to improve security of food production. J Exp Bot 63:1075–1079

Mba C, Tohme J (2005) Use of AFLP markers in surveys of plant diversity. Methods Enzymol 395:177–201

Meksem K, Ruben E, Hyten D, Triwitayakorn K, Lightfoot DA (2001) Conversion of AFLP bands into high-throughput DNA markers. Mol Genet Genomics 265(2):207–214

Melo ATO, Bartaula R, Hale L (2016) GBS-SNP-CROP: a reference-optional pipeline for SNP discovery and plant germplasm characterization using variable length, paired-end genotyping-by-sequencing data. BMC Bioinformatics 17:29

Miah G, Rafii MY, Ismail MR, Puteh AB, Rahim HA, Islam KN, Latif MA (2013) A review of microsatellite markers and their applications in rice breeding programs to improve blast disease resistance. Int J Mol Sci 14(11):22499–22528

Mignouna HD, Abang Asiedu R (2007) Advances in yam (Dioscorea spp.) genetics and genomics proceedings of the 13th ISTRC symposium, pp 72–81

Mignouna HD, Ellis NTH, Asiedu R, Ng QN (1998) Analysis of genetic diversity in Guinea yams (*Dioscorea* spp) using AFLP fingerprinting. Trop Agric (Trinidad) 75:224–229

Mignouna HD, Mank RA, Ellis THN, van den Bosch N et al (2002a) A genetic linkage map of water yam (*Dioscorea alata* L.) based on AFLP markers and QTL analysis for anthracnose resistance. Theor Appl Genet 105:726–735

Mignouna HD, Mank RA, Ellis THN et al (2002b) A genetic linkage map of Guinea yam (Dioscorea rotundata L.) based on AFLP markers. Theor Appl Genet 105:716–725

Mignouna HD, Abang MM, Fagbemi SA (2003) A comparative assessment of molecular marker assays (AFLP, RAPD and SSR) for white yam (*Dioscorea rotundata* Poir.) germplasm characterisation. Ann Appl Biol 142:269–276

Mochida K, Shinozaki K (2013) Unlocking Triticeae genomics to sustainably feed the future. Plant Cell Physiol 54(12):1931–1950

Moe KT, Gwag JG, Park YJ (2012) Trends in genomics and molecular marker systems for the development of some underutilized crops. Gen Genom 34(5):451–466

Molosiwa OO, Aliyu S, Stadler F, Mayes K, Massawe F, Kilian A, Mayes S (2015) SSR marker development, genetic diversity and population structure analysis of Bambara groundnut [*Vigna subterranea* (L.) Verdc.] landraces. Gen Res Crop Evol 62(8):1225–1243

Narina SS, Buyyarapu R, Kottapalli KR, Sartie AM, Ali MI, Robert A et al (2011) Generation and analysis of expressed sequence tags (ESTs) for marker development in yam (*Dioscorea alata* L.) BMC Genomics 12:100

Naylor RL, Falcon WP, Goodman RM, Jahn MM, Sengooba T, Tefera H, Nelson RJ (2004) Biotechnology in the developing world: a case for increased investments in orphan crops. Food Policy 29:15–44

Ngwira AR, Thierfelder C, Lambert DM (2012) Conservation agriculture systems for Malawian smallholder farmers: longterm effects on crop productivity, profitability and soil quality. Renew Agric Food Syst 28(4):350–363

Nimmakayala P, Tomason YR, Jeong J, Ponniah SK, Karunathilake A, Levi A, Perumal R, Levi UK (2010) Genetic reticulation and interrelationships among citrullus species as revealed by joint analysis of shared AFLPs and species-specific SSR alleles. Plant Gen Res 8(1):16–25

Ojiewo C, Tenkouano A, Oluoch M, Yang R (2010) The role of AVRDC—the world vegetable centre in vegetable value chains. Afr J Hort Sci 3:1–23

Padulosi S, Hoeschle-Zeledon I (2004) Underutilized plant species: what are they? LEISA 20(1):56

Petro D, Onyeka TJ, Etienne S, Rubens S (2011) An intraspecific genetic map of water yam (*Dioscorea alata* L.) based on AFLP markers and QTL analysis for anthracnose resistance. Euphytica 179:405–416

Phung NTP, Mai CD, Mournet P, Frouin J, Droc G, Ta NK, Jouannic S, Lê LT, Do VN, Gantet P, Courtois B (2014) Characterization of a panel of Vietnamese rice varieties using DArT and SNP markers for association mapping purposes. BMC Plant Biol 14:371

Pingali PL (2012) Green revolution: impacts, limits, and the path ahead. Proc Natl Acad Sci U S A 109:12302–12308

Powell W, Machray GC, Provan J (1996) Polymorphism revealed by simple sequence repeats. Trends Plant Sci 1(7):215–222

Rahman MM, Aminul MI, Sofian MA, Amru NB (2014) Tropical legume crop rotation and nitrogen fertilizer effects on agronomic and nitrogen efficiency of rice. Sci World J. doi:10.1155/2014/490841

Ranamukhaarachchi DG, Kane ME, Guy CL, Li QB (2000) Modified AFLP technique for rapid genetic characterization in plants. Biotechniques 29(4):858–859

Raney JA, Reynolds DJ, Elzinga DB, Page J, Udall JA, Jellen EN, Bonfacio A, Fairbanks D, Maughan PJ (2014) Transcriptome analysis of drought induced stress in *Chenopodium quinoa*. Am J Plant Sci 5:338–357

Raveendar S, Lee GA, Jeon YA, Lee YJ, Lee JR, Cho GT, Cho JH, Park JH, Ma KH, Chung JW (2015) Cross-amplification of *Vicia sativa* subsp. sativa microsatellites across 22 other *Vicia* species. Molecules 20(1):1543–1550

Ray DK, Mueller ND, West PC, Foley JA (2013) Yield trends are insufficient to double global crop production by 2050. PLoS One 8:1–8

Ren X, Wang J, Liu L, Sun G, Li C, Luo H, Sun D (2016) SNP-based high density genetic map and mapping of btwd1 dwarfing gene in barley. Sci Rep 6:31741. doi:10.1038/srep31741

Roberts RJ, Vincze T, Posfai J, Macelis D (2003) REBASE: restriction enzymes and methyltransferases. Nucleic Acids Res 31(1):418–420

Rubinstein M, Katzenellenbogen M, Eshed R, Rozen A, Katzir N, Colle M et al (2015) Ultrahigh-density linkage map for cultivated cucumber (*Cucumis sativus* L.) using a single-nucleotide polymorphism genotyping array. PLoS One 10(4):e0124101

Saeed B, Baranwal VK, Khurana P (2016) Comparative transcriptomics and comprehensive marker resource development in mulberry. BMC Genomics 17:98

Sánchez-Sevilla JF, Horvath A, Botella MA, Gaston A, Folta K, Kilian A, Denoyes B, Amaya I (2015) Diversity arrays technology (DArT) marker platforms for diversity analysis and linkage mapping in a complex crop, the octoploid cultivated strawberry (Fragaria × ananassa). PLoS One 10(12):e0144960

Sansaloni C, Petroli C, Jaccoud D, Carling J, Detering F, Grattapaglia D, Kilian A (2011) Diversity Arrays Technology (DArT) and next-generation sequencing combined: genome-wide, high throughput, highly informative genotyping for molecular breeding of Eucalyptus. BMC Proc 5(Suppl 7):54

Saski CA, Bhattacharjee R, Scheffler BE, Asiedu R (2015) Genomic resources for water yam (*Dioscorea alata* L.): analyses of EST-sequences, *de novo* sequencing and GBS libraries. PLoS One 10(7):e0134031

Schlotterer C (2004) The evolution of molecular markers—just a matter of fashion? Nat Rev Genet 5:63–69

Semagn K, Bjørnstad A, Ndjiondjop MN (2006) An overview of molecular marker methods for plants. Afr J Biotechnol 5(25):2540–2568

Shi W, Tao F (2014) Vulnerability of African maize yield to climate change and variability during 1961–2010. Food Secur 6:471–481

Si Y, Zhang C, Meng S, Dane F (2009) Gene expression changes in response to drought stress in *Citrullus colocynthis*. Plant Cell Rep 28:997–1009

Suresh S, Chung JW, Cho GT, Sung JS, Park JH, Gwag JG, Baek HJ (2014) Analysis of molecular genetic diversity and population structure in *Amaranthus* germplasm using SSR markers. Plant Biosyst 148(4):635–644

Tanksley SD, Young ND, Paterson AH, Bonierbale MW (1989) RFLP mapping in plant breeding: new tools for an old science. Nat Biotechnol 7:257–264

Tártara SMC, Manifesto MM, Bramardi SJ, Bertero HD (2012) Genetic structure in cultivated quinoa (*Chenopodium quinoa* Willd.), a reflection of landscape structure in Northwest Argentina. Conserv Genet 13:1027-1038

Thies E (2000) Promising and underutilized species, crops and breeds. GTZ, Eschborn, Germany

Tomar RS (2010) Molecular markers and plant biotechnology. New India Publishing, New Delhi

Troconis-Torres IG, Rojas-López M, Hernández-Rodríguez C, Villa-Tanaca L, Maldonado-Mendoza IE et al (2002) Biochemical and molecular analysis of some commercial samples of chilli peppers from Mexico. J Biomed Biotechnol 2012:873090

Varshney RK, Graner A, Sorrells ME (2005) Genic microsatellite markers in plants: features and applications. Trends Biotechnol 23:48–55

Varshney RK, Chen W, Li Y et al (2012) Draft genome sequence of pigeonpea (*Cajanus cajan*), an orphan legume crop of resource-poor farmers. Nat Biotechnol 30(1):83–89

Vatanparast M, Shetty P, Chopra R, Doyle JJ, Sathyanarayana N, Egan AN (2016) Transcriptome sequencing and marker development in winged bean (*Psophocarpus tetragonolobus*; Leguminosae). Sci Rep 6:29070. doi:10.1038/srep29070

Verde I, Abbott AG, Scalabrin S, Jung S, Shu S, Marroni F et al (2013) The high-quality draft genome of peach (*Prunus persica*) identifies unique patterns of genetic diversity, domestication and genome evolution. Nat Genet 45(5):487–494

Vos P, Hogers R, Bleeker M, Reijans M, van de Lee T, Hornes M et al (1995) AFLP: a new technique for DNA fingerprinting. Nucleic Acids Res 23:4407–4414

Wang Z, Hu H, Goertzen LR, McElroy JS, Dane F (2014) Analysis of the *Citrullus colocynthis* transcriptome during water deficit stress. PLoS One 9(8):e104657

Weising K, Nybom H, Pfenninger M, Wolff K, Meyer W (1994) DNA fingerprinting in plants and fungi. CRC Press, Boca Raton

Williams JT, Haq N (2002) Global research on underutilized crops. An assessment of current activities and proposals for enhanced cooperation. ICUC, Southampton

Williams JGK, Kubelik AR, Livak KJ, Rafalski JA, Tingey SV (1990) DNA polymorphisms amplified by arbitrary primers are useful as genetic markers. Nucleic Acids Res 18:6531–6535

Wong QN, Massawe F, Mayes K, Blythe M (2016) Molecular genetic tools to support genetic improvement of winged bean (*Psophocarpus tetragonolobus*) for food and nutrition security. Acta Hortic 1110:1–6

Wong QN, Tanzi AS, Ho WK, Malla S, Blythe M et al (2017) Development of gene based SSR markers in winged bean (*Psophocarpus tetragonolobus* (L.) DC.) for diver-sity assessment. Genes 8(3):100

Yu H, Li Q (2007) Development of EST-SSRs in the Mediterranean blue mussel, Mytilus galloprovincialis. Mol Ecol Notes 7(6):1308–1310

Zeid M, Belay G, Mulkey S, Poland J, Sorrells ME (2011) QTL mapping for yield and lodging resistance in an enhanced SSR-based map for tef. Theor Appl Genet 122:77–93

Zeid M, Assefa K, Haddis A, Chanyalew S, Sorrells ME (2012) Genetic diversity in tef (*Eragrostis tef*) germplasm using SSR markers. Field Crop Res 127:64–70

Zhao Y, Williams R, Prakash CS, He G (2012) Identification and characterization of gene-based SSR markers in date palm (*Phoenix dactylifera* L.) BMC Plant Biol 12:237

Chapter 4
Transcription Factors Associated with Abiotic Stress and Fruit Development in Oil Palm

Siti Nor Akmar Abdullah, Azzreena Mohamad Azzeme, Mortaza Ebrahimi, Engku Ahmad Khairi Engku Ariff, and Farah Hanan Abu Hanifiah

4.1 Introduction

Palm oil has wide applications in the production of various food and nonfood products. It is the main edible oil consumed globally and serves as an important feedstock for biofuels. Demand for palm oil will continue to increase as the demand for vegetable oils by the year 2020 for both food and biofuel will reach 182 million tons (OECD-FAO 2011) compared to 171 million tons vegetable oils consumed in 2015 (IMARC Group 2016). It is also expected that palm oil and its products will make a major contribution to meet future increasing demand for food in developing countries. Besides palm oil which is obtained from the mesocarp of the oil palm fruits, palm

S.N.A. Abdullah (✉) • F.H.A. Hanifiah
Institute of Plantation Studies, Universiti Putra Malaysia, 43400 UPM, Serdang, Selangor, Malaysia
e-mail: snaa@upm.edu.my

A.M. Azzeme
Institute of Plantation Studies, Universiti Putra Malaysia, 43400 UPM, Serdang, Selangor, Malaysia

Department of Biochemistry, Faculty of Biotechnology and Biomolecular Sciences, Universiti Putra Malaysia, 43400 UPM, Serdang, Selangor, Malaysia

M. Ebrahimi
Agricultural Biotechnology Research Institute of Iran-Isfahan Branch, Isfahan, Iran

E.A.K.E. Ariff
Department of Agriculture Technology, Faculty of Agriculture, Universiti Putra Malaysia, 43400 UPM, Serdang, Selangor, Malaysia

S.N.A. Abdullah et al. (eds.), *Crop Improvement*,
DOI 10.1007/978-3-319-65079-1_4

kernel oil extracted from the kernel is widely used in oleochemical industry. In 2008 the palm kernel oil production was about 48 million tons and the prediction of global demand in 2020 will be double and triple in 2050 (Prokurat 2013).

Oil palm productivity is significantly affected by climate change effects resulting in extreme whether condition including prolong drought period (El Nino) and flood. The oil palm fresh fruit bunch (FFB) yield in the peninsular Malaysia and Sabah (in East Malaysia) decreased by about 20% due to the 1997–1998 El Nino because rainfall of less than 100 mm for more than 2 months can greatly affect the yield (Haniff et al. 2010). The effects of haze which occurs for certain period almost annually in recent years in leading palm oil producing countries (Malaysia and Indonesia) need to be addressed. Abiotic stress may affect fruit development and postharvest quality of oil palm fruits and this is of importance in terms of economic value to the industry.

Stress conditions affect the normal functioning of various cellular processes. The developmental and differentiation processes that are responsive to endogenous signals and environmental cues enable plants which are sessile in nature to adapt and survive in a changing environment. The endogenous and environmental stimuli trigger signaling cascades that converge at the gene regulation level, where transcription factors (TFs) play a major regulatory role. TFs are proteins with specific DNA binding domains that bind to DNA regulatory motifs (6–12 base pairs) in the promoter regions of their target genes resulting in the transcriptional activation or repression of these genes. There have been a lot of efforts to study the interaction between TFs and the specific promoter regulatory motifs to better understand the roles of TFs in cellular processes for developmental and environmental adaptation. TFs play critical role in crop improvement efforts since important loci that contribute to crop yield, quality of the harvested product or adaptation to extreme environmental conditions correspond to TFs responsible for changes in transcriptional activity. For example, the *SNORKEL1* (*SK1*) and *SNORKEL2* (*SK2*) genes that belong to the ethylene response factor (ERF) family of transcription factors confer flooding tolerance in rice (Hattori et al. 2009). In the oil palm, mutations in the *SHELL* gene which affects the DNA-binding domain of its encoded product, a MADS-box transcription factor is a major determinant of oil yield from the fruits (Singh et al. 2013). These are among the many examples that indicate that transcription factors and their DNA binding sites are key contributors to phenotypic variation for traits of economic importance.

4.2 Abiotic Stress in Plants

Abiotic stress is the major limiting environmental factor affecting the yield of most crops (Rossini et al. 2016). Plants have evolved mechanism to sense and respond to abiotic stresses such as high salinity, drought, extreme temperature, heavy metal toxicity, and nutrient deficiency (Zhu 2016). Being immobile, plants are equipped with stress signaling system which plays a central role in enabling the plants to withstand the fluctuation in environmental conditions. Many abiotic stresses like drought, high temperature, and salinity cause disruption in water status consequently leading to detrimental effect on the plants. Climate change in recent years that are affecting rainfall patterns and earth temperature is worsening the abiotic stress experienced by field plants (Misra 2014).

Excessive production of reactive oxygen species (ROS) that are highly reactive and toxic under various stress conditions can destroy proteins, lipids, carbohydrates, and DNA and interfere with biochemical processes in the plants which ultimately results in oxidative stress (Demidchik 2015). Heavy metal stresses are caused by the increased amount of heavy metals found in the soils such as lead (Pb), cadmium (Cd), and mercury (Hg) that have adverse effects on plant growth (Osobamiro and Adewuyi 2015). Deficiencies in heavy metals such as iron (Fe), copper (Cu), zinc (Zn), manganese (Mn), boron (B), nickel (Ni), and molybdenum (Mo) which are beneficial for growth and productivity may also cause stress to the plants (Dalcorso et al. 2014).

Abiotic stresses trigger local and long distant signals which are then synchronized and incorporated into the whole plant process so that the plant can respond and survive (Zwack and Rashotte 2015). Abiotic stress response involves cross-talk with energy signaling pathways where autophagy plays an important role. Autophagy destroys specific compounds essential for efficient growth but require major energy input. Consequently, biological processes that protect plants from the stress conditions are stimulated (Michaeli et al. 2016; Wang et al. 2017). Under nutrient deprivation for example, autophagy participates in recycling of nonessential cytoplasmic components to produce raw materials and energy for overcoming the stress. While under oxidative stress, autophagy is involved in degradation of oxidized proteins.

4.2.1 *Abiotic Stress Affecting Oil Palm Productivity*

The productivity of oil palm is largely affected by environmental factors. Region where drought occurs for 2–4 months, there is a possibility of large fluctuations in yield (Arshad et al. 2012). For the optimal oil palm production, sunshine of at least 5 h per day for most of the months and increase to 7 h per day in certain months are essential as inadequate sunshine decreases net assimilation rate and production of female inflorescence. Fresh fruit production could also be reduced from 20 to 30% in a prolonged drought while a combination of water depravity of 50 mm and temperature increment of 1 °C decreases the production at an average of 2.15 t/ha/year accounting about 40% of the average yield (Salmiyati et al. 2014). The rainfall patterns also can affect the oil palm yield after 6–7 months (Awan and Sap 2006). An average monthly temperature for 8 months of less than 27.83 °C before harvesting led to low yield of less than 1.25 t/ha/month all across Malaysia (Shanmuganathan et al. 2014).

Nutrient is important for the productivity of oil palm. Low phosphate reduced fruit production while low potassium reduced the fruit to mesocarp ratio (Salmiyati et al. 2014). Under water stress condition, the ratio of root to shoot increases. On the other hand, a combination of water and nutrient stresses cause the decrease of water content and chlorophyll a/b, nitrogen, and phosphorus concentration in leaves while increasing leaf relative conductivity and potassium concentration in leaves (Sun et al. 2011).

Some micronutrients are associated with the growth and productivity of oil palm. Deficiency of micronutrients such as B causes hook leaf in oil palm, Mn deficiency results in yellowing of interveinal areas in younger fronds, Zn deficiency causes white streaks on lower and mid-crown fronds. In addition, Fe deficiency causes the youngest frond to look droopy and has diffuse blotchy yellowing with white freckles and Cu deficiency leads to whitish yellow mottling of younger fronds (von Uexküll and Fairhurst 1999). B is important for the production, viability, and germination of pollen as well as the growth of pollen tubes. Cu is required for pollen viability while Mn acts as an activator of several enzymes essential for the biosynthesis of important secondary metabolites (Tengoua et al. 2015). Table 4.1 summarizes the abiotic stresses and their symptoms in oil palm.

4.3 Development and Ripening of Climacteric Fruits

Fruits develop from the ovary of a plant after the flowers undergo fertilization process and the ripe fruits often consumed by human and animals vary in taste and texture. Many fruits are made of three layers which are exocarp (skin), mesocarp (fleshy part), endocarp (shell surrounding the seed) and the seed or kernel. After fertilization, the seed develops and the surrounding tissues expand through active cell division and expansion. Ripening period starts in the fruit once the seed is fully developed (McAtee et al. 2013). Fruit ripening progresses through a complex process that involves gradual change in the physiological and metabolic states culminating in the fruits attaining their characteristic flavors, colors, and textures. The tissue surrounding the seed and skin will undergo color change and soften in texture. Gradual accumulation of the specific product such as sugars or lipids (in oil fruit crops) together with an increase in flavor and aroma volatiles will occur (Klee and Giovannoni 2011).

Fruits can be categorized into two types according to their ripening behaviors which are climacteric and non-climacteric fruits. The ripening process in climacteric fruits is regulated by ethylene, an array of transcription factors, and various developmental cues. Production of ethylene concomitant with the climacteric respiration for climacteric fruits is the main signal for the regulation of ripening for these types of fruits. A high cellular respiration rate accompanied by a burst of ethylene production up to 1000-fold of the basal level occurs in climacteric fruits such as tomato, avocado, banana, and oil palm at the onset of ripening (McAtee et al. 2013). In the oil palm fruits, production of ethylene in the mesocarp is controlled in a simi-

Table 4.1 Abiotic stress and its symptoms in oil palm

Abiotic stress	Symptom	References
Boron deficiency	Cause hook leaf	von Uexküll and Fairhurst (1999)
Manganese deficiency	Cause yellowing interveinal areas on younger fronds	von Uexküll and Fairhurst (1999)
Zinc deficiency	Cause white streaks on lower and mid-crown fronds	von Uexküll and Fairhurst (1999)
Iron deficiency	Cause the youngest frond to look droopy and have diffuse blotchy yellowing with white freckles	von Uexküll and Fairhurst (1999)
Copper deficiency	Cause whitish yellow mottling of younger fronds	von Uexküll and Fairhurst (1999)
Combination of water and nutrient stress	Decrease water content, chlorophyll a/b, nitrogen, and phosphorus concentration in leaves while increasing leaf relative conductivity and potassium concentration in leaves	Sun et al. (2011)
Drought	Fluctuation in yield	Arshad et al. (2012)
Inadequate sunshine	Decrease net assimilation rate and production of female inflorescence	Arshad et al. (2012)
Prolonged drought	Reduce FFB yield	Salmiyati et al. (2014)
Combination of water depravity and temperature increment of 1 °C	Reduce production of up to 2.15 t/ha on average	Salmiyati et al. (2014)
Low phosphate	Reduce fruit production	Salmiyati et al. (2014)
Low potassium	Reduce the fruit to mesocarp ratio	Salmiyati et al. (2014)
Low temperature	Yield of less than 1.25 t/ha/month	Shanmuganathan et al. (2014)

lar way with that observed in the tomato. During the fruit developmental stage, an ethylene autoinhibitory system 1 is activated, and transition to a climacteric-like autocatalytic system 2 occurs at the maturation phase in the oil palm mesocarp with the coordinated transcriptional activation of specific 1-aminocyclopropane-1-carboxylic acid (ACC) oxidase and ACC synthase enzymes involved in ethylene biosynthesis (Tranbarger et al. 2011). Climacteric fruits can ripen even after harvest due to the large amount of ethylene gas produced after harvesting (Giovannoni 2001). Auxin is mainly involved in regulating cell division during fruit set and is believed to play a role in influencing the growth phase together with gibberellins by regulating cell expansion (Csukasi et al. 2011). Current knowledge suggests that abscisic acid (ABA) plays the role of promoting ripening in climacteric and non-climacteric fruits (Leng et al. 2014). However, the roles and interplay of different phytohormones in fruit ripening are still not fully understood.

4.3.1 Fruit Development in Oil Palm

Oil palm (*Elaeis guineensis*), a perennial monocotyledonous species, produces fruit which is a drupe with thick and fleshy mesocarp in a fruit bunch. The oil palm is the highest yielding oil producing crop in the world where the mesocarp can accumulate about 80% (dry mass) of edible oil (Murphy 2009). The complex biological processes of oil palm fruit development, ripening, and maturation can be divided into five different phases from juvenile to complete maturity according to days after pollination (DAP), which are phase 1 (30–60 DAP), phase 2 (60–100 DAP), phase 3 (100–120 DAP), phase 4 (120–140 DAP), and phase 5 (140–160 DAP). The growth of the fruit is biphasic occurring at an early stage (phase 1) then at the later stages (phases 3, 4 and 5), especially in phase 5 when lipid accumulation occurs after a lag of growth in phase 2 which coincides with the presence of the highest amounts of indole-3-acetic acid (IAA) and IAA conjugates. The beginning of phase 4 (120 DAP) is the start of the maturation stage characterized by the increase in fresh mass of oil palm mesocarp and the accumulation of lipid and carotenoid. The phase 5 is where the ripening stage starts with a large increase in ABA and ethylene hormones together with detachment of the cell wall. Accumulation of lipids was observed within subcellular spherical organelles occupying the entire cells while high concentration of carotenoids was found in distinct regions in the cells presumably chromoplasts in the ripe fruits at 160 DAP through microscopic visualization (Tranbarger et al. 2011).

In the oil palm fruit bunches, the fruitlets do not ripen at the same time because of the slight variations in the time of pollination. The ripening of the fruits starts from the bottom and progresses towards the top of the bunch (Sambanthamurthi et al. 2000). The quality of extracted palm oil is affected by the asynchronous ripening of the fruits in the oil palm fruit bunch (Nualwijit and Lerslerwong 2014). Figures 4.1 and 4.2 show the fruit bunches on an oil palm tree and the cross section of an oil palm fruit, respectively.

4.4 Structure of Transcription Factors Involved in Abiotic Stress Response and Fruit Development

TFs are the largest functional class of proteins in plants consisting of about 60 families representing 7% of genes that encode for proteins. The different families of TFs are classified based on their conserved structural domains. They play diverse roles in regulating complex biological processes through interaction with the promoter regions of downstream genes (Riechmann and Ratcliffe 2000). Plants adapt to environmental stresses through induced expression of stress-responsive genes. The expression of these genes is spatially and temporarily regulated at the transcriptional level by TFs. TFs and the gaseous hormone, ethylene, are the main players in regulating the ripening process of climacteric fruit. Through their DNA binding

Fig. 4.1 Oil palm tree with ripe and unripe fruit bunches

Fig. 4.2 Cross section of a ripe oil palm fruit

Table 4.2 Classification of *Arabidopsis* AP2/ERF family of transcription factors

AP2/ERF subfamilies	No. of AP2/ERF domain	Subgroup	No. of genes
AP2	2 1		14 3
DREB	1	A-1 to A-6	56
ERF	1	B-1 to B-6	65
RAV	1		6
Others	1		1

Source: Lata and Prasad (2011)

domains, TFs bind to specific DNA motifs in the promoters of their target genes for regulating and coordinating their expression. Outside the DNA-binding domain of the TFs, there are other functionally important domains that are involved in nuclear localization (nuclear localization signal, NLS), transcriptional activity, and interactions with other proteins (Nakano et al. 2006).

4.4.1 AP2-ERF Superfamily

The APETALA2/ethylene response factor (AP2/ERF), one of the biggest family of plant-specific TFs are involved in diverse biological processes. These include response to abiotic and biotic stresses such as cold temperature, drought, microbial pathogen infection, as well as developmental processes like flowering and fruit ripening. Its functional roles are often associated with hormonal signaling pathways involving different phytohormones such as ethylene and ABA (Gu et al. 2017). There are 147 and 167 AP2/ERF members found in the genome sequence of Arabidopsis and rice, respectively (Sakuma et al. 2002; Nakano et al. 2006). Bioinformatics analysis of barley genome which contains 121 AP2/ERF showed that segmental duplication and tandem repeats lead to the expansion of the TF family members (Guo et al. 2016).

The AP2/ERF TFs contain the characteristic AP2/ERF DNA-binding domain of 60–70 amino acids and are classified into subfamilies based on the number of AP2-domain present (Table 4.2) and their sequence homology (Hao-Li et al. 2010). The AP2 family contains two or more AP2/ERF domains or the absence of the conserved WLG motif within these domains while the ABI3/VP1 family (RAV) has an AP2 and a second domain called the B3 domain. The third family, ERF, has a single AP2 domain while the rest of the genes are categorized as soloist. The ERF family has two main subfamilies: the dehydration-responsive element binding factor (DREB) and the ethylene-responsive factors (ERF) (Nakano et al. 2006).

The ERF and DREB can be differentiated based on the types of amino acid found at the 14th and 19th positions in the AP2-conserved domain. DREB proteins have valine and glutamic acid while ERF proteins have alanine and aspartic acid in 14th and 19th amino acid position, respectively. These two conserved amino acids which are critical for DNA binding are found on the β-sheet in the AP2/ERF domain (Akhtar et al. 2012). DREB protein is an important TFs involved in abiotic stress, particularly in conferring resistance to drought, salt, and cold (Lata and Prasad 2011).

There are two major groups classified under DREB, the cold binding factor (CBF)/DREB1 and DREB2 (Druege 2006). DREB1 can be further divided into six groups, A-1 to A-6 (Sakuma et al. 2002). The DSAW and LWSY motifs found in the AP2/ERF domain and C-terminal region, respectively, are A-1 group member specific signatures (Sakuma et al. 2002). DREB TFs bind to the drought responsive element (DRE) and DRE-like *cis*-elements located in the promoter of SRGs to induce their expression in response to abiotic stresses. The DRE possesses a core sequence A/GCCGAC (Gutha and Reddy 2008; Sharoni et al. 2011; Zhang et al. 2014).

4.4.2 Other Families of Transcription Factors

Transcription factors play an important role in the transcriptional control of gene expression in response to various environmental stresses. Members of the myeloblastosis related proteins (MYB) (Li et al. 2015), myelocytomatosis related proteins (MYC) (Fujita et al. 2006), basic leucine zipper containing domain proteins (bZIP) (Kobayashi et al. 2008), NAC (Chen et al. 2014), and WRKY (Chen et al. 2012) TF families have been implicated in the regulation of stress responses. While MADS-box, NAC, and bZIP are important in developmental processes including fruit ripening.

The bZIP proteins contain bZIP domain that is highly conserved and composed of a basic and a leucine zipper regions. The conserved bZIP domain is a basic sequence-specific DNA binding region. The leucine zipper region contains amphiphatic sequences in a coiled-coil form for dimerization specificity. In Arabidopsis, the bZIP family is well characterized for their role in ABA-dependent signal transduction for abiotic stress tolerance. AREB/ABF (ABA-responsive element binding protein/ABRE binding factor), belonging to group A bZIPs modulate transcription of ABA-dependent genes. They preferentially bind to ABA cis-acting element, ABRE such as the G-box (CACGTG) with the ACGT core motif. Numerous ABA and/or stress-regulated genes possess the (C/T)ACGTGGC consensus sequence in their promoter sequences (Choi et al. 2000). The bZIP family also participates in the fruit development process. Transcriptome analysis carried out on the climacteric banana for example, showed that MabZIP genes were differential expressed at the various stages of fruit development and ripening, and in response to a variety of abiotic stresses in the different tissues (Hu et al. 2016). The mechanism involved in ABA-dependent and ABA-independent pathway is summarized in Fig. 4.3.

The diversification and expansion of the plant MADS-box transcription factors are associated with the evolution of novel structures, like flowers, seeds, and fruits (Theissen et al. 2000). At the N-terminal region of the MADS-box family of transcription factors, the conserved MADS-box domain consisting of 56–60 amino

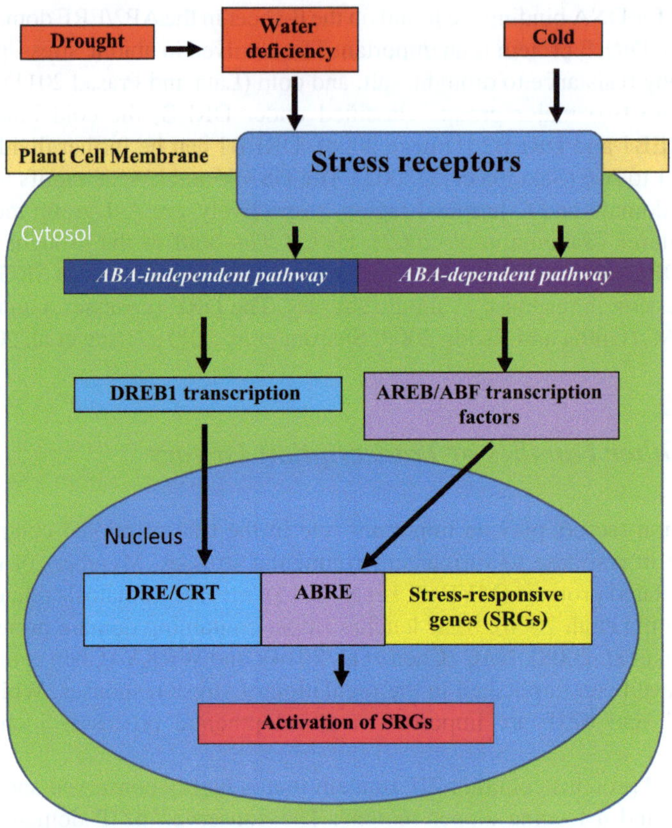

Fig. 4.3 Involvement of ABA-dependent and ABA-independent pathways in response to drought and cold stresses. The stress receptors received drought and cold signals, respectively from surrounding. Activation of ABA-dependent pathway involves AREB/ABF transcription factors. The activation of ABA-independent pathway activates DREB1 transcription factors. DREB1 and AREB/ABA transcription factors can directly bind to DRE/CRT and ABRE element located at stress-responsive genes promoter. Direct interaction of DREB1 with DRE/CRT and AREB/ABF induced expression of stress-responsive genes

acids, involved in DNA binding and dimerization with other MADS-box proteins can be found. Eukaryotic MADS-box is categorized into Type I and Type II (MIKC-type). Genetic and molecular analyses of the MIKC-type categorized their roles in flower organogenesis into five classes, where different combinations are used to specify floral organ identity. The different functional classes identified in Arabidopsis are APETALA1 (AP1), PISTILLATA (PI) and APETALA3 (AP3), AGAMOUS (AG), SEEDSTICK/AGAMOUS-LIKE 1 (STK/AGL11) and SHATTERPROOF (SHP), and SEPALLATA (SEP1, SEP2, SEP3, SEP4) genes in class A, B, C, D, and E, respectively (Grimplet et al. 2016). AG and APETALA1/FRUITFULL (AP1/

FUL) subfamily members are also involved in seed and fruit development (de Folter et al. 2006; Elitzur et al. 2010). A MADS-box transcription factor encoded by *RIN* serves as one of the earliest transcription factors in the induction of ripening in tomato (Vrebalov et al. 2002).

The MYB TF represents 9% of total TFs in *Arabidopsis*. MYB domain that is highly conserved at the N-terminus consists of 1–4 imperfect tandem repeats (MYB repeats). Three regularly distributed tryptophan or phenylalanine residues are found in the MYB repeat of 50–53 amino acids in length. Each repeat contains three α-helices, with the occurrence of helix-turn helix structure formed between the second and third helices. The helix-turn-helix structure may recognize and bind to the specific recognition site C/TAACG/TG (Chen et al. 2014; Li et al. 2015). MYB TFs are involved in plant development and stress tolerance (Mengiste et al. 2003). MYB family members from tomato and apple for example, appeared to be associated with flavonoid and anthocyanin production in ripening fruit (Mes et al. 2008).

The plant-specific NAC TF family encodes 117 genes in the *Arabidopsis* genome. They possess a highly conserved N-terminal DNA-binding NAC domain and a variable C-terminal domain (Chen et al. 2014). The NAC domain consisting of 50–160 amino acids is involved in nuclear localization, DNA binding and dimerization with other NAC domain proteins (Olsen et al. 2005). The diversified C-terminal region which contains simple amino acid repeats and regions rich in serine and threonine, proline and glutamine, or acidic residues serves as the transcriptional repressor and activator (Olsen et al. 2005; Puranik et al. 2012). The NAC proteins are involved in regulating abiotic stress signal (Meng et al. 2009) through binding to the CACG core motif that acts as a drought-responsive NAC recognition sequence (NACRS) (Shiriga et al. 2014). NAC is also involved in plant growth and development, such as vegetative and floral organ formation and fruit development (Ma et al. 2014).

4.5 Roles of Phytohormones as Signaling Molecules for Gene Regulation by Transcription Factors

Phytohormones play significant role in the signal transduction cascade and plant regulatory network throughout their growth period including fruit development and in responding to challenges inflicted by the various abiotic stresses. These phytohormones include ethylene, ABA, IAA, cytokinin (CK), gibberellins (GAs), and jasmonic acid (JA) (Memelink 2009; Kumar et al. 2014). The production and accumulation of these phyhormones are able to generate the network for signal transduction leading to a complex cascade of events that produce the biochemical and physiological changes in plants.

4.5.1 Abscisic Acid (ABA)

ABA is a sesquiterpenoid phytohormone predominantly involved in abiotic stress response. The levels of endogenous ABA have been reported to dramatically increase when plants are exposed to an array of abiotic stresses like saline environment, water deficit, or low temperature (Lim et al. 2015). These stresses are interfering with water availability in plant cells and inducing osmotic stress. Enhancing biosynthesis of ABA than normal level promotes stomatal closure, change in gene expression, and modification of physiological state (Sah et al. 2016). The exogenous application of ABA can elicit the drought response without water deficit where the application can mimic the effect of water stress condition. In rice and *Arabidopsis*, many drought inducible genes were reported to respond to ABA (Park et al. 2001; Xiong et al. 2001; Mahajan and Tuteja 2005; Yanez et al. 2009; Xin et al. 2011).

Exposure to abiotic stresses in plants activates an ABA-dependent signaling system. It induces phosphorylation of protein kinase which then activates the bZIP TFs which immediately bind to ABA-responsive regulatory elements (ABREs) found in the promoter sequence of the target genes (Nakashima et al. 2014). The interaction may induce expression of ABA-responsive genes. Other ABA-dependent TFs like MYC and MYB also function cooperatively in regulating the expression of target genes. Not all drought responses in *Arabidopsis* seemed to be mediated by ABA, since certain genes are induced by drought, salt, and cold in ABA-deficient (aba) and ABA-insensitive Arabidopsis mutants (abi). This led to the discovery of the members of the AP2/ERF family such as DREBs. The DREBs are activated through an ABA-independent pathway (Kizis et al. 2001; Seki et al. 2003; Shinozaki et al. 2003).

4.5.2 Ethylene

Ethylene, a gaseous phytohormone, is involved in various biological processes in plants including fruit ripening, seed germination, cell elongation, cell fate, senescence, abscission, as well as abiotic stress responses (Park et al. 2001). Exposure to different kinds of abiotic stress triggers the biosynthesis of ethylene and this situation may induce reactive oxygen species (ROS). The ROS damage plant cellular system through lipid peroxidation (Wang et al. 2002). The ethylene signaling pathway consists of two major pathways involving the regulation of ethylene biosynthesis and the perception of stress signal by the nucleus. The ethylene transduction signal is perceived by a family of histidine kinase-like receptors: ETHYLENE RESPONSE1 (ETR1), ETR2, ETHYLENE RESPONSE SENSOR1 (ERS1), ERS2, and ETHYLENE INSENSITIVE4 (EIN4). Binding of ethylene to its receptor is facilitated by a copper transporter called RAN-1-delivered copper cofactor (Hall et al. 2012).

When there is no ethylene, the receptors are active and the ethylene response is suppressed by stimulating the negative regulator Constitutive Triple Response1 (CTR1), a Raf-like serine-threonine kinase. Consecutively, the ethylene-signaling pathway is switched off and the ethylene insensitive 2 (EIN2) and EIN3 components are inactivated (Muller and Munne-Bosch 2015). In the presence of ethylene, the interaction between the phytohormone and its receptors induces inactivation of the protein receptor and relieves the CTR1-mediated blockage for further positive downstream signaling event. Hence, it leads to an accumulation of EIN3 and EIN3-like (EIL) transcription factor in the nucleus. The EIN3 and EIL activate expression of targets TF such as ethylene-responsive factor 1 (ERF1). The ERF is the TF responsible in ethylene-responsive activity in plants (Druege 2006; Stepanova and Alonso 2009; Morkunas et al. 2014). The ERF belongs to the AP2/ERF family. Previous reports stated that apart from ethylene, phytohormormone like ABA and jasmonic acids also influence the activation of ERF (Liu et al. 2006; Shinshi 2008).

4.5.3 *Jasmonic Acid (JA)*

JA and its cyclic precursors and derivatives are communally referred as jasmonates (JAs). The phytohormones is a lipid-derived compound known as oxylipins. They regulate various plant response and development, including response to abiotic stress, wounding, fruit ripening, production of viable pollen, root elongation, and defense against insects (Reinbothe et al. 2009). In response to JA signal, the F box protein CORONATINE INSENSITIVE 1-based SCF complexes recruit JASMONATE ZIM-domain (JAZ) repressors for ubiquitination and subsequent degradation through the 26S proteasome pathway, and downstream signaling components essential for various JA responses are subsequently regulated. In the absence of JA, the JAZ proteins associate with the corepressor TOPLESS via the adapter NOVEL INTERACTOR OF JAZ (NINJA), and repress diverse JA response through interaction with various downstream transcription factors (Song et al. 2014). A bHLH TF (e.g., AtMYC2, AtMYC3, AtMYC4) and MYB TF are main targets in the JA signaling pathway and they interact with JAZ family members in the regulation of JA-responsive target genes (Seo et al. 2013).

Previous study has shown that exogenous application of JA is successful in defending plants from drought-induced oxidative damage due to its ability to enhance the activity of antioxidant enzymes (Alam et al. 2014). Several cis-acting elements found in different gene promoters that mediate jasmonate responsiveness are known. The most common regulatory element is the GCC motif (GCCGCC) and the G-box (CACGTG) and other jasmonate-responsive promoter elements were recently discovered, such as JASE1 (CGTCAATGAA) and JASE2 (CATACGTCGTCAA) (Seo et al. 2013).

4.5.4 Auxin

Similar to cytokinin, auxin now has been established as a phytohormonal marker involved in abiotic stresses. A previous study revealed that transcript profiling of plant sampled under various treatments of abiotic stresses like cold, drought, and dehydration induced different kinds of auxin-responsive genes. For example, the expression of *YUCCA7* (*YUC7*) gene that encodes a flavin monooxygenase from the tryptophan dependent auxin biosynthetic pathway is enhanced in elevated endogenous auxin levels. This led to enhanced drought resistance in *Arabidopsis* (Du et al. 2012, 2013). The auxin responses in most tissues are concentration dependent. Hence, when different amounts of exogenous auxins are applied, different tissues respond differently. The endogenous amount of auxin in plant system is highly controlled and regulated (Ludwig-Muller 2011). At low concentrations of auxin, the signaling activity involved the repression of auxin response by AUXIN/INDOLE-3-ACETIC ACID (AUX/IAA) proteins. AUX/IAA proteins suppress AUXIN RESPONSE FACTORS (ARFs), a class of transcription factors that are involved in regulating auxin-responsive gene expression. This activates auxin-regulated gene expression. Under high concentration than maximum requirements, auxin is able to bind to TIR1 (TRANSPORT INHIBITOR RESPONSE 1) and related AUXIN F-BOX (AFB) proteins, namely AFB1, AFB2, and AFB3. The interaction results in the degradation of AUX/IAAs by the 26S proteasome and subsequently the suppression of ARFs is released. ARFs regulate (activate or repress) auxin-responsive gene expression by binding the auxin-responsive element (ARE) often found in the promoters of auxin-responsive genes, resulting in an array of auxin-mediated phenotypic alterations (Popko et al. 2010; Kazan 2013).

4.6 Transcription Factors (TFs) in Abiotic Stress Response in Oil Palm

The vast number of abiotic stress responsive genes found in model plants as well as important crops function to protect the cells from the stress conditions through production of critical metabolic proteins. TFs are essentially master regulators that can control the expression of different genes defined by specific interaction of TF domains and target genes regulatory elements. Under stress conditions, TFs are major players in signal transduction and controlling the expression of the wide variety of stress responsive genes. The many target genes controlled by one transcription factor is referred to as regulons. DREB1/CBF regulons function in ABA-independent manner, while AREB/ABF regulons are involved in ABA-dependent pathways.

4.6.1 EABF and EABF1 That Integrate Multiple Stress Signaling

The action of ABA is responsible for enhancing or suppressing the expression of numerous stress-responsive genes. ABA signaling is regulated by the AREB/ABF proteins mainly from the bZIP families while others belong to the AP2/ERF families of transcription factors (Choi et al. 2000). The isolation of two oil palm ABA-responsive TFs designated as EABF and EABF1 by yeast one-hybrid screening was reported by Omidvar et al. (2012). Multiple sequence alignment and phylogenetic analysis classified EABF and EABF1 as DREB1/CBF (from AP2/ERF family) and bZIP family members, respectively. Using yeast one-hybrid transactivation assay, their roles in the ABRE- and DRE/CRT-mediated expression of target genes were determined. This was further validated by an electrophoretic mobility shift assay which revealed that both EABF and EABF1 could bind to the ABRE and the DRE regulatory elements. *EABF* transcript production was induced by ABA in the fruits and leaves of the oil palm but not in the roots. However, the expression of *EABF1* was induced constitutively in all tissues by ABA.

Besides ABA, ethylene and methyl jasmonate strongly induced the expression of both *EABF* and *EABF1* in the oil palm fruits (Omidvar et al. 2012). In addition, different abiotic stress treatments including drought, cold, and high salts could also strongly induce their expression. This suggests that EABF and EABF1 not only play regulatory roles in response to various abiotic stresses, they could also be involved in integrating multiple stress signaling pathways. Similar observation was reported for ABF from other plant species. For example, in rice, OsABF2 expression was responsive to abiotic stresses such as cold, drought, high salinity, and oxidative stresses, besides ABA (Hossain et al. 2010). The oil palm EABF is induced by ABA, drought, and cold suggesting that it behaves differently from DREB1/CBF. EABF may be involved in both ABA-dependent and ABA-independent pathways based on its strong DRE/CRT-binding property. Nevertheless, recent evidences suggest that the two pathways cross-talk with each other (Roychoudhury et al. 2013).

AREB/ABF-type proteins in rice are phosphorylated and activated by ABA-dependent SnRK2 protein kinases. This subsequently enhanced the expression of the abiotic stress responsive genes. The products of these genes are required in stress responses and tolerance (Todaka et al. 2012). The study by Omidvar et al. (2012) described above provided evidence on the EABF and EABF1 proteins in the ABA signaling pathway and stress response in plants. The genes could serve as potential candidate for genetic modification for improving oil palm crops to be more tolerant to environmental stresses or as diagnostic markers to monitor stress conditions.

4.6.2 Drought Stress Response Mediated by DREB1

Dry condition due to El Nino phenomenon in recent years causes serious adverse impact on palm oil industry (Haniff et al. 2010). However, there is very limited report on molecular-physiological studies on the effects of drought carried out on oil palm. The recently reported EgDREB1 from oil palm demonstrated to have the ability to bind to the DRE/CRT DNA motif through in vivo assay using yeast one-hybrid transactivation assay and in vitro assay using EMSA (Azzeme et al. 2017). EgDREB1 is classified under the A-1 DREB1/CBF subfamily based on phylogenetic analysis and has the conserved amino acid residues, the 14th valine and 19th glutamic acid in the AP2 domain which are critical for binding to the DRE (Akhtar et al. 2012). Besides, it contains the consensus NLS (PKRPAGRTKFRETRHP). The localization of EgDREB1 in the nucleus was validated by the subcellular localization experiment using onion cells. In addition, the PKKPAGR and DSAWR signature sequences characteristic of DREB1 family members were identified. These sequences distinguished DREB1 from the other AP2/ERF proteins (Morran et al. 2011). A cDNA encoding EREBP was isolated from the mesocarp of the oil palm fruit using yeast one-hybrid assay with ERE as a bait. EREBP demonstrated transcriptional activation and transactivation properties in yeast and in vitro. It can bind to ERE in addition to the DRE/CRT motif (Omidvar et al. 2013a).

Gene expression analysis was carried out on both *EgDREB1* and *EREBP* in the leaves and roots of oil palm seedlings subjected to water stress treatments. The treated oil palm seedlings were categorized into different drought severity levels based on physiological parameters and observed physical deteriorations (Azzeme et al. 2016, 2017). *EgDREB1* was expressed under mild drought in oil palm roots suggesting its potential involvement in the initiation of signaling communication from root to shoot. However, in the leaves, *EgDREB1* expression peaks during severe stage of the drought. It was suggested that possibly *EgDREB1* serves in mediating the recovery response of the cells or as a signal in programmed cell death as a way to overcome the abiotic stress (Wituszynska and Karpinski 2013). While *EREBP* transcripts accumulated under mild stress in the leaves at 4.81-fold higher than in the control seedlings and significantly downregulated under moderate stress. The results demonstrated the potential of both genes to serve as expression markers in monitoring drought severity levels in oil palm seedlings using leaves which are more accessible than roots.

Further functional characterization by overexpressing *EgDREB1* in tomato seedlings showed enhanced expression of tomato DRE/CRT and non-DRE/CRT containing genes in cold and PEG treated seedlings, suggesting that EgDREB1 can act as a functional regulator in enhancing tolerance to both cold and drought conditions (Azzeme et al. 2016). EREBP which binds DRE/CRT is also responsive to drought, cold, and high-salinity treatments. Thus, both genes are also candidates for crop improvement for tolerance to the respective stress conditions. The ERE binding ability demonstrated by EREBP suggests that it could mediate both abiotic and biotic stress responses. This was further supported by the ability of different phytohormones including ABA, methyl jasmonate, and salicylic acid to induce its expression. These phytohormones are important signaling molecules in biotic stress response such as during pathogen attack (Abdullah and Akhtar 2016). These results suggest that EREBP like TINY from *Arabidopsis* plays a role in the cross-talk between abiotic and biotic stress responses through DRE- and ERE-mediated signaling pathways (Sun et al. 2008).

4.6.3 CBF as a Key Player Under Cold Stress

Oil palm is a tropical oil crop and cannot withstand cold temperature as its growth and productivity will be badly affected. However, there is interest in expanding the cultivation of this economically important oil crop to regions where there is period of low temperature such as in South China. High throughput gene expression analysis on cold treated oil palm seedlings showed that the biggest number of transcription factors to be upregulated in the leave tissues belong to the AP2/ERF family, which are the CBF followed by the NAC and bZIP family of transcription factors (Lei et al. 2014). Cold tolerance will be enhanced by the induced expression of the CBFs as shown in *Arabidopsis* and other plant species (Lee et al. 2005). Interestingly, RNA-seq analysis of cold stress treated *Elymus nutans* Griseb, a highly cold tolerant alpine perennial forage also found that CBF as the major cold stress responsive genes (Fu et al. 2016) suggesting conservation of the CBF mediated signal transduction in plant species grown in the tropics and those that can withstand low temperature. In addition, MYB15 which a potential suppressor of CBF (Chinnusamy et al. 2003) also showed an increase in expression pattern in the oil palm. Comparison of the AP2 domain in oil palm CBF proteins with those found in temperate plants showed that the sequence is highly conserved implying functional conservation in the DNA binding properties. However, many of the cold-responsive (COR) genes in oil palm do not have a proper DRE motifs, the binding site for CBFs. This might not be critical for the oil palm as it is a tropical crop that is not subjected to cold stress conditions (Lei et al. 2014).

4.7 Transcription Factors in the Development and Ripening of Oil Palm Fruits

4.7.1 Tomato as a Model for Climacteric Fruits

The critical role of ethylene was highlighted from the transcriptome data set where ethylene associated genes showed the highest percentage (37%) among the total differentially expressed genes during development and ripening of tomato (Alba et al. 2005). The mechanisms that coordinate the upregulation or suppression of these genes involved fruit-related TFs and microRNAs targeting some of these TFs, DNA methylation and chromatin remodeling.

ERF family of transcription factors are the mediator for the expression of the ethylene-regulated genes exerting control at the transcriptional level. The ERFs are categorized into different subclasses with contrasting roles as depicted by the differences in their expression profile in wild-type and ripening-impaired mutant tomatoes, Never-ripe (Nr), ripening-inhibitor (rin), and non-ripening (nor). Members of subclass E of ERF were strongly downregulated in the ripening mutants indicating a critical role in the fruit ripening process (Liu et al. 2016). While SlERF1 and SlERF2 could possibly be involved in the regulation of fruit ripening and softening (Pirrello et al. 2006; Li et al. 2007) and members of the Subfamily VII of ERF are mainly associated with fruit ripening (Licausi et al. 2013).

RIN which encodes a MADS-box transcription factor is a key regulatory gene that serves as one of the earliest transcription factors involved in the induction of ripening in tomato. Recently, chromatin immune-precipitation assay showed that a broad range of ripening-associated genes are directly regulated by RIN (Ito 2016). In addition, other MADS-box proteins were identified to be interacting with RIN suggesting their involvement in the regulation of the ripening process. AP2, MYB, NAC, HD-zip, basic helix-loop-helix (bHLH), and auxin response factors (ARFs) are the other TFs involved in regulating fruit ripening (Karlova et al. 2014). SlMYB12, for example, influences flavonoid and anthocyanin accumulation (Adato et al. 2009). AP2a expression is induced by ripening regulatory transcription factors, NOR, CNR, and RIN. AP2a modulates ethylene signaling which leads to a partial negative regulation of the ripening process in tomato (Karlova et al. 2011). Fruit ripening was inhibited in SlNAC1 overexpression lines that have an effect on ethylene synthesis and carotenoid accumulation (Ma et al. 2014).

4.7.2 Regulation of Oil Palm Fruit Ripening

A sharp increase in transcripts for several ERFs and ethylene biosynthetic genes in the oil palm mesocarp during the fruit ripening period resembling the system 2-type ethylene production was observed (Tranbarger et al. 2011). The most predominant ERFs were the type VII, similar to those reported in the ripening of tomato, showed maximum expression at the maturation and ripening phase of the oil palm fruits. Strong conservation amongst the monocots and dicots was shown based on phylogenetic analysis of their signature domains and gene duplication possibly contributes to the multiple type VII ERF expressed in the oil palm mesocarp. In addition, two NAC domain contigs were identified with similarity in sequence and expression at the ripening period as the NAC-NOR gene from tomato, which suggests conservation in protein function.

A transcript of the gene encoding EgAP2-2 with strong similarity with WRI1 peaks at 120 DAP, the onset of lipid accumulation and the maximal expression for many FA biosynthesis genes. The AW-box, the binding site for WRI1 were found in the proximal upstream regions of some genes involved in glycolysis and fatty acid biosynthesis that are expressed in the mesocarp. WRI1 directly activates FA biosynthetic genes and is also involved in regulating the assembly and storage of TAG in oilseed crops (Maeo et al. 2009). WRI1 homologue from oil palm was able to complement the wri1 mutant phenotype of *A. thaliana* (Bourgis et al. 2011; Ma et al. 2013). Together, these findings suggest some common regulatory features between oilseed and oil fruit plants (Tranbarger et al. 2011).

4.7.3 The Role of MADS-Box Transcription Factors

Tranbarger et al. (2011) identified 14 MADS-box domain sequences representing 7 subfamilies during oil palm fruit development. One contig belonging to the AG subfamily showed abundant expression that peaks at the fruit maturation phase (140 DAP). From the GLOBOSA/PISTILLATA (GLO/PI) subfamily, a single contig identical to EgGLO1 (Adam et al. 2006) showed a significant increase in transcript level during the maturation phase and maintains at high level during the ripening period of the oil palm fruits.

Oil palm has three fruit forms based on the shell thickness where a thicker shell corresponds to thinner mesocarp resulting in less oil yield as the oil is extracted from the fleshy mesocarp. This is a monogenic co-dominant inheritance, where tenera, the hybrid (thin shell) is produced from maternal dura (thick shell) cross with paternal pisifera (shelless) palms. A type II MADS-box transcription factor was found to encode the shell thickness gene and different alleles which contain mutations of this gene that associate with the shell thickness trait was determined and used to develop DNA-based marker for accurate screening for dura contamination at the nursery stage. The findings and utility of these DNA-based marker has significant impact to the oil palm industry as planting of the thick shell genotype will result in the loss in revenue amounting to about 30% of the total income as well as wastage of valuable plantation areas to low yielding oil palm trees (Ooi et al. 2016).

4.7.4 An AP2 That Associates with the Ethylene and ABA Signaling Pathways

ABA is an important control factor for fruit ripening besides ethylene in both climacteric and non-climacteric fruits (Leng et al. 2014). Simultaneous increase in ethylene and ABA was observed in the mesocarp tissue of the oil palm during the ripening period (Tranbarger et al. 2011). Identification of ethylene- and/or ABA-regulated transcription factors regulating fruit ripening in oil palm could lead to the discovery of potential candidate genes for improving fruit yield and quality through metabolic engineering and molecular breeding. AP2 from the AP2/ERF family of transcription factors are mainly involved in growth and developmental processes and do not have stress related roles.

A cDNA clone from oil palm mesocarp encoding a TF that interacts with the ERE motif in a yeast one-hybrid assay was identified (Omidvar et al. 2013b). The TF was found to be identical to EgAP2-1 that has been reported to be involved in oil palm embryo development (Morcillo et al. 2007). It contains two highly conserved AP2 DNA binding domains and a linker region that characterized the plant AP2 family belonging to the AP2/ERF superfamily. The EgAP2–1 subcellular localization experiment in onion cells confirmed its nuclear localization even though a nuclear localization signal (NLS) was not found in its amino acid sequence. The ability demonstrated by EgAP2-1 in binding to ERE but not the DRE in vitro, suggested its ERE-mediated regulatory role. It also indicated its potential involvement in regulating ripening but not in controlling abiotic stress response. The *EgAP2-1* was expressed in the mesocarp of the oil palm fruits with an increasing expression towards the ripening phase and no expression was shown in the vegetative tissues including leaves and roots. Its expression was induced in the mesocarp specifically in response to ABA and ethylene but not to different abiotic stresses. Furthermore, EgAP2-1 expression increased at the late stages of fruit ripening in the mesocarp that coincide with the increase in ABA and ethylene in the fruits (Tranbarger et al. 2011) suggesting positive transcriptional regulation through the ABA and/or ethylene signaling pathways. Thus, EgAP2-1 putatively expressed in the fleshy oil palm mesocarp during fruit ripening in an ABA- and/or ethylene-dependent manner. Together the data suggests a possible new regulatory role for AP2 in ABA- and/or ethylene-mediated control of fruit ripening. AP2a has been reported as an important transcription factor that functions downstream of the ripening regulators RIN and CNR in tomato. During tomato fruit development and ripening, it was suggested based on *AP2i* silenced fruit transcriptomic and metabolomics data that AP2a plays an important role in the primary and secondary metabolism, ethylene biosynthesis, and signaling pathways, as well as in chromoplast differentiation (Chung et al. 2010; Karlova et al. 2011).

4.7.5 A Novel CBF with Potential Role in Oil Palm Fruit Ripening

The role of CBF in regulating stress response particularly cold stress is well known, however there is no report on its involvement in developmental processes like fruit ripening. A novel CBF, EgCBF3 that is responsive to cold treatment was isolated from the ripening fruits of the oil palm and proven to be a nuclear localized protein (Ebrahimi et al. 2015). EgCBF3 contains the specific signature sequences characteristic of the A-1 group of CBF, but clustered closely with the GCC binding A-4 group through phylogenetic analysis. In the fruit mesocarp tissues, EgCBF3 was found to show the highest expression at the ripening stage (17 weeks afer anthesis/120 DAF)

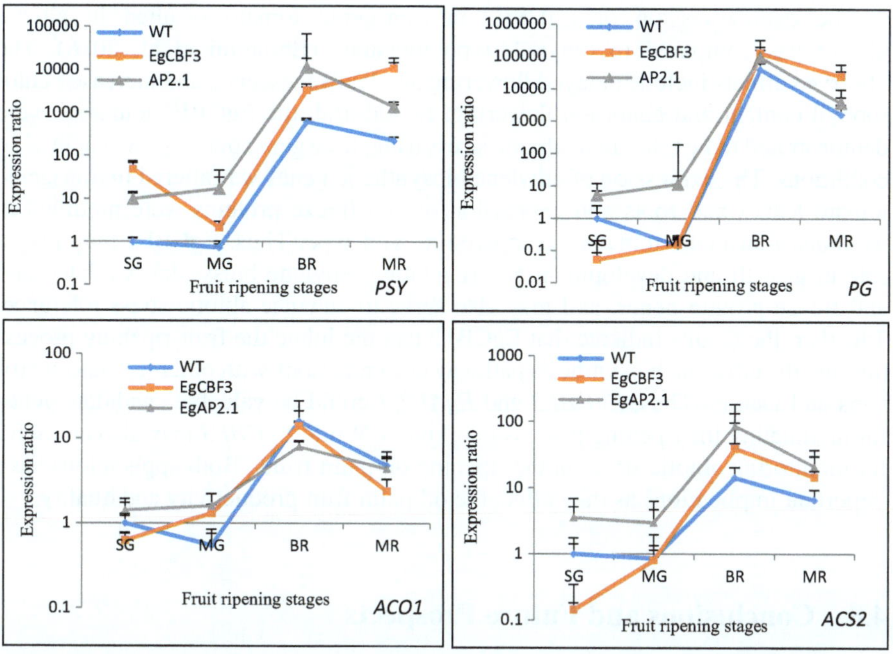

Fig. 4.4 Effects of overexpressing oil palm transcription factors, *EgCBF3* and *EgAP2.1* on fruit ripening genes from tomato. Tomato fruits at different developmental stages were transiently transformed with oil palm transcription factor genes, *EgCBF3* and *EgAP2.1* and the expression of four fruit ripening related genes was analyzed. Phytoene synthase (PSY) represents carotenoid biosynthetic gene, polygalacturonase (PG) represents cell wall degrading gene, ACC oxidase (ACO) and ACC synthase (ACS) represent ethylene biosynthetic genes. Small green (SG), mature green (MG), breaker (BR), mature ripe (MR) represent different developmental stages of tomatoes

while a low level of expression was observed in the vegetative roots and leaves tissues. Like *EgAP2-1*, expression of *EgCBF3* appeared to correlate with the elevated levels of ethylene and ABA that were detected in the oil palm mesocarp tissue during the ripening period (Tranbarger et al. 2011). The expression was also affected by abiotic stresses including drought and high salt. Functional studies using EMSA and yeast one-hybrid assay validated its ability to bind to DRE/CRT promoter motif. Transient overexpression of EgCBF3 in tomato as a model system, modified the expression of ABA and ethylene biosynthetic genes, the cell wall degrading enzyme, polygalacturonase and the ethylene-responsive, fruit-ripening gene (E8) suggesting its role in regulating the expression of ripening-related genes in addition to its role in mediating abiotic stress response. Figure 4.4 compares the effects of overexpressing oil palm transcription factor genes, EgCBF3 and EgAP2.1 on fruit ripening genes from tomato. Zhao et al. (2009) reported on the expression pattern of *LeCBF1* in mature tomato green fruits that coincide with endogenous ethylene production. Interestingly, a CBF-like gene in the climacteric tomato fruit also transactivates the ripening related gene, polygalacturonase 1 promoter. The transactivation was enhanced by the application of exogenous ethylene (Tacken et al. 2010).

The overexpression of *EgCBF3* in transgenic tomato resulted in altered phenotypes compared to the wild-type tomatoes (Ebrahimi et al. 2016). The observed effects include delayed flowering and leaf senescence and increased chlorophyll content and abnormal flowering. In addition, the EgCBF3 tomatoes were demonstrated to be tolerant to abiotic stress using transgenic plants grown in in vitro conditions. The expression of ethylene biosynthetic genes was altered in transgenic tomato leaves and roots and expression of anti-freeze proteins were modified in wounded transgenic tomatoes compared to wild type. Thus, EgCBF3 may play a role in growth and development and regulating ethylene biosynthesis-related and anti-freeze protein genes, and may also serve to enhance abiotic stress tolerance. Together, the results indicate that EgCBF3 can modulate the fruit-ripening process through the ethylene biosynthesis pathway or interactions with other regulatory proteins and factors. Thus, *EgCBF3* and *EgAP2-1* could be valuable candidate genes for modulating the ripening process in oil palm. While *EgCBF3* may also be useful for monitoring abiotic stress in the ripening oil palm fruits. Both applications have important implications as they effect the oil palm fruit productivity and quality.

4.8 Conclusions and Future Prospects

Regulation of gene expression at the transcriptional level where TFs play a major role is of great importance. TFs are master regulators of biosynthetic pathways and stress responses in plants. However, studies on their biological activities and the functions of their regulons are still at its infancy in the oil palm compared to other economically important crops like rice, wheat, rapeseed, and maize. Nevertheless, a number of transcription factors with potential roles in abiotic stress response and fruit ripening have been identified in the oil palm. The roles of the transcription factors and their potential regulons can be inferred based on their expression in specific tissue under different treatments such as the association of oil palm CBFs with cold stress and EgDREB1 with drought stress while others like the ABF and ABF1 which are responsive to ABA are implicated in a variety of stress responses. Studies thus far suggest evolutionary conservation in oil palm with regard to the ripening process where the involvements of ERF, AP2, and MADS-box were observed to be similar to that reported for the climacteric tomato fruit while the role of WRI1 in regulating fatty acid synthesis and oil deposition is similar to that found for the oilseed crop. New findings for plants on the role of oil palm EgCBF3 in fruit ripening worth a more in-depth study. While the oil palm MADS-box gene that regulates shell thickness, a major determinant of oil yield in oil palm is already at the advanced stage on the verification of its utility for selection of planting materials in the nursery. More detailed functional studies that were carried out validated the nuclear localization and interaction of some of the oil palm transcription factors with the specific promoter motif.

Limited land for expansion especially in Malaysia and the uncertainty of climate that leads to extreme weather for a prolong period manifested as El Nino and La

Nina are among the pressing issues in oil palm cultivation. Reliance on good plantation practices alone is insufficient. Complementation with sustainable technical inputs with strong scientific basis is required for overcoming the various problems more effectively. Genomic resources in the form of genome sequence and germplasm materials should be tapped to look for suitable alleles and variants of genes associating with the traits of economic importance. The molecular markers which enable selection of planting materials at the nursery stage clearly is cost effective and in the field, performance of oil palm trees and the effects of abiotic stresses can also be monitored using easily accessible tissues such as the leaves if suitable expression markers can be obtained. However, the knowledge on the roles of the genes of economic interest and the effects of sequence variants on gene function are very limited in the oil palm.

The biological complexity with regard to the roles of the different transcription factors is not only confined to the level of DNA–protein interaction as often the transcription factors also interact with other proteins that modulate their transcriptional regulatory activities. Clearly other techniques such as yeast two-hybrid that validate protein–protein interaction and chromatin immunoprecipitation sequencing (ChIP-seq), one of the more recent development which allows identification of the entire regulons of a transcription factor will be of great value. Such studies if complemented with functional characterization in planta using model plant systems or via editing of the oil palm genome would provide vital information on the reliability and utility of the targeted transcription factor genes for oil palm improvement.

References

Abdullah SNA, Akhtar MS (2016) Plant necrotrophic fungal pathogen interaction: mechanism and mode of action. In: Hakeem KR, Akhtar MS, Siti Nor Akmar A (eds) Plant, soil and microbes. Springer, New York, pp 29–53

Adam H, Jouannic S, Morcillo F, Richaud F, Duval Y, Tregear JW (2006) MADS box genes in oil palm (Elaeis guineensis): patterns in the evolution of the SQUAMOSA, DEFICIENS, GLOBOSA, AGAMOUS, and SEPALLATA subfamilies. J Mol Evol 62:15–31

Adato A, Mandel T, Mintz-Oron S, Venger I, Levy D, Yativ M, Domínguez E, Wang Z, De Vos RC, Jetter R, Schreiber L, Heredia A, Rogachev I, Aharoni A (2009) Fruit-surface flavonoid accumulation in tomato is controlled by a SlMYB12-regulated transcriptional network. PLoS Genet 5:e1000777. doi:10.1371/journal.pgen.1000777

Akhtar M, Jaiswal A, Taj G, Jaiswal JP, Qureshi MI, Singh NK (2012) DREB1/CBF transcription factors: their structure, function and role in abiotic stress tolerance in plants. J Genet 91:385–395

Alam MM, Nahar K, Hasanuzzaman M, Fujita M (2014) Exogenous jasmonic acid modulates the physiology, antioxidant defense and glyoxalase systems in imparting drought stress tolerance in different Brassica species. Plant Biotechnol Rep 8:279–293

Alba R, Payton P, Fei Z, McQuinn R, Debbie P, Martin G, Tanksley S, Giovannoni J (2005) Transcriptome and selected metabolite analyses reveal multiple points of ethylene control during tomato fruit development. Plant Cell 17:2954–2965

Arshad CM, Armanto ME, Zain AM (2012) Evaluation of climate suitability for oil palm (Elaeis guineensis Jacq.) cultivation. J Environ Sci Eng B 1:272–276

Awan AM, Sap MNM (2006) A framework for predicting oil-palm yield from climate data. Proceedings of the postgraduate annual research seminar 2006, pp 360–366

Azzeme AM, Abdullah SNA, Aziz MA, Wahab PEM (2016) Oil palm leaves and roots differ in physiological response, antioxidant enzyme activities and expression of stress-responsive genes upon exposure to drought stress. Acta Physiol Plant 38:1–12

Azzeme MA, Abdullah SNA, Aziz MA, Wahab PEM (2017) Oil palm drought inducible DREB1 induced expression of DRE/CRTandnon-DRE/CRT-containing genes in lowland transgenic tomato under cold and PEG treatments. Plant Physiol Biochem 112:129–151

Bourgis F, Kilaru A, Cao X, Ngando-Ebongue G-F, Driraf N, Ohlrogged JB, Arondela V (2011) Comparative transcriptome and metabolite analysis of oil palm and date palm mesocarp that differ dramatically in carbon partitioning. Proc Natl Acad Sci U S A 108:12527–12532

Chen L, Song Y, Li S, Zhang L, Zou C, Yu D (2012) The role of WRKY transcription factors in plant abiotic stresses. Biochim Biophys Acta 1819:120–128

Chen X, Wang Y, Lv B, Li J, Luo L, Lu S, Zhang X, Ma H, Ming F (2014) The NAC family transcription factor OsNAP confers abiotic stress response through the ABA pathway. Plant Cell Physiol 55:604–619

Chinnusamy V, Ohta M, Kanrar S, Lee B-h, Hong X, Agarwal M, Zhu J-K (2003) ICE1: a regulator of cold-induced transcriptome and freezing tolerance in Arabidopsis. Genes Dev 17:1043–1054

Choi HI, Hong JH, Ha JO, Kang JY, Kim SY (2000) ABFs, a family of ABA-responsive element binding factors. J Biol Chem 275:1723–1730

Chung MY, Vrebalov J, Alba R, Lee J, McQuinn R, Chung JD, Klein P, Giovannoni (2010) A tomato (Solanum lycopersicum) APETALA2/ERF gene, SlAP2a, is a negative regulator of fruit ripening. Plant J 64:936–947

Csukasi F, Osorio S, Gutierrez JR, Kitamura J, Giavalisco P, Nakajima M, Fernie AR, Rathjen JP, Botella MA, Valpuesta V, Medina-Escobar N (2011) Gibberellin biosynthesis and signalling during development of the strawberry receptacle. New Phytol 191:376–390

Dalcorso G, Manara A, Piasentin S, Furini A (2014) Nutrient metal elements in plants. Metallomics 6:1770–1788

Demidchik V (2015) Mechanisms of oxidative stress in plants: from classical chemistry to cell biology. Environ Exp Bot 109:212–228

Druege U (2006) Ethylene and plant responses to abiotic stress. In: Khan NA (ed) Ethylene action in plants. Springer-Verlag, New York, pp 81–118

Du H, Wu N, Fu J, Wang S, Li X, Xiao J, Xiong L (2012) A GH3 family member, OsGH3-2, modulates auxin and abscisic acid levels and differentially affects drought and cold tolerance in rice. J Exp Bot 63:6467–6480

Du H, Liu H, Xiong L (2013) Endogenous auxin and jasmonic acid levels are differentially modulated by abiotic stresses in rice. Front Plant Sci 4:1–10

Ebrahimi M, Abdullah SNA, Aziz MA, Namasivayam P (2015) A novel CBF that regulates abiotic stress response and the ripening process in oil palm (Elaeis guineensis) fruits. Tree Genet. Genomes 11(3):56

Ebrahimi M, Abdullah SNA, Aziz MA, Namasivayam P (2016) Oil palm EgCBF3 conferred stress tolerance in transgenic tomato plants through modulation of the ethylene signaling pathway. J plant physiol 202:107–120

Elitzur T, Vrebalov J, Giovannoni JJ, Goldschmidt EE, Friedman H (2010) The regulation of MADS-box gene expression during ripening of banana and their regulatory interaction with ethylene. J Exp Bot 61:1523–1535

de Folter S, Shchennikova AV, Franken J, Busscher M, Baskar R, Grossniklaus U, Angenent GC, Immink RG (2006) A Bsister MADS-box gene involved in ovule and seed development in petunia and Arabidopsis. Plant J 47:934–946

Fu J, Miao Y, Shao L, Hu T, Yang P (2016) De novo transcriptome sequencing and gene expression profiling of Elymus nutans under cold stress. BMC Genomics 17:1–19

Fujita M, Fujita Y, Noutoshi Y, Takahashi F, Narusaka Y, Yamaguchi-Shinozaki K, Shinozaki K (2006) Crosstalk between abiotic and biotic stress responses: a current view from the points of convergence in the stress signaling networks. Curr Opin Plant Biol 9(4):436–442

Giovannoni J (2001) Molecular biology of fruit maturation and ripening. Annu Rev Plant Physiol Plant Mol Biol 52:725–749

Grimplet J, Martinez-Zapater JM, Carmona MJ (2016) Structural and functional annotation of the MADS-box transcription factor family in grapevine. BMC Genomics 17:1–23

Gu C, Guo ZH, Hao PP, Wang GM, Jin ZM, Zhang SL (2017) Multiple regulatory roles of AP2/ERF transcription factor in angiosperm. Bot Stud 58:6

Guo B, Wei Y, Xu R, Lin S, Luan H, Lv C, Zhang X, Song X, Xu R (2016) Genome-wide analysis of APETALA2/ethylene-responsive factor (AP2/ERF) gene family in barley (Hordeum vulgare L.) PLoS One 11:e0161322. doi:10.1371/journal.pone.0161322

Gutha LR, Reddy AR (2008) Rice DREB1B promoter shows distinct stress-specific responses and the overexpression of cDNA in tobacco confers improved abiotic and biotic stress tolerance. Plant Mol Biol 68:533–555

Hall BP, Shakeel SN, Amir M, UlHaq N, Qu X, Schaller GE (2012) Histidine kinase activity of the ethylene receptor ETR1 facilitates the ethylene response in Arabidopsis. Plant Physiol 159:682–695

Haniff MH, Mohammad AT, Noor MRM, Din AK, Latiff J, Sani ARA, Abdullah R (2010) Impact of El Niño occurrence on oil palm yield in Malaysia. Planter 86:873–852

Hao-Li M, Han-lin Z, Huai-yu Z, Jie Z (2010) Cloning and expression analysis of an AP2/ERF gene and its responses to phytohormones and abiotic stresses in rice. Ric Sci 17:1–9

Hattori Y, Nagai K, Furukawa S, Song X, Kawano R, Sakakibara H, Wu J, Matsumoto T, Yoshimura A, Kitano H, Matsuoka M, Mori H, Ashikari M (2009) The ethylene response factors SNORKEL1 and SNORKEL2 allow rice to adapt to deep water. Nature 460:1026–1030

Hossain MA, Lee Y, Cho JI, Ahn CH, Lee SK, Jeon JS, Kang H, Lee CH, An G, Park PB (2010) The bZIP transcription factor OsABF1 is an ABA responsive element binding factor that enhances abiotic stress signaling in rice. Plant Mol Biol 72:557–566

Hu W, Wang L, Tie W, Yan Y, Ding Z, Liu J, Li M, Peng M, Xu B, Jin Z (2016) Genome-wide analyses of the bZIP family reveal their involvement in the development, ripening and abiotic stress response in banana. Sci Rep 6:1–15

IMARC Group. (2016) Global vegetable oil market stimulated by demand for biofuel and potential opportunities in emerging markets. https://imarcgroup.wordpress.com/2016/09/23/global-vegetable-oil-market/. Accesed 6 June 2017

Ito Y (2016) Regulation of tomato fruit ripening by MADS-box transcription factor. Jpn Agric Res Quart 50:33–38

Karlova R, Rosin FM, Busscher-Lange J, Parapunova V, Do PT, Fernie AR, Fraser PD, Baxter C, Angenent GC, de Maagd RA (2011) Transcriptome and metabolite profiling show that APETALA2a is a major regulator of tomato fruit ripening. Plant Cell 23:923–941

Karlova R, Chapman N, David K, Angenent GC, Seymour GB, de Maagd RA (2014) Transcriptional control of fleshy fruit development and ripening. J Exp Bot 65:4527–4541

Kazan K (2013) Auxin and the integration of environmental signals into plant root development. Ann Bot 112:1655–1665

Kizis D, Lumbreras V, Pages M (2001) Role of AP2/EREBP transcription factors in gene regulation during abiotic stress. FEBS Lett 498:187–189

Klee HJ, Giovannoni JJ (2011) Genetics and control of tomato fruit ripening and quality attributes. Annu Rev Genet 45:41–59

Kobayashi F, Maeta E, Terashima A, Kawaura K, Ogihara Y, Takumi S (2008) Development of abiotic stress tolerance via bZIP-type transcription factor LIP19 in common wheat. J Exp Bot 59:891–905

Kumar R, Khurana A, Sharma AK (2014) Role of plant hormones and their interplay in development and ripening of fleshy fruits. J Exp Bot 65:4561–4575

Lata C, Prasad M (2011) Role of DREBs in regulation of abiotic stress responses in plants. J Exp Bot 62:4731–4748

Lee BH, Henderson DA, Zhu JK (2005) The Arabidopsis cold-responsive transcriptome and its regulation by ICE1. Plant Cell 17:3155–3175

Lei X, Xiao Y, Xia W, Mason AS, Yang Y, Ma Z, Peng M (2014) RNA-seq analysis of oil palm under cold stress reveals a different C-repeat binding factor (CBF) mediated gene expression pattern in *Elaeis guineensis* compared to other species. PLoS One 9:1–20

Leng P, Yuan B, Yangdong G (2014) The role of abscisic acid in fruit ripening and responses to abiotic stress. J Exp Bot 65:4577–4588

Li C, Ng CK-Y, Fan L-M (2015) MYB transcription factors, active players in abiotic stress signaling. Environ Exp Bot 114:80–91

Li Y, Zhu B, Xu W, Zhu H, Chen A, Xie Y, Shao Y, Luo Y (2007) LeERF1 positively modulated ethylene triple response on etiolated seedling, plant development and fruit ripening and softening in tomato. Plant Cell Rep 26:1999–2008

Licausi F, Ohme-Takagi M, Perata P (2013) APETALA2/ethylene responsive factor (AP2/ERF) transcription factors:mediators of stress responses and developmental programs. New Phytol 199:639–649

Lim CW, Baek W, Jung J, Kim J-H, Lee SC (2015) Function of ABA in stomatal defense against biotic and drought stresses. Int J Mol Sci 16:15251–15270

Liu Y, Zhao T, Liu J, Liu W, Liu Q, Yan Y, Zhou H (2006) The conserved Ala37 in the ERF/AP2 domain is essential for binding with the DRE element and the GCC box. FEBS Lett 580:1303–1308

Liu M, Gomes BL, Mila I, Purgatto E, Peres LE, Frasse P, Maza E, Zouine M, Roustan JP, Bouzayen M, Pirrello J (2016) Comprehensive profiling of ethylene response factor expression identifies ripening-associated ERF genes and their link to key regulators of fruit ripening in tomato. Plant Physiol 170:1732–1744

Ludwig-Muller J (2011) Auxin conjugates: their role for plant development and in the evolution of land plants. J Exp Bot 62:1757–1773

Ma W, Kong Q, Arondel V, Kilaru A, Bates PD, Thrower NA, Benning C, Ohlrogge JB (2013) Wrinkled1, a ubiquitous regulator in oil accumulating tissues from Arabidopsis embryos to oil palm mesocarp. PLoS One 8:1–13

Ma N, Feng H, Meng X, Li D, Yang D, Wu C, Meng Q (2014) Overexpression of tomato SlNAC1transcription factor alters fruit pigmentation and softening. BMC Plant Biol 14:351

Maeo K, Tokuda T, Ayame A, Mitsui N, Kawai T, Tsukagoshi H, Ishiguro S, Nakamura K (2009) An AP2-type transcription factor, WRINKLED1, of Arabidopsis Thaliana binds to the AW-box sequence conserved among proximal upstream regions of genes involved in fatty acid synthesis. Plant J 60:476–487

Mahajan S, Tuteja N (2005) Cold, salinity and drought stresses: an overview. Arch Biochem Biophys 444:139–158

McAtee P, Karim S, Schaffer R, David K (2013) A dynamic interplay between phytohormones is required for fruit development, maturation, and ripening. Front Plant Sci 4:79. doi:10.3389/Fpls.2013.00079

Memelink J (2009) Regulation of gene expression by jasmonate hormones. Phytochemistry 70:1560–1570

Meng C, Cai C, Zhang T, Guo W (2009) Characterization of six novel NAC genes and their responses to abiotic stresses in Gossypium hirsutum L. Plant Sci 176:352–359

Mengiste T, Chen X, Salmeron J, Dietrich R (2003) The BOTRYTIS SUSCEPTIBLE1 gene encodes an R2R3MYB transcription factor protein that is required for biotic and abiotic stress responses in Arabidopsis. Plant Cell 15:2551–2565

Mes PJ, Boches P, Myers JR, Durst R (2008) Characterization of tomatoes expressing anthocyanin in the fruit. J Am Soc Hort Sci 133:262–269

Michaeli S, Galili G, Genschik P, Fernie AR, Avin-Wittenberg T (2016) Autophagy in plants–what's new on the menu? Trends Plant Sci 21:134–144

Misra AK (2014) Climate change and challenges of water and food security. Int J Sustainable Built Environ 3:153–165

Morcillo F, Gallard A, Pillot M, Jouannic S, Aberlenc-bertossi F, Collin M, Verdeil JL, Tregear JW (2007) EgAP2-1, an AINTEGUMENTA-like (AIL) gene expressed in meristematic and proliferating tissues of embryos in oil palm. Planta 226:1353–1362

Morkunas I, Mai VC, Waskiewicz A, Formela M, Golinski P (2014) Major phytohormones under abiotic stress. In: Ahmad P, Wani MR (eds) Physiological mechanisms and adaptation strategies in plants under changing environment, vol 2. Springer, New York, pp 87–135

Morran S, Eini O, Pyvovarenko T, Parent B, Singh R, Ismagul SE, Shirley N, Langridge P, Lopato S (2011) Improvement of stress tolerance of wheat and barley by modulation of expression of DREB/CBF factors. Plant Biotechnol J 9:230–249

Muller M, Munne-Bosch S (2015) Ethylene response factors: a key regulatory hub in hormone and stress signaling. Plant Physiol 169:32–41

Murphy DJ (2009) Oil palm: future prospects for yield and quality improvements. Lipid Technol 21:257–260

Nakano T, Suzuki K, Fujimura T, Shinsh H (2006) Genome-wide analysis of the ERF gene family in Arabidopsis and rice. Plant Physiol 140:411–432

Nakashima K, Yamaguchi-Shinozaki K, Shinozaki K (2014) The transcriptional regulatory network in the drought response and its crosstalk in abiotic stress responses including drought, cold and heat. Front Plant Sci 5:1–7

Nualwijit N, Lerslerwong L (2014) Post harvest ripening of oil palm fruit is accelerated by application of exogenous ethylene. Songklanakarin J Sci Technol 36:255–259

Olsen AN, Ernst HA, Leggio LL, Skriver K (2005) NAC transcription factors:structurally distinct, functionally diverse. Trends Plant Sci 10:79–87

Omidvar V, Abdullah SNA, Ho CL, Mahmood M, Al-Shanfari AB (2012) Isolation and characterization of two ABRE-binding proteins: EABF and EABF1 from the oil palm. Mol Biol Rep 39:8907–8918

Omidvar V, Abdullah SNA, Ho CL, Mahmood M (2013a) Isolation and characterization of an ethylene-responsive element binding protein (EgEREBP) from oil palm (Elaies guineensis). Aust J Crop Sci 7:219–226

Omidvar V, Abdullah SNA, Ebrahimi M, Ho CL, Mahmood M (2013b) Gene expression and functional characterization of the EgAP2-1 transcription factor during oil palm fruit ripening and in response to ethylene and ABA treatments. Biol Plant 57:646–654

Ooi LCL, Low ETL, Abdullah MO, Nookiah R, Ting NC, Nagappan J, Manaf MAA, Chan KL, Halim MA, Azizi N, Omar W, Murad AJ, Lakey N, Ordway JM, Favello A, Budiman MA, Brunt AV, Beil M, Leininger MT, Jiang N, Smith SW, Brown CR, Kuek ACS, Bahrain S, Hoynes-O'Connor A, Nguyen AY, Chaudhari HG, Shah SA, Choo YM, Sambanthamurthi R, Singh R (2016) Non-tenera contamination and the economic impact of SHELL genetic testing in the Malaysian independent oil palm industry. Front Plant Sci 7:1–13

Osobamiro MT, Adewuyi GO (2015) Levels of heavy metals in the soil: effects of season, agronomic practice and soil geology. J Agric Chem Environ 4(4):109

Organization for Economic Co-operation and Development-Food and Agricultural Organization of the United Nations (OECD-FAO) (2011) Agriculture outlook 2011–2020.

Park JM, Park C, Lee S, Ham B, Shin R, Paek K (2001) Overexpression of the tobacco TSi1 gene encoding an EREBP/AP2-type transcription factor enhances resistance against pathogen attack and osmotic stress in tobacco. Plant Cell 13:1035–1046

Pirrello J, Jaimes-Miranda F, Sanchez-Ballesta M, Bouzayen M (2006) Sl-ERF2, a tomato ethylene response factor involved in ethylene response and seed germination. Plant Cell Physiol 47:1195–1205

Popko J, Hansch R, Mendel RR, Polle A, Teichmann T (2010) The role of abscisic acid and auxin in the response of poplar to abiotic stress. Plant Biol 12:242–258

Prokurat S (2013) Palm oil - strategic source of renewable energy in Indonesia and Malaysia. J Modern Sci 3:425–443

Puranik S, Sahu PP, Srivastava PS, Prasad M (2012) NAC proteins:regulation and role in stress tolerance. Trends Plant Sci 17:369–381

Reinbothe C, Springer A, Samol I, Reinbothe S (2009) Plant oxylipins: role of jasmonic acid during programmed cell death, defence and leaf senescence. FEBS J 276:4666–4681

Riechmann J, Ratcliffe O (2000) A genomic perspective on plant transcription factors. Curr Opin Plant Biol 3:423–434

Rossini MA, Maddonni GA, Otegui ME (2016) Multiple abiotic stresses on maize grain yield determination: additive vs multiplicative effects. Field Crop Res 198:280–289

Roychoudhury A, Paul S, Basu S (2013) Cross-talk between abscisic acid-dependent and abscisic acid-independent pathways during abiotic stress. Plant Cell Rep 32:985–1006

Sah SK, Reddy KR, Li J (2016) Abscisic acid and abiotic stress tolerance in crop plants. Front Plant Sci 7:1–26

Sakuma Y, Liu Q, Dubouzet JG, Abe H, Shinozaki K, Yamaguchi-Shinozaki K (2002) DNA-binding specificity of the ERF/AP2 domain of Arabidopsis DREBs, transcription factors involved in dehydration- and cold-inducible gene expression. Biochem Biophys Res Commun 290:998–1009

Salmiyati HA, Idayu I, Supriyanto E (2014) Oil palm plantations management effects on productivity fresh fruit bunch (FFB). APCBEE Procedia 8:282–286

Sambanthamurthi R, Sundram K, Tan Y (2000) Chemistry and biochemistry of palm oil. Prog Lipid Res 39:507–558

Seki M, Kamei A, Yamaguchi-Shinozaki K, Shinozaki K (2003) Molecular responses to drought, salinity and frost: common and different paths for plant protection. Curr Opin Biotechnol 16:194–199

Seo JS, Koo YJ, Jung C, Yeu SY, Song JT, Kim J, Choi Y, Lee JS, Choi YD (2013) Identification of a novel jasmonate-responsive element in the AtJMT promoter and its binding protein for AtJMT repression. PLoS One 8:1–14

Shanmuganathan S, Narayanan A, Mohamed M, Ibrahim R, Haron K. (2014) A hybrid approach to modelling the climate change effects on Malaysia's oil palm yield at the regional scale. Paper presented at the first international conference on soft computing and data mining (SCDM 2014), Universiti Tun Hussein Onn Malaysia (UTHM), Johor, Malaysia

Sharoni AM, Nuruzzaman M, Satoh K, Shimizu T, Kondoh H, Sasaya T, Choi I, Omura T, Kikuchi S (2011) Gene structures, classification and expression models of the AP2/EREBP transcription factor family in rice. Plant Cell Physiol 52:344–360

Shinozaki K, Yamaguchi-Shinozaki K, Seki M (2003) Regulatory network of gene expression in the drought and cold stress responses. Curr Opin Plant Biol 6:410–417

Shinshi H (2008) Ethylene-regulated transcription and crosstalk with jasmonic acid. Plant Sci 175:18–23

Shiriga K, Sharma R, Kumar K, Yadav SK, Hossain F, Thirunavukkarasu N (2014) Genome-wide identification and expression pattern of drought-responsive members of the NAC family in maize. Meta Gene 2:407–417

Singh R, Low EL, Ooi LC, Ong-Abdullah M, Ting N, Nagappan J, Nookiah R, Amiruddin MD, Rosli R, Manaf MAA, Chan KL, Halim MA, Azizi N, Lakey N, Smith SW, Budiman MA, Hogan M, Bacher B, Brunt AV, Wang C, Ordway JM, Sambanthamurthi R, Martienssen RA (2013) The oil palm SHELL gene controls oil yield and encodes a homologue of SEEDSTICK. Nature 500:340–344

Song S, Qi T, Wasternack C, Xie D (2014) Jasmonate signaling and crosstalk with gibberellin and ethylene. Curr Opin Plant Biol 21:112–119

Stepanova AN, Alonso JM (2009) Ethylene signaling and response: where different regulatory modules meet. Curr Opin Plant Biol 12:548–555

Sun S, Yu J-P, Chen F, Zhao T-J, Fang X-H, Li Y-Q, Sui S-F (2008) TINY, a dehydration-responsive rlement (DRE)-binding protein-like transcription factor connecting the DRE- and ethylene-responsive element-mediated signaling pathways in Arabidopsis. J Biol Chem 283:6261–6271

Sun CX, Cao HX, Shao HB, Lei XT, Xiao Y (2011) Growth and physiological responses to water and nutrient stress in oil palm. Afr J Biotechnol 10:10465–10471

Tacken E, Ireland H, Gunaseelan K, Karunairetnam S, Wang D, Schultz K, Bowen J, Atkinson RG, Johnston JW, Putterill J, Hellens RP, Schaffe RJ (2010) The role of ethylene and cold temperature in the regulation of the apple POLYGALACTURONASE1 gene and fruit softening. Plant Physiol 153:294–305

Tengoua FF, Hanafi MM, Idris AS, Syed-Omar SR (2015) Screening for optimum concentrations of boron, copper and manganese for the growth of three-month old oil palm seedlings in solution culture. Pertanika J Trop Agric Sci 38:113–126

Theissen G, Becker A, Di Rosa A, Kanno A, Kim JT, Munster T, Winter KU, Saedler H (2000) A short history of MADS-box genes in plants. Plant Mol Biol 42:115–149

Todaka D, Nakashima K, Shinozaki K, Yamaguchi-Shinozaki K (2012) Toward understanding transcriptional regulatory networks in abiotic stress responses and tolerance in rice. Rice 5:1–9

Tranbarger TJ, Dussert S, Joet T, Argout X, Summo M, Champion A, Cros D, Omore A, Nouy B, Morcillo F (2011) Regulatory mechanisms underlying oil palm fruit mesocarp maturation, ripening, and functional specialization in lipid and carotenoid metabolism. Plant Physiol 156:564–584

von Uexküll HR, Fairhurst TH (1999) Some nutritional disorders in oil palm. Better Crops Int 13:16–21

Vrebalov J, Ruezinsky D, Padmanabhan V, White R, Medrano D, Drake R, Schuch W, Giovannoni J (2002) A MADS-box gene necessary for fruit ripening at the tomato ripening-inhibitor (Rin) locus. Science 296:343–346

Wang KL-C, Li H, Ecker JR (2002) Ethylene biosynthesis and signaling networks. Plant Cell 14 Suppl:S131–S151

Wang W, Xu M, Wang G, Galili G (2017) Autophagy: an important biological process that protects plants from stressful environments. Front Plant Sci 7:2030

Wituszynska W, Karpinski S (2013) Programmed cell death as a response to high light, UV and drought stress in plants. In: Vahdati K, Leslie C (eds) Abiotic stress-plant responses and applications in agriculture. InTech, Rijeka, Croatia, pp 207–246

Xin H, Qin F, Tran LP (2011) Transcription factors involved in environmental stress responses in plants. In: Ahmad P, Prasad MNV (eds) Environmental adaptation and stress tolerance of plants in the era of climate change. Springer, New York, pp 279–295

Xiong L, Gong Z, Rock CD, Subramaniam S, Guo Y, Xu W, Galbraith D, Zhu J (2001) Modulation of abscisic acid signal transduction and biosynthesis by an Sm-like protein in Arabidopsis. Dev Cell 1:771–781

Yanez M, Caceres S, Orellana S, Bastias A, Verdugo I, Ruiz-Lara S, Casaretto JA (2009) An abiotic stress-responsive bZIP transcription factor from wild and cultivated tomatoes regulates stress-related genes. Plant Cell Rep 28:1497–1507

Zhang P, Yang P, Zhang Z, Han B, Wang W, Wang Y, Cao Y, Hu T (2014) Isolation and characterization of a buffalograss (Buchloe dactyloides) dehydration responsive element binding transcription factor, BdDREB2. Gene 536:123–128

Zhao D, Shen L, Fan B, Yu M, Zheng Y, Lv S, Sheng J (2009) Ethylene and cold participate in the regulation of LeCBF1 gene expression in postharvest tomato fruits. FEBS Lett 583:3329–3334

Zhu JK (2016) Abiotic stress signaling and responses in plants. Cell 167:313–324

Zwack PJ, Rashotte AM (2015) Interactions between cytokinin signalling and abiotic stress responses. J Exp Bot 66:4863–4871

Chapter 5
Sensing Crop Diseases

Thangavel Lakshmipriya, Subash C.B. Gopinath, and Uda Hashim

5.1 Introduction

Production of good quality and nutritious crops is important for human and animal health and it is also one of the important determinants of farmers' income. The yield of crops is influenced by different factors where generation of disease free crops in particular can determine good crop harvest. Crops are widely infected by different pathogens which can lead to great loss in yield and reduce the benefits that can be obtained from crops for human and animals (Fig. 5.1; Table 5.1). Like in human and animals, pathogens are also severely infecting plants. The exposed plant leaves are often vulnerable to pathogen attack but in certain cases other parts of the plant can also be infected depending on the type of pathogens.

Crops are widely infected by a variety of diseases caused by different kinds of organism including viruses and bacteria (Fig. 5.2). There are many plant pathogenic viruses such as *Lettuce mosaic virus* that infects lettuce (Mazier et al. 2004), *Citrus psorosis virus* that causes damage to the bark of the citrus plant, *Tomato bushy stunt virus* that infects eggplant (Robinson and Harrison 1982), and *Grapevine fanleaf virus* that causes disease on the grapevine leaves (Andret-Link et al. 2004). Pathogenic bacterial species directly infect different parts of the plant. Bacteria can survive in various tissues of the plant including the vegetative tissues such as leaves and stem as well as the reproductive tissues like fruits (de Nadra 2007; Pennisi 2015). Examples of common plant pathogenic bacteria are *Pseudomonas syringae* pv. *syringae* (causes infection on beans), *Pseudomonas syringae* pv. *pisi* (peas), *Ralstonia solanacearum* (potato and tomato) and *Erwinia* spp. (apples and pears).

T. Lakshmipriya • S.C.B. Gopinath (✉) • U. Hashim
Institute of Nano Electronic Engineering, Universiti Malaysia Perlis, 01000 Kangar, Perlis, Malaysia

School of Bioprocess Engineering, Universiti Malaysia Perlis, 02600 Arau, Perlis, Malaysia
e-mail: subash@unimap.edu.my

S.N.A. Abdullah et al. (eds.), *Crop Improvement*,
DOI 10.1007/978-3-319-65079-1_5

Fig. 5.1 Diseases on crop plants. The appearances of infection in different crops are displayed

Table 5.1 Pathogens and their host plants

Plant	Pathogen	Type of pathogen	References
Grapevine	*Xylella fastidiosa*	Bacteria	Minsavage et al. (1994)
Onion	*Sclerotium cepivorum*	Fungi	Haq et al. (2003)
Olive	*Pseudomonas savastanoi* pv. *savastanoi*	Bacteria	Bertolini et al. (2003)
Potato	*Candidatus liberibacter solanacearum*	Bacteria	Li et al. (2006)
Citrus	*Citrus tristeza virus*	Virus	Saponari et al. (2007)
Tomato	*Pepino mosaic virus*	Virus	Gutiérrez-Aguirre et al. (2009)
Rice	*Burkholderia glumae*	Bacteria	Fang et al. (2009)
Almond	*Candidatus phytoplasma prunorum*	Bacteria	Yvon et al. (2009)
Sweet orange	*Candidatus Liberibacter asiaticus*	Bacteria	Rigano et al. (2014)

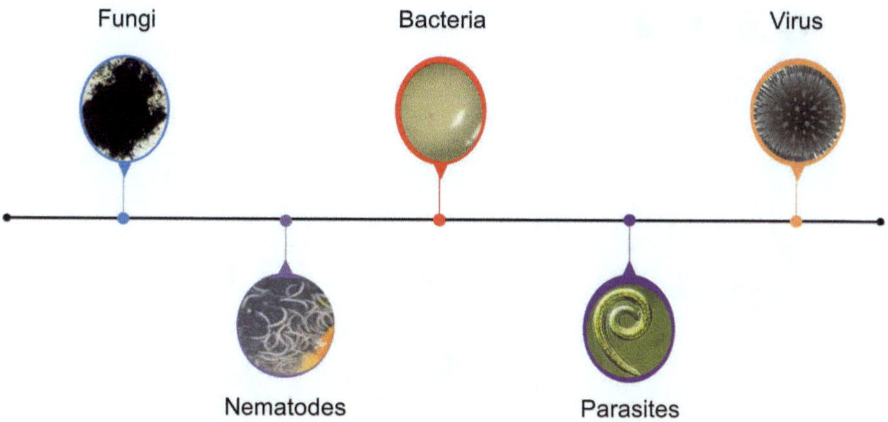

Fig. 5.2 Causative agents for the infections on crop plants. Different types of organism are displayed

Degree of infection by the bacteria varies and it can be accelerated by additional factors such as temperature, humidity level and seasonal variations.

5.2 Biosensors

Determining the occurrence of different pathogens and disease-causing agents on plants can be done by different ways. Biosensor is one of the most important tools that can help to identify diseases in plants, animals and human. Application of biosensor in the field of agriculture is necessary for improving crop yield and consequently the economic outcome. A conventional biosensing system has a transducer with electrical/non-electrical output elements and operates with the assistance of different probes (Fig. 5.3). In general, a biosensor can work on solid substrate or it can be solution-based. Furthermore, biosensors are widely categorized as label-based and label-free biosensors. These sensing systems will have probe to capture or diagnose the disease-causing agent(s) in the sample to be tested. Different probes have been produced in the past and among these probes, aptamer, antibody and DNA are attested to be the predominant molecules in several instances (Gopinath 2006; Gopinath et al. 2006a, 2006b, 2012; Matsushita and Penmetcha 2009; Gopinath and Kumar 2013; Lakshmipriya et al. 2013; Suenaga and Kumar 2014; Samy et al. 2015).

These probes are capable in detecting diseases in plants and they have specific secondary or tertiary structures. To form the proper structures, the probes undergo appropriate folding processes to interact with the target molecules (Fig. 5.4). With these structure forming molecules, surface chemical linker is necessary to immobilize the molecules on the solid sensing surface. The common linkers are 3-aminopropyltriethoxysilane, glutaraldehyde, N,N'-carbonyldiimidazole and thiol

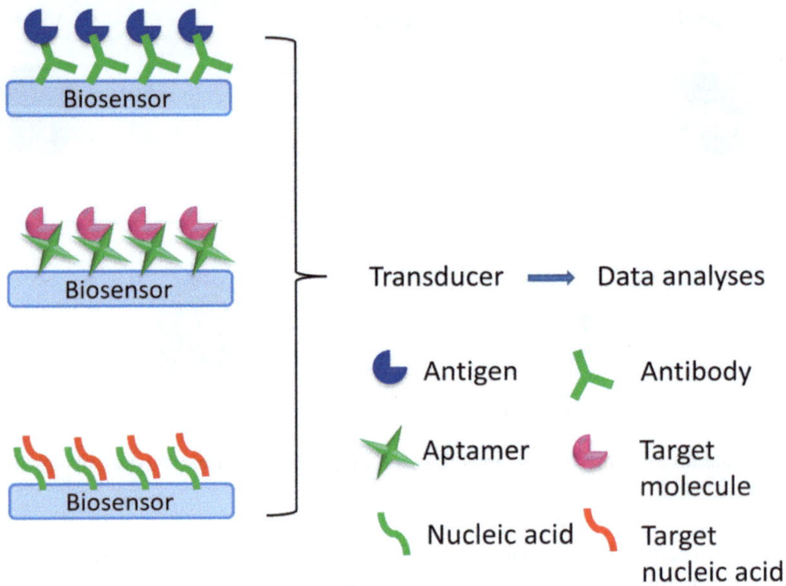

Fig. 5.3 Elements for biosensor. Probe, transducer and output are indicated

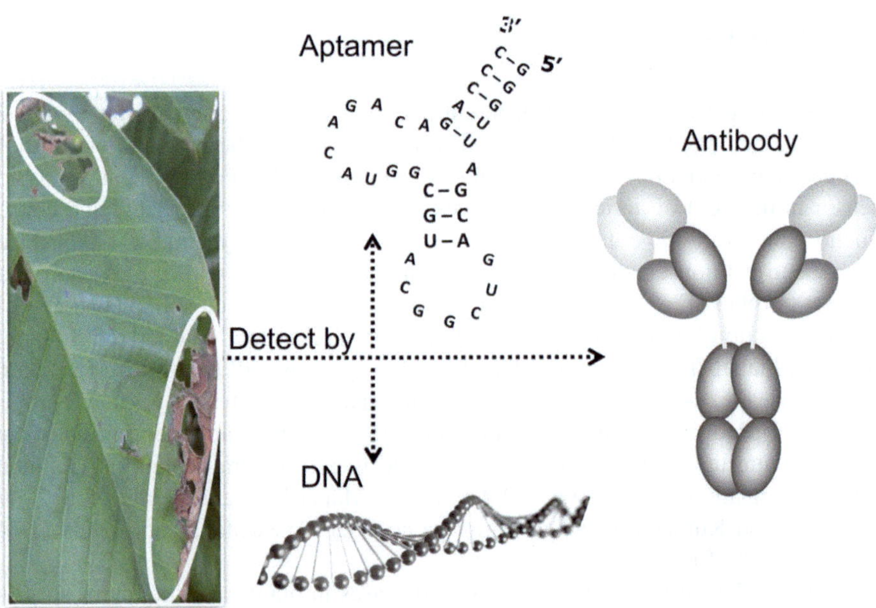

Fig. 5.4 Probes for the detection of crop diseases. Aptamer, antibody and nucleic acids are the potential probes

compounds. The choice of linker depends on the substrate to be used on the sensing surface. On the other hand, solution-based sensing may not rely on these chemical linkers. However, solution-based sensors may need radioactive or fluorescent compounds such as fluorescein, rhodamine and cyanine as labelling methods.

Label-free method is the general choice as it offers several advantages. These include no requirement for chemical modification on the molecules thus maintaining the intactness of the molecular structure with no alteration on the original affinity of the probe to the analyte. Even though the sensing offers the choice either to be done on solid or in solution, the biological reaction must be carried out under wet condition. The genuine and high-performance sensor is usually expected to have higher sensitivity and specificity. In this chapter, we discuss on the detection of plant diseases using the above-mentioned probes in order to promote higher crop yield.

5.3 Detection of Crop Diseases by Aptamer

5.3.1 Production of Aptamer

Aptamer is one of the probes that have been used to identify plant diseases. Aptamer is a single strand nucleic acid [deoxyribonucleic acid (DNA) or ribonucleic acid (RNA)], called 'artificial antibody', generated by the process called 'SELEX' (systematic evolution of ligands by exponential enrichment) (Fig. 5.5). In the SELEX process, initially we need to synthesize the random nucleic acid library and these molecules can be amplified further by polymerase chain reaction (PCR) in the case of DNA library and in vitro transcription followed by PCR in the case of RNA library. In the preparation of the initial nucleic acid library, both the length and number of molecules must be considered. Once prepared, the library will be complexed with the desired target molecule followed by the separation of bound molecule to the target from the unbound one. The separation can be done using solid media such as on functionalized titre plate, through column matrix, on the affinity beads or by any other affinity processes. Then, the bound molecules will be amplified to be used for the next round of selection steps. After several iterative rounds of selection process in similar manner, the high-affinity molecules will be cloned into the vector system and sequenced (Gopinath 2006, 2007, 2008; Chandra and Gopinath 2009; Gopinath et al. 2012; Lakshmipriya et al. 2013).

5.3.2 Aptasensors for the Detection of Crop Diseases

Aptamers have been selected as probes against a wide range of targets from small molecules to whole cells. The selected aptamers have several advantages over antibodies including higher sensitivity and specificity, easier and cheaper to prepare,

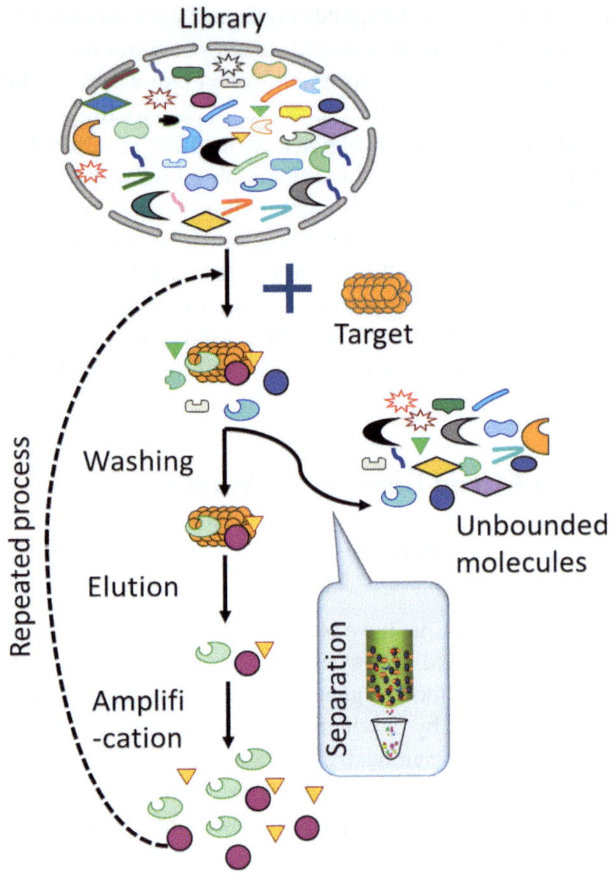

Fig. 5.5 SELEX process. This process involves iterative rounds of complex formation, separation and amplification

non-immunogenic and no batch variation. Because of these positive characteristics, aptamer-mediated applications are widespread in all fields including for environmental monitoring, medical diagnosis, drug discovery and delivery and food quality checking (Gopinath et al. 2016). Aptamer probes are given great attention in the biosensing field due to their high sensitivity in detecting the target molecule. For plant diseases, this characteristic can help to detect the diseases at earlier infection stages. In the field of agriculture, aptamers are used to detect pathogens, such as viruses and bacteria on the plants and it also helps to detect metal ions in soils and water where plants are grown, which have an indirect influence on the susceptibility of plants to infection.

On the sensing surface, aptamer can be immobilized directly through the chemical linker or extending aptamer sequences at either the 5′ or 3′ end. Furthermore, the extended aptamer may be immobilized on the sensing surface by complexing with oligonucleotide before being attached on the sensing surface. Alternatively, the

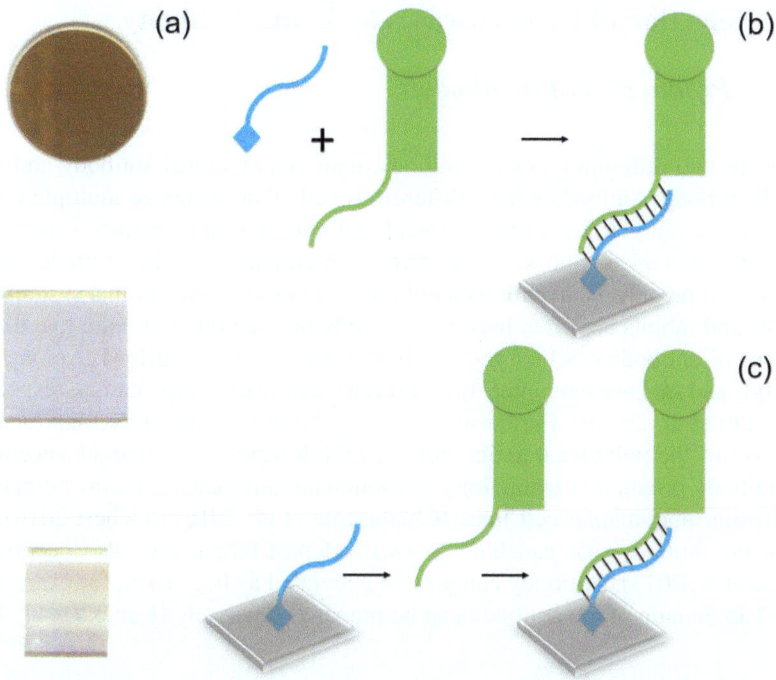

Fig. 5.6 Immobilization of aptamer on the sensing surface. (**a**) Different types of sensing surfaces. On these surfaces, two different strategies are possible. (**b**) Strategy with pre-complexed molecules (oligonucleotide and extended aptamer) (**c**) strategy with post-complexed molecules

basic oligonucleotide can attach first on the surface, followed by the duplex formation of extended aptamer (Fig. 5.6). In the latter case, consideration on the length of the extended portion for duplex formation is necessary. This length should be able to withstand the experimental temperature. Usually only oligonucleotides of 20 mer and above in length can withstand high temperature. Moreover, it is better to avoid A-U base pairing in the case of RNA-extension, while G-C pair base pairing is advisable and ideal. Usually, A-T has two hydrogen bonding, whereas G-C has three hydrogen bonding, which shows obvious greater stability of the G-C complementation. In addition, peptide aptamer has also been used as an alternative to the nucleic acid-based aptamer. These are small combinatorial proteins which can bind to target molecules at specific sites. They are only about 5–20 amino acid long embedded as a loop within a stable protein scaffold and are often produced using yeast-two-hybrid system.

Ochratoxin A is a toxin produced by the species of *Aspergillus* and *Penicillin*. These fungi commonly infect grains and grapes. Using an aptamer generated against Ochratoxin A, a sandwich assay has been designed on amorphous silicon (Costantini et al. 2016) and was demonstrated to be a potential way to detect crop diseases. Lopez-Ochoa et al. (2006) found the aptamer to block the replication of geminivirus. This is an indirect way of aptamer detection, rather than direct detection that is often described.

5.4 Detection of Crop Diseases by Using Antibody

5.4.1 Production of Antibody

There are two different types of antibody, namely polyclonal antibody and monoclonal antibody. Antibodies from different B cells that recognize multiple epitopes on the same antigen are called a polyclonal antibody. In contrast, a monoclonal antibody is obtained from a single antibody producing B cell and binds with one unique epitope only. Conventional antibodies are produced in animal system where rodents and rabbits are often used for antibody production. There are five different isotypes of antibodies, which include IgA, IgD, IgE, IgG and IgM. Among these, IgD, IgE and IgG are monomer; IgA and IgM are dimer and pentamer, respectively (Gopinath et al. 2014b). Furthermore, IgG is the most abundantly found antibody and most of the polyclonal antibodies are IgG. Recently, with the advancement in the antibody production technology, recombinant antibodies can now be manufactured using mammalian cell lines (Chudasama et al. 2016; Reichert 2016) while bispecific antibody that can bind to two different targets can also be produced (Spiess et al. 2013). Antibody can also be generated as fragments, for example, the Fc or Fab domain of the antibody can be prepared separately (Kaneko et al. 2006).

5.4.2 Immunosensors for the Detection of Crop Diseases

Detection of pathogens by antibodies can be done at high efficiency for determining the different diseases. Immunoassay-based sensing systems in particular, which include the enzyme-linked immunosorbent assay (ELISA) and western blot, are commonly used to identify diseases with the help of antibodies. Antibody is a 'Y' shaped molecule that can be easily generated for a particular target of interest. With any probe(s), the detection system can be designed against a single target (monoplex) or multiple targets (muliplex). Using multiplex detection, chilli vein-banding mottle virus, *Acidovorax avenae* subsp. *citrulli*, melon yellow spot virus, and watermelon silver mottle virus have been detected with the corresponding partner antibodies. Researchers tagged antibody to the fluorescent coated magnetic microsphere for the detection of virus (Charlermroj et al. 2013). Peters et al. (2007) detected the *Dickeya dianthicola* and *Pectobacterium atrosepticum* in the extract of the potato plant. They conjugated the alexa flour to the antibody generated against the bacteria to be detected. Also biotinylated antibody has been used for the detection of food borne pathogens *Salmonella* and *Escherichia coli* (Wang et al. 2011). The common strategy for the detection of virus using antibody is displayed in Fig. 5.7. As stated above, ELISA is one of the potential sensing strategies for the detection of not only the plant pathogens, but it is also highly suitable for any kinds of antigen from the disease-causing agents. ELISA has been formulated by two different methods (direct and indirect ELISAs) (Fig. 5.8). In direct ELISA, the primary (first) antibody

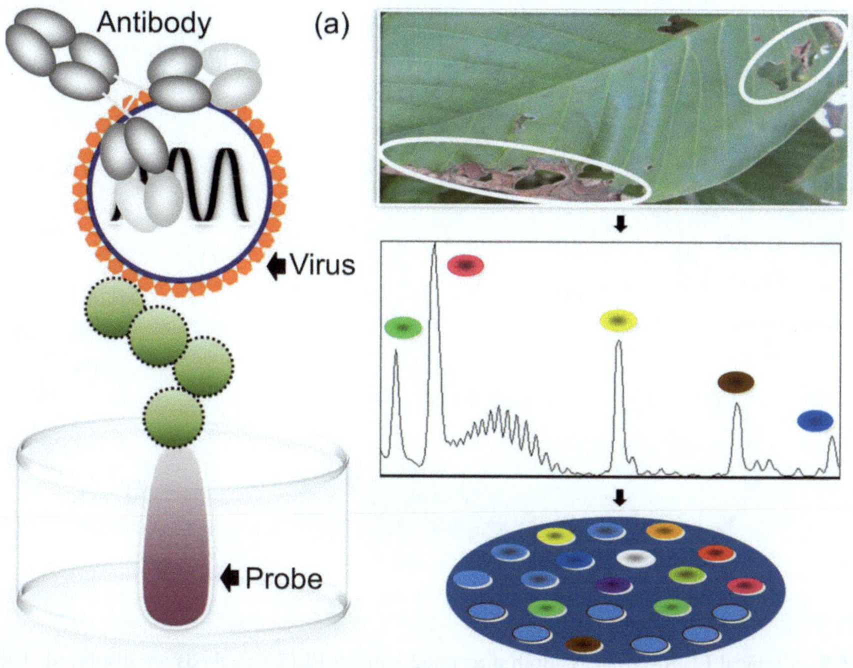

Fig. 5.7 (**a**) A common strategy for the detection of pathogens using antibody. (**b**) Design to detect infections on plant by different pathogens. This strategy is ideal for array-based detections

is labelled and this can be achieved using the commercially available kit. Whereas indirect ELISA, which is widely used, involved labelling the secondary antibody.

5.5 Detection of Crop Diseases by Using DNA

5.5.1 *Synthesis of Nucleic Acids*

Nucleic acids (DNA or RNA) are commonly prepared either chemically or enzymatically. Conventional synthetic methods have limitations on the length of DNA or RNA during preparation; however, the advancement in technology brought the choice to prepare the longer size nucleic acids. Through enzymatic synthesis, single stranded DNA may be prepared by reverse transcription and asymmetric PCR. Double stranded DNA can be prepared by PCR from the original template. RNA can be prepared by transcription reaction from the DNA template. DNA or RNA can also be extracted from the plants or microbes or animal system using the appropriate reagents. Furthermore, the extracted nucleic acids may be fragmented by the restriction enzymes.

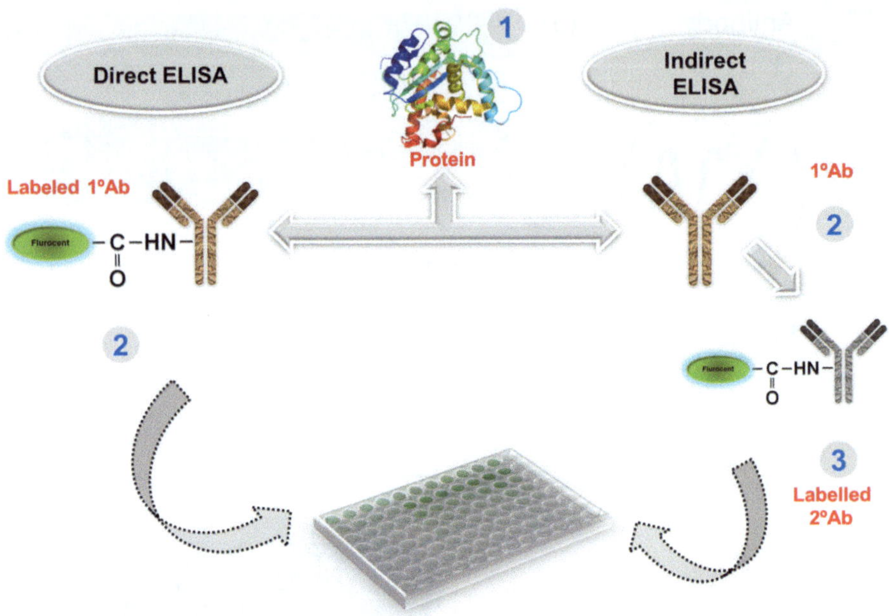

Fig. 5.8 A typical ELISA strategy. Both direct and indirect ELISA methods are displayed. The sequential steps involved in these methods are indicated by numbers

5.5.2 Nucleic Acid Sensors for the Detection of Crop Diseases

In general, biomolecules are identified by the partner molecules, such as DNA, antibody and aptamer. DNA probes have been used with various sensors to identify diseases by different sensing methods, such as PCR amplification, fluorescence-based detection system, Raman spectroscopy, Surface Plasmon Resonance (SPR) and Biacore-based SPR system. Among the probes, identification by DNA is sensitive and specific. PCR-based detection of plant diseases shows accurate result. Suitable primers can be ideally designed for the particular DNA of interest from the microorganism (Hatziloukas et al. 1995). The specificity of the detection mainly depends on the suitable complementation. Papadakis et al. (2015) used multiplex PCR for the simultaneous detection of bacterial pathogens in plants. They detected the bacteria *Ralstonia solanacearum*, *Pseudomonas syringae*, and *Xanthomonas campestris* pv. *vesicatoria* from tomato plants. Postinikova et al. (2008) detected the bacteria *Rhodococcus fascians, Pseudomonas Acidovorax facilis, syringae* pv. *syringae, Xylella fastidiosa and Ralstonia solanacearum* by DNA amplification-based detection with the assistance of mass spectrometry. Matsushita and Penmetcha (2009) analysed the non-coding RNA (*Chrysanthemum* stunt viroid) in *Cyrysanthemum* known for causing stunt disease affecting the flowers. In their study, they revealed the role of *Chrysanthemum* stunt viroid for the infection, replication and disease spread.

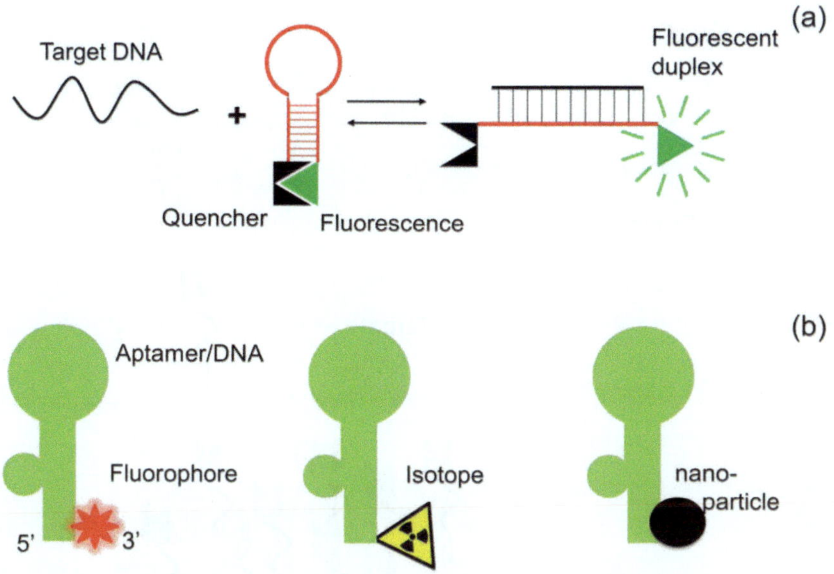

Fig. 5.9 Labelling on DNA. (**a**) Molecular beacon type. (**b**) Different methods of labelling on nucleic acid

Nucleic acid is easy to be labelled with different compounds or materials such as fluorescent, radioactive materials and nanomaterials. Detection of diseases by molecular beacon is one of the efficient ways as shown for human viral detection (Fig. 5.9a; Yamamoto and Kumar 2000). Other than fluorescent labelling, radioisotope labelling is one of the powerful and sensitive strategies. Currently, nanoparticle-based capturing of nucleic acid is getting popular for high-performance detection (Fig. 5.9b; Gopinath et al. 2014a). Among the different nanoparticles, gold nanoparticle has been highly studied for sensing applications due to its potential characteristics, such as easier to tailor, cheaper, suitable for non-fouling, variable in size and ideal with biologically relevant molecules. Other than gold nanoparticle, silica nanoparticles have also been recommended for sensing purposes due to their unique features. The above-mentioned labelling techniques are promising for the array-based screening, especially for high-throughput analysis, where several types of nucleic acid targets need to be screened. By tagging the above-mentioned labels, it will be ideal to screen several numbers of samples (Fig. 5.10).

5.6 Conclusions and Future Prospects

Based on the above discussions, there are three major probes that have been proposed in the past for detecting plant diseases, and these probes are aptamer, antibody and nucleic acid. These probes offer different advantages as demonstrated

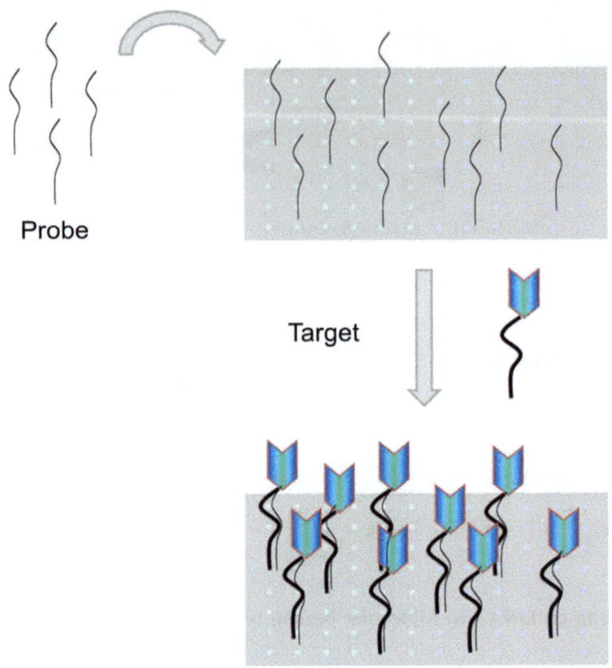

Fig. 5.10 Array-based techniques using nucleic acid sequences as the probe and target

previously. By complementing different probes, there will be a further development in the high-performance sensing systems. Moreover, these are well suited for the commonly available sensing strategies (Fig. 5.11). With these technologies, making sensing systems for important plant pathogen such as *Ganoderma boninense* that causes devastating disease to the oil palm, a major plantation crop will find the way for industrial application. Furthermore, detection of the availability of metals or other micronutrients in the plant environments will optimize the conditions necessary to protect plants from the pathogens that cause the diseases.

Further, there are additional developments such as early diagnosis, multiple target detections, molecular discrimination and high-throughput screening, considered to be the immediate necessities for the crop improvements. Among the three probes discussed in this overview, aptamers are more relevant to fulfil all these tasks and to be associated with the sensing systems. However, for high-throughput screening all three probes can be used. It is wise to generate complementing strategies, for example, aptamer-antibody or nucleic acid-antibody complemented probes, in the sensing set-up that will pave the way to overcome all current issues or necessities.

Acknowledgement Lakshmipriya thank Universiti Malaysia Perlis for providing the post-doctoral position.

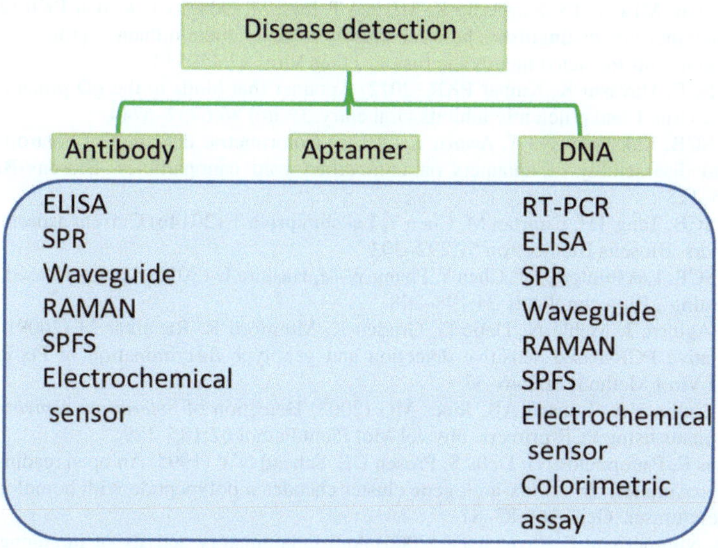

Fig. 5.11 Potential techniques suitable for different probes

References

Andret-Link P, Laporte C, Valat L, Ritzenthaler C, Demangeat G, Vigne E, Laval V, Pfeiffer P, Stussi-Garaud C, Fuchs M (2004) Grapevine fanleaf virus: still a major threat to the grapevine industry. J Plant Pathol 86:183–195

Bertolini E, Penyalver R, García A, Olmos A, Quesada JM, Cambra M, López MM (2003) Highly sensitive detection of *Pseudomonas savastanoi pv. savastanoi* in asymptomatic olive plants by nested-PCR in a single closed tube. J Microbiol Methods 52:261–266

Charlermroj R, Himananto O, Seepiban C, Kumpoosiri M, Warin N, Oplatowska M, Gajanandana O, Grant IR, Karoonuthaisiri N, Elliott CT (2013) Multiplex detection of plant pathogens using a microsphere immunoassay technology. PLoS One 8:e62344. doi:10.1371/journal.pone.0062344

Chudasama V, Maruani A, Caddick S (2016) Recent advances in the construction of antibody-drug conjugates. Nat Chem 8:114–119

Costantini F, Sberna C, Petrucci G, Reverberi M, Domenici F, Fanelli C, Manetti C, De Cesare G, Derosa M, Nascetti A, Caputo D (2016) Aptamer-based sandwich assay for on-chip detection of Ochratoxin A by an array of amorphous silicon photosensors. Sens Actuators B 230:31–39

Fang Y, Xu LH, Tian WX, Huai Y, Yu SH, Lou MM, Xie GL (2009) Real-time fluorescence PCR method for detection of *Burkholderia glumae* from rice. Rice Sci 16:157–160

Gopinath SCB (2006) An efficient RNA aptamer against human influenza B virus Hemagglutinin. J Biochem 139:837–846

Gopinath SCB (2007) Antiviral aptamers. Arch Virol 152:2137–2157

Gopinath SCB (2008) Anti-coagulant aptamers. Thromb Res 122:838–847

Gopinath SCB (2009) Mapping of RNA – protein interactions. Anal Chim Acta 636:117–128

Gopinath SCB, Kumar PKR (2013) Aptamers that bind to the hemagglutinin of the recent pandemic influenza virus H1N1 and efficiently inhibit agglutination. Acta Biomater 9:8932–8941

Gopinath SCB, Balasundaresan D, Akitomi J, Mizuno H (2006a) An RNA aptamer that discriminates bovine factor IX from human factor IX. J Biochem 140:667–676

Gopinath SCB, Misono TS, Kawasaki K, Mizuno T, Imai M, Odagiri T, Kumar PKR (2006b) An RNA aptamer that distinguishes between closely related human influenza viruses and inhibits haemagglutinin-mediated membrane fusion. J Gen Virol 87:479–487

Gopinath SCB, Hayashi K, Kumar PKR (2012) Aptamer that binds to the gD protein of herpes simplex virus 1 and efficiently inhibits viral entry. J Virol 86:6732–6744

Gopinath SCB, Lakshmipriya T, Awazu K (2014a) Colorimetric detection of controlled assembly and disassembly of aptamers on unmodified gold nanoparticles. Biosens Bioelectron 51:115–123

Gopinath SCB, Tang TH, Citartan M, Chen Y, Lakshmipriya T (2014b) Current aspects in immunosensors. Biosens Bioelectron 57:292–302

Gopinath SCB, Lakshmipriya T, Chen Y, Phang W-M, Hashim U (2016) Aptamer-based 'point-of-care testing'. Biotechnol Adv 34:198–208

Gutiérrez-Aguirre I, Mehle N, Delić D, Gruden K, Mumford R, Ravnikar M (2009) Real-time quantitative PCR-based sensitive detection and genotype discrimination of Pepino mosaic virus. J Virol Methods 162:46–55

Haq MA, Collin HA, Tomsett AB, Jones MG (2003) Detection of *Sclerotium cepivorum* within onion plants using PCR primers. Physiol Mol Plant Pathol 62:185–189

Hatziloukas E, Panopoulos NJ, Delis S, Prosen DE, Schaad NW (1995) An open reading frame in the approximately 28-kb tox-argk gene cluster encodes a polypeptide with homology to fatty acid desaturases. Gene 166:83–87

Kaneko Y, Nimmerjahn F, Ravetch JV (2006) Anti-inflammatory activity of immunoglobulin G resulting from Fc sialylation. Science 313:670–673

Lakshmipriya T, Fujimaki M, Gopinath SCB, Awazu K (2013) Generation of anti-influenza aptamers using the systematic evolution of ligands by exponential enrichment for sensing applications. Langmuir 29:15107–15115

Li W, Hartung JS, Levy L (2006) Quantitative real-time PCR for detection and identification of *Candidatus Liberibacter* species associated with citrus huanglongbing. J Microbiol Methods 66:104–115

Lopez-Ochoa L, Ramirez-Prado J, Hanley-Bowdoin L (2006) Peptide aptamers that bind to a geminivirus replication protein interfere with viral replication in plant cells. J Virol 80:5841–5853

Matsushita Y, Penmetcha KKR (2009) In vitro-transcribed chrysanthemum stunt viroid RNA is infectious to chrysanthemum and other plants. Phytopathology 99:58–66

Mazier M, German-Retana S, Flamain F, Dubois V, Botton E, Sarnette V, Le Gall O, Candresse T, Maisonneuve B (2004) A simple and efficient method for testing Lettuce mosaic virus resistance in in vitro cultivated lettuce. J Virol Methods 116:123–131

Minsavage GV, Thompson CM, Hopkins DL, Leite RMVBC, Stall RE (1994) Development of a polymerase chain-reaction protocol for detection of Xylella-fastidiosa in plant-tissue. Phytopathology 84:456–461

de Nadra MM (2007) Nitrogen metabolism in lactic acid bacteria from fruits: a review. Commun Curr Res Educ Top Trends Appl Microbiol 1:500–510

Papadakis G, Skandalis N, Dimopoulou A, Glynos P, Gizeli E (2015) Bacteria murmur: application of an acoustic biosensor for plant pathogen detection. PLoS One 10:e0132773. doi:10.1371/journal.pone.0132773

Pennisi E (2015) Microbiology. Leaf bacteria fertilize trees, researchers claim. Science 348:844–845

Peters J, Sledz W, Bergervoet JHW, Van Der Wolf JM (2007) An enrichment microsphere immunoassay for the detection of *Pectobacterium atrosepticum* and *Dickeya dianthicola* in potato tuber extracts. Eur J Plant Pathol 117:97–107

Postinikova E, Baldwin C, Whitehouse CA, Sechler A, Schaad NW, Sampath R, Harpin V, Li F, Melton R, Blyn L, Drader J, Hofstadler S, Schneider WL (2008) Identification of bacterial plant pathogens using multilocus polymerase chain reaction/electrospray ionization-mass spectrometry. Phytopathology 98:1156–1164

Reichert JM (2016) Antibodies to watch in 2016. MAbs 8:197–204

Rigano LA, Malamud F, Orce IG, Filippone MP, Marano MR, Amaral AMD, Castagnaro AP, Vojnov AA (2014) Rapid and sensitive detection of *Candidatus* Liberibacter asiaticus by loop mediated isothermal amplification combined with a lateral flow dipstick. BMC Microbiol 14:86
Robinson IK, Harrison SC (1982) Structure of the expanded state of tomato bushy stunt virus. Nature 297:563–568
Samy RP, Stiles BG, Chinnathambi A, Zayed ME, Alharbi SA, Franco OL, Rowan EG, Kumar AP, Lim LHK, Sethi G (2015) Viperatoxin-II: a novel viper venom protein as an effective bactericidal agent. FEBS Open Bio 5:928–941
Saponari M, Manjunath K, Yokomi RK (2007) Quantitative detection of Citrus tristeza virus in citrus and aphids by real-time reverse transcription-PCR (TaqMan(R)). J Virol Methods 147:43–53
Spiess C, Merchant M, Huang A, Zheng Z, Yang NY, Peng J, Ellerman D, Shatz W, Reilly D, Yansura DG, Scheer JM (2013) Bispecific antibodies with natural architecture produced by co-culture of bacteria expressing two distinct half-antibodies. Nat Biotechnol 31:753–758
Suenaga E, Kumar PKR (2014) An aptamer that binds efficiently to the hemagglutinins of highly pathogenic avian influenza viruses (H5N1 and H7N7) and inhibits hemagglutinin-glycan interactions. Acta Biomater 10:1314–1323
Wang H, Li Y, Wang A, Slavik M (2011) Rapid, sensitive, and simultaneous detection of three foodborne pathogens using magnetic nanobead-based immunoseparation and quantum dot-based multiplex immunoassay. J Food Prot 74:2039–2047
Yamamoto R, Kumar PKR (2000) Molecular beacon aptamer fluoresces in the presence of Tat protein of HIV-1. Genes Cells 5:389–396
Yvon M, Thébaud G, Alary R, Labonne G (2009) Specific detection and quantification of the phytopathogenic agent '*Candidatus* phytoplasma prunorum'. Mol Cell Probes 23:227–234

Chapter 6
Development of Genomic Resources and Assessing Their Potential for Accelerated *Acacia* Breeding

Wikneswari Ratnam, Chee Yen Choong, and Muhammad Asif Javed

6.1 Introduction

Acacia is a woody tree with wide distribution in all the continents except Europe and Antarctica (Maslin 2001). It comprises three subgenera—the pantropical subgenera *Acacia* and *Aculeiferum* distributed from Central and South America to Southeast Asia including Africa, and the subgenus *Phyllodineae* occurring primarily in Australia (Ross 1981). There are more than 1200 reported species of *Acacia*, and 955 of these are found only in Australia (Maslin 2001). Some tropical species (*A. auriculiformis, A. mangium* and *A. crassicarpa*) and temperate species (*A. melanoxylon, A. saligna* and *A. mearnsii*) from the subgenus *Phyllodineae* and some species from the subgenus *Acacia* (*A. senegal* and *A. nilotica*) have been planted commercially (Doran and Turnbull 1997; CAB International 2000). *Acacia* is a legume (subfamily Mimosoideae), and it is well known in forming an important symbiotic relationship with *Rhizobium* and *Bradyrhizobium*, the nitrogen fixing bacteria (Doyle 1994). This symbiotic association enables *Acacia* trees to grow well on poor soils. They are widely grown for fuel wood, soil rehabilitation and amenity plantings. Besides, the seeds of many species are taken as vegetable protein and also main source of Arabica gum (Maslin 2001). The increasing demand for paper source together with a decline in fibre supply from forests has forced the pulp and paper industry to look for more sustainable fibre sources to supplement the natural forest based resources.

W. Ratnam (✉) • C.Y. Choong
Faculty of Science and Technology, Universiti Kebangsaan Malaysia,
43600 UKM, Bangi, Selangor, Malaysia
e-mail: wicki@ukm.edu.my

M.A. Javed
Faculty of Forestry and Range Management, University of Agriculture Faisalabad,
38000 Faisalabad, Pakistan

S.N.A. Abdullah et al. (eds.), *Crop Improvement*,
DOI 10.1007/978-3-319-65079-1_6

In developing countries where forest resource is limited, paper is mainly produced from annual plants (non-wood fibre). Globally pulp is produced from eucalypt, however, it is (wood fibre) a less likely choice in many countries due to high water demand of the species coupled with climate change associated with water shortage. Nevertheless, wood pulp is still very much necessary for feasible pulp processing (Jahan et al. 2008). Energy crisis is faced by many countries due to unstable supply of fuel wood and price increase in natural gas dependence on wood biomass, particularly in poor rural areas. Furthermore, natural forestland is declining rapidly due to the competition with other land uses and deforestation. It is becoming more difficult to continuously supply pulpwood from natural forest to support increasing growth of pulp and paper industry. Therefore, plantation for fast growing tree species needs to be established to counteract the declining supply of pulpwood from the forest. Similarly, planting of forest species thriving on poor lands generally not suitable for planting other crops can increase the planted area for increased production of pulpwood. *Acacia mangium*, *A. auriculiformis* and their hybrid seem to be good candidates for pulpwood and softwood production in Malaysia as they grow well on poor soils, and *Acacia* plantations have been established in Peninsular Malaysia and Sabah.

Lignin is present as one of the essential wood components comprising 10–30% of the wood mass. It reduces pulp yield, increases cost of pulp production, risk environment safety due to increase chemical processing, and reduces paper whitishness (Chiang et al. 1998). Understanding lignin biosynthesis and its content will help in selecting planting materials with low lignin content (Whetten et al. 1998). Conventional breeding in *Acacia* is always hampered by long reproductive cycle of the plant, coupled with the difficulty in controlled crossing and seed germination for hybrid progenies (Ng et al. 2009). Therefore, the *Acacia* breeding, as much as the breeding for other forest trees, has been mostly done on recurrent selection from open-pollinated families (Butcher and Moran 2000). Marker-assisted selection (MAS) breeding is a promising approach to select efficiently superior planting materials with better wood and pulp properties for forest plantation establishment in relatively shorter period compared to conventional breeding.

Our work focuses on *A. mangium*, *A. auriculiformis* and their hybrid (*A. mangium* × *A. auriculiformis*) as these *Acacia* species have been established in plantations in Malaysia. We aim to produce elite planting materials for pulpwood production by understanding the molecular basis of lignin biosynthesis in these *Acacia* species through genomics approaches (Yong et al. 2011; Ong and Wickneswari 2011, 2012; Wong et al. 2011, 2012). At the same time we developed various DNA markers for linkage mapping and QTL analysis. All these information and efforts are crucial for the molecular breeding of the *Acacia* species.

6.2 Expressed Sequence Tag Analysis

The expressed sequence tags (ESTs) of plants available in the public databases are mainly from non-woody species (*Arabidopsis*, soybean, rice and maize), with the exception of the model woody tree species *Populus*. In order to understand the lignin and cellulose biosynthesis in *Acacia*, a cDNA library was constructed on RNA extracted from the inner bark tissues of the *Acacia* hybrid (Yong et al. 2011). A total of 3182 high quality ESTs were obtained from the cDNA library, and this represents 1982 unique transcripts consisting 663 contigs and 1319 singletons. A total of 1053 unigenes (derived from 1750 ESTs) were found to match with protein sequences in the public database, and 867 of these unigenes were of significant match (E-value ≤ e^{-10}). These significantly matched unigenes of known functions were subjected to functional analysis, and this revealed 231 genes grouped into 11 distinct functional classes (Fig. 6.1).

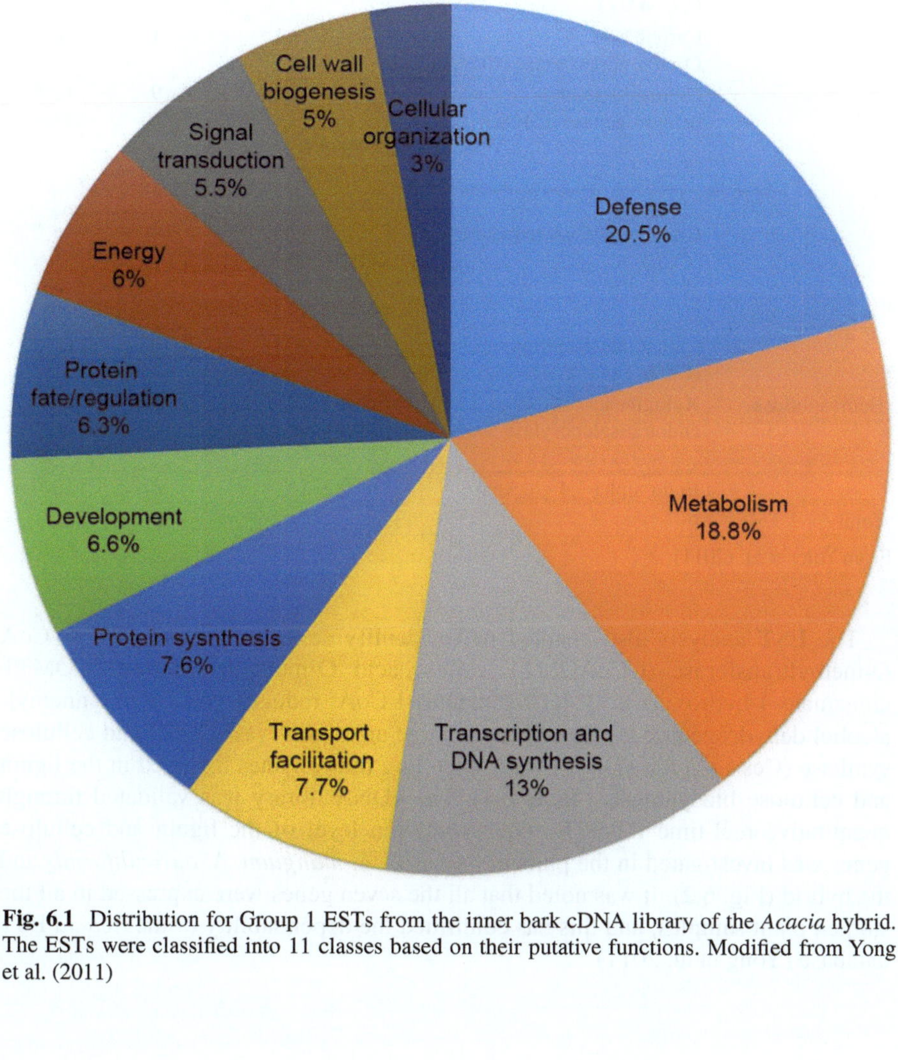

Fig. 6.1 Distribution for Group 1 ESTs from the inner bark cDNA library of the *Acacia* hybrid. The ESTs were classified into 11 classes based on their putative functions. Modified from Yong et al. (2011)

Table 6.1 The *Acacia* ESTs related to wood formation genes. E-value was obtained from BLASTX analysis

Component	Putative function	Accession number of EST	Number of ESTs	E-value
Cellulose	Cellulose synthase (*CesA*)	GR482640.1	1	$1e^{-28}$
Lignin	Phenylalanine ammonia-lyase (*PAL*)	GR482512.1, GR483108.1, GR483107.1, GR481639.1, GR481574.1	5	$2e^{-42}$
	Caffeoyl-CoA O-methyltransferase (trans-caffeoyl-CoA 3-O-methyltransferase) (*CCoAMT*)		1	$3e^{-14}$
	Caffeic acid O-methyltransferase (*COMT*)	GR482772.1	1	$7e^{-35}$
	Cinnamyl-alcohol dehydrogenase (*CAD*)	GR482657.1, GR482473, GR482003.1, GR481889.1, GR482389.1	9	$1e^{-50}$
	Cinnamoyl-CoA reductase (*CCR*)	GR482692.1, GR481799.1, GR482550.1, GR482839.1	4	$1e^{-49}$
	Cinnamate-4-hydroxylase (*C4H*)	GR482704.1, GR481724.1	4	$8e^{-47}$
Hemicellulose	Xyloglucan endotransglucosylase	GR481522, GR481593, GR480896	6	$3e^{-61}$ $6e^{-30}$
	Endo-alpha-1,4 glucanase	GR482784	1	$1e^{-28}$
Total			32	

From Yong et al. (2011)

The EST analysis also enabled us to identify seven unigenes, caffeoyl-CoA O-methyltransferase (CCoAOMT), caffeic acid O-methyltransferase (COMT), cinnamate-4-hydroxylase (C4H), cinnamoyl-CoA reductase (CCR), cinnamyl-alcohol dehydrogenase (CAD), phenylalanine ammonia-lyase (PAL) and cellulose synthase (CesA) in the cDNA library encoding the enzymes involved in the lignin and cellulose biosynthesis (Table 6.1). The cDNA library was validated through quantitative real-time RT-PCR. The expression level of the lignin and cellulose genes was investigated in the phloem tissues of *A. mangium*, *A. auriculiformis* and the hybrid (Fig. 6.2). It was noted that all the seven genes were expressed in all the studied plant samples, and this has confirmed the dependability of the *Acacia* EST database (Yong et al. 2011).

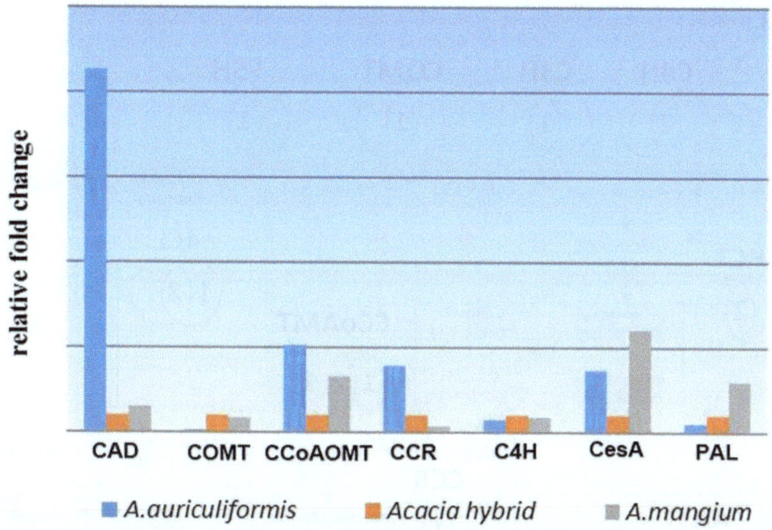

Fig. 6.2 Relative expression levels of seven wood formation genes in the inner bark tissue of *A. mangium*, *A. auriculiformis* and the hybrid determined using RTqPCR. CAD: cinnamyl-alcohol dehydrogenase, COMT: caffeic acid O-methyltransferase, CCoAOMT: caffeoyl-CoA O-methyltransferase, CCR: cinnamoyl-CoA reductase, C4H: cinnamate-4-hydroxylase, PAL:phenylalanine ammonia-lyase and CesA: cellulose synthase. Modified from Yong et al. (2011)

6.3 Transcriptome Sequencing

The general monolignol biosynthesis and wood-related pathways are well studied and characterized. The genes in these pathways, however, are poorly characterized in the *Acacia* species. Currently, RNA-seq on short-read platform has frequently been used as a rapid approach to generate comprehensive transcriptome data for informative sequence discovery. We sequenced the transcriptomes of *A. mangium* and *A. auriculiformis* from non-normalized cDNA libraries of pooled inner bark tissues using paired-end libraries and an Illumina GAII machine (Wong et al. 2011). The de novo assembly analysis on the transcriptome sequences produced a total of 35,759 and 42,217 contigs with an average 498 and 496 bp in length for *A. mangium* and *A. auriculiformis*, respectively. The de novo assemblies on the *A. mangium* and *A. auriculiformis* transcriptome data had a total length of 17,838,260 and 21,022,649 bp, respectively. The largest contig was 15,262 bp in length. All the ten monolignol biosynthesis pathway genes were detected in the transcriptome data using BLASTX (Basic Local Alignment Search Tool for Protein). Further analysis revealed 21 and 19 isoforms of lignin genes from *A. mangium* and *A. auriculiformis* (Fig. 6.3).

A few main classes of transcription factors involved in the lignin and wood formation were also identified in both the *Acacia* species (Wong et al. 2011). Basically, these transcription factors were present in similar number of contigs, though the

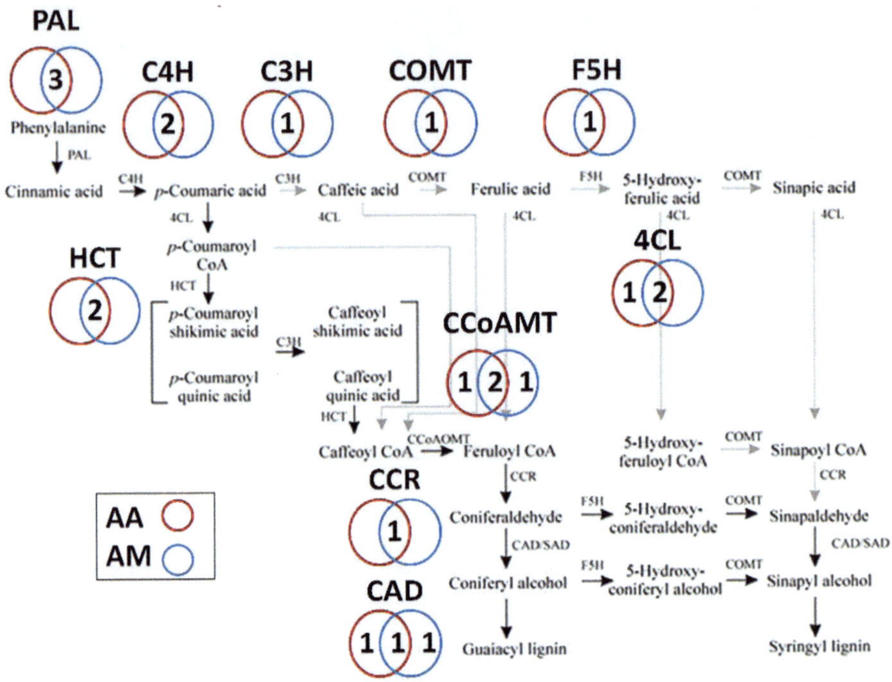

Fig. 6.3 Monolignol biosynthesis pathway isoforms of *A. mangium* and *A. auriculiformis*. The number of isoforms found in *A. auriculiformis* (in *red circle*) and *A. mangium* (in *blue circle*) based on BLASTX (E-value ≤ $1E^{-10}$) and conserved motifs. The number of orthologous isoforms shared by both species is indicated in the overlapping region. *Phenylalanine ammonia lyase* (PAL), *Cinammate 4-hydroxylase* (C4H),*4-coumarate 3-hydroxylase* (C3H), *caffeic acid O-methyltransferase*(COMT), Ferulate 5-hydroxylase (F5H), 4-coumarate:CoA ligase (4CL), *hydroxycinnamoyl-CoA shikimate/quinatehydroxy cinnamoyltransferase* (HCT), *caffeoyl CoA 3-O-methyltransferase* (CCoAOMT), *cinnamyl alcohol dehydrogenase* (CAD) and *cinnamoyl Co-A reductase* (CCR). From Wong et al. (2011)

NAC (NAM, ATAF and CUC) family was found to be substantially more abundant in *A. auriculiformis* compared to *A. mangium*. Furthermore, nine *A. mangium* and eight *A. auriculiformis* contigs were identified as class III HD-ZIP (homeodomain-leucine zipper), a member of homeobox (HB) family. Besides, some members of the R2R3-MYB (R2R3 myeloblastosis) family were found to be involved in the control of lignin deposition and secondary wall formation by interacting with other MYB genes, activated by NAC transcription factor master switches and AC element binding (Perez-Rodriguez et al. 2010). The AC element is a cis-acting element, and is normally found in most promoters of the genes involved in monolignol biosynthesis. We managed to identify five contigs, three in *A. mangium* and two in *A. auriculiformis*, which are homologous to MYB regulating wood-related pathways in other plants. Apart from MYB, the NtLIM1 (tobacco LIM protein) in tobacco has also been proven to bind AC elements, and its inhibition has reduced lignin content (Zhong and Ye 2009). One of the *A. mangium* contigs obtained was homologous to the NtLIM1 with identity more than 80%. We carried out phylogenetic analysis on the predicted MYB proteins with wood-related MYBs of other plants, and it revealed that all these MYBs fall into three distinct groups (Wong et al. 2011). One *A. mangium* MYB protein belonging to group one was clustered together with AmMYB308, EgMYB1, ZmMYB31 and ZmMYB42, indicating an important role of MYB in regulating the monolignol biosynthesis pathway. EgMYB1 is found to bind AC element and repress the monolignol biosynthesis pathway (Kawaoka et al. 2000). AmMYB308, ZmMYB31 and ZmMYB42 were shown to have effect on lignin contents by regulating the expression of lignin genes (Newman et al. 2004; Bomal et al. 2008). Two *A. auriculiformis* and one *A. mangium* proteins are homologues of the *Arabidopsis* group two MYBs, namely MYB61 and Pine MYB8. Pine MYB8 is homologous to MYB61, and its overexpression has caused ectopic lignin deposition, though its exact function is not clear (Zhong et al. 2007, 2008). Only one *A. mangium* MYB protein belongs to group three which is homologous to MYB20 and MYB43 of *Arabidopsis*. It has been shown that NAC master switches activate MYB20, MYB42 and MYB43 to regulate downstream MYB proteins in wood-related pathways (Patzlaff et al. 2003). On the other hand, MYB85 has shown to induce secondary wall biosynthetic genes (Legay et al. 2010). Pine MYB1 is another member of this group with the ability to bind AC elements (Tamagnone et al. 1998) and is involved in secondary cell wall deposition (Zhong et al. 2008).

We have produced the first comprehensive de novo assembly transcriptome-wide analyses for *A. auriculiformis* and *A. mangium*. The comprehensive and high quality assemblies of the *Acacia* transcriptomic data have enabled us to identify the major genes involved in the lignin biosynthesis and wood-related pathways in the *Acacia* species. We have demonstrated that transcriptome sequencing is a cost-effective approach for gene discovery, regulatory sequence identification, and informative marker development in a non-model woody tree (Wong et al. 2011).

6.4 MicroRNA Analysis

Lignin after cellulose is the second most abundant biopolymer in wood, amounting 15–35% of the wood dry weight. It is an integral component of wood, and is essential for plant structure and defence. However, lignin is a highly undesirable component in the pulping process as it reduces the yield and increases the cost of pulp production. High lignin level could also limit the carbon availability for cell division and growth (Kirst et al. 2004). This is due to a higher carbohydrate consumption during lignin synthesis which consequently may have caused a slower growth rate (Kirst et al. 2004). The effect of low lignin on the growth of aspen and loblolly pine has been well established based on evidences from several model plant species (Hu et al. 1999; Wu et al. 1999; Li et al. 2003; Kirst et al. 2004). Similar trend has also been demonstrated in our study on *A. mangium*. We recorded that *A. mangium* tree (accession Am54) with a larger diameter at breast high (DBH) of 58.0 cm had a lower lignin content (21%) whereas tree (accession Am48) with smaller DBH (35.5 cm) had a higher lignin content (41%). Therefore, the amount of lignin may have directly influenced the growth of *Acacia* plant. Subsequently, we generated two small RNA libraries using Illumina GAII sequencing from these two *A. mangium* accessions (Am48 and Am54) with contrasting lignin content (Ong and Wickneswari 2011). A total of 10 million short sequence reads were obtained from the secondary xylem of Am48 (high lignin content), and 14 million sequence reads from Am54 (low lignin content). It was observed that the small RNAs of *A. mangium* consisted of a set of 12 highly conserved miRNA families, 82 novel miRNAs and a large proportion of non-conserved small RNAs with low expression level (Ong and Wickneswari 2011). Further analysis revealed that the predicted target genes of those conserved and non-conserved miRNAs with differential expression are of transcription factors associated with the gene regulation in the pathway of lignin biosynthesis. It is believed that some of these small RNAs are involved in epigenetic silencing. Majority of the genes up-regulated in xylem tissue are involved in lignin biosynthesis and secondary wall formation while those genes up-regulated in bark tissue have broader functions (Prassinos et al. 2005). Differential expression analysis of the small RNAs in secondary xylem tissues with contrasting lignin content (accessions Am48 and Am54) suggests that a miRNA cascade plays an interconnected role in regulating the lignin biosynthetic pathway in the *Acacia* species.

In an effort to unravel the role of various miRNA as the potential regulatory sequences of secondary cell wall biosynthesis in *A. mangium*, the differential expression of six highly conserved miRNA families (amg-miR156, amg-miR159, amg-miR166, amg-miR168, amg-miR172 and amg-miR394) were investigated (Ong and Wickneswari 2012). Among the highly conserved miRNA families only amg-miR166 was observed with a strong differential expression between xylem and phloem tissues (Ong and Wickneswari 2012). The functional characterization of amg-miR166 targeting various tissues had enabled us to reveal four gene families [ATHB8 (*Arabidopsis* homeobox), ATHB15, HD-ZIP III and REVOLUTA] involved in the xylem development. Apparently, these four gene families vary in their functions. In the –psRNA target analysis, however, the nine different members of HD-ZIP III had conserved miRNA target sequences. It was observed that the precursor structure of amg-miRNA166 had undergone exhaustive sequence variation, and this can be seen even within members of the same family (Ong and Wickneswari 2012). Gene expression analysis also showed that three genes in lignin biosynthesis pathways (CAD, C4H and CCoAOMT) were up-regulated in compression wood whereas a miRNA cascade was down-regulated (Fig. 6.4). It is most likely that an evolutionary mechanism for miRNA has targeted on the length of their 5′ and 3′ UTRs (untranslated regions) and their cellular role. These findings can be effectively used to further understand the small RNA roles in lignin biosynthesis, and this information is useful to develop gene constructs for silencing genes involved in monolignol biosynthesis without affecting much the fitness and viability of plants.

6.5 Development of DNA Markers for *Acacia*

The development of polymerase chain reaction (PCR) based DNA markers such as simple sequence repeat (SSR), single nucleotide polymorphism (SNP) and cleaved amplified polymorphic sequence (CAPS) has shortened the time needed to generate

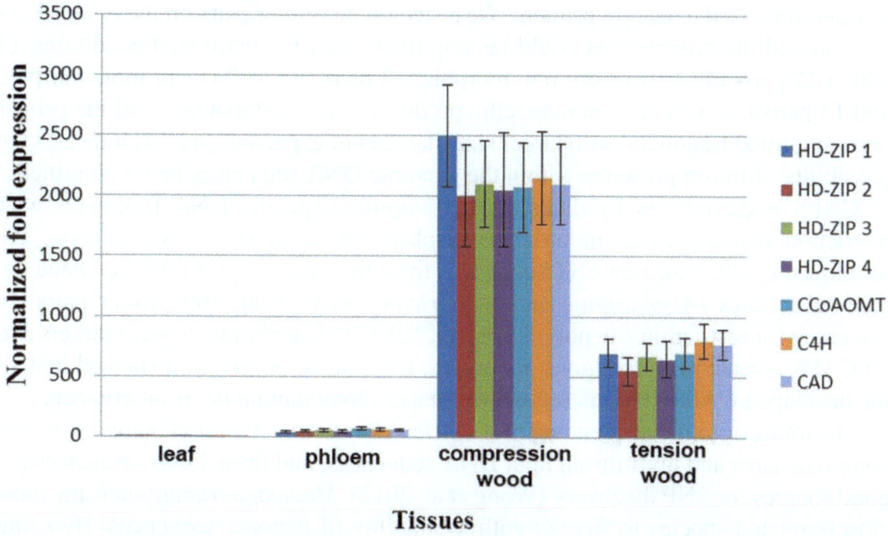

Fig. 6.4 Expression of various HD-ZIP III transcription factors and lignin genes in various tissues of *A. mangium* through RT-qPCR analysis. Actin gene was used to normalize cDNA amounts from various tissues. The data presented are average of three different trees with three replicates for each tree. Modified from Ong and Wickneswari (2012)

genetic linkage maps (Marques et al. 1998). The EST database of the *Acacia* hybrid (Yong et al. 2011) and the transcriptomic data of *A. mangium* and *A. auriculiformis* (Wong et al. 2011) are good sources and references for the development of DNA markers. We have made use of the *Acacia* genomics information to develop SSR, CAPS and SNP markers for the *Acacia* species.

SSR markers derived from EST sequences are normally found to be viable source of polymorphism and they are highly transferable among closely related species. The EST database from the *Acacia* hybrid was searched for SSR sequences. From 6415 non-redundant EST sequences, 1042 (16.42%) ESTs contained SSR motifs. These results clearly demonstrated that ESTs are a valuable source for mining and harvesting SSR sequences. Non-redundant ESTs are unique sequences; they are expected to provide more informative SSR markers for the mapping of non-redundant genes (or unigenes). We found no mononucleotide repeats in the EST database, and hexanucleotide repeats were the least abundant. This result may be due to variation in the quality of sequence data being analysed and the different criteria used for mining SSR motifs in the EST databases. For every class of SSR the frequency of SSR decreases with increasing repeat length. The heptanucleotide repeats found had insufficient length to be of use. The SSRs identified from the EST database composed mainly of di, tri, tetra, hexa and penta nucleotide with decreasing proportions of 47%, 40%, 8%, 4% and 1%, respectively. We found that poly TA was the most common among the dinucleotide repeats, while poly AAG was the most common among trinucleotide repeats. Only 143 SSRs from the screening of the EST database fulfil the criteria for use as molecular markers. Of these, 126 potential SSRs with di- and trinucleotide repeats were used to design primers. We tested on 53 primer pairs on the *Acacia* species, and all the primer pairs could be amplified under the optimized conditions, of which 25 pairs (47.17%) were polymorphic, 13 pairs (24.53%) were monomorphic and 15 pairs (28.3%) gave non-specific products. We noted that some of the primer pairs amplified fragments with larger size than that of expected, and this indicates the possibility of intron presence within the genomic DNA sequences being amplified.

CAPS is carried out by digesting the amplified specific DNA fragments with restriction endonucleases in order to display restriction site polymorphism. We developed CAPS based on EST sequences from the *Acacia* hybrid. We screened the parental species (*A. mangium* and *A. auriculiformis*) using 309 primer pairs. In screening for restriction site polymorphism, 128 CAPS are homozygous markers and 29 CAPS revealed heterozygous genotypes. Qualitative observation showed that all the developed CAPS markers segregated in a co-dominant mode of inheritance.

The transcriptomic sequencing in *A. auriculiformis* and *A. mangium* has provided comprehensive and high-throughput DNA sequences, and these DNA sequences are good sources for SNP discovery (Wong et al. 2012). *Medicago truncatula* is the most closely related species to *Acacia* with availability of genome sequences. By using *Medicago truncatula* as a genome reference, most of the *Acacia* SNPs were converted to successful assays. A total of 37,786 SNPs were identified through the alignment of 7839 de novo transcriptome contigs from the parents of the *Acacia* hybrid (Wong et al. 2012). This assessment revealed that the SNP assay failure was mainly caused by the presence of exon-intron boundary within the PCR amplified segment. Nevertheless, further SNP filtering and improvement in assay design had yielded a higher assay success rate (92.4%) and conversion rate (57.4%) based on the genotyping of 768 SNPs. Finally, we identified 258 and 319 polymorphic SNPs in the *A. auriculiformis* and *A. mangium* germplasms, respectively.

6.6 Development of Mapping Populations

Acacia has displayed an excellent growth for pulp yield. However, the plant is susceptible to wilt and root rot diseases (Potter et al. 2006). *A. mangium* is found to be highly susceptible to heart rot disease. On the other hand, *A. auriculiformis* is not suitable for timber production due to crooked-twisted trunk (Kojima et al. 2009). The *Acacia* hybrid has been proven to be more superior than its parents, but it still carries inferior traits from the parents—high lignin content, poor stem form, low wood density and short fibre length. Besides, the production of viable seeds and germination of seeds from natural or controlled pollination are always problematic (Kijkar 1992). Breeding programme to produce elite *Acacia* planting materials with better traits such as fast growth, low lignin content, high wood density, straight stem form, disease resistance, adaptability to different soil types, and high pulp quality is important for the sustainability of pulp and paper industry (Hardiyanto 2014). Conventional breeding in forest species is always hampered by long life cycle, the difficulty in hybridization and large land requirement for progeny evaluation. Undoubtedly, MAS breeding is a promising approach for selecting superior (better wood and pulp quality) materials for planting. Development of DNA markers based on the parents of a mapping population is an effective way to obtain enough informative markers to establish a linkage mapping of the plant. Therefore, a first step towards MAS is to construct a genetic linkage map based on a segregating family for the desired traits.

We have established two *Acacia* mapping populations (one is for wood density and the other one is for fibre length) for linkage mapping and QTL analysis. To develop the mapping populations, parents (*A. mangium* and *A. auriculiformis*) were selected based on contrasting traits for wood density and fibre length. Flower phenological studies revealed that the flowering period in *A. mangium* pollen donor tree used for wood density cross was 3–4 months more than the *A. mangium* tree used in fibre length cross. Likewise, the flowering period for *A. auriculiformis* female tree in the wood density cross was more than *A. auriculiformis* female tree in the fibre length cross. *A. auriculiformis* female tree in wood density cross produced flowers almost throughout the whole year compared to *A. auriculiformis* female tree of fibre length cross which only produced flowers in certain months of a year. Furthermore, staminate flowers were found in fibre length *A. auriculiformis* female, which reduced the number of stigmas available for pollination work. However, both the species were synchronous in flowering time which facilitated the pollination procedure. Pollinations carried out using emasculation resulted into 2.7–3.2% pods. Some of the crosses from both methods appeared to be successful at the early stage, but after several days floral buds shrivelled and fell off. This could be due to pollen viability and incompatibility or environmental factors such as extreme changes in the weather condition and floral damage due to insects. A total of 90 pods were harvested from the trees in the wood density cross. In case of fibre length cross, a total of 52 pods were harvested. The number of seeds per pod ranged from 1 to 12 respectively with an average of four seeds per pod. All harvested pods were used for clonal propagation.

All the seeds in a pod were germinated to produce enough progenies because the density and quality of a linkage map is affected by the mapping population size. Controlled pollinations between female (*A. auriculiformis*) and male (*A. mangium*) parents generated 174 F_1 seedlings for wood density cross whereas 214 F_1 seedlings were generated for fibre length cross. The number of F_1 progenies produced for wood density cross was higher than the fibre length cross because the availability of flowers in the donor tree for the wood density cross was more promising compared to the donor tree in the fibre length cross. *Acacia* is a cross pollinated plant species, therefore seed progenies were screened by isozyme markers to select true hybrid and to find pollen contamination using SSR and SNP markers (Asif et al. 2015; Asif et al. 2017). Non-hybrid progenies detected were discarded and pure hybrids were further multiplied through shoots of the progenies. Progenies from the two mapping populations were planted in the field for phenotypic and genotypic evaluations in Peninsular Malaysia and East Malaysia, i.e., (a) Forest Research Institute Malaysia field station at Segamat, Johor; (b) Asiaprima RCF Sdn. Bhd. field trial site at Lancang, Pahang and (c) Borneo Tree Seeds and Seedlings Supplies Sdn, Bhd. field trial site at Bintulu, Sarawak.

6.7 Linkage Mapping

We intended to do linkage mapping on the two *Acacia* mapping populations (one is for wood density and the other one is for fibre length) using the SNP markers. Due to the small number of SNP markers (only 108) that were available for fibre length cross, linkage mapping for the fibre length population was terminated. Before performing linkage analysis, genotype data for all individuals was screened for

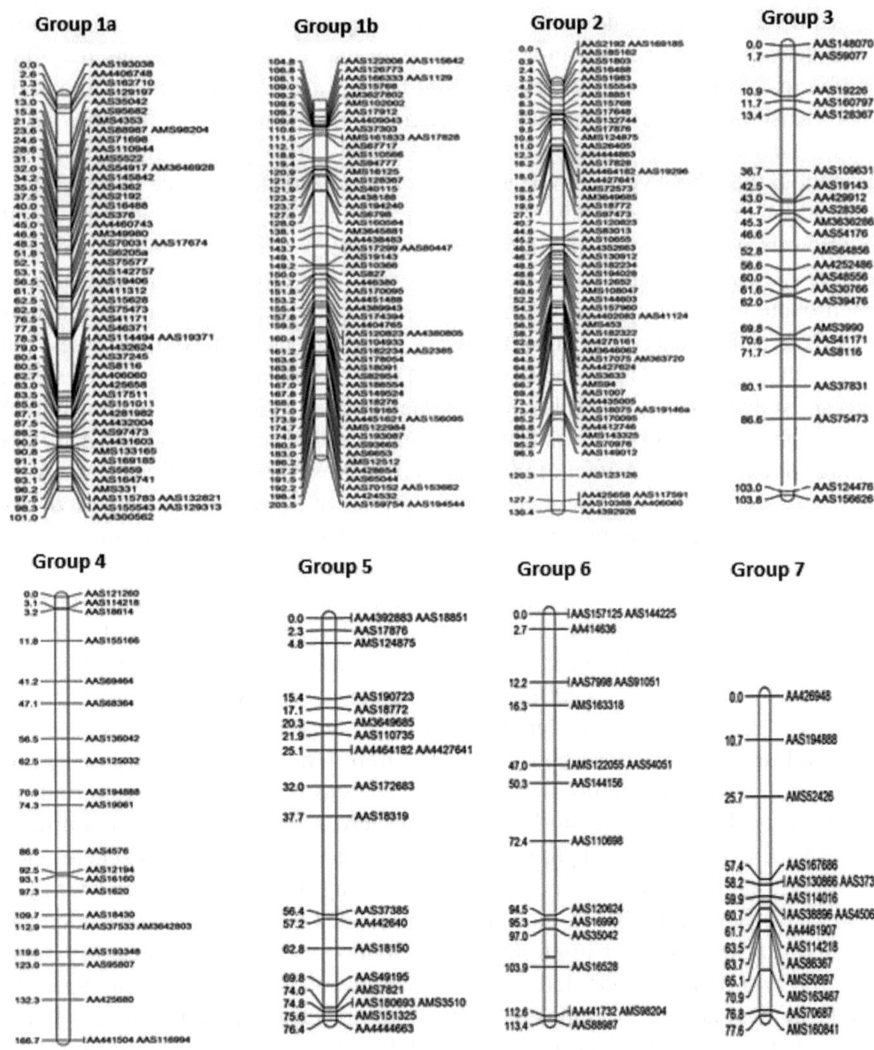

Fig. 6.5 Linkage map for *A. auriculiformis* × *A. mangium* wood density cross

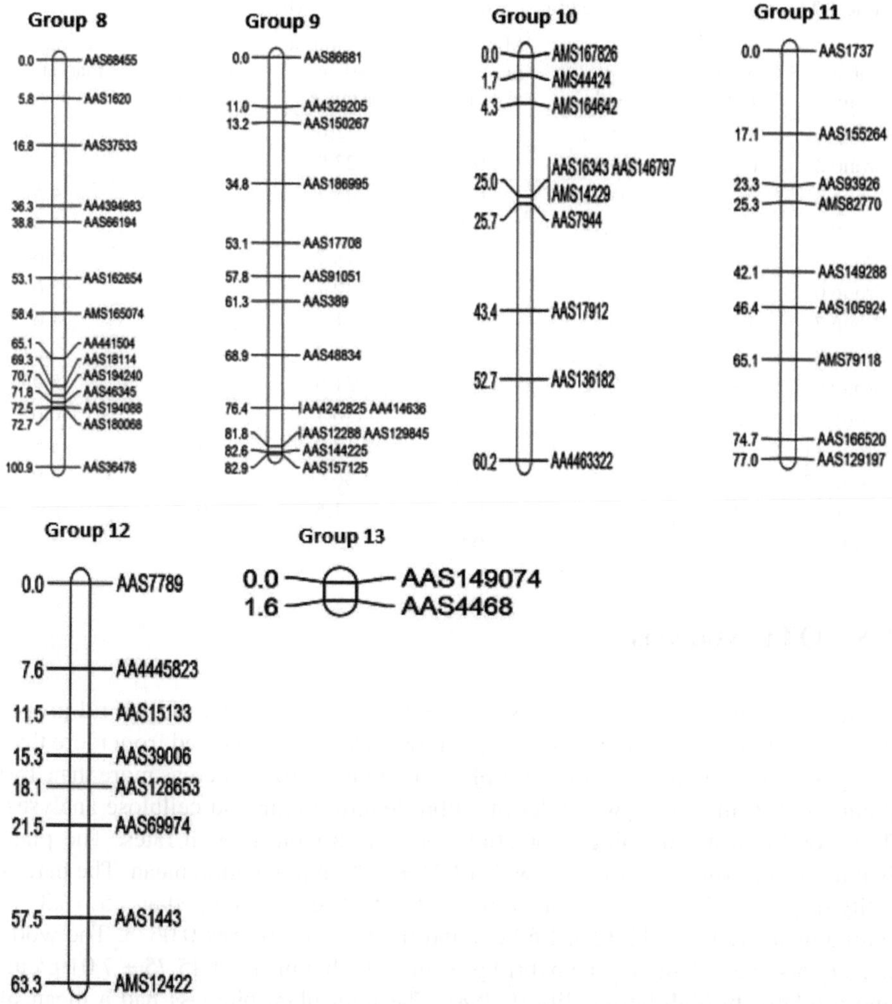

Fig. 6.5 (continued)

heterozygosity, similarity and segregation distortion. Markers exhibiting highly significant segregation distortion were excluded from the linkage mapping. A total of 396 demonstrably heterozygous SNP markers were available to establish the linkage map for wood density. After linkage analysis, 333 of these SNP markers were successfully mapped onto 13 linkage groups corresponding to the *Acacia* haploid genome ($2n = 26$; $n = 13$) (Fig. 6.5). The map spanned 1205.82 cM with average inter marker distance of 3.62 cM (Table 6.2). The genetic length of the linkage groups ranged from 1.6 (linkage group 13) to 203.48 cM (linkage group 1), with an average of 3.62 cM. Linkage group 1 was the most saturated, having 117 markers with an average marker density of 1.74 cM. On the other hand, the linkage group 13 had the least number of markers (2).

Table 6.2 Wood density integrated linkage map of the *Acacia* hybrid

Linkage group	Number of markers	Map length	Minimum marker distance	Maximum marker distance	% Marker distortion	Average marker distance
Group 1	117	203.48	0.1	13.6	21.37	1.74
Group 2	60	130.35	0.1	23.8	56.67	2.17
Group 3	23	103.84	0.4	23.3	8.69	4.51
Group 4	22	166.74	0.1	34.4	40.90	7.58
Group 5	21	76.4	0.8	18.7	0	3.64
Group 6	17	113.37	0.8	30.7	17.65	6.67
Group 7	16	25.72	0.2	31.7	93.75	1.61
Group 8	14	100.9	0.2	28.2	21.43	7.21
Group 9	14	82.92	0.3	21.6	0	5.93
Group 10	10	60.2	0.7	20.7	20	6.02
Group 11	9	77	2	18.7	55.56	8.56
Group 12	8	63.3	2.8	36	12.5	7.9
Group 13	2	1.6	0	1.6	0	0.8
Total	333	1205.82	0.66	23.31	26.81	3.62

6.8 QTL Analysis

Phenotypic data and wood samples were collected from 3-year-old hybrid progenies from both the *Acacia* mapping populations. Data was collected from more than 500 plants from both populations for plant height and DBH whereas more than 150 plants were sampled for wood density, fibre length, lignin and cellulose analyses. The wood density mapping population showed variable growth rates. The plant height ranged from 6.2 to 18.3 m with 12.21 ± 1.79 m population mean. The heritability of plant height was estimated to be 0.0655. The DBH ranged at 3.5–16.3 cm with a mean value of 11.32 ± 1.69 cm and the heritability was 0.0625. The wood density also varied among the hybrid progenies with a mean of 45.75 ± 7.01 g/cm^3 and a relatively high heritability (0.288). The total plant biomass had a mean of 180.36 ± 11.05 Kg/tree with a relatively low heritability (0.045). All the four traits were normally distributed and suitable for QTL analysis.

Analysis of variance was performed for DBH and plant height to test the significance of variation among the hybrid progenies. Both traits were significantly different in both populations suggesting suitability of the traits for QTL analysis. Both traits were positively correlated with a high correlation (r = 0.86). All the components of the QTL analysis were performed with the MapQTL 6.0. The detection of QTLs was done based on a non-parametric method using Kruskal-Wallis (KW) test. The KW test ranks all individuals according to the quantitative traits, and it classifies the traits by marker genotype. A segregating QTL (with big effect) is normally linked closely to the tested markers, and this results in large differences in the average rank of marker genotype classes. A test statistic based on the ranks in the marker genotype classes is calculated and indicated by "K" in the KW test statistics

(Lehmann 1975). The test is normally performed on many linked and unlinked loci. Therefore, the use of a stringent significance level for the individual tests is adopted so that an overall significance level at 0.05 (K*) is obtained. However, the best will be at a level of at least 0.005 (K**).

Markers with a *P*-value <0.05 in KW test are taken as QTL candidates (Table 6.3). Using this method for DBH, 30 QTLs in wood density mapping population across the 13 linkage groups were detected. Similarly for plant height a total of 26 QTLs for wood density population were detected. For wood density, QTLs for DBH were detected on linkage groups 1, 4, 6, 7, 9 10 and 11. However, majority of the QTLs were present on linkage groups 1 (14 QTLs), 6 (6 QTLs) and 11 (4 QTLs). It is noted that the 14 QTLs on linkage group 1 were present in close proximity to each other. The other 10 QTLs were present in less than 1 cM distance apart. Similarly in case of linkage group 10, five QTLs were present in less than 1 cM. In case of linkage group 11, all four QTLs were present in a close proximity at a distance of less

Table 6.3 QTL analysis for four main traits

Trait	Linkage group	Locus	Position	K	Degree of freedom	Significance level
Plant height	1	m_AAC438876_1	130.731	4.397	1	b
	4	m_AAC444886_MtS_1	0.0	4.817	1	b
	4	m_AAC438188	11.753	4.96	1	b
	4	m_AAS10050-Mt_1a	20.686	4.381	1	b
	4	m_AMS1251-Mt2_1	149.657	4.568	1	b
	4	m_AAC428654_1	151.302	5.974	1	b
DBH	5	m_AAS9580-Mt7_1	43.823	4.214	1	b
	5	B_AAS1620	68.822	5.118	1	b
	10	m_AAS8668-Mt_1	0.0	4.213	1	b
Wood density	1	b_AAS1707-Mt5	28.063	3.883	1	b
	1	m_AAC444030-Mt5_2	28.063	3.883	1	b
	1	b_AAS4117-Mt1	43.871	4.358	1	b
	1	m_AAS679-Mt8_4	68.247	4.02	1	b
	1	b_AAS19424-Mt0	72.067	5.118	1	b
	1	m_AAC445148-Mt8_1	86.649	5.883	1	b
	1	m_AAS827_2	89.919	5.883	1	b
	1	m_AAS10366_1	90.717	5.523	1	b
	1	b_AAS19143	90.747	5.883	1	b
Biomass	5	b_AAS1620	68.822	6.86	1	c
	5	m_AAS16160_1	72.965	3.704	1	a
	5	m_AAS12194_1a	73.57	3.704	1	a
	10	m_AAS8668-Mt1_1	0.0	3.948	1	b
	10	m_AAC432920-Mt5_1	11.119	2.827	1	a

[a]Indicates significance level 0.05
[b]Indicates significance level 0.005
[c]Indicates significance level 0.001
K is Kruskal-Wallis test statistics

than 1 cM. The QTLs detected shown to contribute 7–8% of DBH total phenotypic standard deviation. A range of 7–14% of phenotypic variance had been reported for QTLs related to different traits in the forest species *Populus* (Pakull et al. 2009).

The highest K value (6.4**) was observed for a marker present on linkage group 1 at 9.183 cM map position followed by 5.7** at linkage group 7 (77.396 cM), and two significant markers with similar K value of 5.2** were present in linkage group 1 at 28.063 cM. The QTLs associated with the highest K value showed homology with genes from *M. truncatula* genome i.e., amino transferase family III, glucan endo-1,3-beta-glucosidase and chalcone reductase. BLAST search for plausible function of these genes revealed that these QTLs were related to plant growth, stress response, disease resistance and growth hormones. The class III amino transferase family is an important subfamily which has multiple genes randomly localized across the plant genome. Amino transferases are pyridoxal 5′-phosphate-dependent enzymes, and they are involved in development and plant growth, and response to abiotic stresses (Sun et al. 2014). On the other hand, glucan endo-1,3-beta-glucosidase belongs to glycosyl hydrolases family 17, and it is involved in carbohydrate metabolic process through hydrolase activity by hydrolyzing O-glycosyl compounds. This enzyme is believed to be an important plant defence-related substance against fungal pathogens. Chalcone synthase or naringenin-chalcone synthase is an enzyme ubiquitous to higher plants, and it belongs to a family of polyketide synthase enzymes (PKS) known as type III PKS. The type III PKS is associated with the chalcone production, and chalcone is a class of organic compounds for natural defence mechanisms and also as synthetic intermediates. Another gene villin-4 associated with DBH in *Acacia* is found to be involved in root hair growth in *Arabidopsis* (Zhang et al. 2011). CCR4-NOT (carbon catabolite-repression 4-Not) transcription complex subunit 1 belongs to the CCR4-NOT complex, and it is an evolutionarily conserved protein complex involved in the mRNA decay and transcriptional control. These genes in rice have shown to be related with growth and development and in nutrient responses (Cai et al. 2011).

For plant height trait, 26 significant QTLs for wood density population were identified. The K values ranged from 2.733 to 12.145 across six linkage groups. Two major QTL blocks were detected on linkage groups 1 and 4, and the QTLs with high K value were located on linkage group 4. Three major QTLs located on linkage group 1 were related to genes involved in protein synthesis and protein folding. Plant height QTLs on linkage group 4 had K values ranging from 2.87 to 12.15. These QTLs were present within a small block ranging from 47.36 to 116.19 cM. A major QTL with K value of 12.15 is related to molybdenum cofactor sulfurase gene homologue from *M. truncatula*. This gene is essential for plant nitrogen assimilation, phytohormone synthesis, purine degradation and sulphite detoxification. Two more major genes, sucrose-phosphate synthase and cellulose synthase, were mapped on linkage group 4. Sucrose-phosphate synthase a gene reported in rice controlling height by accumulating high level of sucrose in leaves compared to control imparting taller plants. Similarly, cellulose synthase gene which is part of starch pathway has been reported to be related to plant growth.

A significant positive correlation was observed between DBH and plant height. The two QTLs (one on linkage group 4 and another one on linkage group 9) affecting both the traits were successfully mapped onto two same genomic regions. The presence of the QTLs for both DBH and plant height detected in the same genomic regions may be due to the tight linkage of two genes or to one gene with a pleiotropic effect on these two traits.

6.9 Conclusions and Future Prospects

The genomics information generated from the EST analysis, microRNA analysis and transcriptome sequencing in the *Acacia* species is enormous. This allows us to understand better the lignin biosynthesis and wood formation in the *Acacia* species. At the same time, it also shows that the wood formation is a complicated process involving regulation of various factors and elements. The DNA sequences generated from these *Acacia* genomic studies have enabled us to develop various types of DNA markers (SSR, CAPS and SNP) for many uses including linkage mapping and QTL analysis. It is no doubt that the linkage map and QTLs established is an important step to accelerate the molecular breeding of the *Acacia* hybrid.

The availability of this first composite genetic linkage map and its effective utilization through the identification of QTLs for various important traits have proven that there is future potential of these studies to find genes controlling more traits of economic importance in the *Acacia* species. Nevertheless, we identified most of the traits studied have low heritability and the QTLs only explained small magnitude of genetic variance. Thus, environment seems to play an important role in determining these traits in the *Acacia* species. Therefore, this has opened up opportunities and challenges for future research in *Acacia*. We think that there is a need to verify these detected QTLs in an independent population. At the same time identifying and validating the functions of the genes in the detected QTLs in relation to the phenotypic traits are important. The stability of QTLs should also be tested under different environmental conditions. The true heritability of phenotypic traits could be determined through large progeny sets in multi-environmental trials (genotype × environment). We also feel that more wood-related traits such as lignin and cellulose contents, yield and disease resistance should be screened. The understanding of the molecular basis of economically important traits can help in manipulation of different genes controlling these traits in order to breed new desirable genotypes, particularly those which are disease resistant and can withstand changing climatic conditions. Ultimately, there is a need to evaluate the feasibility of molecular breeding in commercial *Acacia* genetic improvement programmes.

References

Asif MJ, Ariffin MAT, Yit HM, Wong M, Abdullah MZ, Muhammad N, Wickneswari R (2015) Utilization of STMS markers to verify admixture in clonal progenies of *Acacia* mapping populations and relabeling using assignment tests. J For Sci 61:200–209

Asif MJ, Mohd Zaki A, Norwati M, Wickneswari R (2017) Detecting mislabeling and identifying unique progeny in Acacia mapping population using SNP markers. J Forest Res doi:10.1007/s11676-017-0405-8s

Bomal C, Bedon F, Caron S, Mansfield SD, Levasseur C, Cooke JE, Blais S, Tremblay L, Morency MJ, Pavy N, Grima-Pettenati J, Seguin A, Mackay J (2008) Involvement of *Pinus taeda* MYB1 and MYB8 in phenylpropanoid metabolism and secondary cell wall biogenesis: a comparative in planta analysis. J Exp Bot 59(14):3925–3939

Butcher PA, Moran GF (2000) Genetic linkage mapping in *Acacia mangium*. 2. Development of an integrated map from two outbred pedigrees using RFLP and microsatellite loci. Theor Appl Genet 101:594–605

CAB International (2000) Forestry compendium global module. [CD-ROM]. CAB International, Wallingford

Cai H, Song Z, Lian X (2011) Sequence and expression analysis of CCR4-associated factor 1 (OsCAF1) gene family in rice. J Plant Mol Biol Biotechnol 2(1):47–64

Chiang VL, Puumal RJ, Takeuchi H, Eckert RE (1998) Comparison of softwood and hardwood kraft pulping. TAPPI J 71:173–176

Doran JC, Turnbull JW (1997) Australian trees and shrubs: species for land rehabilitation and farm planting in the tropics. ACIAR monograph no. 24. Australian Centre for International Agricultural Research, Canberra

Doyle JJ (1994) Phylogeny of the legume family: an approach to understanding the origins of nodulation. Annu Rev Ecol Syst 25:325–349

Hardiyanto EB (2014) Challenges for Acacia breeders. In: Sustaining the future of Acacia plantation forestry. Hue, pp. 2–4

Hu WJ, Harding SA, Lung J, Popko JL, Ralph J, Stokke DD, Tsai CJ, Chiang VL (1999) Repression of lignin biosynthesis promotes cellulose accumulation and growth in transgenic trees. Nat Biotechnol 17(8):808–812

Jahan MS, Rowshan S, Arjumand R (2008) Alkaline pulping and bleaching of *Acacia auriculiformis* grown in Bangladesh. Turk J Agric For 32:339–347

Kawaoka A, Kaothien P, Yoshida K, Endo S, Yamada K, Ebinuma H (2000) Functional analysis of tobacco LIM protein Ntlim1 involved in lignin biosynthesis. Plant J 22(4):289–301

Kijkar S (1992) Handbook. Vegetative propagation of *Acacia mangium* × *Acacia auriculiformis*. ASEAN-Canada Forest Tree Seed Centre, Saraburi, Thailand, pp 9–12

Kirst M, Myburg AA, de Leon JPG, Kirst ME, Scott J, Sederoff R (2004) Coordinated genetic regulation of growth and lignin revealed by quantitative trait locus analysis of cDNA microarray data in an interspecific backcross of *Eucalyptus*. Plant Physiol 135:2368–2378

Kojima M, Yamamoto H, Okumura K, Ojio Y, Yoshida M, Okuyama T, Ona T, Matsune K, Nakamura K, Ide Y, Marsoem N, Sahri MH, Hadi YS (2009) Effect of the lateral growth rate on wood properties in fast-growing hardwood species. J Wood Sci 55:417–424

Legay S, Sivadon P, Blervacq AS, Pavy N, Baghdady A, Tremblay L, Levasseur C, Ladouce N, Lapierre C, Seguin A, Hawkins S, Mackay J, Grima Pettenati J (2010) EgMYB1, an R2R3 MYB transcription factor from eucalyptus negatively regulates secondary cell wall formation in *Arabidopsis* and poplar. New Phytol 188(3):774–786

Lehmann EL (1975) Nonparameterics. McGraw-Hill, New York

Li L, Zhou YH, Cheng XF, Sun JY, Marita JM, Ralph J, Chiang VL (2003) Combinatorial modification of multiple lignin traits in trees through multigene cotransformation. Proc Natl Acad Sci U S A 100:4939–4944

Marques CM, Araujo JA, Ferreira JG, Whetten R, O'Malley DM, Liu BH, Sederoff R (1998) AFLP genetic maps of *Eucalyptus globulus* and *E. tereticornis*. Theor Appl Genet 96:727–737

Maslin BR (2001) Introduction to *Acacia* flora of Australia. A Mimosaceae, *Acacia*. Part 1 Melbourne: Australia biological re-sources study (ABRS)/commonwealth scientific and industrial research organization (CSIRO). ACIAR, Canberra

Newman LJ, Perazza DE, Juda L, Campbell MM (2004) Involvement of the R2R3-MYB, AtMYB61, in the ectopic lignification and dark-photomorphogenic components of the det3 mutant phenotype. Plant J 37(2):239–250

Ng CH, Lee SL, Ng KKS, Muhammad N, Ratnam W (2009) Mating system and seed variation of *Acacia* hybrid (*A. mangium x A. auriculiformis*). J Genet 88:25–31

Ong SS, Wickneswari R (2011) Expression profile of small RNAs in *Acacia mangium* secondary xylem tissue with contrasting lignin content – potential regulatory sequences in monolignol biosynthesis pathway. BMC Genomics 12(Suppl 3):S13

Ong SS, Wickneswari R (2012) Characterization of microRNA expressed during secondary wall biosynthesis in *Acacia mangium*. PLoS One 7(11):e49662

Pakull B, Groppe K, Meyer M, Markussen T, Fladung M (2009) Genetic linkage mapping in aspen (*Populus tremula* L. and *Populus tremuloides* Michx.) Tree Genet Genomes 5:505–515

Patzlaff A, Newman LJ, Dubos C, Whetten RW, Smith C, McInnis S, Bevan MW, Sederoff RR, Campbell MM (2003) Characterisation of Pt MYB1, an R2R3-MYB from pine xylem. Plant Mol Biol 53(4):597–608

Perez-Rodriguez P, Riano-Pachon DM, Correa LG, Rensing SA, Kersten B, Mueller-Roeber B (2010) PlnTFDB: updated content and new features of the plant transcription factor database. Nucleic Acids Res 38(suppl_1):D822–D827

Potter K, Rimbawanto A, Beadle C (2006) Heart rot and root rot in tropical *Acacia* plantations. In: Proceedings of a workshop, Yogyakarta, Feb 7–9, pp 124–127

Prassinos C, Ko JH, Yang J, Han KH (2005) Transcriptome profiling of vertical stems segments provides insights into the genetic regulation of secondary growth in hybrid aspen trees. Plant Cell Physiol 46(8):1213–1225

Ross JH (1981) An analysis of the African *Acacia* species: their distribution, possible origins and relationships. Bothalia 13:389–413

Sun YJ, Zhu L, Guo JS, Zhou C (2014) Heterologous expression of EsABA1 enhances salt tolerance with increased accumulation of endogenous ABA in transgenic tobacco. Turk J Bot 38:1067–1079

Tamagnone L, Merida A, Parr A, Mackay S, Culianez-Macia FA, Roberts K, Martin C (1998) The AmMYB308 and AmMYB330 transcription factors from antirrhinum regulate phenylpropanoid and lignin biosynthesis in transgenic tobacco. Plant Cell 10(2):135–154

Whetten RW, MacKay JJ, Sederoff RR (1998) Recent advances in understanding lignin biosynthesis. Annu Rev Plant Physiol Plant Mol Biol 49:585–609

Wong MML, Cannon CH, Wickneswari R (2011) Identification of lignin genes and regulatory sequences involved in secondary cell formation in *Acacia auriculiformis* and *Acacia mangium* via *de novo* transcriptome sequencing. BMC Genomics 12:342

Wong MML, Cannon CH, Wickneswari R (2012) Development of high-throughput SNP-based genotyping in *Acacia auriculiformis* × *A. mangium* hybrids using short-read transcriptome data. BMC Genomics 13:726

Wu RL, Remington DL, MacKay JJ, McKeand SE, OMalley DM (1999) Average effect of a mutation in lignin biosynthesis in loblolly pine. Theor Appl Genet 99:705–710

Yong SYC, Choong CY, Cheong PL, Pang SL, Nor Amalina R, Harikrishna JA, Mat-Isa MN, Hedley P, Milne L, Vaillancourt R, Wickneswari R (2011) Analysis of ESTs generated from inner bark tissue of an *Acacia auriculiformis* × *Acacia mangium* hybrid. Tree Genet Genomes 7:143–152

Zhang Y, Xiao Y, Du F, Cao L, Dong H, Ren H (2011) Arabidopsis VILLIN4 is involved in root hair growth through regulating actin organization in a Ca^{2+}dependent manner. New Phytol 190:667–682

Zhong R, Ye ZH (2009) Transcriptional regulation of lignin biosynthesis. Plant Sig Behav 4(11):1028–1034

Zhong R, Richardson EA, Ye ZH (2007) The MYB46 transcription factor is a direct target of SND1 and regulates secondary wall biosynthesis in *Arabidopsis*. Plant Cell 19(9):2776–2792

Zhong R, Lee C, Zhou J, McCarthy RL, Ye ZH (2008) A battery of transcription factors involved in the regulation of secondary cell wall biosynthesis in *Arabidopsis*. Plant Cell 20:2763–2782

Chapter 7
Engineering of Secondary Metabolites in Tissue and Cell Culture of Medicinal Plants: An Alternative to Produce Beneficial Compounds Using Bioreactor Technologies

Mortaza Ebrahimi and Arash Mokhtari

7.1 Introduction

More than 50,000 of plant species all around the world are used for medicinal purposes (http://www.fao.org) and it is believed that one-third of the flora on earth have therapeutic properties (Maitra and Satya 2014). Each of these plants has the potential of producing 5000–25,000 different metabolites (Trethewey 2004). Plant secondary metabolites comprise a wide variety of organic compounds such as terpenes, phenolics, and alkaloids that show different biological activities. Their uses include pharmaceuticals (*Atropa belladonna, Digitalis lanata, Taxus* sp., *Hypericum perforatum*), agrochemicals (*Calceolaria andina*), colors (*Lythospermum erithroryzum*), flavors (*Mentha* spp., *Camellia sinensis* and *Vanilla planifolia*), fragrances (*Lavandula* spp.), biopesticides (*Chrysanthemum cinenariaefolium*), and food and food additives (*Coffea arabiga, Vanilla tahitiensis* and *Piper* spp.). Until now, the FDA has approved several plant-derived active metabolites such as taxol/paclitaxel (*Taxus* spp.), topotecan and irinotecan (*Camptotheca acuminata*), etoposide and teniposide (*Podophyllum peltatum*), and vinblastine and vincristine (*Catharanthus roseus*) as anticancer compounds. These chemicals are structurally too complex to be developed by semi- or total-chemical synthesis. So, the only feasible choice was extraction of these compounds from their natural sources (Gomez-Galera et al. 2007). Herbal remedy is increasing throughout the world. Around 25% of the new pharmaceutical drugs in the market are based on plant originated molecules (Raskin et al. 2002; Terryn and Van Montagu 2006). The 16.50 billion USD worth of worldwide plant-derived

---

M. Ebrahimi (✉)
Agricultural Biotechnology Research Institute of Iran-Isfahan Branch, Isfahan, Iran
e-mail: m.ebrahimi@abrii.ac.ir

A. Mokhtari
Agricultural Biotechnology Research Institute of Iran-Isfahan Branch, Agricultural Research, Education and Extension Organization (AREEO), Tehran, Iran

S.N.A. Abdullah et al. (eds.), *Crop Improvement*,
DOI 10.1007/978-3-319-65079-1_7

secondary metabolite industries in the 1990s extended to about 70 billion USD in 2010 (Chatterjee 2010). It is estimated that plant-derived medicines hit around 35.4 billion USD between 2013–2020 (Giri and Zaheer 2016) and that the global market of medicinal plants will reach to five trillion USD by the end of 2050 (Kalia 2005). The demand for medicinal herbs, even in developed countries, was fascinating. Since 2010, the yearly growth rate of phyto-pharmaceuticals in European and North American market were 15–20% and 25–30%, respectively (Chatterjee 2010).

Because of global warming, the calamitous environmental conditions worsen throughout the world. This situation resulted in more biotic and abiotic stresses and destruction of natural areas for most of the economically valuable plant species. Besides, over-harvesting of plants with economically valuable active compounds from natural resources is one of the main reasons for plant extinction. Populations of medicinal plants on their natural habitats are dwindling because of indiscriminate harvesting for medicinal uses and food, overgrazing, landslides and degradation, climate changes, and the other anthropogenic activities. Between 4000 and 10,000 species of medicinal plants are categorized as endangered (Canter et al. 2005) and unfortunately the number is increasing. For example, *Panax ginseng* has nearly disappeared in their natural growing area. It is reported that *Dioscorea balcanica, Podophyllum hexandrum*, and *Pilocarpus jaborandis* are now regarded as extinct species (Nosov 2012). Up to 80–90% of commercial supply of medicinal plants are collected from natural resources (Nosov 2012). Cultivation of most of medicinal plants is unprofitable. Only 10% of 1200–1300 conventional medicinal and aromatic plant species in Europe are cultivated on approximately 70,000 hectares (Nosov 2012). Slow growth rate, low yields, difficulties in growing many of these medicinal plants in commercial scale, and fluctuations of metabolite concentrations due to varietal, physiological, developmental stage, geographical and seasonal differences are examples of problems to produce secondary metabolites via field cultivation of medicinal plants (Rao and Ravishankar 2002; Gomez-Galera et al. 2007). Production of 1 kg of taxol, an anticancer drug, needs to collect around 10,000 kg of dry bark of *Taxus* sp. (Vidensek et al. 1990). These reasons and additional concerns regarding the instability of secondary metabolites related to species and environmental conditions are motivating factors for those searching for an efficient alternative method.

Plant cell and tissue culture introduced as new sources of plant originated chemicals (Rao and Ravishankar 2002). This approach has been attempted since the late 1950s to produce beneficial secondary metabolites. The following results stimulated new studies to develop efficient in vitro systems in many countries. But, many barriers and problems still exist and need to be overcome for commercialization of the products. Production of active metabolites using in vitro systems is influenced by the fact that most of the secondary metabolites are produced in low quantities. In addition, the slow growth rate of plant cells and tissues, and differences between the metabolomes of the original tissue in the whole plant with the tissue cultured cells and tissues are major considerations. These can lead to a high product price or unsuccessful production of the metabolites. Different techniques have been attempted to increase the efficiency of secondary metabolite production using in vitro methods, and further elucidation of metabolic pathways and dedicated molecular mechanisms is going to facilitate manipulation of plant cells as green factories.

There are several reports indicating elevated level of terpenoid indole alkaloids (TIAs) in tissue culture of genetically modified periwinkle cells, tissue, and plants. Because of the two important anti-tumor compounds, vinblastine and vincristine obtained from Madagascar Periwinkle (*Catharanthus roseus* (L.)), it is one of the most extensively studied medicinal plants. But these bisindole alkaloids accumulated in small quantities (0.001%) in the plant leaves (O'Keefe et al. 1997). The researchers are seeking to increase these metabolite using tissue culture techniques as an alternative, alongside genetic manipulation for enhanced concentrations of vinblastine and vincristine. Overexpression of the *strictosidine synthase* (*Str*) gene in cell lines of periwinkle resulted in tenfold STR activity with higher level of strictosidine, ajmalicine, catharanthine, serpentine, and tabersonine than wild type (Canel et al. 1998). In another experiment, the cell suspension culture of transgenic cell lines in periwinkle overexpressing *tryptophan decarboxylase* (*Tdc*) gene accumulated more TIAs (serpentine, catharanthine, strictosidine) compared to wild type (Whitmer et al. 2002).

7.2 Secondary Metabolisms and Economically Active Metabolites of Interest

Schematic representation of the primary and secondary metabolisms in plants is demonstrated in Fig. 7.1. Plants are able to synthesize the primary complex molecules like carbohydrates using sunlight, CO_2 and H_2O in a process known as photosynthesis. Primary metabolism is the generator of chemical energy in the form of ATP for the ongoing cellular processes. Secondary metabolites are by-products of primary metabolism. The biochemist Albrecht Kossel was the first person who used the term "secondary metabolites" to describe the cellular chemical components that originated from "primary metabolites" (Talapatra and Talapatra 2015). These substances serve as defense compounds, repellants, and attractants *in planta*, and display several biological activities to be used by human (Table 7.1). Plant secondary metabolites have been classified structurally into several major groups. Also, these compounds can be categorized according to their precursors or the metabolic pathways. Bourgaud et al. (2001) subdivided these compounds into three groups termed terpenes and steroids, phenolics, and alkaloids (Bourgaud et al. 2001). Complexity of the secondary metabolites makes it difficult to place them under these three groups. In another classification, they have been divided into the following five groups: phenylpropanoids, alkaloids, polyketides, isoprenoids, and flavonoids (Verpoorte et al. 2000). Some of secondary metabolites are present in all plants, while the others are synthesized in specific species or tissues and sometimes in response to external stimuli.

The mevalonic acid (MVA), shikimic acid, 1-deoxy-D-xylulose phosphate (DXP), and polyketide pathways are examples of the well-known secondary metabolic pathways. The biosynthesis of terpenoids, steroids, and related natural molecules are related to MVA and DXP pathways, while shikimic acid pathway is correlated with the biosynthesis of various aromatic amino acids and structural carbon containing metabolites. Biosynthesis of secondary metabolites in higher plants

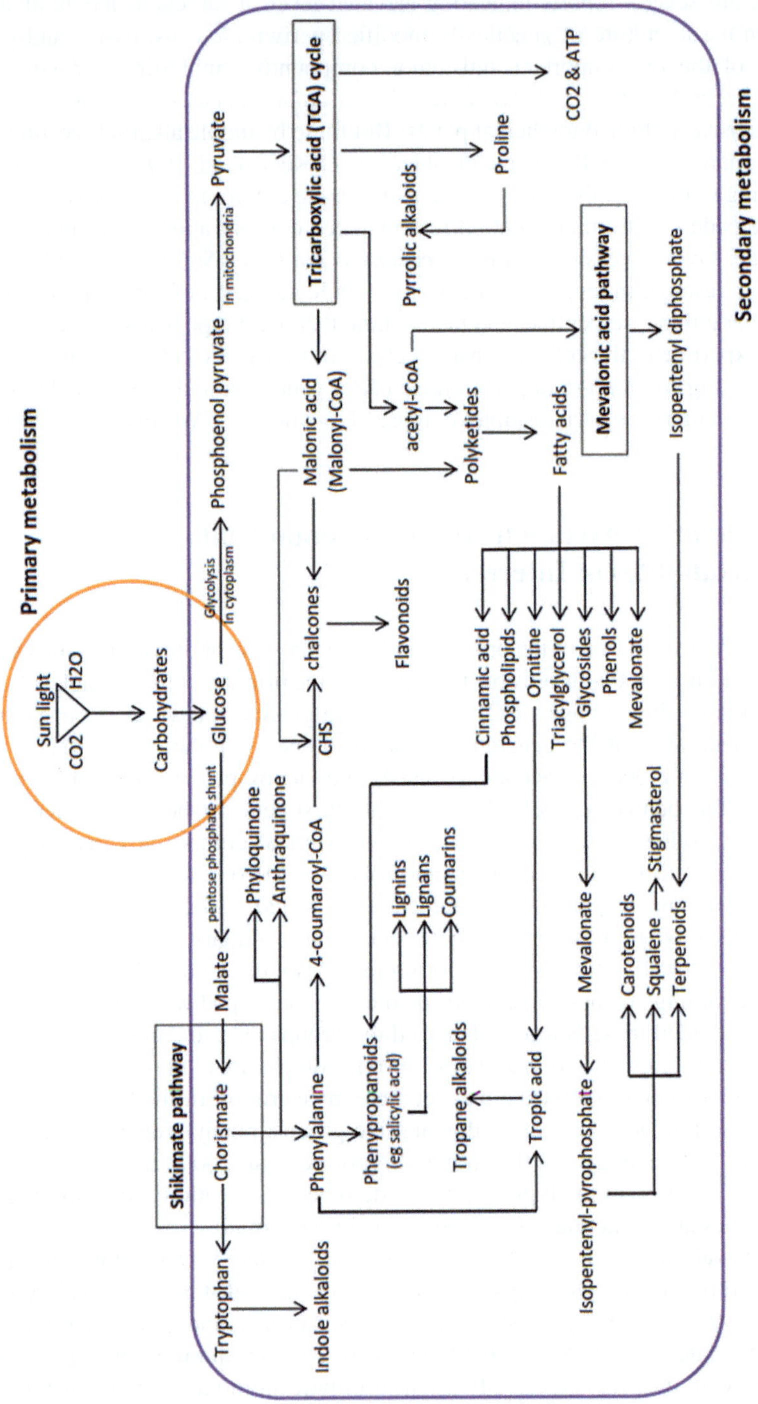

Fig. 7.1 A simple schematic representation of the major secondary metabolites in plants. *CHS* chalcone synthase

Table 7.1 Examples of plants derived biologically active compounds

Active compound(s)	Type	Application	Plant	References
Reserpine	*Alkaloids*	Anti-hypertensive, psychotropic	*Rauwolfia serpentine L.*	Muller et al. (1952)
Stevioside, rebaudioside	*Terpenes*	Natural sweeteners	*Stevia rebaudiana Bertoni*	Ishima and Katayama (1976)
Codeine, morphine	*Alkaloids*	Analgesic, antitussive	*Papaver somniferum*	Erdelsky (1978)
Ajmalicine	*Alkaloids*	Antihypertensive	*Catharanthus roseus*	Kurz et al. (1981)
Atropine	*Alkaloids*	Anti-cholinergic	*Atropa belladonna*	Yamada and Hashimoto (1982), Rajput (2013)
Atropine, Hyoscyamine, scopolamine	*Alkaloids*	Anti-cholinergic	*Datura* spp.	Yamada and Hashimoto (1982)
Cocaine	*Alkaloids*	Local analgesic	*Erytrhoxylum coca and Erythroxylum novogranatense*	Plowman and Rivier (1983)
Capsaicine	*Phenolic*	Analgesic	*Capsicum* spp.	Tominack and Spyker (1987), Anand and Bley (2011)
Vinblastine, vincristine	*Alkaloids*	Antineoplastic	*Catharantus roseus L.*	Schiel Om Berlin (1987)
Psoralen, Xhanothotoxin	*Coumarins*	Anticoagulant	*Apium graveolens*	Zobel et al. (1991)
Shikonin	*Terpenoids*	Anticancer	*Lithospermum erythrorhizon*	Shimomura et al. (1991)
Ajmaline, Ajmalicine	*Alkaloids*	Antihypertensive	*Rauwolfia* spp.	Wink and Roberts (1998)
Hyperforin	*Phloroglucinol*	Anti-depressant	*Hypericum perforatum*	Wentworth et al. (2000)
Glycyrrhizic acid	*Terpenoids*	Natural sweeteners	Glycyrrhiza spp.	Kitagawa (2002)
Trans-resveratrol	*Phenolics*	Anticancer, Antidiabetes	*Polygonum cuspidatum*	Burns et al. (2002), Counet et al. (2006)

(continued)

Table 7.1 (continued)

Active compound(s)	Type	Application	Plant	References
Taxol	*Diterpene alkaloid*	Antineoplasic	*Taxus* spp.	Croteau et al. (2006)
Artemisinin	*Sesquiterpene*	Anti-malarial	*Artemisia anua*	Weathers et al. (2011)
Digoxin, digitoxin	*Steroids*	Cardiotonic	*Digitalis* spp.	Calderón-Montaño et al. (2013)
Pilocarpine	*Alkaloids*	Cholinergic	*Pilocarpus jaborandi L.*	Santos and Moreno (2013)
Diosgenin	*Steroids*	Contraceptive	*Dioscorea* spp.	Talapatra and Talapatra (2015)
Genistein	*Flavonoids*	Anticancer	*Glycine max*	Talapatra and Talapatra (2015)
Quinine	*Alkaloids*	Anti-malarial	*Cinchona* spp.	Talapatra and Talapatra (2015)
Nicotine	*Alkaloids*	Neuroactive, insecticide	*Nicotiana* spp.	Talapatra and Talapatra (2015)

take places in separate compartments under the influence of the conventional acetate-mevalonic acid and non-mevalonic acid pathways. The biosynthesis of sterols, sesquiterpenes, and ubiquinones is related to acetate-mevalonic acid pathway that operates mainly in the cytosol and mitochondria, while the non-mevalonic acid pathway takes place in the plastid, operates the biosynthesis of carotenoids, hemiterpenes (C5), monoterpenes (C10), sesquiterpenes (C15), and diterpenes (C20) (Singh and Sharma 2014). Biosynthesis of most of the chemicals is not dedicated to a single route. For example, C6 from polyketide pathway and C3–C6 from shikimic acid pathway come together to form flavonoids.

7.2.1 Terpenes and Steroids

Terpenes and steroids constitute the largest groups of secondary metabolites in the plant kingdom with cyclic structure and cyclopentane perhydrophenanthrene backbone, respectively. Terpenes are metabolites of interest because of their roles in plant defense system under biotic and abiotic stresses, and their beneficial biological activities in the medical and industrial sectors as scent, flavours, fragrances, spices, sweetener, anti-inflammatory, anti-parasitic, and anti-tumor. These active metabolites show diversity in the number of C5-units: hemiterpenes (C5), monoterpenes (C10), sesquiterpenes (C15), diterpenes (C20), triterpenes (C30), tetraterpenes (C40), and polyterpenes (>C40) (Ashour et al. 2010; Alvarez 2014). The mevalonate and methylerythritol phosphate (MEP) pathways are two compartmentalized pathways involved in the first step of the terpenoids biosynthesis. Brassinosteroids, carotenoids, and saponins are groups of plant steroidal compounds. Ergosterol, stigmasterol, β-sitosterol, campesterol, and stigmasterol are examples of plant steroids with health-promoting properties as anti-atherosclerotic, anti-inflammatory, cholesterol lowering, and anti-oxidative activities (Alvarez 2014). The terpenes such as amarogentin (*Swertia japonica*), hernandulcin (*Lippia dulcis*), shikonin (*Lithospermum erythrorhizon*), digitoxin (*Digitalis purpurea*), ginsenosode (*Panax ginseng*), and astragaloside (*Astragalus membranaceus*) have received much attention for production using transformed root culture (Bajaj and Ishimaru 1999).

7.2.2 *Phenolics*

The aromatic compounds like lignin, coumarins, isoflavonoids, stilbenes, proanthocyanidines, and condensed tannins are phenolic secondary metabolites involved in several aspects of plant growth and developments. As indicated in Fig. 7.1, most of phenolic compounds originated from phenylpropanoid pathway. Other alternative pathways also drive the biosynthesis of some phenolics. It is believed that the polyketide pathway mediates the synthesis of several phenolic compounds. Tetrahydrocannabinoids is catalyzed from terpenoids (Alvarez 2014). The flavonoids and stilbenes are the two main subgroups of phenolics. Most of the fruits and vegetables contain flavonoids. There are several substances categorized as flavonoids: anthocyanidins, flavanones, naringen, silymarin, hesperitin, pelargonidins, cyanidins, delphinidins, rutin, and kaempferol (Gibbs 2007; Alvarez 2014). These compounds are very beneficial as anticancer, natural anti-inflammatory, potent antioxidant, scavenger of free radicals, and plant UV-protectants. These compounds are subjected to degradation during storage and cooking and it is believed that only 9% of the people receive enough flavonoids through consuming fruits and vegetables (Gibbs 2007). Stilbenes are another small group of phenolics that participate in plant defense system and display multiple beneficial function as medicine. A number of plant families like *Liliaceae, Fagaceae, Vitaceae,* and *Mirtaceae* produce stilbenes (Giorcelli et al. 2004). The 3,5,4′-trihydoxystilbene, also known as resveratrol, is the most common stilbene and its *trans* isomer has beneficial effects on some of the serious human diseases (Bradamante et al. 2004; Halls and Yu 2008). It is desirable to manipulate for resveratrol in plants, because of its potential nutraceutical property and as a cosmetic ingredient or for enhancing plant resistance against diseases. Engineering for enhanced level of resveratrol in plant is simply achievable by ectopic expression of *stilbene synthase* (STS). It has been demonstrated that introduction of STS gene can result in increased level of resveratrol in several plant species (Fischer et al. 1997; Liu et al. 2006; Nicoletti et al. 2007; Hanhineva et al. 2009). The Anthraquinone (*Rubia tinctorum*), Xanthone (*Swertia japonica*), Lignan (*Linum flavum*), Rosmarinic acid (*Ocimum basilicum*), Anthocyanin (*Lobelia chinensis*), and Rutin (*Fagopyrum esculentum*) are examples of phenolic compounds in medicinal plants that have been tested for production from transformed root culture (Bajaj and Ishimaru 1999).

7.2.3 Alkaloids

Alkaloids are nitrogen-containing compounds (tropane-, morphinan-, indole-, pyrrolidine-, piperidine-, pyrrolizidine-, quinoline-, isoquinoline-, aporphine-, imidazole-, diazocin-, purine-, steroidal-, amino-, and terpenoid indole alkaloids) capable to form complex with metal ions and salts with acids (Alvarez 2014). Biosynthesis of alkaloids is performed mainly in mevalonic acid (MVA)/deoxy-D-xylulose P (DXP) and shikimic acid pathways (Talapatra and Talapatra 2015). While the nitrogen atoms in alkaloids are provided from L-amino acids.

People of 1200–1400 B.C. took advantage of opium as medicine. The latex of *Papaver somniferum* or opium is a rich source of codeine and morphine. These compounds are alkaloids. Alkaloids that are biosynthesized in different organs provide protective advantages for plants. Ruminants keep off browsing the herbs containing alkaloids. Alkaloids can cause severe damage to liver. Pyrrolizidine alkaloids have been recognized as the main cause of the cirrhosis and failure of the liver. These compounds are hepatotoxic, mutagenic, and carcinogenic. Some of the *Crotalaria* spp. (*Leguminosae*) and *Heliotropium* spp. (*Boraginaceae*) contain pyrrolizidine alkaloids. The calabash-curare alkaloids have strong paralysis properties and was used as dart poisons for hunting. With more than 21,000 structures, alkaloids comprise the biggest group of secondary metabolites. The purine alkaloids are synthesized from small molecules such as amino acids L-aspartic acid, L-glutamine, and L-glycine. The tea (*Camellia sinensis*), coffee (*Coffea* spp.), cacao (*Theobroma cacao*), guarana (*Ilex paraguariensis*), and cola (*Cola nitida*) are the rich source of purine alkaloids (Springob and Kutchan 2009). Caffeine is one of the major purine alkaloids that we receive daily by drinking coffee. This drink originated from 1000 A.D. in Ethiopia. Caffeine is an analgesic and defensive agent against herbivores in plants (Hollingsworth et al. 2002) and as inhibitor of germination of coffee seedling (Friedmann and Waller 1985). The alkaloids (S)-hyoscyamine and (S)-scopolamine are tropane alkaloids occurring only in *Solanaceae* family. These alkaloids originated from the amino acids ornithine and/or arginine. The *Solanaceae*, *Brassicaceae*, *Erythroxylaceae*, *Euphorbiaceae*, *Rhizophoraceae*, and *Proteaeceae* families are well known for having tropane alkaloids. Some of the alkaloids with medicinal properties are given in Table 7.1. These compounds are of interest to human societies because of their prominent role in human health care. For example, atropine is an anti-cholinergic agent and has an antidote activity against intoxication. Besides, before surgery the patient will receive atropine to decrease salivation and respiratory secretion (Springob and Kutchan 2009). Hyoscyamine and scopolamine have been used as analgesic and sleeping poisons, aphrodisiacs and psychoactive drugs. Some of the plant species like *Atropa belladonna, Datura* species, *Hyoscyamus niger*, and *Mandragora officinarum* are recognized as lethal plants because of these toxic alkaloids.

7.3 Tissue Culture of Medicinal Plants

The levels of secondary metabolite found in plants are subjected to variation due to unfavorable environmental conditions. Production of active metabolites in some medicinal plants is usually a time-consuming process, which sometimes take around 20 years. In ginseng, the root growth rate in the plantation is around 1 gram/year, while the rate for tissue cultured roots was reported to be up to 2 gram dry weight/liter/day (Nosov 2012). Besides, the quantity and quality of the medicinal plant active metabolites are influenced by climate and growing conditions in the field or natural habitats. It has been shown that composition of ginsenoside in ginseng roots harvested from plantation is different from wild ginseng (Li and Mazza 1999). Tissue culture of some plants exhibits an advantage over field grown materials. For example, the rhizomes of the field grown mountain arnica lack the smell and taste which are present in tissue cultured plants (Ellenberger 1998). One alternative method to produce the economically bioactive compounds is the cell and hairy root culture system. Plant cell and tissue cultures hold great promise for controlled production of myriad of useful secondary metabolites on demand. Transgenic hairy root cultures have revolutionized the role of plant tissue culture in secondary metabolite production. They are unique in their genetic and biosynthetic stability, faster in growth, and more easily maintained. Using this methodology, a wide range of chemical compounds has been synthesized. Theoretically, in vitro grown plant cells and tissues have the potential to serve as a renewable source of economically valuable bioactive compounds. The following are some of the potential advantages of cell and tissue cultures compared to extraction of useful metabolites from field grown herbs:

- Continuous production of active metabolites independent of the environmental variations
- Production of uniform and pure metabolites by Good Manufacturing Practice (GMP) in a predictable system that guarantee a product free of agrochemicals throughout the cultivation process
- The possibility for enrichment of target metabolites and an increase in biomass production
- Easier and simple extraction procedures that can help to reduce aggressive solvents

- Authoritative and reliable identity of particular species of medicinal plant
- Simplicity to monitor genotype variability
- Easy to handle the cell lines to produce new compounds

Feasibility of in vitro techniques for production of the pharmaceuticals falls in doubt by the failure of Charles Pfizer Co. first attempts in the 1950s. The first successful strive for production of substances through tissue culture technique was reported in 1956 (Chandra and Chandra 2011). Later on in 1978, additional progresses were made in other countries particularly in Germany and Japan which led to the use of tissue culture as a sustainable source for production of secondary metabolites. Despite the significant progress, plant cell culture is still a complex process with relatively long period compared to bacterial culture and it is sensitive to mechanical damage. All these factors are influencing production of bioactive metabolites in in vitro condition. The difficulty of in vitro techniques is not the only limiting factor for commercialization of products. Sometimes, the products from this technology failed because of regulatory reasons. This was the reason that curbed the production of sanguinarine and vanilla in the USA, despite the successful achievement of in vitro production of these compounds.

In addition to the bioactive secondary metabolites, production of foreign proteins such as antibodies, vaccines, enzymes, and therapeutics medicinal proteins have been achieved using plant tissue culture techniques. The plant systems provide a less expensive and often clean source for these groups of chemicals. This is a low contamination system with reduced risk of mammalian pathogens and viruses. But significant challenges related to plant-based expression systems still remain to be overcome.

Production of plant metabolites via tissue culture system has been used for a limited number of medicinal plants mainly due to insufficient production of metabolites and high culture cost to meet economic production. Nosov (2012) provided a table indicating some of the successful examples of metabolites produced from medicinal plants by in vitro techniques in commercial scale. For example, Mitsui Petrochemical Industries in Japan applied a submerged fermentation processes to produce the bioactive metabolites berberine, arbutin, ginseng, and shikonin. These chemicals are produced in bioreactors with 4000–20,000 L capacity. Ginsenosides has been produced from tissue culture of *Panax ginseng* in bioreactor. It was reported that 27% of in vitro-derived cell dry weight contained Ginsenosides compared with 4.5% in whole plants (Chandra and Chandra 2011). In Japan, the Mitsui Petrochemical Industries Group started to produce berberine in bioreactor. They obtained a high yielding line with more than 0.1 g/L/day berberine. Selection of high yielding cell lines of *Coptis japonica* resulted in accumulation of 7 g/L of the alkaloid berberine in suspension cell culture. Until now, several metabolites of different groups have been synthesized using in vitro culture of plant cells. Production of phenolic compounds (hydroxycinnamic acids, coumarins, lignans, stilbenes, flavonoids, tannins, isoflavonoids, anthraquinones, benzoquinones, naphthaquinones), alkaloids (acridine, quinolizidine, pyridine), isoprenoids (sesquiterpenes, mono-, di-, tri-, and tetraterpenes), and some other minor groups (betalains, polyacetylenes, thiophenes, alliin, pyrethrins, nonproteinogenic amino acids) have been reported (Nosov 2012).

Several factors are affecting production of plant-derived compounds using tissue culture systems. The cell line, medium, temperature, pH, light, inoculum, gas composition, mixing technique, elicitors, bioreactor type, and growing mode are the main elements determining the level of secondary metabolites and cell biomass. Tissue culture-mediated secondary metabolite production is affected by several limiting factors such as slow growth rate and doubling time (approx. 30–40 h), low specific production rates, instability of cell lines, shear sensitivity, high cost of bioreactors, and accumulation of the active metabolites in the cells. All these factors should be overcome by applying appropriate strategies.

The bioreactor culture system is the promising way of commercial production of secondary metabolites. The engineering of adventitious and hairy roots represents one of today's fastest growing areas in pharmaceutical and nutraceutical economy of the world. Efficient nutrients supply to the cells, possibly combined with the application of mechanical stimulation to direct cellular activity as well as differentiation and function of adventitious and hairy roots in the bioreactor, are the significant steps towards development of high-tech products (Sivakumar 2006). Currently, metabolic engineering in hairy roots opened a new opportunity to improve the metabolic pathway of desired molecules in bioreactors. Most of the bioactive molecules, such as camptothecin, vinblastine, and ginsenosides, are of root origin. Harvesting roots is destructive for the plants and hence there has been increasing interest in developing in vitro cultures of adventitious and hairy roots from various medicinal plant species. The gap between discovery and commercialization can be bridged when biotechnologists and bioengineers join forces and integrate their research disciplines to develop bioreactor technology for the production of bioactive compounds (Sivakumar 2006). For example, constitutive expression of *hyoscyamine 6β-hydroxylase* from *Hyoscyamus niger* into hyoscyamine-rich *Atropa belladonna* plants using *A. rhizogenes* resulted in induction of hairy roots with enhanced levels of hydroxylase activity and up to fivefold higher concentrations of scopolamine than control (Hashimoto et al. 1993).

7.4 Enhance Metabolic Output of Cellular Lines Using Conventional Approaches

Several methods have been used to increase metabolic output and levels of valuable molecules using in vitro culture of medicinal plants. The biomass growth rate and level of bioactive compounds are the two main determinants for economical production of secondary metabolites in in vitro condition. Different factors are impacting these parameters. Generally, the majority of cell lines in callus or suspension cultures do not fully differentiate, and yield low amounts of secondary metabolites. Strain improvement is often carried out which include selection of a mother plant with high contents of desired bioactive products for callus induction to obtain high-producing cell lines. Appropriate techniques must be adopted for selecting those cell lines with higher yield.

Various physicochemical aspects including media composition, plant growth regulators (PGRs), temperature, light, pH, and aeration influence the culture productivity of plant cells. The type and concentration of the carbon source have important effects on cell growth and yield of secondary metabolites. Available concentration of nitrogen was also found to affect the contents of proteinaceous or amino acid products in cell suspension cultures. General plant tissue culture medium including MS, LS, or B5 usually have both NO^{3-} and NH^{4+} ions as sources of nitrogen. The ratio of the NH^{4+}/NO^{3-} and the total content of nitrogen are significantly affective for successful plant tissue culture systems. Phosphate level in the medium may have a major impact on the production of secondary metabolites in plant cell cultures. Higher concentrations of phosphate ion can enhance the cell growth with negative influence on secondary product accumulation. The effects of PGRs on secondary metabolite production are variable. Important biochemical changes are induced by some PGRs. The PGRs have remarkable consequences on physiological and biochemical processes and gene regulation as well as plant growth and development. Secondary metabolite production was strongly affected by the presence of 6-benzylaminopurine (BA), kinetin (Kin), or other cytokinines such as 4-chloro-2-diphenylurea in the medium (Karuppusamy 2009). Production of secondary metabolites in in vitro culture of *Mentha piperita* was monitored only by the addition of cytokinin, which resulted in about 40% increase in the total production of essential oils (Santoro et al. 2013). Gibberellic acid is also effective on plant cell cultures. Gibberellic acid increases secondary metabolite production in *Echinacea purpurea* hairy roots (Abbasi et al. 2012). A considerable rise in the contents of phenols and flavonoids in culture of *Stevia rebaudiana* was proven in response to a combination of BA either with gibberellic acid (GA3) or indol-3-acetic acid (IAA) compared to singly applied of PGRs indicating synergistic effects of PGRs (Radić et al. 2016).

The biosynthetic pathway is also controlled by physical conditions of cell culture like temperature and pH. Each plant species may favor a specific temperature for optimum in vitro metabolic activity usually in the range of 17–25 ° C. In the case of *Eleutherococcus senticosus,* the low (12 and 18 °C) and high (30 °C) temperatures resulted in significant reduction in fresh weight, dry weight, total phenolics, flavonoids, and total eleutheroside accumulation, while low temperature increased eleutheroside E accumulation in somatic embryos (Abdullah et al. 2006). It has been shown that the pH value of *Stevia rebaudiana*'s nod culture is a determinant of leaf metabolite, polyphenolics levels, and their distribution between different tissues (Radić et al. 2016). The inoculum density is another determinant for plant cell growth and accumulation of secondary metabolites. The flavonoid accumulation in cell suspension cultures of *Glycyrrhizin inflata* was shown to be correlated with the inoculum density as well as sucrose and nitrogen concentrations (Yang et al. 2009).

Elicitation is one of the most useful strategies for elevating the in vitro production of secondary metabolites. Elicitors can be categorized into biotic or abiotic factors. Endophytes are microbes that may stimulate host plant growth, improve nutrient supply, and protect host plants from biotic and abiotic stresses. There are several reports of endophyte-induced biosynthesis of secondary metabolites in host plants. Two fungal endophytes, *Curvularia* sp. *CATDLF5* and *Choanephora infundibulifera*

CATDLF6 isolated from the leaves of *Catharanthus roseus* were found to elevate vindoline content by 229–403% (Pandey et al. 2016). Abiotic elicitors have various effects on the production of secondary metabolites and can be divided into three types on the basis of physical (light, ultrasound, UV, thermal stress, salt stress, drought, and osmotic stress), chemical (heavy metals, nanoparticles mineral salts, amino acids, and gaseous toxins), and hormonal (salicylic acid and jasmonates). Elicitation effects of 1,2,4,5-tetraoxane and 2,5-diphenylthiophene in shoot cultures of two *Nepeta* species have been studied (Dmitrović et al. 2016). Addition of 150 mg/L of tryptophan resulted in elevated level of thymol to 390%, compared to mother plant (Abedaljasim et al. 2016). The other strategy such as permeabilization (Hussain et al. 2012), immobilization (Rao and Ravishankar 2002; Hussain et al. 2012), addition of precursors (Masoumian et al. 2011; Dheeranupattana and Natthiya 2013; Paek et al. 2014), biotransformation (Sabir et al. 2011; Banerjee et al. 2012), and hairy root and organ culture (Srivastava and Srivastava 2008; Hussain et al. 2012; Paek et al. 2014) have been studied extensively for production of different bioactive compounds from alkaloids, phenolics, and terpenoids.

7.5 Metabolome Differences Between Whole Plant and Tissue Cultured Materials

The properties of major metabolites found in the whole (organized) plants are generally similar to those in cell, tissue, and organ culture systems. The in vitro cultured cells often exhibit the same features of primary metabolic processes as those found in the whole plant cells. Differences that occur between the two systems can lead to altered secondary metabolite profiles. A cell suspension culture is submersed in a liquid medium containing the necessary nutrient factors. The different kinds of nutrients can enter the cells either through diffusion or by active uptake. Plant growth regulators (PGRs) which are incorporated exogenously to most plant cell cultures cause a special metabolic and cell differentiation changes in the in vitro cultured cells.

Although the plants have all the genetic information for the synthesis of PGRs, but in the whole plant not all the cells synthesize the PGRs. Growth and developmental processes are regulated by PGR gradient across the plant organs. The cell cultures mostly need exogenous PGR. A comparison study of the metabolome for primary metabolites extracted from the leaves of *Arabidopsis thaliana* and *Arabidopsis*'s T87 cell suspension culture indicated that some metabolites that were produced in the leaves were missing in the T87 cell culture and vice versa (Fukusakia et al. 2006). Primary metabolites are essential for the biosynthesis of the secondary; it is therefore expected that each variation in primary metabolites would be reflected in the secondary metabolite profiles. In fact, whole plant secondary metabolism exhibits much more variation compared to the plant organ or undifferentiated cultures. Production of secondary metabolities in cell suspension cultures have been widely published and it was proposed as a technology to overcome problems of variable product quantity and quality from whole plants due to the effects of different environmental factors, such as climate, diseases, and pests. Despite the differences, the in vitro systems using cell culture were successful in producing bioactive secondary metabolites using cell culture.

In the cell suspension culture of *Anthriscus sylvestris*, trace amounts of deoxy-podophyllotoxin could be detected. Koulman et al. (2003) carried out feeding experiments using deoxy-podophyllotoxin, yatein, and anhydropodorhizol. Their results showed that only deoxypodophyllotoxin was converted into podophyllotixin, yielding significantly higher concentration than measured in whole plants. Another research indicated that the roots of *A. sylvestris* produced a diverse range of lignins, but cell suspension cultures have a hydroxylating activity specific for deoxypodophyllotoxin. Hence suspension culture is a useful bioconversion catalyst for the synthesis of podophyllotoxin from deoxypodophyllotoxin (Koulman et al. 2003). Induction of these properties in plant cells would result in an interesting new source for podophyllotoxin. The presence of total phenols, leucocyanidins and (−)-epicatechin was compared between the mature plants, seedling and callus cultures of *Theobroma cacao* by Jalal and Collin (1979). They reported some compounds were found in both callus and intact plants and some were dedicated to callus, only. For example, the callus was found to contain two new flavonoid glycosides, phenolic acid complexes, p-coumaric and caffeic acids (Jalal and Collin 1979). In *Papaver bracteatum*, 2–3 seasons is necessary to accumulate thebaine (Razdan 2003). Novel synthetic pathways can be recovered from deviant and mutant cell lines which can lead to the production of new compounds, not previously found in whole plants. In some examples the cultured cell produces novel secondary metabolites that are not found in the whole plant. The *Lithospermum erythrorhizon* cell culture synthesized the rosmarinic acid which has not been detected in intact plant (Fukui et al. 1984). This shows that a novel unexpected compound can sometimes be observed after modifications of a metabolic pathway. Breitenbach et al. (2014) investigated the carotenoids contents of engineered rice callus. Manipulation of carotenoid pathway resulted in formation of a novel carotenoid *4-keto-α-carotene* in rice callus. In further assessment, they applied a bacterial *ketolase* gene in callus. The transgenic callus was able to produce astaxanthin, a non-plant carotenoid. The authors suggested that astaxanthin could have been produced from the combination of *ketolase* gene products and the native endogenous enzymes (Breitenbach et al. 2014).

Generally, cultured plant cells produce lower quantities and different features of secondary metabolites when compared with the whole plant. Also, the quantitative and qualitative profiles may change over the time. Inadequate production of secondary metabolites in cell culture systems is often caused by the lack of cell differentiation. Regardless of the disappointing statements, there are several examples indicating the potential of tissue culture over whole plant biosynthesis of some bioactive compounds. In the cell suspension culture of *Lithospermum erythrorizon,* 20% of the cell's dry weight contained shikonin compared to 1.5% in the whole plant (Takahashi and Fujita 1991). The same results have been reported for a few compounds such as Ginsenoside (*Panax ginseg*), Anthraquinones (*Morindaci trifo-*

lia, Galiumverum, Galiumaparine), Ajmalicine (*Catharanthus roseus*), Rosmarinic acid (*Coleus blumeii*), Ubiquinone-10 (*Nicotiana tabacum*) (Zhong 2001), Diosgenin (*Dioscorea deltoids*) (Rokem et al. 1984), Anthocyanin (*Vitis* sp., *Euphorbia milli, Perilla frutescens*), Benzyl isoquinolone alkaloids (*Coptis japonica*), Berberine (*Thalictrum minor, Coptis japonica*), Nicotine (*Nicotiana tabacum*), Bisoclaurine (*Stephania cepharantha*), and Tripdiolide (*Triptreyqium wilfordii*) (Ulbrich et al. 1985; Zhong 2001) in which the tissue cultured cells accumulated more metabolites compared to the whole plant.

7.6 Genetic Transformation and Manipulation of Metabolic Pathway

The in vitro techniques can serve as one of the alternative tools to produce valuable active metabolites, but the yield and stability of the product in in vitro condition are the main concern. Examples of successful production of economically valuable secondary metabolites are limited to 16 plant species for production of 13 bioactive compounds of interest (Nosov 2012). It seems that our failure to set up in vitro-mediated production of valuable metabolites in other medicinal plants could be partly due to the shortage or lack of our knowledge on the molecular mechanisms of different metabolic pathways. Manipulation of the metabolic pathway can be achieved after elucidation of the functional and regulatory proteins involved in the biosynthesis of the specific compound and even the related pathways. The quality, yield, stability, and the processing steps of a specific metabolite are subjected to change through metabolite engineering. Suppression of unwanted metabolite (e.g., toxic compounds) or introduction of a new pathway for de novo synthesis of dedicated metabolite (e.g., biofuel) are the other advantages of genetic engineering. The traditional limitations on in vitro culture-mediated production of plant secondary metabolites can be overcome using genetic engineering.

Besides our understanding on plant metabolome, availability of in vitro regeneration methods and genetic manipulation systems are prerequisites to engineer metabolic pathways of secondary products in plants. Several plant transformation methods have been grouped into two main categories: (a) direct gene delivery such as biolistics, protoplast fusion, microinjection, and electroporation; and (b) indirect transformation system using natural genetic engineers mainly *A. tumefaciens* and *A. rhizogenes* (Zarate and Verpoorte 2007). Alongside limitations and attractive applications provided from all these methods, transformation procedure using natural genetic engineers is most preferred due to the simplicity, inexpensive handling, ability for transformation of long DNA fragments, low frequency of transgene rearrangement, preferential integration of DNA segments into transcriptionally active sites, low rate of transgene silencing, and integration of several single copy of transgene. These factors make the *Agrobacterium* as attractive engineers (Kumar et al. 2013).

Extensive attempts have been made to manipulate for production of secondary metabolites using tissue culture several years ago (Bajaj and Ishimaru 1999). Production of nicotine from transgenic roots of *Nicotiana rustica* expressing *ornithine decarboxylase* gene was reported by Hamill et al. (1990). Overexpression of *hyoscyamine 6β-hydroxylase* (H6H) from *Hyoscyamus niger* in hairy roots of *Atropa belladonna* resulted in an enhanced level of scopolamine (Hashimoto et al. 1993). The transgenic hairy roots showed fivefold concentration of this tropane alkaloid in genetically engineered root cultures. In another attempt, Kang et al. (2011) reported elevated level of hyoscyamine and scopolamine in *Scopolia parviflora* root culture. Co-expression of two key enzymes *putrescine N-methyl transferase* (PMT) and H6H in hairy roots not only increased the level of tropane alkaloids, but also showed a potential role in root development (Kang et al. 2011). Tanshinones produced in *Salvia miltiorrhiza* are health-promoting diterpenoids with many important applications. According to Shi et al. (2016) two genes including *geranylgeranyl diphosphate synthase* (*SmGGPPS*) and *1-deoxy-d-xylulose-5-phosphate synthase* (*SmDXSII*) were introduced into hairy roots of *S. miltiorrhiza*. Overexpression of *SmGGPPS* and *SmDXSII* in hairy roots produced higher content of tanshinone than control and single-gene transformed lines. Tanshinone production in the double-gene transformed line *GDII10* reached 12.93 mg/g dry weight, which is the highest tanshinone content that has been achieved through genetic engineering. Furthermore, transgenic hairy root lines showed higher antioxidant and antitumor activities than control lines. Also, contents of chlorophylls, carotenoids, indole acetic acid, and gibberellins were significantly increased in the transgenic *Arabidopsis thaliana* plants (Shi et al. 2016).

As mentioned earlier, plant natural products are divided into three main categories known as terpenes and steroids, phenolics, and alkaloids. By far, with around 40,000 compounds, isoprenoids represent the largest secondary metabolites in plants. These valuable terpenoids are produced in low quantities in specialized cells such as glandular trichomes and laticifers. The squalene epoxide and several prenyl diphosphates, including isopentenyl diphosphate (IDP), dimethyl allyl diphosphate (DMADP), geranyl diphosphate (GDP), farnesyl diphosphate (FDP), and geranylgeranyl diphosphate (GGDP), are the precursors that catalyze the production of isoprenoids through two different pathways located in the cytosol or plastids. The two main precursors, IDP and DMADP together originated from mevalonic acid (MVA) pathway, in the cytosol and from pyruvate and a non-mevalonate or methylerythritol phosphate (MEP) pathways in the plastids. The process consists of a dephosphorylation step and is often followed by cyclization (Staniek et al. 2013). The resulted isoprenoid structures are subjected to hydroxylation process catalyzed by specialized enzymes termed cytochrome P450. These enzymes have been recognized as candidate gene for enhanced production of triterpenoides using genetic engineering (Fukushima et al. 2011). The triterpene glycyrrhizin is a natural sweetener derived from the rhizomes of licorice (*Glycyrrhiza* spp.) plants. Due to overharvesting from its natural resources, licorice is under extinction and has received considerable attention for production through in vitro techniques. Several enzymes mediate the biosynthesis of glycyrrhizin from 2,3-oxidosqualene, a common precursor of triterpenes and phytosterols (Abe et al. 1993). It has been shown that Cytochrome P450s are critical enzymes in the biosynthesis of terpenoids and other natural products in plants. Seki et al. (2008) reported that CYP88D6 (a cytochrome P450 mono-oxygenase) catalyzes the biosynthesis of 11-oxo-β-amyrin, an intermediate between β-amyrin and glycyrrhizin. This group also isolated a second P450 (CYP72A154) responsible for C-30 oxidation of 11-oxo-β-amyrin in the glycyrrhizin pathway to produce glycyrrhetinic acid (Seki et al. 2011). However, tissue cultures attempts for production of glycyrrhizin in licorice cell culture were unsuccessful (Hayashi et al. 1988, 1990; Ayabe et al. 1990). Seki et al. (2011) showed that β-amyrin synthase (*bAS*), *CYP88D6* and *CYP72A154* were expressed in both stolons and roots of licorice with no transcript detection in the leaves. These enzymes play critical function on the biosynthesis of glycyrrhizin in the roots and stolons. Although the first attempts on establishments of hairy root culture of licorice failed to produce glycyrrhizin, the ongoing researches on metabolite engineering of hairy roots were promising. Toivonen and Rosenqvist (1995) were not able to report glycyrrhizin production on transformed hairy roots (Toivonen and Rosenqvist 1995). They applied *A. rhizogenesis* to induce hairy roots on explants without engineering for any enzymes responsible for glycyrrhizin biosynthetic pathway. Instead, Ri-mediated transformation of *G. uralensis* using *A. rhizogenesis* harboring a cassette with constitutive expression of *Squalene Synthase Gene* (*GuSQS1*) resulted in enhanced level of glycyrrhizin (Lu et al. 2008). One of the transgenic lines was able to produce 2.5 mg/g dry weight glycyrrhizin which was 2.6-fold higher than control. Despite the success, the resulted quantity was not high enough to meet the prerequisites of an economically valuable amount of glycyrrhizin using hairy root culture. By recent gene discovery efforts and characterization, it is now expected that engineering for a single gene in biosynthetic pathway of glycyrrhizin should not be enough to yield a high-producing line (Seki et al. 2008, 2011). As mentioned above, Cytochrome P450s have critical functions in different steps of converting 2,3-oxidosqualene to glycyrrhizin.

The recent findings using functional genomics revealed several Cytochrome P450s. This family consists of specialized enzymes responsible for the biosynthesis of terpenoids. Taxol or paclitaxel is another example from diterpenes with high anticancer activity. Extraction of taxol from its natural source faces serious limitations. At least 2–4 mature trees must be cut for extraction of adequate amount of the drug for treatment of a single patient (Schoendorf et al. 2001; Jennewein et al. 2001). The tissue culture has been introduced as an alternative for production of taxol. Elucidation of the taxol biosynthetic pathway and the other rate-limiting steps lead to the development of high yielding recombinant plant or microbial cells. The taxadiene synthase was the first cloned enzyme that mediates for cyclization of the isoprenoid precursor geranylgeranyl diphosphate (GGDP) to taxa-4(5), 11(12)-diene (Wildung and Croteau 1996). The following reactions are mainly catalyzed by Cytochrome P450 enzymes (Walker and Croteau 2001). Although several attempts have been made for ectopic expression of the taxol related biosynthesis genes in microbial host like *E. coli* (Walker and Croteau 2000a, b), but the main conventional source of this drug in the market originated from plant cell culture technologies optimized by the companies such as ESCA genetics (United States), Phyton Catalytic (United States/Germany), Nippon Oil (Japan), and Samyang Genex Co. (Korea) (Nosov 2012). Among the enzymes involved in the biosynthesis of terpenoid, cytochrome P450 hydroxylases are the big challenge for functionalization in microbial systems (Khosla and Keasling 2003). They are decorators of the carbon skeleton in terpenoid. Until now, about 14 cytochrome P450s genes were reported to be involved in the biosynthesis of taxol (Morant et al. 2003). Difficulties of engineering for plant terpenoid biosynthetic pathways related genes in microbial cells motivated the use of the native plant species for production of some economically valuable bioactive compounds. Several genes have been cloned and identified with potential role in taxol biosynthesis such as *Geranylgeranyl diphosphate synthase* (Hefner et al. 1998), *Taxadiene synthase* (Wildung and Croteau 1996), *Taxane 5α-hydroxylase* (Jennewein et al. 2004), *Taxa-4(20), 11(12)-dien-5α-ol-O-acetyltransferase* (Walker and Croteau 2000a), *Taxane 10β-hydroxylase* (Schoendorf et al. 2001), *Taxane 13α-hydroxylase* (Jennewein et al. 2001), *10-Deacetylbaccatin III-10-O-acetyltransferase* (Walker and Croteau 2000a), *Taxane 2α-O-benzoyltransferase* (Walker and Croteau 2000b), *Phenylalanine aminomutase* (Walker et al. 2004), *Baccatin III:3-amino-3-phenylpropanoyltransferase* (Walker et al. 2002), and *3′-N-debenzoyl-2′-deoxytaxol N-benzoyltransferase* (Walker et al. 2002).

Hypericum perforatum or St John's wort is another interesting candidate with potentials for in vitro-mediated production of economically valuable medicinal compounds through genetic engineering of metabolic pathways. *H. perforatum* is one of the ancient medicinal herbs with two thousand years old history in remedy of human being (Zobayed et al. 2005). This plant contains several valuable chemical groups such as phenolics, flavonoids, naphthodianthrones, and phloroglucinols with approved bioactivity (Barnes et al. 2001). It has been reported that the hypericins content in the field grown plants varies and the difference can be up to 50-fold depending on the season (Southwell and Bourke 2001). This fluctuation in metabolites is a critical issue when the active metabolite is recorded in different phytopharmaceutical preparations. It can result in up to 13- to 17-fold differences in hypericin and pseudohypericin amounts, respectively (Greeson et al. 2001). So, the cell and tissue cultures of this plant have been attempted for stable and large-scale production of pharmaceutically important compounds. However, the in vitro system was not successful due to the low performance and unreliable yield of both bioactive compounds and the biomass.

The hypericins and hyperforins are the two main bioactive compounds of *H. perforatum* (Barnes et al. 2001) belonging to naphthodianthrones and phloroglucinols, respectively (Schröder 1997). It has been shown that type III polyketide synthases (PKSs) are the key enzymes in the biosynthesis of these compounds (Klingauf et al. 2005; Karioti and Bilia 2010). The PKSs multi-gene family mediates the biosynthetic reactions for most of the plant secondary metabolites as well as naphthodianthrones, phloroglucinols, xanthones, and flavonoids. In higher plants the PKSs are categorized in three groups: *chalcone synthase* (CHS-type), *stilbene synthase* (STS-type), and *coumaroyl triacetic acid synthase* (CTAS-type). The PKSs responsible for the biosynthesis of hypericins and hyperforins is the CHS-type (Flores-Sanchez and Verpoorte 2009).

Biosynthesis of hypericin is related to the metabolism of anthranoid. The type III PKS-mediated cyclization process is performed using one molecule of *Acetyl-CoA* and seven molecules of *malonyle-CoA*. The resulted octaketide participates in the following cyclization and decarboxylation process to produce emodin anthrone as an intermediate (Kirakosyan et al. 2004; Zobayed et al. 2005). This compound serves as a substrate for the biosynthesis of hypericin. In the following, another key enzyme *Hypericum perforatum phenolic oxidative coupling protein* (*Hyp-1*) catalyzes the ongoing reactions to produce hypericin (Lin-Fang et al. 2014). It is believed that *Hyp-1* is not a limiting enzyme. In contrast to PKSs with differential expression limited to flowers and leaves (He et al. 2012), maximum expression of *Hyp-1* was reported in all tissues as well as the roots, the non-hypericin producing organ in *H. perforatum* (Kosuth et al. 2011). So, it is axiomatic that hypericin should not be detectable in the roots. Despite the contradictory reports (Cui et al. 2010), this finding is consistent with Gaid et al. (2016) who indicated that hypericin was not detectable in hairy root culture of *H. perforatum*.

Hyperforin is the other important bioactive compound in *H. perforatum*. It has been shown that expression of *HpPKS1* and *HpPKS2* are related to the biosynthesis of hyperforin and hypericin, respectively (Karppinen and Hohtola 2008). Karppinen and Hohtola (2008) showed that the maximum transcript level of *HpPKS1* is in the flowers, while the *HpPKS2* demonstrated high expression level in the flower buds and leaves. The minimum expression was detected in the roots, for both of these genes. The *HpPKS1* mediates the biosynthesis of hyperforin by cyclization of one molecule of isobutyryl-CoA with three molecules of malonyl-CoA to form phlorisobutyrophenone. Then phlorisobutyrophenone is converted to dimethylallyl-phlorisobutyrophenone in a reaction catalyzed by dimethylallyltranstransferase (MAT) and dimethylallyl diphosphate (DMAPP). De Novo sequencing of transcriptome indicated that MAT is expressed in almost all tissues of *H. perforatum* (He et al. 2012). These findings corroborate that inadequate production of hyperforin and hypericin in hairy root culture of *H. perforatum* should be partly due to the lack of expression of *HpPKS1* and *HpPKS2*.

As mentioned earlier, metabolite engineering has been introduced as a promising tool to overcome low performance, instability, and unreliable yield of secondary metabolites using plant cell and tissue cultures. But, implementation of this method for manipulation of metabolic pathways in medicinal plants encounters significant difficulties related to the efficiency of transformation and accurate assessment of the consequences on the metabolic pathways following genetic engineering at the cellular and subcellular level. Manipulation of *H. perforatum* for enhanced production of bioactive metabolites has so far been hampered due to the lack of an efficient transformation system. Until now, there are few reports on establishment of transgenic events in *H. perforatum* (Di Guardo et al. 2003; Vinterhalter et al. 2006; Franklin et al. 2007; Bertoli et al. 2008; Santarem et al. 2008, 2010; Tusevski et al. 2013, 2014). *Agrobacterium*-mediated transformation of *H. perforatum* was shown to be affected by factors such as early oxidative burst and enhanced level of antimicrobial factors which kills ~99% of agrobacteria after 12 h of cocultivation (Pitzschke 2013). Besides, the low level accumulation of foreign proteins and even plant originated secondary metabolites, and silencing and instability of the transgene in engineered lines, are the problems that need to be overcome in metabolic pathway engineering of medicinal plants.

7.7 Regulatory Proteins Involved in Secondary Metabolism

Manipulation of secondary metabolites is influenced by the fact that there are competing pathways with potential intervention effects as they compete for the same substrate or intermediates with the pathway of interest. Metabolic engineering using a single structural gene can stimulate a contest in accessibility of cofactors and development of feedback-inhibition mechanisms by accumulation of end products. The turnover number (k_{cat} values) of specialized enzymes is significantly lower than their related corresponding enzymes of core metabolism (Bar-Even and Tawfik 2013). Incidentally, the other rate-limiting enzymatic steps and compartmentation related issues of pathways are the problems that need to be overcome in plant metabolite engineering. A general believe is that most of the metabolic enzymes in plant are promiscuous (Bar-Even and Tawfik 2013). These enzymes can often function on different substrates. It seems that successful engineering for plant secondary metabolites using single structural genes should coincide with the manipulation of regulatory proteins located upstream of the metabolic pathways. According to Sato et al. (2001), the suggestion was "fortification of multiple steps by over-expression of multiple biosynthetic genes, manipulation of regulatory genes that control expression of multiple pathway enzymes, or both."

The enzyme activity and specificity of the substrates are not the only factors affecting biosynthesis of secondary metabolites in plants. Availability of cofactors, negative feedback, intracellular chemical flux, source of chemicals, and pH are among the parameters influencing secondary metabolism. Although modification of a single gene helps in identifying the limiting steps involved in the biosynthesis of certain metabolites, but it does not result in an enhanced level of secondary metabolites. As the biosynthetic pathways are usually complicated, application of regulatory genes seems more helpful than a single gene. General regulation of the biosynthetic pathway is more trustable to increase production of bioactive compounds (Verpoorte et al. 2000). Manipulation for master transcription factors can lead to the regulation of the entire pathway or critical steps in the biosynthesis of dedicated metabolites. The early reports on feasibility of metabolite engineering using regulatory genes have been published by several researchers including Lloyd et al. (1992) and Martin (1996) for enhanced production of anthocyanins in maize and Arabidopsis. Anthocyanin is the end products of the flavonoid biosynthetic pathway. It has been shown that expression of the structural genes involved in the biosynthesis of these pigments is directly regulated by a combination of two regulatory proteins with homology to c-MYB and bHLH transcription factors (Mol et al. 1998). Ectopic expression of *R* (a MYB protein) and *C1* (a bHLH protein) resulted in accumulation of anthocyanins in in vitro cultured maize cells (Bruce et al. 2000). The biosynthesis of flavonoid as well as terpenoid indole alkaloids (TIA) correlated with ethylene- and jasmonate-mediated transcriptional machinery in plants (Zhou et al. 2010; Zhu et al. 2015).

Hormone-elicited transcriptional machinery plays a major role on secondary metabolism in plants. Plant hormones are considered as conserved elicitors of expression for a wide range of transcription factors. Ethylene serves as a signal to trigger specific biological responses. Accumulation of secondary metabolites in tomato fruit is partly correlated with ethylene-mediated transcriptional machinery. The signals are perceived by a multi-gene family of membrane-localized ethylene receptors (ETRs) (Wang et al. 2002) that negatively regulate ethylene responses through *Constitutive Triple Response1* (*CTR1*), a putative MAP-kinase kinase kinase (MAPKKK) (Kieber et al. 1993; Huang et al. 2003). So far, six members of ETRs (SlETR1, SlETR2, SlETR3 or SlNR, SlETR4, SlETR5, and SlETR6) with ethylene binding ability (Zhang et al. 2010) and four CTRs (SlCTR1, SlCTR2, SlCTR3, and SlCTR4) with differential expression have been isolated from tomato (Lin et al. 2008). There are several positive regulators (EIN2, EIN3, EIN5, EIN6, and EIL1) downstream of the CTRs. Existence or absence of ethylene is a determining factor that mediates the ongoing regulatory events. The ethylene downstream responsive signals will be inactivated in the absence of ethylene. This process is mediated by binding of ETRs to CTR1 and their interaction with EIN2. While presence of ethylene inactivates the ETRs followed by deactivation of CTR1, resulting in the EIN2 functioning as a positive regulator of the ethylene pathway (Klee 2004; Joo et al. 2007; Li and Guo 2007) through EIN3 and other Ethylene Insensitive-Like Proteins (EILs) transcription factors (Roman et al. 1995). These transcription factors can bind to the primary ethylene response element (PERE) on the promoter of ethylene responsive factors (ERFs) (Solano et al. 1998). Altered level of the genes with GCC-box and DRE/CRT containing promoters are the consequences of ETRs expression (Joo et al. 2007; Li and Guo 2007). This process indicates the importance of precise choices of regulatory proteins to manipulate for secondary metabolism in plants. Here, selection of AP2/ERF related proteins could be helpful.

Jasmonate is another master regulator of transcriptional machinery of secondary metabolism. This metabolite is an effective elicitors originated from oxidation of unsaturated fatty acids in plant cells. Jasmonate is directly correlated with stress in plants. Stress can result in subcellular damage followed by impairs electron transport system. These events lead to enhanced level of reactive oxygen species (ROS) (Price et al. 1989). Treatment of the adventitious root culture with higher levels of sucrose increased the total phenolics in *H. perforatum* (Tian and Russell 1999). These stress-induced metabolic processes can be partly dictated by jasmonate-elicited transcription machinery. It is speculated that the resulted osmotic stress of sucrose treatment on adventitious roots was followed by oxidative stress and other subsequent events, including lipid peroxidation and accumulation of secondary metabolite. Expression of Octadecanoid-Responsive Catharanthus AP2/ERF-domain (ORCAs) is induced by methyl jasmonate (MeJA). These transcription factors regulate the biosynthesis of several TIAs (Vom Endt et al. 2002). There is a direct target site (G-box) for MYC2, a basic helix-loop-helix (bHLH) factor in *ORCA3* gene in *Arabidopsis, N. tabaccum,* and *C. roseus* (Geyter et al. 2012). MYC2 is a jasmonate-elicited transcription factor that regulates the secondary metabolism. Our knowledge on jasmonate-mediated secondary metabolism in *Arabidopsis* indicates that although an enhanced level of jasmonate can promote JA-responsive gene expression by MYC2 transcription factors, but overexpression of these transcription factors does not theoretically means that expression of JA-responsive gene will be elevated. In this regards, several repressor and co-repressor proteins have to be released off from their binding to the transactivation domain of MYC protein. JA ZIM domain (JAZ) is a repressor protein that directly binds to MYC. The other protein is a Skp-Cullin-F-box-type E3 ubiquitin ligase complex (SCF + COI1(CORONATIVE INSENSITIVE 1)). Existence of JA mediates degradation of this complex from JAZ protein through 26S proteasome. This release can promote expression of JA-responsive gene by MYC protein (Geyter et al. 2012).

Transcriptional reprogramming of secondary metabolism in plants is triggered by JA through recruitment of transcription factors from several families. Geyter et al. (2012) have published a review that summarizes the jasmonate-elicited regulatory proteins involved in metabolism of plant bioactive compound. Several examples of regulatory genes from AP2/ERF, bHLH, R2R3-MYB, WRKY, NAC, DOF, HD-ZIP, and TFIIIA zinc finger were shown to be recruited by JA signaling with regulatory role to navigate the metabolism in plants.

7.8 Transporters and Secretion of Secondary Metabolites to Culture Medium or Subcellular Compartments

Production of valuable bioactive compounds using cell culture systems or transgenic plants overexpressing biosynthetic enzyme genes have been reported (Yazaki 2004: Rischer et al. 2013) with limited successes. Some secondary metabolites, especially alkaloids (e.g., berberine and benzylisoquinoline), are toxic to prokaryotic and eukaryotic cells. One of the mechanisms in plants to escape from such cytotoxicity is excretion of those chemicals from the cytosol to the apoplast or the other subcellular compartments such as vacuole (Shitan 2016). Transporter proteins provide such opportunities. Tissue culture-mediated production of metabolites offers the advantages that made it economically competitive for the extraction of chemicals from the whole plant systems. The total cost of secondary metabolite purification can be reduced if the desired compound is released into the liquid culture medium. Recovery of chemicals from plant biomass requires a series of complicated operations that increases the total cost of production greater than secreted one. According to Shitan (2016) the secondary metabolites transporter proteins can be roughly classified into 4 families: the ABC (ATP-binding cassette) transporter, nitrate-peptide transporter (NRT), multidrug and toxic compound extrusion (MATE), and purine permease (PUP) families. Some examples of current understanding of these transporters are described as follows. *CrTPT2* is a plasma membrane-localized G-type ABC transporter responsible for the efflux of catharanthine to the leaf surface, where it plays a role in protection of plants against herbivores (Yu and luca 2013). In *Nicotiana tabacum*, nicotine is biosynthesized in roots and translocated to aerial parts via the xylem (Shoji and Hashimoto 2013). The MATE transporters *jasmonate-inducible alkaloid transporter 1* (JAT1), JAT2, and MATE1 show nicotine transport activities, probably acting as nicotine/proton antiporters and localize to the vacuolar membrane. JAT1 and JAT2 likely transport nicotine into leaf vacuoles (Morita et al. 2009); MATE1 and MATE2 likely transport nicotine into root vacuoles (Shoji et al. 2009). Biochemical analysis of *VvABCC1* of grapevine (*Vitis vinifera*) proved that this C-type ABC transporter mediates cotransport of glucosylated anthocyanidin, *malvidin 3-O-glucoside* with glutathione (GSH) in an ATP-dependent manner (Francisco et al. 2013). As transporters for terpenoids, a G-type ABC transporters have been reported so far. *Nicotiana plumbaginifolia pleiotropic drug resistance1* (*NpPDR1/NpABC1*) was first isolated as a plasma membrane-localized protein induced by treatment with sclareolide, a close analog of the endogenous antifungal diterpene compound sclareol. *NpPDR1* is preferentially expressed throughout the root, in the leaf glandular trichomes, and in the flower petals and is proposed to transport sclareol (Jasinski et al. 2001).

Accumulation of desirable metabolite(s) can be problematic and limiting factors that have to be overcomed by appropriate strategy. So, targeting of a specific metabolite to vacuole, other subcellular compartments or even to the culture medium following constitutive expression of specialized enzyme(s) or regulatory proteins in a transgenic medicinal plant is often essential for modulation of a particular pathway.

7.9 Conclusions and Future Prospects

Global warming is going to change the life on earth. Our planet is getting warmer and this phenomenon is threatening our resources. How can we provide food and the other necessaries for ~9 billion people in the year 2050? Besides food, most of the chemicals used in medicine and industries are of plant origin. It is axiomatic that most of the plant diversity and suitable lands will be lost because of climate changes. With this condition, invention and developments of new tools are promising. The bioreactor targeted production of plant secondary metabolites is a modern approach that has to be further developed using new strategies as an alternative for whole plant harvested in nature. We have witnessed a continuous intensive effort of several scientists from different laboratories on this field. The emergence of OMICS as state-of-the-art tools in biotechnology is giving indications of future successes in precise manipulation of metabolic pathway(s) in medicinal plant. These will lead to discovery of key nucleotide targets for modification using powerful genome editing clean technologies such as CRISPR/Cas9 system.

References

Abbasi BH, Stiles AR, Saxena PK, Liu CZ (2012) Gibberellic acid increases secondary metabolite production in *Echinacea purpurea* hairy roots. Appl Biochem Biotechnol 168(7):2057–2066

Abdullah MS, Ali MB, Yu KW, Hahn EJ, Paek KY (2006) Effect of temperature on secondary metabolites production and antioxidant enzyme activities in *Eleutherococcus senticosus* somatic embryos. Plant Cell Tiss Org Cult 85(2):219–228

Abe I, Rohmer M, Prestwich GD (1993) Enzymatic cyclization of squalene and oxidosqualene to sterols and triterpenes. Chem Rev 93:2189–2206

Abedaljasim MJ, Al-Jibouri AAS, Abass Ali AJ, Majeed DM (2016) Improvement of phenols production by amino acids in callus cultures of *Verbascum thapsus* L. Am J Plant Sci 7:84–91

Alvarez MA (2014) Plant biotechnology for health: from secondary metabolites to molecular farming. Spring, New York. doi:10.1007/978-3-319-05771-2

Anand P, Bley K (2011) Topical capsaicin for pain management: therapeutic potential and mechanisms of action of the new high-concentration capsaicin 8% patch. Br J Anaesth 107(4):490–502

Ashour M, Wink M, Gershenzon J (2010) Biochemistry of terpenoids: monoterpenes, sesquiterpenes and diterpenes. In: Wink M (ed) Annual plant reviews: biochemistry of plant secondary metabolism, vol 40, 2nd edn. Wiley, New York

Ayabe S, Takano H, Fujita T, Furuya T, Hirota H, Takahashi T (1990) Triterpenoid biosynthesis in tissue cultures of *Glycyrrhiza glabra* var. glandulifera. Plant Cell Rep 9:181–184

Bajaj YPS, Ishimaru K (1999) Genetic transformation of medicinal plants. In: YPS B (ed) Transgenic medicinal plants, Biotechnology in Agriculture and Forestry, vol 45. Springer, New Delhi, pp 1–29

Banerjee S, Singh S, Rahman LU (2012) Biotransformation studies using hairy root cultures–a review. Biotechnol Adv 30:461–468

Bar-Even A, Tawfik DS (2013) Engineering specialized metabolic pathways—is there room for enzyme improvements? Curr Opin Biotechnol 24:310–319

Barnes J, Anderson LA, Phillipson JD (2001) St. John's wort (*Hypericum perforatum* L.): a review of its chemistry, pharmacology, and clinical properties. J Pharm Pharmacol 53:583–600

Bertoli A, Giovannini A, Ruffoni B, Guardo AD, Spinelli G, Mazzetti M et al (2008) Bioactive constituent production in St John's wort in vitro hairy roots Regenerated plant lines. J Agric Food Chem 56:5078–5082. doi:10.1021/jf0729107

Bourgaud F, Gravot A, Milesi S, Gonntier E (2001) Production of plant secondary metabolites: a historical perspective. Plant Sci 161:839–851

Bradamante S, Barenghi L, Villa A (2004) Cardiovascular protective effects of resveratrol. Cardiovasc Drug Rev 22(3):169–188

Breitenbach JP, Bai C, Rivera SM, Anela RC, Capell T, Christou P, Zhu C, Sandmann G (2014) A novel carotenoid, 4-keto-α-carotene, as an unexpected by-product during genetic engineering of carotenogenesis in rice callus. Phytochemistry 98:85–91

Bruce W, Folkerts O, Garnaat C, Crasta O, Roth B, Bowen B (2000) Expression profiling of the maize flavonoid pathway genes controlled by estradiol-inducible transcription factors CRC and P. Plant Cell 12:65–80

Burns J, Yokota T, Ashihara H, Lean ME, Crozier A (2002) Plant foods and herbal sources of resveratrol. J Agric Food Chem 50:3337–3340

Calderón-Montaño J, Burgos-Morón E, Lopez-Lazaro M (2013) The cardiac glycosides digitoxin, digoxin and ouabain induce a potent inhibition of glycolysis in lung cancer cells. WebmedCentral CANCER 4:WMC004323

Canel C, Lopes-Cardoso MI, Whitmer S, van der Fits L, Pasquali G, van der Heijden R, Hoge JHC, Verpoorte R (1998) Effects of over-expression of strictosidine synthase and tryptophan decarboxylase on alkaloid production by cell cultures of *Catharanthus roseus*. Planta 205:414–419

Canter PH, Thomas H, Ernst E (2005) Bringing medicinal plants into cultivation: opportunities and challenges for biotechnology. Trends Biotechnol 23(4):180–185

Chandra S, Chandra R (2011) Engineering secondary metabolite production in hairy roots. Phytochem Rev 10:371–395

Chatterjee SK (2010) Status of medicinal plants in global perspective: biodiversity destruction and conservation. In: Proceedings of botanicals in integrated health care convention 2010, 26–28 December 2010. The Agriculture Horticultural Society of India, Kolkata, India, pp 32–36

Counet C, Callemien D, Collins S (2006) Chocolate and cocoa: new sources of trans-resveratrol and trans-ieced. Food Chem 98:649–657

Croteau R, Ketchum RE, Long RM, Kaspera R, Wildung MR (2006) Taxol biosynthesis and molecular genetics. Phytochem Rev 5:75–97

Cui XH, Chakrabarty D, Lee EJ, Paek KY (2010) Production of adventitious roots and secondary metabolites by *Hypericum perforatum* L. in a bioreactor [J]. Bioresour Technol 101(12):4708–4716

Dheeranupattana S, Natthiya C (2013) Effects of sodium acetate and sucrose on in vitro alkaloid production from *Stemona* Sp. culture. Asian J Plant Sci 12(2):92–96

Di Guardo A, Cellarova E, Koperdáková J, Pistelli L, Ruffoni B, Allavena A et al (2003) Hairy root induction and plant regeneration in *Hypericum perforatum* L. J Genet Breed 57:269–278

Dmitrović S, Skorić M, Boljević J, Aničić N, Božić D, Mišić D, Filipović V, Opsenica D (2016) Elicitation effects of a synthetic 1,2,4,5-tetraoxane and a 2,5-diphenylthiophene in shoot cultures of two *Nepeta* species. J Serb Chem Soc 81(9):999–1012

Ellenberger A (1998) Assuming responsibility for a protected plant: Weleda's endeavour to secure the Firm's supply of *Arnica Montana*: first international symposium on the conservation of medicinal plants in trade in Europe. TRAFFIC Europe, Kew, pp 127–130

Erdelsky K (1978) In: Thorpe T (ed) Fourth international congress plant tissue and cell culture. Canada, University of Calgary

Fischer R, Budde I, Hain R (1997) Stilbene synthase gene expression causes changes in flower colour and male sterility in tobacco. Plant J 11(3):489–498

Flores-Sanchez IJ, Verpoorte R (2009) Plant polyketide synthases: a fascinating group of enzymes. Plant Physiol Biochem 47:167–174

Francisco RM, Regalado A, Ageorges A et al (2013) ABCC1, an ATP binding cassette protein from grape berry, transports anthocyanidin 3-O-glucosides. Plant Cell 25:1840–1854

Franklin G, Oliveira M, Dias ACP (2007) Production of transgenic *Hypericum perforatum* plants via particle bombardment-mediated transformation of novel organogenic cell suspension cultures. Plant Sci 172:1193–1203. doi:10.1016/j.plantsci.2007.02.017

Friedmann J, Waller GR (1985) Caffeine hazards and their prevention in germinating seeds of coffee (*Coffea arabica* L.) J Chem Ecol 9:1099–1106

Fukui H, Yamazaki K, Tabata M (1984) Two phenolic acids from *Lithospermum erythrorhizon* cell suspension cultures. Phytochemistry 23(10):2398–2399

Fukusakia E, Jumtee K, Bamba T, Yamaji T, Kobayashi A (2006) Metabolic fingerprinting and profiling of *Arabidopsis thaliana* leaf and its cultured cells T87 by GC/MS. Z Naturforsch C 61(3–4):267–272

Fukushima EO, Seki H, Ohyama K, Ono E, Umemoto N, Mizutani M, Saito K, Muranaka T (2011) CYP716A subfamily members are multifunctional oxidases in Triterpenoid biosynthesis. Plant Cell Physiol 52(12):2050–2061

Gaid MP, Haas BT, Scholl S, Beerhues L (2016) Hyperforin production in *Hypericum perforatum* root cultures. J Biotechnol 222:47–55

Geyter ND, Gholami A, Goormachtig S, Goossens A (2012) Transcriptional machineries in jasmonate-elicited plant secondary metabolism. Trends Plant Sci 17(6):349–359

Gibbs T (2007) Phytonutrients - the natural drugs of the future. In: Zhu YZ, Tan BKH, Bay BH, Liu CH (eds) Natural products: essential resources for human survival. World Scientific Publishing Co. Pte. Ltd, Hackensack, pp 1–26

Giorcelli AF, Sparvoli MF, Tava A, Balestrazzi A, Vrhovsek U, Calligari P, Bollini R, Confalonieri M (2004) Expression of the stilbene synthase (StSy) gene from grapevine in transgenic white poplar results in high accumulation of the antioxidant resveratrol glucosides. Transgenic Res 13:203–214

Giri CC, Zaheer M (2016) Chemical elicitors versus secondary metabolite production in vitro using plant cell, tissue and organ cultures: recent trends and a sky eye view appraisal. Plant Cell Tissue Org J 126(1):1–18

Gomez-Galera S, Pelacho AM, Gene A, Capell T, Christou P (2007) The genetic manipulation of medicinal and aromatic plants. Plant Cell Rep 26:1689–1715. doi:10.1007/s00299-007-0384-x

Greeson JM, Sanford B, Monti DA (2001) St. John's wort (*Hypericum perforatum*): a review of the current pharmacological, toxicological and clinical literature. Psychopharmacology (Berl) 153:402–414

Halls C, Yu O (2008) Potential for metabolic engineering of resveratrol biosynthesis. Trends Biotechnol 26:77–81

Hamill JD, Robins RJ, Parr AJ, Evans DM, Furze JM, Rhodes MJC (1990) Over-expressing a yeast ornithine decarboxylase gene in transgenic roots of *Nicotiana rustica* can lead to enhanced nicotine accumulation. Plant Mol Biol 15:27–38

Hanhineva K, Kokko H, Siljanen H, Rogachev I, Aharoni A, Kärenlampi SO (2009) Stilbene synthase gene transfer caused alterations in the phenylpropanoid metabolism of transgenic strawberry (*Fragaria x ananassa*). J Exp Bot 60(7):2093–2106

Hashimoto T, Yun DJ, Yamada Y (1993) Production of tropane alkaloids in genetically engineered root cultures. Phytochemistry 32:713–718

Hayashi H, Fukui H, Tabata M (1988) Examination of triterpenoids produced by callus and cell suspension cultures of *Glycyrrhiza glabra*. Plant Cell Rep 7:508–511

Hayashi H, Fukui H, Tabata M (1990) Biotransformation of 18-β-Glycyrrhetinic acid by cell suspension cultures of *Glycyrrhiza glabra*. Phytochemistry 29:2149–2152

He M, Wang Y, Hua W, Zhang Y, Wang Z (2012) De novo sequencing of *Hypericum perforatum* transcriptome to identify potential genes involved in the biosynthesis of active metabolites. PLoS One 7(7):e42081

Hefner J, Ketchum REB, Croteau R (1998) Cloning and functional expression of a cDNA encoding geranylgeranyl diphosphate synthase from *Taxus canadensis* and assessment of the role of this prenyltransferase in cells induced for Taxol production. Arch Biochem Biophys 360:62–74

Hollingsworth RG, Armstrong JW, Campbell E (2002) Caffeine as a repellent for slugs and snails. Nature 417:915–916

Huang Y, Li H, Hutchison CE, Laskey J, Kieber JJ (2003) Biochemical and functional analysis of CTR1, a protein kinase that negatively regulates ethylene signaling in Arabidopsis. Plant J 33:221–233

Hussain MS, Fareed S, Ansari S, Rahman MA, Ahmad IZ, Saeed M (2012) Current approaches toward production of secondary plant metabolites. J Pharm Bioallied Sci 4(1):10–20

Ishima N, Katayama O (1976) Sensory evaluation of stevioside as a sweetener. Rep Natl Food Res Inst 31:80–85

Jalal MAF, Collin HA (1979) Secondary metabolism in tissue cultures of *Theobroma cacao*. New Phytol 83(2):343–349

Jasinski M, Stukkens Y, Degand H et al (2001) A plant plasma membrane ATP binding cassette-type transporter is involved in antifungal terpenoid secretion. Plant Cell 13:1095–1107

Jennewein S, Rithner C, Williams R, Croteau R (2001) Taxol biosynthesis: taxane 13a-hydroxylase is a cytochrome P450-dependent monooxygenase. Proc Natl Acad Sci U S A 98:13595–13600

Jennewein S, Wildung MR, Chau M, Walker K, Croteau R (2004) Random sequencing of an induced Taxus cell cDNA library for identification of clones involved in Taxol biosynthesis. Proc Natl Acad Sci U S A 101:9149–9154

Joo S, Kim WT, Louis S (2007) A gaseous plant hormone ethylene: the signalling pathway. J Plant Biol 50:109–116

Kalia AN (2005) Worldwide trade scenario of medicinal and aromatic plants. In: Tyagi CS, Verma PK, Hooda JS, Yadav OP, Goyal RK (eds) Course compendium winter school on advances in medicinal, aromatic and under-utilized plants research, 29 September–19 October 2005 at CCSHAU. Hisar, Haryana, India, pp 271–273

Kang YM, Park DJ, Min JY, Song HJ, Jeong MJ, Kim YD, Kang SM, Karigar CS, Choi MS (2011) Enhanced production of tropane alkaloids in transgenic Scopolia parviflora hairy root cultures over-expressing putrescine N-methyl transferase (PMT) and hyoscyamine-6β-hydroxylase (H6H). In Vitro Cell Dev Biol Plant 47:516–524

Karioti A, Bilia AR (2010) Hypericins as potential leads for new therapeutics. Int J Mol Sci 11:562–594

Karppinen K, Hohtola A (2008) Molecular cloning and tissue-specific expression of two cDNAs encoding polyketide synthases from *Hypericum perforatum*. J Plant Physiol 165:1079–1086

Karuppusamy S (2009) A review on trends in production of secondary metabolites from higher plants by in vitro tissue, organ and cell cultures. J Med Plants Res Vol 3(13):1222–1239

Khosla C, Keasling JD (2003) Metabolic engineering for drug discovery and development. Nat. Rev. Drug Des Discov 2:1019–1025

Kieber JJ, Rothenberg M, Roman G, Feldman KA, Ecker JR (1993) CTR1, a negative regulator of the ethylene response pathway in Arabidopsis, encodes a member of the Raf family of protein kinases. Cell 72:427–441

Kirakosyan A, Sirvent TM, Gibson DM (2004) The production of hypericins and hyperforin by in vitro cultures of St. John's Wort (*Hypericum perforatum*) [J]. Biotechnol Appl Biochem 39(1):71–81

Kitagawa I (2002) Licorice root. A natural sweetener and an important ingredient in Chinese medicine. Pure Appl Chem 74(7):1189–1198

Klee HJ (2004) Ethylene signal transduction. Moving beyond arabidopsis. Plant Physiol 135:660–667

Klingauf P, Beuerle T, Mellenthin A, El-Moghazy SA, Boubakir Z et al (2005) Biosynthesis of the hyperforin skeleton in *Hypericum calycinum* cell cultures. Phytochemistry 66:139–145

Kosuth J, Smelcerovic A, Borsch T, Zuehlke S, Karppinen K, Spiteller M et al (2011) The hyp-1 gene is not a limiting factor for hypericin biosynthesis in the genus hypericum. Funct Plant Biol 38:35–43. doi:10.1071/FP10144

Koulman A, Kubbinga ME, Batterman S, Woerdenbag HJ, Pras N, Woolley JG, Quax WJ (2003) A phytochemical study of Lignans in whole plants and cell suspension cultures of *Anthriscus sylvestris*. Planta 69(8):733–738

Kumar N, Gulati A, Bhattacharya A (2013) L-glutamine and L-glutamic acid facilitate successful agrobacterium infection of recalcitrant tea cultivars. Appl Biochem Biotechnol 170:1649–1664. doi:10.1007/s12010-013-0286-z

Kurz WG, Chatson KB, Constabel F et al (1981) Alakloid production in *Catharanthus roseus* cell cultures. Planta Med 42:22–31

Li H, Guo H (2007) Molecular basis of the ethylene signaling and response pathway in Arabidopsis. J Plant Growth Regul 26:106–117

Li TSC, Mazza G (1999) Hortscience 34:85–87

Lin Z, Alexander L, Hackett R, Grierson D (2008) LeCTR2, a CTR1-like protein kinase from tomato, plays a role in ethylene signaling, development and defense. Plant J 54:1083–1093

Lin-Fang H, Zeng-Hui W, Shi-Lin C (2014) Hypericin: chemical synthesis and biosynthesis. Chin J Nat Med 12(2):0081–0088

Liu S, Hu Y, Wang X, Zhong J, Lin Z (2006) High content of resveratrol in lettuce transformed with a stilbene synthase gene of *Parthenocissus henryana*. J Agric Food Chem 54(21):8082–8085

Lloyd AM, Walbot V, Davis RW (1992) Arabidopsis and nicotiana anthocyanin production activated by maize regulators R and C1. Science 258:1773–1775

Lu HY, Liu JM, Zhang HC, Yin T, Gao SL (2008) Ri-mediated transformation of *Glycyrrhiza uralensis* with a squalene synthase gene (GuSQS1) for production of glycyrrhizin. Plant Mol Biol Rep 26:1–11

Maitra S, Satya P (2014) Value added products from medicinal plants. In: Sharangi AB, Datta S (eds) Value addition of horticultural crops: recent trends and future directions. Springer, India, pp 113–124

Martin C (1996) Transcription factors and the manipulation of plant traits. Curr Opin Chem Biol 7:130–138

Masoumian M, Arbakariya A, Syahida A, Maziah M (2011) Effect of precursors on flavonoid production by *Hydrocotyle bonariensis* callus tissues. Afr J Biotechnol 10(32):6021–6029

Mol J, Grotewold E, Koes R (1998) How genes paint flowers and seeds. Trends Plant Sci 3:212–217

Morant M, Bak S, Moller BL, Werck-Reichhart D (2003) Plant cytochromes P450: tools for pharmacology, plant protection and phytoremediation. Curr Opin Biotechnol 14:151–162

Morita M, Shitan N, Sawada K et al (2009) Vacuolar transport of nicotine is mediated by a multi-drug and toxic compound extrusion (MATE) transporter in *Nicotiana tabacum*. Proc Natl Acad Sci U S A 106:2447–2452

Muller JM, Schlittler E, Bein HJ (1952) Reserpin, der sedative Wirkstoffaus *Rauwofia serpentina* Benth. Experientia 8:338

Nicoletti I, De Rossi A, Giovinazzo G, Corradini D (2007) Identification and quantification of stilbenes in fruits of transgenic tomato plants (*Lycopersicon esculentum* Mill.) by reversed phase HPLC with photodiode Array and mass spectrometry detection. J Agric Food Chem 55(9):3304–3311

Nosov AM (2012) Application of cell Technologies for production of plant derived bioactive substances of plant origin. Appl Biochem Microbiol 48(7):609–624

O'Keefe BR, Mahady GB, Gills JJ, Beecher CWW (1997) Stable vindoline production in transformed cell cultures of *Catharanthus roseus*. J Nat Prod 60:261–264

Paek KY, Murthy HN, Zhong JJ (2014) Production of biomass and bioactive compounds using bioreactor technology. Springer, Dordrecht. doi:10.1007/978-94-017-9223-3

Pandey SS, Singh S, Vivek Babu CS, Shanker K, Srivastava NK, Shukla AK, Kalra A (2016) Fungal endophytes of *Catharanthus roseus* enhance vindoline content by modulating structural and regulatory genes related to terpenoid indole alkaloid biosynthesis. Sci Rep 25(6):26583

Pitzschke A (2013) Agrobacterium infection and plant defense—transformation success hangs by a thread. Front Plant Sci 4(519):1–12. doi:10.3389/fpls.2013.00519

Plowman T, Rivier L (1983) Cocaine and cinnamoylcocaine content of *Erythroxylum* species. Ann Bot 51:641–659

Price AH, Atherton NM, Hendry GAF (1989) Plants under drought stress generate activated oxygen. Free Radic Res Commun 8:61–66

Radić S, Vujčić V, Glogoški M, Radić-Stojković M (2016) Influence of pH and plant growth regulators on secondary metabolite production and antioxidant activity of *Stevia rebaudiana* (Bert). Period Biol 118(1):9–19

Rajput H (2013) Effects of Atropa belladonna as an anti-cholinergic. Nat Prod Chem Res 1:1. 1000104

Rao SR, Ravishankar GA (2002) Plant cell cultures: chemical factories of secondary metabolites. Biotechnol Adv 20:101–153

Raskin I, Ribnicky DM, Komarnytsky S, Ilic N, Poulev A, Borisjuk N, Brinker A, Moreno DA, Ripoll C, Yakoby N, O'Neal JM, Cornwell T, Pastor I, Fridlender B (2002) Plants and human health in the twenty-first century. Trends Biotechnol 20:522–531

Razdan MK (2003) Introduction to plant tissue culture. Science Publishers. p 375. ISBN, 1578082374, 9781578082377

Rischer H, Hakkinen ST, Ritala A et al (2013) Plant cells as pharmaceutical factories. Curr Pharm Des 19:5640–5660

Rokem JS, Schwarzberg J, Goldberg I (1984) Autoclaved fungal mycelia increase diosgenin production in cell suspension cultures of *Dioscorea deltoidea*. Plant Cell Rep 3:159–160

Roman G, Lubarsky B, Kieber JJ, Rothenberg M, Ecker JR (1995) Genetic analysis of ethylene signal transduction in *Arabidopsis thaliana*: five novel mutant loci integrated into a stress response pathway. Genetics 139:1393–1409

Sabir F, Sangwan RS, Singh J, Misra LN, Pathak N, Sangwan NS (2011) Biotransformation of withanolides by cell suspension cultures of *Withania somnifera* (Dunal.) Plant Biotechnol Rep 5:127–134

Santarem ER, Zamban DC, Felix LM, Astarita LV (2008) Secondary metabolism of *Hypericum perforatum* induced by agrobacterium rhizogenes. In Vitro Cell Dev Biol Anim 44:S52

Santarem E, Silva T, Freitas K, Sartor T, Astarita L (2010) Agrobacterium rhizogenes and salicylic acid trigger defense responses in *Hypericum perforatum* shoots. In vitro cell Dev. Biol Anim 46:S159–S160

Santoro MV, Nievas F, Zygadlo J, Giordano W, Banchio E (2013) Effects of Growth Regulators on Biomass and the Production of Secondary Metabolites in Peppermint (*Mentha piperita*) Micropropagated in Vitro. Am J Plant Sci 04:49–55

Santos AP, Moreno PRH (2013) Alkaloids derived from histidine: imidazole (Pilocarpine, pilosine). In: Ramawat KG, Merillon J-M (eds) Handbook of natural products. Chapter 6. Springer. Disponível Em: http://www.springerreference.com/docs/html/chapterdbid/363118.html

Sato F, Hashimoto T, Hachiya A, Tamura KI, Choi KB, Morishige T, Fujimoto H, Yamada Y (2001) Metabolic engineering of plant alkaloid biosynthesis. Proc Natl Acad Sci U S A 98:367–372

Schiel Om Berlin J (1987) Large scale fermentation and alkaloid production of cell suspension cultures of *Catharanthus roseus*. Plant Cell Tiss Org Cult 8:153–161

Schoendorf A, Rithner CD, Williams RM, Croteau R (2001) Molecular cloning of a cytochrome P450 taxane 10β-hydroxylase cDNA from Taxus and functional expression in yeast. Proc Natl Acad Sci U S A 98:1501–1506

Schröder J (1997) A family of plant-specific polyketide synthases facts and predictions. Trends Plant Sci 2:373–378

Seki H, Ohyama K, Sawai S, Mizutani M, Ohnishi T, Sudo H, Akashi T, Aoki T, Saito K, Muranaka T (2008) Licorice beta-amyrin 11-oxidase, a cytochrome P450 with a key role in the biosynthesis of the triterpene sweetener glycyrrhizin. Proc Natl Acad Sci U S A 105(37):14204–14209

Seki H, Sawai S, Ohyama K, Mizutani M, Ohnishi T, Sudo H, Fukushima EO, Akashi T, Aoki T, Saito K, Muranaka T (2011) Triterpene functional genomics in licorice for identification of CYP72A154 involved in the biosynthesis of glycyrrhizin. Plant Cell 23(11):4112–4123

Shi M, Luo X, Li L, Huang S, Zhang T, Wang H, Kai G (2016) Enhanced DiterpeneTanshinone accumulation and bioactivity of transgenic *Salvia miltiorrhiza* hairy roots by pathway engineering. J Agric Food Chem 64(12):2523–2530

Shimomura K, Sudo H, Saga H, Kamada H (1991) Shikonin production and secretion by hairy root cultures of *Lithospermum erythrorhizon*. Plant Cell Rep 10:282–285

Shitan N (2016) Secondary metabolites in plants: transport and self-tolerance mechanisms. Biosci Biotechnol Biochem 80(7):1283–1293

Shoji T, Hashimoto T (2013) Smoking out the masters: transcriptional regulators for nicotine biosynthesis in tobacco. Plant Biotech 30:217–224

Shoji T, Inai K, Yazaki Y et al (2009) Multidrug and toxic compound extrusion-type transporters implicated in vacuolar sequestration of nicotine in tobacco roots. Plant Physiol 149:708–718

Singh B, Sharma RA (2014) Plant terpenes: defense responses, phylogenetic analysis, regulation and clinical applications. 3 Biotech 5(2):129–151. doi:10.1007/s13205-014-0220-2

Sivakumar G (2006) Bioreactor technology: a novel industrial tool for high-techproduction of bioactive molecules and biopharmaceuticals from plant roots. Biotechnol J 1:1419–1427

Solano R, Stepanova A, Chao Q, Ecker JR (1998) Nuclear events in ethylene signaling: a transcriptional cascade mediated by ETHYLENE-INSENSITIVE3 and ETHYLENE-RESPONSE-FACTOR1. Genes Dev 12:3703–3714

Southwell IA, Bourke CA (2001) Seasonal variation in hypericin content of *Hypericum perforatum* L. (St. John's wort). Phytochemistry 56:437–441

Springob K, Kutchan TM (2009) Introduction to the different classes of natural products. In: Osbourn AE, Lanzotti V (eds) Plant-derived natural products. Synthesis, function, and application. Springer, New York. doi:10.1007/978-0-387-85498-4

Srivastava S, Srivastava AK (2008) Hairy root culture for mass-production of high-value secondary metabolites. Crit Rev Biotechnol 27(1):29–43

Staniek A, Bouwmeester H, Fraser PD, Kayser O, Martens S, Tissier A, van-der Krol S, Wessjohann L, Warzecha H (2013) Natural products – modifying metabolite pathways in plants. Biotechnol J 8:1159–1171

Takahashi D, Fujita Y (1991) In: Plant cell culture in Japan. Komamine A, Misawa M, Dicosmo F (eds). Cosmetic Materials. 72–78

Talapatra SK, Talapatra B (2015) Chemistry of plant natural products: stereochemistry, conformation, synthesis, biology, and medicine. Springer, Heidelberg. doi:10.1007/978-3-642-45410-3

Terryn N, Van Montagu M, Inzé D, Goosens A (2006) Functional genomic approaches to study and engineer secondary metabolites in plant cell cultures. In: Bogers LJ, Craker LE, Jange D (eds) Medicinal and aromatic plants. Springer, Netherlands, pp 291–300

Tian HQ, Russell SD (1999) Culture-induced changes in osmolality of tobacco cell suspensions using four exogenous sugars. Plant Cell Tiss Org Cult 55:9–13

Toivonen L, Rosenqvist H (1995) Establishment and growth characteristics of glycyrrhiza glabra Hairy root cultures. Plant Cell Tiss Org Cult 41:249–258

Tominack RL, Spyker DA (1987) Capsicum and capsaicin–a review: case report of the use of hot peppers in child abuse. J Toxicol Clin Toxicol 25(7):591–601

Trethewey RN (2004) Metabolite pro fi ling as an aid to metabolic engineering in plants. Curr Opin Plant Biol 7:196–201

Tusevski O, Stanoeva JP, Stefova M, Kungulovski D, Pancevska NA, Sekulovski N et al (2013) Hairy roots of *Hypericum perforatum* L.: a promising system for xanthone production. Cent Eur J Biol 8:1010–1022. doi:10.2478/s11535-013-0224-7

Tusevski O, Petreska Stanoeva J, Stefova M, Pavokovic D, Gadzovska Simic S (2014) Identification and quantification of phenolic compounds in *Hypericum perforatum* L. transgenic shoots. Acta Physiol Plant 36:2555–2569. doi:10.1007/s11738-014-1627-4

Ulbrich B, Weisner W, Arens H (1985) In: Neumann KH, Reinhard E (eds) Primary and secondary metabolism of plant cell cultures. Springer, Berlin, pp 293–303

Verpoorte R, van der Heijden R, Memelink J (2000) Engineering the plant cell factory for secondary metabolite production. Transgenic Res 9:323–343

Vidensek N, Lim P, Campbell A, Carlson C (1990) Taxol content in bark, wood, root, leaf, twig, and seedling from several *taxus* species. J Nat Prod 53:1609–1610

Vinterhalter B, Ninkovic S, Cingel A, Vinterhalter D (2006) Shoot and root culture of *Hypericum perforatum* L. transformed with Agrobacterium rhizogenes A4M70GUS. Biol Plant 50:767–770. doi:10.1007/s10535-006-0127-9

Vom Endt D, Kijne JW, Memelink J (2002) Transcription factors controlling plant secondary metabolism: what regulates the regulators? Phytochemistry 61(2):107–114

Walker K, Croteau R (2000a) Molecular cloning of a 10-deacetylbaccatin III-10-O-acetyl transferase cDNA from Taxus and functional expression in Escherichia coli. Proc Natl Acad Sci U S A 97(2):583–587

Walker K, Croteau R (2000b) Taxol biosynthesis: molecular cloning of a benzoyl- CoA:taxane 2alpha-O-benzoyltransferase cDNA from Taxus and functional expression in Escherichia coli. Proc Natl Acad Sci U S A 97(25):13591–13596

Walker K, Croteau R (2001) Taxol biosynthetic genes. Phytochemistry 58(1):1–7

Walker K, Fujisaki S, Long R, Croteau R (2002) Molecular cloning and heterologous expression of the C-13 phenylpropanoid side chain-CoA acyltransferase that functions in Taxol biosynthesis. Proc Natl Acad Sci U S A 99:12715–12720

Walker K, Klettke K, Akiyama T, Croteau R (2004) Cloning, heterologous expression, and characterization of a phenylalanine aminomutase involved in taxol biosynthesis. J Biol Chem 279:53947–53954

Wang KLC, Li H, Ecker JR (2002) Ethylene biosynthesis and signaling networks. Plant Cell 14:131–152

Weathers PJ, Arsenault PR, Covello PS, McMickle A, Teoh KH, Reed DW (2011) Artemisinin production in *Artemisia annua*: studies in planta and results of a novel delivery method for treating malaria and other neglected diseases. Phytochem Rev 10(2):173–183

Wentworth JM, Agnostini M, Love J, Schwabe JW, Chatterjee VKK (2000) St John's wort, a herbal antidepressant, activates the steroid X receptor. J Endocrinol 166:R11

Whitmer S, van der Heijden R, Verpoorte R (2002) Effect of precursor feeding on alkaloid accumulation by a tryptophan decarboxylase over-expressing transgenic cell line T22 of *Catharanthus roseus*. J Biotechnol 96:193–203

Wildung MR, Croteau R (1996) A cDNA clone for taxadiene synthase, the diterpene cyclase that catalyzes the committed step of taxol biosynthesis. J Biol Chem 271(16):9201–9204

Wink M, Roberts MW (1998) Alkaloids: biochemistry, ecology, and medicinal applications. Plenum Press, New York

Yamada Y, Hashimoto T (1982) Production of tropane alkaloids in cultured cells of hyoscyamus niger. Plant Cell Rep 1:101e103

Yang Y, He F, Yu L, Ji J, Wang Y (2009) Flavonoid accumulation in cell suspension cultures of *Glycyrrhiza inflata* Batal under optimizing conditions. Z Naturforsch C 64:68–72

Yazaki K (2004) Natural products and metabolites. In: Christou P, Klee H (eds) Handbook of plant biotechnology, vol 2. Wiley, Chichester, pp 811–857

Yu F, Luca DV (2013) ATP-binding cassette transporter controls leaf surface secretion of anticancer drug components in *Catharanthus roseus*. Proc Natl Acad Sci U S A 110:15830–15835

Zarate R, Verpoorte R (2007) Strategies for the genetic modification of the medicinal plant *Catharanthus roseus* (L.) G. Don. Phytochem Rev 6:475–491

Zhang S, Li N, Gao F, Yang A, Zhang J (2010) Overexpression of TsCBF1 gene confers improved drought tolerance in transgenic maize. Mol Breeding 26:455–465

Zhong JJ (2001) Biochemical Engineering of the Production of Plant-Specific Secondary Metabolites by Cell Suspension Cultures. In: Scheper T (ed) Advances in biochemical engineering, Biotechnology, vol 72. Springer, Heidelberg, pp 1–26

Zhou ML, Hou HL, Zhu XM, Shao JR, Wu YM, Tang YX (2010) Molecular regulation of terpenoid indole alkaloids pathway in the medicinal plant, *Catharanthus roseus*. J Med Plants Res 4(25):2760–2772

Zhu J, Wang M, Wen W, Yu R (2015) Biosynthesis and regulation of terpenoid indole alkaloids in *Catharanthus roseus*. Pharmacogn Rev 9(17):24–28

Zobayed SMA, Afreen F, Kozai T (2005) Temperature stress can alter the photosynthetic efficiency and secondary metabolite concentrations in St John's wort. Plant Physiol Biochem 43:977–984

Zobel AM, Wang J, March RE, Brown SA (1991) Identification of eight coumarins occuring with psoralen, xanthoxin, and bergapten on leaf surfaces. J Chem Ecol 17(9):1859–1870

Chapter 8
Bioreactor Technology for Sustainable Production of Valuable Plant Metabolites: Challenges and Advances

Morteza Khanahmadi and Kee Yoep Paek

8.1 Introduction

Human started a new phase of development through domestication of animals and their controlled propagation for food, clothes, and transportation. The subsequent phase in human advancement was agriculture, i.e., propagation of plants under some degree of control for various needs. Incognizant propagation of microorganisms mainly for enhancement of quality or shelf life of foods may have started at about the same time. Through advances in science and technology, the degree of control on the mentioned propagation processes gradually enhanced. In the case of microorganisms, appropriate chambers and vessels having different volumes and equipped with various sensors and other monitoring and control instruments were developed for controlled cultivation. A large part of construction knowhow of the mentioned vessels and their instrumentation were derived from the "chemical reactors," the equipment in which chemical "reactions" takes place. Hence, the former equipped vessels were subsequently called "bioreactors." The purpose of a wide range of microorganism cultivation processes is to "ferment" various substrates, i.e., to convert their sugars into alcohol. Therefore the bioreactors used for "fermentation" processes were named "fermentor," the term which is sometimes wrongly used synonymous to bioreactor for other microbial processes. By progress in science and technology the idea of animal and plant cells cultivation in equipped vessels for various proposes emerged and realized. These vessels are also named bioreactor.

M. Khanahmadi (✉)
Agricultural Biotechnology Research Institute of Iran-Isfahan Branch, Agricultural Research, Education and Extension Organization (AREEO), Isfahan, Iran
e-mail: khanahmadi@abrii.ac.ir

K.Y. Paek
Department of Horticultural Science, Chungbuk National University, Cheongju, South Korea

S.N.A. Abdullah et al. (eds.), *Crop Improvement*,
DOI 10.1007/978-3-319-65079-1_8

Cultivation of living organisms in bioreactors is carried out for various purposes. They may be cultured for investigation of their growth and product formation kinetics under the influence of different levels of effective factors. The investigations are usually performed in small bench scale bioreactors. The target of cultivation in larger bioreactors is usually a product which could be the organism itself, i.e., the biomass, or otherwise one or more of its valuable metabolites. In some cases, the aim of cultivation is to destroy or remove an unwanted compound from the cultivation medium or form the gas stream flowing through the bioreactor.

In the case of plants as well as animals, despite the fact that biomass and most of their metabolites are more or less valuable, production in bioreactor is not profitable as they are produced less costly in the nature and farms. The photosynthetic ability of plants provides them with free substrate from nature, while expensive substrate is required for cultivation of plant cells and organs in bioreactors. Photo-bioreactors are an exception in this regard but they have their own restrictions including slow growth rate, the need for CO_2, and shallow light penetration depth which all lead to low volumetric productivity, and consequently high capital cost. Hence, production of main plant products such as carbohydrates, proteins, and fat is almost not economically feasible in bioreactors.

Many of primary metabolites such as ethanol, lactic acid, and amino acids are produced in bioreactors at commercial scale; however they are produced by cultivation of fast growing microorganisms rather than slow growing plant cells. Some of these compounds are also produced via chemical synthesis. Nevertheless, the case may change for plant secondary metabolites, which are compounds synthesized by specific plants for special purposes. These include for combating viral, microbial, and pest attacks or to cope with drought, cold, heat, and salinity stresses and also for attracting pollinating insects. Most of plant secondary metabolites are not producible through natural metabolic paths of microorganisms. Furthermore, these compounds usually have too complex skeletons and numerous stereo specific atoms, so their chemical synthesis cannot outcompete the natural biosynthesis. Accordingly, plant secondary metabolites and plant specific complex primary metabolites are candidates for production via cultivation of plant cells and tissues in bioreactor.

8.2 Valuable Plant Metabolites

A wide range of plant metabolites have been used in various applications as natural colorants, odorants, flavorings, pesticides, and drugs. The market is increasing and more plant metabolites will gain commercial value as people tend to replace synthetic products with natural ones.

Global natural food colors & flavors market is expected to reach USD 7.79 billion by 2020 growing at a compound annual growth rate (CAGR) of 7.0% from 2015 to 2020. Some of plant flavors such as vanillin, decalactone (peach aroma), furaneol (strawberry aroma), raspberry ketone, cis-3-hexeno l (grassy-green aroma), and nootkatone (grapefruit aroma) are candidates for production in bioreactors at commercial scale. The two latter compounds also act as pheromone and insecticide/repellant, respectively.

The natural food coloring market is growing at 10–15% annually and many plant colorants are already produced in commercial scale using bioreactor technology. These include Astaxantin produced by the fungus *Xanthophyllomyces dendrorhous* and algae *Haematococcus lacustris*, and *H. pluvialis*; Betacarotene by the fungus *Blakeslea trispora,* and *Phycomyces Blakesleeanus* and algae *Dunaliella salina,* and *D. bardwil;* Lycopene by genetically modified (GM) fungus *Fusarium sporotrichioides and* GM bacteria *Erwnia uredovors.* Anthocyanins are also produced by plant cell suspension culture in bioreactors (Nosov 2012).

Plant metabolites have gained a significant share of biological drugs. In 2006, around 2500 biological drug were under development, 900 in preclinical trial stage and 1600 in clinical trial stage from which 700, 350, and 300 were devoted, respectively, to cancer, AIDs, and infectious diseases. Hepatitis, heart diseases, and diabetes were next in rank. Biopharmaceuticals market will reach USD 70 billion in 2010 (Luitjens and Pralong 2011).

8.3 Economical Challenges

Despite the growing market of plant metabolites, only a few of them are produced in bioreactors at commercial scale and most of these are produced by microorganisms rather than the plant cells and tissues. In fact, anthocyanins, berberine, rosmarinic acid, shikonine, ginsenosides, paclitaxel, and a few others are the only reported plant cell culture metabolites which are successfully commercialized. For a sustainable market, the plant metabolites produced in bioreactors should be cheaper than the same metabolites cultivated in farms or harvested from the nature. Here we try to give a preliminary economic analysis of the metabolite production in plant cell and tissue culture bioreactors and estimate the conditions for successful competition of bioreactor with farms in metabolite production. Most of the raw data are valid for Iran and so the quantitative results may be different for other countries. However, the overall results may clarify economic challenges of metabolite production in plant cell and tissue culture bioreactors.

A simplified overall view of metabolite production process is depicted in Fig. 8.1. The process is divided into two major units, which are the bioreactor unit and the extraction unit. The product of the bioreactor unit is dried biomass, whether cells or tissues, which then goes through the extraction unit for separation and purification of the target metabolite. It should be noted that a metabolite production factory may merely contain bioreactor unit and the produced dry biomass could be sold to existing factories that are designed for extraction and purification of metabolites from herbs and other plants. In this way, we will focus on bioreactor unit and directly compare production costs of dry biomass from this unit with that of dried plants harvested from farms or nature. One point that should be mentioned is that in some cases the target metabolite(s) may be secreted from the growing biomass into the

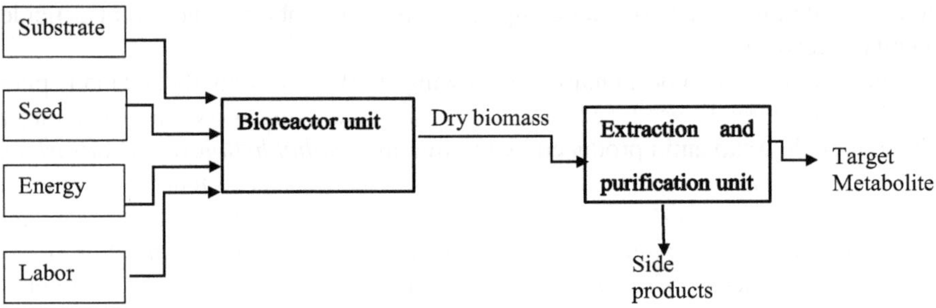

Fig. 8.1 Scheme of metabolite production process using bioreactor technology

medium within the bioreactor. This will omit extraction and may reduce purification costs. In most cases, this appears to be the feature that favor bioreactor production rather than the farm.

The major items that contribute to the biomass production costs in a bioreactor unit are assumed to be the substrate, seed culture, energy, labor, maintenance, and also investment costs, i.e., depreciation and loan interest. We estimate production costs using the following figures and assumptions:

- Substrate: Referring to the literature, Murashige and Skoog medium (MS) is one of the most common substrates used for both cell suspension culture and tissue cultures such as adventitious and hairy roots. Usually, half strength MS fortified with 30 g/L sucrose and traces of some other compounds is used as the culture medium. According to local market prices, the cost of such medium is roughly USD 0.2/L.
- Biomass yield: Referring to many published research results, 10 g of dry biomass per L of half strength MS medium is an acceptable record. The same figure is used in the present study.
- Culture time: Depending on the genotype, culture conditions, and suitable growth stage for attaining maximum production of target metabolite, the optimum culture time may vary from 10 days to 3 months. A culture time of 1 month is assumed as the base value in the present cost estimation.
- Capital investment: Capital investment is estimated based on a simplified scheme of a bioreactor unit (Fig 8.2). The unit is composed of a bioreactor, a compressor, a filter cartridge for sterilization of air stream, a feed mixing and storage tank, a steam sterilization heat exchanger for sterilization of culture medium, a biomass separation and drying unit, a boiler for steam supply, a chiller for control of bioreactor in desired temperature, and eventually a monitoring and control unit. The bioreactor is assumed to be of bubble column type with a useful volume of 20 cubic meters. Based on this capacity, 200 kg of dry biomass will be produced in each 1 month batch, i.e., 2500 kg dry biomass per year. Due to the slow growth of plant cells, an aeration rate of 0.1 volume air per volume culture media per minute (VVM) is reported to be appropriate for plant cell bioreactors. Based on these figures, appropriate size of the devices is calculated and capital investment

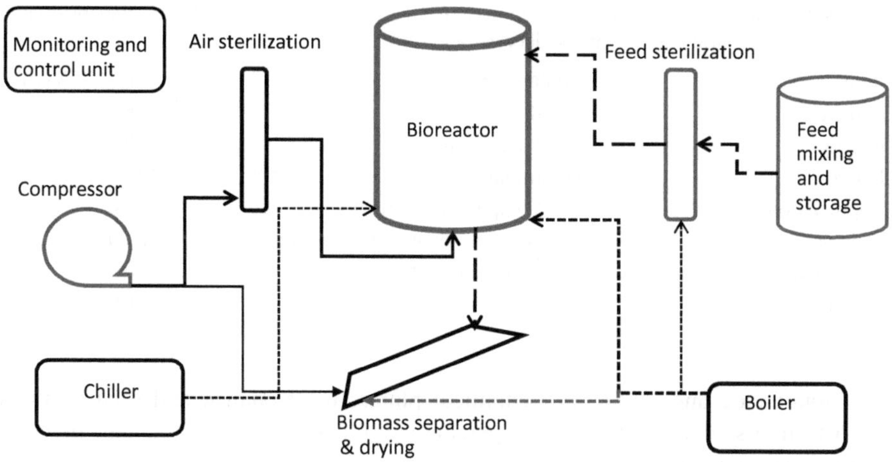

Fig. 8.2 A simplified scheme of the bioreactor unit

Table 8.1 A rough capital cost estimation for a 20 m^3 stainless steel bubble type bioreactor unit and its peripheral equipment and accessories

Item	Value(USD)
Bioreactor	14,500
Boiler	14,000
Compressor	5500
Chiller	8000
Sterilization units	9000
Biomass separation and drying	6000
Instrumentation and control	15,000
Piping and wiring	11,000
Sum of equipment prices	83,000
Installation and startup	16,600
Building	17,400
Capital cost	117,000

of the unit is estimated using local prices of the devices and equipment (Table 8.1). Referring to the table, total capital investment of the bioreactor unit will be around USD 117,000.

– Energy: A power of 2.4 kW is required to supply an aeration rate of 0.1 VVM to the 20 cubic meter bioreactor. In this way annual aeration energy cost will be USD 600 based on local electricity price. As a rough estimate, total energy expenses are taken to be two times of aeration energy cost, i.e., USD 1200/year.
– Labor: Sum of work-hours for the plant is assumed to be equal to 200 h/month which corresponds to a labor cost of USD 11,500/year according to local salary.

Table 8.2 Estimation of the total production costs for a 20 m^3 stainless steel bubble type bioreactor unit and its peripheral equipment and accessories

Item	Estimation basis	Value (USD/year)
Substrate	USD 0.2/L	48,000
Energy	USD 0.029/kwh	1200
Labor	200 h/month	11,500
Depreciation	10% of capital investment/year	11,700
Interest	10 years payback and interest rate of 20%	11,700
Maintenance and repair	5% of capital investment/year	5850
Total production costs		90,000

- Maintenance and repair: For chemical plants, maintenance and repair annual costs are usually taken to a fixed percentage of capital investment. This percentage is taken to be 5% in the present study.
- Depreciation: Operational life of the plant is assumed to be 10 years. In this way, the annual depreciation rate will be USD 11,700/year.
- Interest: Assuming that the total capital is loaned with a 10 years payback period, the interest cost will be 11,700/year according to local interest rates.

Sum of the estimated costs gives the biomass production cost to be USD 90,000/year (Table 8.2).

Taking into account that 200 kg biomass dry weight is produced in each 1 month batch, the annual biomass production by the 20 cubic meter bioreactor unit will be 2400 kg/year. In this way, total cost of biomass production in the bioreactor will be USD 37.5/kg of dry biomass. This rough value gives a base for the quantitative analysis of bioreactor vs. farm.

Local price of many farmed medical herbs with an annual yield of 2–3 t dry biomass per hectare is around USD 5/kg. This figure may fluctuate temporarily due to high demand or low supply; however it cannot be increased permanently since the farmers will dedicate more area to it which eventually decreases the price. It is seen that the biomass produced in the bioreactor is much more expensive than that cultivated in the farm. Having said that, the conditions may change in the following cases:

- For slow growing or poor growing plants which have a low annual biomass yield. In this case expenses of farming the plant biomass may be well above biomass cultivation in bioreactor. For example, ginseng is a slow growing plant with medical properties that its price may reach USD 50–150 per kg (fresh weight) in Korea. Its cultivation in bioreactor is now commercialized.
- When concentration of the target metabolite(s) in the biomass produced within the bioreactor is sufficiently higher than its concentration in biomass harvested from farm or the nature. In such a cases bioreactor could compete with farm if the biomass is priced based on the target metabolite(s) concentration. Based on above computations the biomass produced in bioreactor could compete that produced in farm if concentration of target metabolite in it was eight times higher and the biomass pricing was linearly proportional to the concentration of target metabolite (Fig. 8.3). It should be noted that the extraction and purification

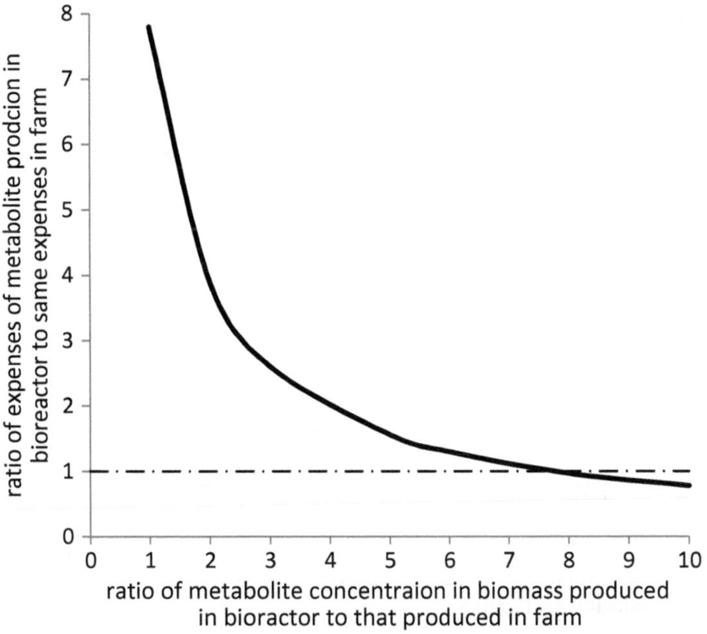

Fig. 8.3 Effect of increase in target metabolite concentration on metabolite production costs in bioreactor compared with costs of metabolite production by farming

expenses per unit weight of the target metabolite will decrease as its concentration increases in the biomass. In this way, relation of biomass price to its target metabolite concentration will grow steeper than linear, and hence the biomass produced in bioreactor could compete that of farm in lower target metabolite concentration ratios, e.g., five instead of eight.

There are several target metabolites which are commercially produced in bioreactor despite cheap mother plant biomass. For example, Nauttermann, a German company, produces rosmarinic acid in commercial scale by cultivation of *Coleus blumei* cells in bioreactor. Concentration of rosmarinic acid in the cultured cells reach to 27% of dry weight, nearly an order of magnitude higher than that found in herbs such as *Salvia officinalis* which are natural resources of the mentioned metabolite.

Referring to Table 8.2, substrate cost is the main determining item of production costs and any reduction in it will improve bioreactor competitiveness (Fig 8.4). It is seen that even with free substrate the biomass produced in bioreactor is more expensive than that harvested from the farm. This is due to other production costs which are independent of substrate cost.

A similar trend will be seen if duration of each bioreactor run could be reduced without any decrease in product (Fig. 8.5). In this way, the batch time is reduced and annual production is increased proportionally without the need for extra energy or capital investment. However, annual amount of consumed substrate will increase proportional to increase in number of bioreactor runs and this prevents biomass production costs to fall below the farm biomass price.

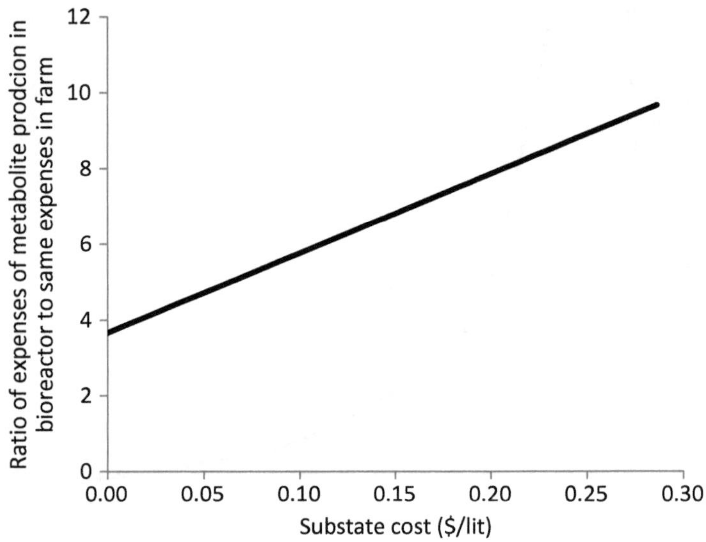

Fig. 8.4 Effect of substrate cost on metabolite production costs in bioreactor compared with costs of metabolite production by farming

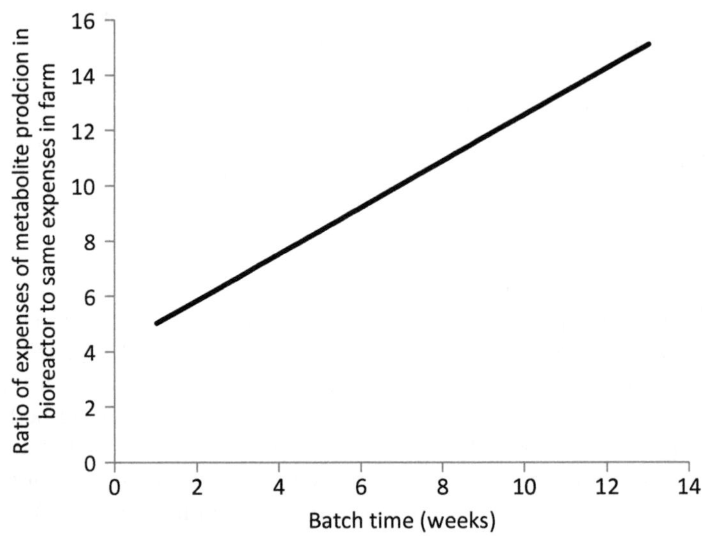

Fig. 8.5 Effect of cultivation time on metabolite production cost in bioreactor compared with cost of metabolite production by farming

In conclusion, a combination of increase in metabolite concentration in biomass, reduction of substrate cost, and reduction of batch time should be considered to make production of plant metabolites in bioreactor economically feasible. Reduction in substrate cost could be achieved by using cheaper ingredients or enhancing the amount of biomass produced per unit weight of substrate. Reduction of capital cost by the use of cheaper equipment or increase in volumetric productivity of bioreactor is another means for cutting the cost. Sustainable plant metabolite production in bioreactor will be achievable through research for these goals.

Referring to the literature, research is ongoing in various aspects of metabolite production by the use of bioreactor technology. Some articles introduce new metabolites or new plant sources of a known metabolite. Others depict efforts made to enhance productivity or improve machinery in different steps of metabolite production commercialization. These steps could be defined as follows (Davies and Deroles 2014):

- Step I-Selection: searching natural resources for plants with superior genetic capability for production of target metabolite.
- Step II-Genetic manipulation: mutation or metabolic engineering for production of elite cell lines with enhanced productivity of target metabolite.
- Step III-Lab scale optimization of culture conditions: determination of values corresponding to the best productivity of target metabolite for measurable independent parameters such as temperature, pH, medium composition, and elicitors. A large number of laboratory scale optimization experiments can be carried out simultaneously using simple laboratory glasswares. In this way optimization time is shortened despite the slow growth rate of plant cells.
- Step IV-Bench & pilot scale optimization: influence of physical design and operational mode of bioreactor on the process productivity are tested on bench and pilot scale bioreactors. Furthermore, optimization of some parameters such as aeration rate and mixing intensity which depend on vessel size and shape is performed in this step. These bioreactors include provisions for continuous monitoring trend of change in various parameters and also adjusting them in appropriate values. However, they are expensive and a few of them are available in each laboratory and hence the number of experiments is limited.

The ultimate goal of research activity in any of these steps is to make commercial production of a metabolite technically and economically feasible. While most achievements of these researches are product specific, a few others have more general application. Physical design of bioreactor is a subject with general application and any progress in it for a specific process is extendable to other processes. Various bioreactor designs which are used for plant cell and tissue cultures and advancement made in them are reviewed in the rest of the chapter.

8.4 Advances in Bioreactor Design

The bioreactors used for valuable plant metabolite production could be grouped based on various aspects. Based on the state of the growth environment, they are assorted into liquid phase bioreactors, gas phase bioreactors, and combined gas and liquid phase bioreactors. In the latter case the growing body stands in gas phase but is submerged into liquid medium intermittently or at least one time. They may be also divided into suspended cell culture bioreactors and tissue culture bioreactors according to structure of the growing body. Different tissues are cultured in bioreactors for propagation or metabolite production. For metabolite production, adventitious roots and hairy roots are the most frequently used tissues. Design of plant bioreactors originated from the knowhow of custom microbial fermentors. However, several inherent differences between them raised the need for advancement of the present designs and investigation for new ones.

8.4.1 Liquid Phase Bioreactors

Custom liquid phase fermentors also known as submerged fermentors are designed to provide the suspended microbial cells with a controlled temperature and medium pH and supply of sufficient oxygen. Aeration and mixing are routine operations in the fermentors. Mixing is needed for keeping the culture homogenous, i.e., providing the same conditions in all points within the culture volume. Homogeneity is a prerequisite for effective control and predictable performance of a fermentor. Furthermore, mixing prevents settlement of the cells and enhances heat transfer which is important in removing heat of metabolism. Mixing also improves oxygen supply which is a challenge in liquid state cultures due to trace solubility of oxygen in water. Oxygen supply is performed by blowing sterilized air into liquid culture in the form of tiny bubbles. Volumetric rate of air, which is usually expressed as VVM, size of bubbles and extent of mixing are major factors influencing oxygen supply. On the other hand, upward motion of air bubbles causes a degree of mixing which alleviate and in some designs eliminate the need for mechanical mixing. The bioreactors which are mixed merely by aeration, e.g., air lift fermentors, are called pneumatically driven for distinguishing them from mechanically driven ones, e.g., stirred tank fermentors. Both aeration and mechanical mixing are energy consuming and, moreover, exert shear stress on suspended cells which can injure or tore their walls and membranes. So they should be kept at minimum levels.

Compared with most of microorganisms, lower shear stress levels may harm plant cells since they are generally weaker against shear stress due to their larger size and having rigid fragile wall and large vacuoles. On the other hand, appropriate shear stress intensity prevents plant cells from being aggregated into lumps which adversely affect performance by restricting oxygen and nutrient transfer to the interior of the cells. Various effects of shear stress on growth and metabolite production of cell is reported in the literature. As mentioned before, compared with most microorganisms, plant cells are slow growing, so their biological oxygen demand is substantially lower and hence nondestructive shear stress may be sufficient to supply appropriate oxygen. A typical plant culture biological oxygen demand (BOD) is estimated to be 100 μmol oxygen per g dry weight of biomass per hour, it is one-tenth of a typical yeast BOD. In this way for a relatively high plant tissue concentration of 10 g-DW/L and an oxygen conversion rate of 50%, aeration rate of 0.004 VVM would be sufficient (Florez et al. 2016). In practice, typical flow rates of 0.1 VVM are used in plant cultures, it is an order of magnitude lower than that used for fast growing microorganisms. It should be noted that secretion of some compounds in plant cell cultivations may increase the liquid medium viscosity which in turn amplifies shear stress in a given mixing intensity.

Slowness of plant cell growth augments challenge of contamination for all kind of plant cell and tissue cultures. Long cultivation time of plant cell cultures provide enough time for fast growing microorganisms to propagate from a few cells to large populations which eventually contaminates and destroys the culture and wastes valuable medium and large amount of energy and labor.

Taking these facts into account a variety of new designs and design improvement are proposed and tested for suspended plant cell growth and it is claimed that many of the large scale cultivation problems are already solved (Nosov 2012).

8.4.1.1 Mechanically Driven Liquid Phase Bioreactors

Stirred tank bioreactor: These bioreactors include a cylindrical vessel into which an impeller and an air sparger are mounted. A major challenge in this kind of bioreactors is appropriate design of impeller to provide sufficient mixing without injuring the plant cells. No large scale use of stirred tank bioreactors for plant cell culture is reported probably due to the mentioned challenge. However, many of them which are a few litter in size are used in both cell and tissue cultures for research proposes in recent decades. Several different designs of impeller and mixing system are developed by researchers (Mamun et al. 2015). Successful production of transgenic tobacco cells in a 60 L stirred tank bioreactor was reported recently (O'Neill et al. 2015).

Rotating drum bioreactors: As its name indicate, mixing in these bioreactors is done by rotation of horizontally mounted bioreactor vessel. This design is not common in liquid state culture of microorganisms and was developed for solid state fermentation, i.e., cultivation of microorganisms over moist solid particles. A laboratory scale case study showed that it was able to maintain higher cell viability than stirred tank bioreactors (Tanaka et al. 1983) Large scale plant cell culture by use of rotating drum bioreactors has not been not reported.

8.4.1.2 Pneumatically Driven Liquid Phase Bioreactors

Bubble column bioreactors: It is a primitive pneumatically driven bioreactor composed of a cylindrical vessel and an air sparger mounted at its bottom. Conic vessels are also common. This type of bioreactor is successfully used in various scales for cultivation of plant cells and tissues. A 500 L one was successfully used for production of *Stevia rebaudiana* shoots (Akita et al. 1994).

Airlift bioreactor: This advanced design of bubble column bioreactor consists of two coaxial vertical cylinders and the air sparger is mounted at the bottom of an internal cylinder which is named riser or draft tube. The liquid medium is lifted by air bubbles in draft tube and returns downward through the space between the two cylinders. This medium circulation improves mixing and removes dead zones.

The air lift bioreactor was tested at laboratory scale for several plant cell cultures with or without modifications such as the use of conical draft tube (Townsley et al. 1983) or meshed draft tube (Kusakari et al. 2012).

Combined liquid and gas phases bioreactors: Growing plant cells and tissues submerged in liquid medium suffers from two disadvantages. Uniform and sufficient oxygen supply becomes challenging when population of cells or tissues increases and a dense culture occurs. The other challenge is absorption of excessive water into plant tissues which is called hyperhydricity. This phenomenon reduces mechanical strength and may cause malfunction or damage to the tissue due to oxygen depletion, formation of reactive oxygen species, and oxidative stress induction (Ziv 2005). Combined designs for bioreactors are proposed to circumvent these challenges.

Temporary immersed bioreactors: This design consists of two vessels one of which acts as a bioreactor and the other as a medium reservoir. The liquid medium is exchanged periodically between the two vessels through connecting tubes by the use of compressed air or a pump. In each period the plant cells and tissues remain immersed for several minutes then the medium is withdrawn and the tissues are left in direct contact with the gas phase for the remaining time of the period which may last a few hours. This design was tested for small scale propagation of various plant tissues and elimination of hyperhydricity was reported (Alvard et al. 1993) Mass propagation of tropical crops are reported in 5–10 L temporary immersion bioreactors and 46% reduction costs compared with standard procedure was reported for sugarcane propagation (Gonzalez 2005). Recently, a modification was reported in which the liquid medium was transferred between the bioreactor and the reservoir by changing the reservoir elevation. In this way, the gas and the liquid phases are de-coupled which increases process flexibility and reduces operational cost especially in processes where the gas phase is enriched with oxygen or carbon dioxide (Florez et al. 2016).

Large scale temporary immersion bioreactor system has not been reported yet and it seems that some challenges will occur as the system is enlarged. Fast filling and draining of a large bioreactor, e.g., 10 m^3, needs large compressor or pump and may harm cultured biomass, on the other hand slow filling and draining will cause different immersion time for biomass that is located at different heights in bioreactor. This in turn may lead to nonuniform growth and performance of biomass. Furthermore, fresh weight of biomass in dense cultures may reach several 100 kg per m^3 of bioreactor volume. In tall bioreactors, such a big biomass weight could injure the biomass at the bottom layers during the draining period.

8.4.2 Novel Bioreactors

Novelties in design of liquid phase bioreactors are experimented to respond to special needs of some cultures.

8.4.2.1 Diffusion Bioreactors

Delivery of oxygen to liquid medium could be performed by diffusion through a submerged membrane instead of bubbling and mixing. This is advantageous for shear sensitive cells and tissues. Diffusers with different configurations such as coil, disc, and basket made from hydrophobic polypropylene or silicon membranes were tested. Gentle rotation of diffuser could homogenize bioreactor contents and prevents cell settling. Membrane clogging due to cell adhesion is a possible danger. Much lower foams was formed in diffusion bioreactor compared with conventional stirred tank due to the absence of bubbles and damaged cell debris (Piehl et al. 1988).

8.4.2.2 Perfusion Bioreactors

In some plant cell cultures the target metabolite(s) exudes from cells into the medium. In this case the precious biomass cells may be used for production of the metabolite in prolonged culture by perfusion, i.e., continuous feeding of the bioreactor with fresh medium. However accumulation of metabolite may harm the cells or suppress the metabolite production. Furthermore, the metabolite may react with medium constituents and/or degrade with prolonged cultivation times. In such cases the metabolite could be extracted from the medium continuously or intermittently while keeping the biomass alive. For this, the medium is separated from suspended biomass by sedimentation, centrifuging or filtration, then it is treated with resins or other means to separate the metabolite. The remaining medium may be then recycled to the bioreactor or rejected. Several perfusion processes for metabolite production by the use of small scale bioreactors has been reported. Adding resin to culture medium of *Catharanthus roseus* cell cultures enhanced ajmalicine production by 70% and all of it was absorbed by resins and recovered (Lee Parsons and Shuler 2002). Instead of adding resins directly into bioreactor, resins were packed in a separate column and culture medium was recirculated through the column. By this method a 20-fold increase in sanguinarine productivity of *Eschscholtzia californica* suspension cell culture was achieved (Klvana et al. 2005). In some cases a compatible organic solvent is added directly to the bioreactor to improve extraction of target metabolite form the growing biomass. Up to 70% of thiophenes was extracted from hairy roots of *Tagetes patula* into hexane in a bubble column two phase bioreactor (Buitelaar et al. 1991). Overall yield of tabersonine and lochnericine was increased fourfold by use of silicone oil as extractive solvent in a bubble column bioreactor in which hairy root of *Catharanthus roseus* was cultured (Tikhomiroff et al. 2002).

8.4.3 Operational Improvements

Fed-batch operations: Conventional bioprocesses are usually run in batch mode. In this mode of operation the bioreactor is filled with the sterile culture medium, then the seed culture is added to bioreactor and conversion of substrate to biomass and various metabolites will start. The bioprocess is ended after reaching to the desired conversion of substrate to the target bioproduct(s). No more substrate is added to the bioreactor and the culture volume remains constant during the batch process. In this way, concentration of substrates is deceased and that of bioproducts is increased gradually during the bioprocess. Cells are in contact with high concentration of substrate at the beginning which may be inhibiting to cell growth. On the other hand, there is very low concentration of substrate and high concentration of bioproduct at the final stages of the bioprocess both of which slows down and eventually ceases the bioprocess.

The same as chemical processes, yield of many bioprocesses will improve if batch mode is replaced with fed-batch one. In fed-batch mode the bioreactor is partially filled with sterile culture medium having appropriate substrate concentration. The bioprocess will start by adding seed culture to bioreactor. When substrate decreases to rate limiting concentration or metabolites are increased to inhibiting concentrations, a defined volume of concentrated substrate is fed to bioreactor and mixed with culture medium. This will increase substrate concentration and reduces inhibitor concentration both of which boost bioprocess rate. This intermittent feeding could be repeated several times until the culture reaches to the desired volume. Several successful fed-batch plant cultures are reported. Fed-batch cultivation improved production of caftaric acid derivatives in adventitious root culture of *E. angustifolia* in a 5 L balloon type bioreactor (Baque et al. 2012). Fed batch increased biomass and cichoric acid by 1.5- and 1.37-fold in *Echinacea purpurea* adventitious root cultivation in air lift bioreactors (Wu et al. 2007).

8.4.4 Gas Phase Bioreactors

Gas phase bioreactors are always filled with a sterile airstream which may be enriched with oxygen, carbon dioxide, or other gases. The sterile liquid medium is sprayed or trickled over the biomass which is held within the bioreactor usually by means of a support. The medium exchanges substrate with metabolites from the biomass as it flows downward as a thin layer over the tissues. The used medium is collected at the bioreactor bottom from where it could be recycled to feed container (batch mode), or discarded after or without recovery of valuable metabolites and replenished by fresh medium (continuous mode). When the liquid layer is absent oxygen is directly absorbed by tissue from the gas phase. Otherwise oxygen dissolves into the thin liquid layer from where it is absorbed by the tissue. The liquid spraying is usually intermittent to prevent a permanent liquid layer which may restrict gas exchange and cause ethylene accumulation (Weathers and Zobel 1992). Furthermore, oxygen supply is not a challenge in gas phase bioreactors. Some hairy roots which are produced by solid phase culture will not grow in liquid phase and the gas phase cultivation is the only choice for them (Weathers et al. 2008).

Gas phase bioreactors are suitable for plant tissues and organs as they form porous structures which are not flooded with liquid and so the gas could flow through the bulk of biomass along with the liquid. Meshes or other kinds of frame are generally used to distribute biomass within bioreactor volume and strengthen its porous structure. Even with the use of frames, keeping the tissues from being injured under high wet biomass weight seems to be a big challenge in the use of large volume gas phase bioreactors. Uniform distribution of liquid within the bulk of biomass is another challenge in this regard. Hence, scale up of gas phase bioreactors seems to be practical by modular approach, i.e., by the use of several small bioreactors instead of one large one.

Modular approach is used in other processes where construction of large equipment is not feasible. Membrane systems are well known cases which are scaled up by increasing the number of modules instead of their volume. A clear example is sea water desalination plant which consists of a huge number of reverse osmosis modules. Modular systems are more flexible as their capacity is changed simply by adding new modules or removing some of them. Another important advantage of modular systems over conventional large volume systems comes from independent operation of each module. In this way occurrence of a problem in one module cause only a little product loss. This is of crucial importance in plant culture processes which are amenable to contamination. Contamination of one large bioreactor will result in the loss of the entire product while contaminations of one of several small bioreactors will result in the loss of a few percent of the product.

Despite the mentioned advantages, commercialization of modular systems is challenging due to two disadvantages in using them compared with their conventional counterparts. The first disadvantage is that they are more labor intensive, e.g., handling of one large bioreactor needs less labor than several small ones. The second is that they are more capital intensive, e.g., ten of 1 m^3 bioreactors cost much more than one 10 m^3 bioreactor with the same degree of instrumentation. This is due to the fact that the vessel surface to volume ratio decreases as its volume increase and so less material is needed to construct a large volume vessel than several small vessels with the same total volume. Furthermore, one set of instruments is needed for monitoring and control of a large well mixed bioreactor while the same set is needed for each of the small ones.

Having said that, modular plant culture systems have the chance to be commercialized due to the fact that small vessels should bear smaller forces and so they could be constructed from cheaper materials such as plastics. In fact disposable bioreactors are being developed which open the way for commercialization of plant culture bioreactors and specially the gas phase ones.

Gas phase bioreactors and disposable bioreactor technology are more novel than liquid phase ones and so they are reviewed here with more details.

8.4.4.1 Mist/Spray Bioreactors

Some plants grow over rocks adjacent to waterfalls by absorbing water from misted air via their hanging roots. This observation triggered the idea of aeroponics, i.e., soilless cultivation of plants on a support and misting the nutrient solution over their suspended roots. The first aeroponic system was developed in 1942 for pineapple (Carter 1942) and then it was employed in many research and production cases. Mist production in earlier aeroponic systems was provided by the use of nozzles. Depending on nozzle configuration and liquid pressure, the size of liquid droplets could fall in the same rank as liquid spray, i.e., larger than 60 μ, or mist, 60–30 μ or fog, i.e., smaller than 30 μ. Working with nozzles have two drawbacks, i.e., considerable energy consumption to pressurized liquid and frequent fouling.

Ultrasonic misting was used in aeroponic systems from the beginning of the 1990 to circumvent the mentioned drawbacks. No considerable pressure was required and fouling is not a challenge. Very uniform fog of nutrient solution with a mean droplet diameter of 10 μm which is produced by ultrasonic device that penetrates into the root network and is absorbed by the roots. In this way, the excess volume of medium which should be sprayed for uniform distribution of nutrients is reduced considerably and hence lesser energy is consumed for spraying and recirculation of medium. Furthermore, thinner or no liquid layer forms over the root surface and so oxygen transfer is enhanced (Weathers and Zobel 1992). The mist particles has large surface to volume ratios and therefore are readily saturated with oxygen. They transport oxygen to tissues and enhance their growth rate. In mist bioreactors, the biomass could occupy 37% of the bioreactor volume before onset of oxygen limitation, this figure is as low as 6% in liquid phase bioreactors (Stiles and Liu 2013). Packing fractions as high as 71% has been reported for mist bioreactors. In liquid phase cultures, such high biomass densities are difficult to achieve due to the occurrence of high pressure drops (Weathers et al. 2008). Packing fraction corresponds to the productivity of a bioreactor and so its increase will reduce production costs.

Uniform distribution of seed biomass is important in mist bioreactors since more compact regions entrap more mist and grow faster. This in turn will intensify the compactness heterogeneity (Williams and Doran 2000). The heterogeneity is not desirable and may reduce volumetric yield of bioreactor and synchronous growth of biomass. A mathematical model was suggested for this phenomenon (Wyslouzil et al. 1997).

Liquid particles should enter into the bioreactor from the top and move downward through the bed of biomass by gravity and maybe gas flow. Upward flow cause entrapment of particles in bottom layer of biomass which results in nonuniform growth and necrosis of biomass (Chatterjee et al. 1997). Two ultrasonic mist designs are used to satisfy this constraint. In earlier design the mist is produced in a separate vessel by a conventional ultrasonic mist generator and is swept into top of bioreactor by air flow through a connecting pipe. In recent designs an ultrasonic atomizer is mounted on the top of bioreactor and the extra vessel is eliminated. Both of the designs were investigated in laboratory scale. Compared to bubble bioreactor, mist bioreactor produced more artemisicin from *Artemisia annua* hairy roots (Kim et al.

2002). *Tagetes patula L.* hairy roots had the highest growth rate and metabolite production rate in mist bioreactor compared with that of bubble bioreactors and trickling bed bioreactors (Suresh et al. 2005). Bubble column, mixed, spray and mist bioreactors were investigated for *Azadirachtin* production by hairy root cultivation of *Azadirachta indica* and the mist bioreactor was reported to give superior results (Srivastava and Srivastava 2012).

8.4.4.2 Trickle-bed Bioreactors

This kind of bioreactor is filled with an inert packing which has appropriate shape to produce a highly porous bed. Nutrient solution which is evenly distributed over the top of the bed trickles down and nourishes the biomass which is growing on the surface of the packing. It is a gas phase bioreactor and air flow through the bed could be from the bottom to the top or vice versa. This kind of bioreactor is conventional in biological waste treatment processes and is known as trickling filter; however it is reported to be suitable for plant cell culture as well. A 14 L tricking bioreactor was investigated for cultivation of *Hyoscyamus muticus* hairy roots. It was filled with Intalox which is a conventional packing in chemical industry. The bioreactor was operated as bubble column in initial stages until the roots were immobilized onto the bioreactor matrix. Then it was shifted to trickling mode and the nutrient solution was sprayed on top of the bed. Biomass growth rate was higher than that in flasks and a final biomass of 752 g fresh weight per L (36 g-DW/L) was achieved which is considerably higher than liquid phase bioreactor volumetric yields. The presence of rigid packing that supports biomass seems to permit scale up of this type of gas phase bioreactors to large volumes. It is claimed that trickling bed bioreactors as large as 10,000 L are feasible. Lower energy consumption for liquid spray is another advantage of this bioreactor compared with other gas phase bioreactors which use finer sprays (Ramakrishnan and Curtis 2004).

8.4.5 Disposable Bioreactors

An important recent advance in plant cell and tissue culture is the use of disposable bioreactors in which expensive stainless steel vessel are replaced with a cheap bag or vessel made from plastic. Various disposable sensors, mixers, connections, and tubing are being developed which make further reductions in capital investment. Capital investment needed for production of a drug by disposable bioreactors maybe an order of magnitude lesser than that needed for production of the same product by conventional stainless steel units. Short construction and installation time is the second advantage of disposable bioreactors over conventional units which their respective time may be more than 5 years. The long duration increases the investment risk (Lee et al. 2011). Furthermore, as mentioned before, modular approach with all of its advantages may become economically feasible by the use of cheap disposable bioreactors.

Another advantage of disposable bioreactors comes from the fact that cleaning is needed between consecutive batches of conventional bioreactors. This is a costly and time-consuming operation especially in pharmaceutical and food industries in which strict standards should be met (Weathers et al. 2010). Disposable bioreactors need no cleaning as they are single use. Furthermore, disposable bags are easily compacted and sterilized by gamma irradiation or autoclaving. Sterilized and ready-to-use bags are offered by manufacturing companies.

Disposable bioreactors are used for both liquid phase and gas phase cultivations. Wave bioreactors are familiar liquid phase bioreactors which are used for plant cell cultures as well as animal and microbial cell cultures. The sterile bag is partially filled with culture medium and is put on a smoothly oscillating plate. The bag includes a perfusion system for gas or substrate injection. Wave bioreactors are commercialized and Wave Biotech Company offers them. Bubble column and mechanically mixed disposable bioreactors are also reported. PBS Biotech Company produced fully integrated and pre-configured 500 L modular bioreactors with disposable vertical bags which are equipped with wheels which are driven by magnetic coupling between the wheel and housing unit or by buoyant force of gas bubbles that are introduced from the sparger at the bottom and are captured in the air cups on the wheel periphery. PBS bioreactors are presently used for animal cell cultures. PBS bioreactors as large as 5000 L are expected to become available (Loffelholz et al. 2014).

Size of the disposable bag is limited by the weight of the contained liquid. This limitation is circumvented by putting it in a rigid case or cage. A cage bioreactor is used by Protalix Company in Israel for production of Taliglucerase alfa, a medical recombinant protein commercially known as Elelyso, by cultivation of carrot cells (Weathers et al. 2010). Mibelle Biochemistry, another health care company, uses disposable bioreactors for production of valuable metabolites via plant cell and hairy root cultures. Many of the reported gas phase bioreactors consist of disposable bags or vessels. ROOTec Bioactives Ltd., a biotech company which was recently acquired by Green2Chem S.A, offered a mist bioreactor having a disposable vessel.

Rigid disposable vessels are made from polystyrene or polycarbonate. Flexible bags may be made from mono layer or multilayer films. In each case the layer in contact with culture medium which should be inert, sealable and flexible, have only trace amounts of extractable compounds, and possess acceptable resistance to chemicals. Diffusion and torsion is made from polypropylene, polyethylene, or ethylene vinyl alcohol. Multilayer bags consist of an inner contact layer; a middle layer which acts as a barrier for gases and vapors and is usually made from ethylene vinyl acetate copolymer or poly vinyl dichloride; and outer layer which improves mechanical stability may be the same as the connecting layer or otherwise made from poly acetylene or poly ethylene terephthalate. The layers are bonded to each other by poly olefin gluing layers. Mono or multilayer films are shaped into bags by one of the welding methods namely ultrasonic, laser, pulse, or constant heating. Standard bags are cubic but bags with cylindrical shape are also produced (Vanhamel and Masy 2011). Conventional sterilize able thermoplastic tubes with appropriate welding and sealing property and chemical compatibility are used in disposable bioreactors (Lehmann et al. 2014). Aseptic tube welders and disconnecting devices are manufactured by several companies for welding of two thermoplastic tubes or disconnection of a tube without disruption of sterility. Peristaltic pumps, diaphragm pumps, and syringe pumps are useable for pumping liquid in the disposable systems (Rothe and Eibl 2011).

Disposable fittings and connectors with various shapes and sizes are developed by the manufacturers. Some of them are pre-sterilized and offered as ready-to-use aseptic packs. "Steam in place" connectors, i.e., sterilize able with steam, are offered for connection of disposable parts to conventional stainless steel systems. Special fittings are developed for aseptic connection of conventional pH and DO probes into disposable bioreactors; Kleenpak sterile connector offered by Thermo Fisher Scientific Company is one of the first ones (Rothe and Eibl 2011).

Most of the present disposable bioreactors rely on conventional systems for mixing, monitoring, and control. This make bioreactor periphery pell-mell and full of various connections and furthermore prevents complete elimination of two costly and time-consuming operation, i.e., cleaning in place (CIP) and sterilization in place (SIP). Attempts are in progress to produce disposable sensors, mixers, and other peripheral devices. The target is to make fully disposable plug and play bioreactor systems. Plug and play systems reduce installation time and contamination risk and are applicable by non-expert operators. Low shear disposable wheels developed by PBS Company remove the need for conventional mechanical mixers where insertion of their shaft into bioreactor increases contamination risk and the shear stress created by them restricts bioreactor scale up.

Disposable sensors for temperature, pressure, pH, and dissolved oxygen concentration (DO) are already offered by vendors. More limited models of disposable sensors are also released for flow rate, conductivity, carbon dioxide concentration, cell density, glucose, lactate, glutamine, and glutamate. Disposable sensors for pH, DO, and DCO2 are optic based and use dye which is sensitive to measured parameters. Increase of viscosity and rheology change from Newtonian to non-Newtonian disturb their measurement. Furthermore, the dye may be destructed by phytochemical reactions and so optical sensors are not recommended for light requiring cultivations. Disposable sensors for pressure, temperature and conductivity use semiconductors, and cell density sensors are capacitance based. Sensors which measure chemical compounds use ion selective electrodes which are based on enzymatic reactions (Lehmann et al. 2014).

8.5 Conclusions and Future Prospects

Bioreactors are potential alternatives of farms and nature for production of valuable plant metabolites. To actualize the potential, production costs should be reduced by conventional optimization experiments, metabolic engineering tools, and bioreactor design innovations. Along with enhancements in conventional liquid phase bioreactors which are suited for large scale cultivation of suspended cell cultures, advances in gas phase bioreactors are ongoing. The latter bioreactors seem to be more appropriate for hairy roots cultures and can lead to higher volumetric productivities.

Modular approach is a more convenient way for scale up of gas phase bioreactors and development of disposable bioreactor technology may make this approach economically feasible. Both rigid and flexible disposable vessels, tubing, and connections are already used for bioreactors and various disposable sensors are under development. Being single use, they should be made as cheap as possible to cut production costs.

References

Akita M, Shigeoka T, Koizumi Y, Kawamura M (1994) Mass propagation of shoots of Stevia rebaudiana using a large scale bioreactor. Plant Cell Rep 13(3):180–183

Alvard D, Cote F, Teisson C (1993) Comparison of methods of liquid medium culture for banana micropropagation. Plant Cell Tiss Org Cult 32(1):55–60

Baque MA, Moh SH, Lee EJ, Zhong JJ, Paek KY (2012) Production of biomass and useful compounds from adventitious roots of high-value added medicinal plants using bioreactor. Biotechnol Adv 30(6):1255–1267

Buitelaar RM, Langenhoff AAM, Heidstra R, Tramper J (1991) Growth and thiophene production by hairy root cultures of Tagetes patula in various two-liquid-phase bioreactors. Enzyme Microb Technol 13(6):487–494

Carter W (1942) A method of growing plants in water vapor to facilitate examination of roots. Phytopathology 732:623–625

Chatterjee C, Correll MJ, Weathers PJ, Wyslouzil BE, Walcerz DB (1997) Simplified acoustic window mist bioreactor. Biotechnol Tech 11(3):155–158

Davies KM, Deroles SC (2014) Prospects for the use of plant cell cultures in food biotechnology. Curr Opin Biotechnol 26:133–140

Florez SL, Curtis MS, Shaw SE, Hamaker NK, Larsen JS, Curtis WR (2016) A temporary immersion plant propagation bioreactor with decoupled gas and liquid flows for enhanced control of gas phase. Biotechnol Prog 32:337–345

Gonzalez EJ (2005) Mass propagation of tropical crops in temporary immersion systems. In: Hvoslef-Eide AK, Preil W (eds) Liquid culture systems for in vitro plant propagation. Springer, Dordrecht, pp 197–211

Kim YJ, Weathers PJ, Wyslouzil BE (2002) Growth of Artemisia annua hairy roots in liquid and gas phase reactors. Biotechnol Bioeng 80(4):454–464

Klvana M, Legros R, Jolicoeur M (2005) In situ extraction strategy affects benzophenanthridine alkaloid production fluxes in suspension cultures of Eschscholtzia californica. Biotechnol Bioeng 89(3):280–289

Kusakari K, Yokoyama M, Inomata S, Gozu Y, Katagiri C, Sugimoto Y (2012) Large-scale production of saikosaponins through root culturing of Bupleurum falcatum L. using modified airlift reactors. J Biosci Bioeng 113(1):99–105

Lee Parsons CWT, Shuler ML (2002) The effect of ajmalicine spiking and resin addition timing on the production of indole alkaloids from Catharanthus roseus cell cultures. Biotechnol Bioeng 79(4):408–415

Lee B, Langer E, Zheng R (2011) Next-generation single-use bioreactor technology and the future of biomanufacturing: a summary from the manufacturer's and user's perspective. In: Ebil R, Ebil D (eds) Single use technology in biopharmaceutical manufacture. John Wiley & Sons, Hoboken, pp 53–67

Lehmann N, Dittler I, Lamasa M, Ritala A, Rischer H, Eibl D, Oksman-Caldentey KM, Ebil R (2014) Disposable bioreactors for cultivation of plant cell cultures. In: Paek KY, Murthy HN, Zhong JJ (eds) Production of biomass and bioactive compounds using bioreactor technology. Springer, Dordrecht, pp 17–46

Loffelholz C, Kaiser SC, Werner S, Eibl D (2014) CFD as a tool to characterize single use bioreactors. In: Ebil R, Ebil D (eds) Single-use technology in biopharmaceutical manufacture. Wiley, Hoboken, pp 263–279

Luitjens A, Pralong A (2011) Going fully disposable - current posibilities: a case study from crucell. In: Ebil R, Ebil D (eds) Single use technology in biopharmaceutical manufacture. John Wiley & Sons, Chicester, pp 53–67

Mamun NHA, Egertsdotter U, Aidun CK (2015) Bioreactor technology for clonal propagation of plants and metabolite production. Front Biol 10(2):177–193

Nosov AM (2012) Application of cell technologies for production of plant-derived bioactive substances of plant origin. Appl Biochem Microbiol 48(7):609–624

O'Neill KM, Schilthuis AM, Leiter CA, Neihaus KM, Judge NA, Twiddy E, O'Brien AD, Curtis WR (2015) Scale-up of transgenic tobacco cells that express intimin of enterohemorrhagic Escherichia coli O157: H7 for use as a transitional platform for an oral cattle vaccine. In Vitro Cell Dev Biol Plant 51(3):315–323

Piehl G-W, Berlin J, Mollenschott C, Jr L (1988) Growth and alkaloid production of a cell suspension culture of Thalictrum rugosum in shake flasks and membrane-stirrer reactors with bubble free aeration. Appl Microbiol Biotechnol 29(5):456–461

Ramakrishnan D, Curtis WR (2004) Trickle-bed root culture bioreactor design and scale-up: growth, fluid dynamics, and oxygen mass transfer. Biotechnol Bioeng 88(2):248–260

Rothe S, Eibl D (2011) Systems for coupling and sampling. In: Ebil R, Ebil D (eds) Single use technology in biopharmaceutical manufacture. John Wiley & Sons, Chicester, pp 53–67

Srivastava S, Srivastava AK (2012) In vitro azadirachtin production by hairy root cultivation of Azadirachta indica in nutrient mist bioreactor. Appl Biochem Biotechnol 166(2):365–378

Stiles AR, Liu C-Z (2013) Hairy root culture: bioreactor design and process intensification. In: Biotechnology of hairy root systems. Springer, Berlin, pp 91–114

Suresh B, Bais HP, Raghavarao K, Ravishankar GA, Ghildyal NP (2005) Comparative evaluation of bioreactor design using Tagetes patula L. hairy roots as a model system. Process Biochem 40(5):1509–1515

Tanaka H, Nishijima F, Suwa M, Iwamoto T (1983) Rotating drum fermentor for plant cell suspension cultures. Biotechnol Bioeng 25(10):2359–2370

Tikhomiroff C, Allais S, Klvana M, Hisiger S, Jolicoeur M (2002) Continuous selective extraction of secondary metabolites from Catharanthus roseus hairy roots with silicon oil in a two liquid phase bioreactor. Biotechnol Prog 18(5):1003–1009

Townsley PM, Webster F, Kutney JP, Salisbury P, Hewitt G, Kawamura N, Choi L, Kurihara T, Jacoli GG (1983) The recycling air lift transfer fermenter for plant cells. Biotechnol Lett 5(1):13–18

Vanhamel S, Masy C (2011) Production of disposable bags: a manufacturer's report. In: Eibl R, Eibl D (eds) Single-use technology in biopharmaceutical manufacture. John Wiley & Sons, Chicester, pp 113–134

Weathers PJ, Zobel RW (1992) Aeroponics for the culture of organisms, tissues and cells. Biotechnol Adv 10(1):93–115

Weathers P, Liu C, Towler M, Wyslouzil B (2008) Mist reactors: principles, comparison of various systems, and case studies. Elect J Integ Biosci 3(1):29–37

Weathers PJ, Towler MJ, Xu J (2010) Bench to batch: advances in plant cell culture for producing useful products. Appl Microbiol Biotechnol 85(5):1339–1351

Williams GRC, Doran PM (2000) Hairy root culture in a liquid dispersed bioreactor: characterization of spatial heterogeneity. Biotechnol Prog 16(3):391–401

Wu C-H, Murthy HN, Hahn E-J, Paek K-Y (2007) Enhanced production of caftaric acid, chlorogenic acid and cichoric acid in suspension cultures of Echinacea purpurea by the manipulation of incubation temperature and photoperiod. Biochem Eng J 36(3):301–303

Wyslouzil BE, Whipple M, Chatterjee C, Walcerz DB, Weathers PJ, Hart DP (1997) Mist deposition onto hairy root cultures: aerosol modeling and experiments. Biotechnol Prog 13(2):185–194

Ziv M (2005) Simple bioreactors for mass propagation of plants. In: Hvoslef-Eide AK, Preil W (eds) Liquid culture systems for in vitro plant propagation. Springer, Dordrecht, pp 79–93

Chapter 9
Cisgenesis and Intragenesis as New Strategies for Crop Improvement

Mahdi Moradpour and Siti Nor Akmar Abdullah

9.1 Introduction

Crop genetic improvement can be achieved through three different ways: sexual crosses between selected genotypes of plants, mutation breeding and adding novel genetic material resulting in genetic modification of crops (Joshi et al. 2009). The transfer of genetic information between species through crossing followed by repeated backcrossing is called introgression.

Conventional breeding concentrates on transfer and introgression of valuable genes found particularly in wild genetic material into the commercial varieties. Introgressive breeding combines genes and genetic elements from a donor with recipient plant. Subsequent repeated backcrossing with one of the parents together with selection produces the desirable traits. An example of a collection of introgression lines (ILs) is the development of a new melon IL in the background of Charentais genetic. The genomic regions associated with flowering and fruit quality traits that have been identified formed the basis for the genetic collection (Perpiñá et al. 2016).

M. Moradpour • S.N.A. Abdullah (✉)
Institute of Plantation Studies, Universiti Putra Malaysia, 43400 UPM, Serdang, Selangor, Malaysia
e-mail: snaa@upm.edu.my

S.N.A. Abdullah et al. (eds.), *Crop Improvement*,
DOI 10.1007/978-3-319-65079-1_9

Conventional plant breeding is a deep-rooted method for plant improvement that is increasingly becoming advanced as it infuses with new developed technologies. It has benefitted from the employment of molecular technologies for different tasks to boost its programmes without combining with alien genes into the end product of the bred plant variety (Acquaah 2015). Existing variability within a species or its close relatives that are sexually compatible is useful for genetic improvement through conventional breeding (Acquaah 2015). Breeding efficiency for nutritional and/or quality enrichment and biotic and/or abiotic stresses depends on the variability for the traits of interest. Mutation breeding is the key for the production of exploitable and utilizable variation (Ankita et al. 2015). However, the majority of the mutations are receding. In seed propagated crops, this feature does not cause any problem where homozygous recessive trait or genotypes can be obtained. Nevertheless, mutations have played a crucial role in the generation of numerous successful planted cultivars globally.

The final product from conventional breeding is a plant containing a gene of interest from the donor plant in combination with the unwanted genes leading to linkage drag (Jacobsen and Schouten 2007). These problems will not arise if specific target genes are transferred utilizing newly developed breeding techniques such as transgenesis and cis/intragenesis. Transgenesis modifies the genetic of a receiver plant with one or more genes of interest from an unrelated donor plant or from any species including prokaryotic organisms (Schouten et al. 2006a). It allows stable integration of transferred entire gene(s) or partial gene(s) derived from a foreign organism. The foreign gene is introduced into the recipient genome followed by a selection method to confirm insertion of the foreign gene (Joshi 2010). Gene insertion is an atypical process and does not occur in nature. In this process, the genetic material such as gene sequences from different sources containing coding sequence is fused to the promoter from a different gene or produced synthetically in the sense or antisense orientation (Ankita et al. 2015). The gene construct is introduced into the host plant cell via direct transfer such as using biolistic gun or indirectly through *Agrobacterium*-mediated techniques (Hansen 2000). Transgenesis dramatically increased the promises for crop improvement through expanding the pool of traits that could be integrated into the plant genome. It breaks the reliance of breeders on existing biodiversity or on the genes of close relatives of a particular crop for additional genetic diversity because the transformation technology allows for the introduction of exogenous genes including those that do not exist naturally in the target crop (D'Halluin and Ruiter 2013).

However, food derived from genetically modified crops (GM crops) received wide public opposition because people are concerned about the insufficiency of existing regulation, in particular the unnaturalness of the genetic modification approaches. Hence, scientists are now considering insertion of genes derived from the plant itself or its closed relatives through two strategies which are closely related known as cisgenic and intragenic modifications. In principle, the techniques confined the transfer of genes to closely related or sexually compatible species of plants and do not involve any foreign DNA (Mielby et al. 2013; Hunter et al. 2014).

9.2 Cisgenesis

Schouten et al. (2006a) defined a cisgenic plant as the alteration of a receiver plant genome using native gene or genes from a closely related species. The native gene contains intron and flanking regions such as its own regulatory sequences in the sense orientation identical to that found in the donor plant. In principle, cisgenic crop is generated through transferring a native and entire copy of a natural gene complete with its own regulatory regions and maintaining its natural genetic elements (Schouten et al. 2006a, b). More than one cisgenes can be inserted into the cisgenic plants. In cisgenesis, foreign genes including the vector backbone genes and the selectable marker genes are not found in the final product.

The difference between cisgenesis and conventional breeding is that cisgenic crops contain only the desired gene(s) and there are no other genes being transferred (Espinoza et al. 2013). Indeed, in this way, cisgenesis can overcome a major bottleneck for crop improvement faced by traditional breeding due to the linkage drag issues (Schouten et al. 2006a).

Unlike transgenesis, in the cisgenic approach, plants receive genes only from crossable species via genetic engineering and those imported genes are under the control of their own native regulatory components in their natural orientation (Schouten et al. 2006a). Moreover, external genetic material like selectable marker genes is absent. A major worry in transgenesis is the rapid spread of transgenes to environment through the flow of pollens (Holme et al. 2013), while in the event of cisgenesis, the cisgenes are taken from the close wild relatives. Thus, cisgenesis is as safe as conventional breeding. It integrates the conventional breeding with advanced biotechnology and intensely accelerate the process of breeding by reduction of the undesirable effects of linkage drag (Rommens et al. 2007).

A variety of cisgenic crops with different traits (Table 9.1) have been developed such as cisgenic durum wheat with enhanced baking quality (Gadaleta et al. 2008), cisgenic potato resistant to late blight (Haverkort et al. 2009), cisgenic apple with scab resistance (Vanblaere et al. 2011), cisgenic grapevine with fungal disease resistance (Dhekney et al. 2011), cisgenic poplar with different growth types (Han et al. 2011) and cisgenic barley with enhanced grain phytase activity (Holme et al. 2012). Figure 9.1 illustrates a schematic image of different approaches applied for genetic modification in plants based on the definitions of transgenesis and cis/intragenesis.

9.3 Intragenesis

In intragenesis, different plant genetic components are recombined in vitro to produce an expression gene construct that is introduced into a plant within the same sexually compatible gene pool (Rommens 2004; Rommens et al. 2007). Therefore, in intragenesis the source of the genes for the genetic modification is from the same or a crossable species. Unlike cisgenes, intragenes are hybrid genes. Selected

Table 9.1 List of crops developed with different traits through cisgenesis or intragenesis

	Crops	Intra-/ cisgenesis	Type	Gene	Trait	References
Seed propagated self-pollinating crops	Durum wheat (*Triticum turgidum* var. durum)	(Cis)	Introduction of gene from related species	*1Dy10*	Improved baking quality	Gadaleta et al. (2008)
	Barley (*Hordeum vulgare*)	(Cis)	Overexpression	*HvPAPhy_a*	Improved grain phytase activity	Holme et al. (2012)
	Wheat (*Triticum aestivum*)	(Cis)	Introduction of genes only from closely related species	Wheat class I chitinase gene	Fungal pathogens resistance	Maltseva et al. (2014)
Outcrossing species	Perennial ryegrass (*Lolium perenne*)	(Intra)	Overexpression	Lpvp1	Drought tolerance	Bajaj et al. (2008)
	Alfalfa (*Medicago sativa*)	(Intra)	Silencing	Comt	Reduced levels of lignin	Weeks et al. (2008)
Woody species	Poplar (*Populus* spp.)	(Cis)	Overexpression	Genes involved in growth	Different growth types drought	Han et al. (2011)

Crops with commercially widespread clones	Potato (*Solanum tuberosum*)	(Intra)	Silencing	*GBSS*	High amylopectin	de Vetten et al. (2003)
	Potato (*Solanum tuberosum*)	(Intra)	Silencing	*Ppo*	Preventing black spot bruise	Rommens et al. (2004)
	Strawberry (*Fragaria* spp.)	(Intra)	Overexpression	*PGIP*	Grey mould resistance	Schaart (2004)
	Potato (*Solanum tuberosum*)	(Intra)	Silencing	*Ppo, R1, PhL*	Limiting degradation of starch. Limiting acryl-amide formation	Rommens et al. (2006)
	Potato (*Solanum tuberosum*)	(Intra)	Silencing	*StAs1, StAS2*	Limiting acrylamide formation	Rommens et al. (2008)
	Potato (*Solanum tuberosum*)	(Cis)	Introduction of genes from related species	R-genes	Late blight resistance	Haverkort et al. (2009)
	Melon (*Cucumis melo* L.)	(Cis)	Introduction of genes from related species	*At1/At2*- glyoxylate aminotransferase	Downy mildew resistance	Benjamin et al. (2009)
	Potato (*Solanum tuberosum*)	(Intra)	Silencing	*StAs1*	Limiting acrylamide formation	Chawla et al. (2012)
	Apple (*Malus domestica*)	(Cis)	Gene from related species	*HcrVf2*	Scab resistance	Joshi et al. (2009), Vanblaere et al. (2011), Krens et al. (2015), Würdig et al. (2015), Chizzali et al. (2016)
	Grapevine (*Vitis vinifera*)	(Cis)	Introduction of gene from related species	VVTL-1	Fungal disease resistance	Dhekney et al. (2011), Dalla Costa et al. (2016)

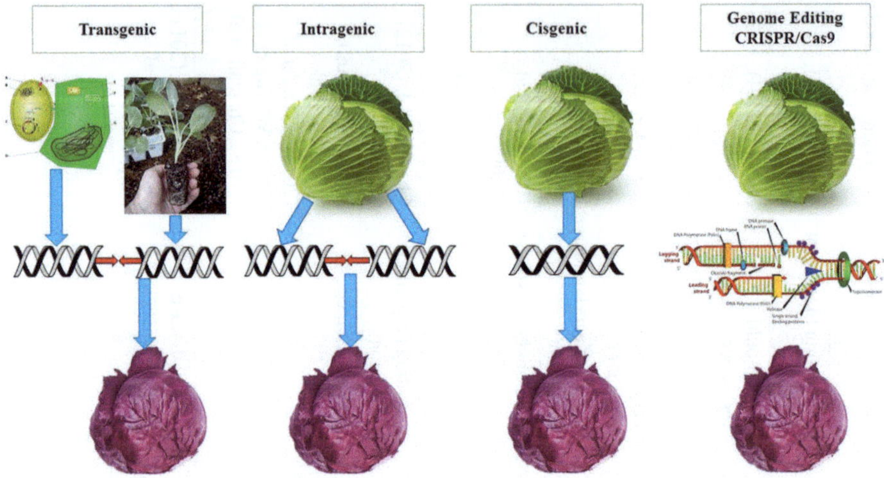

Fig. 9.1 Transgenic, intragenic, cisgenic and genome editing for genetic modification of crop plants

genetic elements from different genes can be combined in intragenesis (Rommens et al. 2007). Therefore, intragenesis approach which allows the genetic elements to be rearranged is obviously less restrictive in comparison to cisgenesis. A list of crops that have been commercially developed with different traits through intragenesis is given in Table 9.1.

It is essential for *Agrobacterium*-mediated transformation to use sequences of T-DNA border from sexually compatible DNA pool (P-DNA borders). Methods such as PCR analyses and intensive database searches can be used to determine successful transfer of the targeted plant DNAs. The plant phenotype obtained through intragenesis is not achievable through traditional breeding because the expression profile of the freshly assembled gene may differ from that observed naturally (Schaart and Visser 2009; Devi et al. 2013).

In intragenesis, antisense or RNA interference (RNAi) can be employed for silencing the gene(s) (Schaart and Visser 2009). Intragenic potatoes with enhanced processing qualities was produced by reducing black spot bruise through silencing the polyphenol oxidase gene (*PPO*) (Rommens et al. 2004). Reduction of cold-induced sweetening was also achieved through gene silencing to limit acrylamide formation (Rommens et al. 2006). Intragenesis approach was employed to develop non-browning apples through RNAi silencing of the apple *PPO* gene. Moreover, overexpression of polygalacturonase inhibiting protein to reduce fungal polygalacturonase effect was done through intragenic strawberry to increase resistance to grey mould (Schaart 2004).

9.4 Development of Cis/Intragenics: General Outline

In cisgenic and intragenic plants, integration of genes taken from a related and sexually compatible species into a desirable background is accomplished in the same way as in transgenesis with a few extra steps (Ankita et al. 2015). Prerequisites of cis/intragenic approach include the plant sequence information, the isolation and characterization of genes of interest from crossable relatives, cis/intragenic vectors within the P-DNA borders.

The major steps involve isolation of the DNA that codes for the protein, insertion of DNA into a plasmid vector, transformation using an appropriate method, elimination of selectable marker genes and, finally, selection of recombinants containing the inserts. The same traditional transformation methods are used to produce cisgenic plants with many unknown effects in the final cisgenic organism and marketed product due to the gene insertion (Hunter 2014).

Agrobacterium-mediated transformation has been used more frequently to produce cisgenic plants. The practice of biolistic transformation technique has been less documented (Romano et al. 2003; Akhond and Machray 2009; Lusser et al. 2012). Comparison of two direct DNA transfer methods showed that both biolistic transformation and protoplast transformation are suitable for transferring cisgene into poplars (Nietsch et al. 2017).

A critical requirement to produce environmental friendly cisgenic plants is the elimination of selectable marker genes. The final cisgenic plant will contain the gene of interest from cross compatible species. Additionally, foreign genes such as vector backbone and selectable marker genes should not be present in the primary intra/cisgenic transformants or their progenies (Holme et al. 2013). Different approaches allowing the construction of marker-free genetically modified plants (GM plants) have been described in the literature (Dalla Costa et al. 2016), principally based on the target plant and the efficiency of the transformation technique used (Krens et al. 2003).

9.5 Special Techniques to Generate Cis/Intragenic Plants

To consider cisgenesis as one of the potential technologies for the future of agriculture, the Commission Implementing Regulation (EU) No 503/2013 (European Parliament 2013) clarifies that the applicant shall therefore aim to develop genetically modified organisms (GMOs) without the use of antibiotic resistance marker genes arguing that it is now possible to develop GMOs without the use of antibiotic resistance marker genes (Breyer et al. 2014). Hence, the availability of efficient systems to avoid or eliminate selectable marker genes is becoming progressively more important. The transfer of an uncomplicated vector containing only the promoter, the target gene and the terminator is the basic design (de Vetten et al. 2003). Nevertheless, this method needs intense PCR screening for positive transformation events, and is hence costly, time-consuming and feasible only for biolistic transformation technique (Vidal et al. 2006) or effective *Agrobacterium*-mediated gene transfer protocols combined with highly transformable species (Malnoy et al. 2010; Ballester et al. 2008; Petri et al. 2011).

9.5.1 Transformation Without Selection

Cis/introgenic plants must be free from other added sequences such as selectable marker genes. A transformation procedure without selectable marker genes was developed to produce intragenic potato enriched in amylopectin. This procedure was also employed to produce cisgenic potatoes with late blight resistance trait (de Vetten et al. 2003). A technique was developed to produce potatoes free from selectable marker with improvement in processing characteristics but the selectable marker gene was transiently expressed initially (Rommens 2004; Rommens et al. 2006).

Isopentenyl transferase is encoded by the bacterial *ipt* gene that catalyses cytokinin isopentenyl adenosine formation in plants (Barry et al. 1984). Introduction of the *ipt* gene into the vector construct harbouring the target gene enables identification of the transformants based on cytokinin-induced abnormality (Rommens et al. 2004). Nevertheless, Richael et al. (2008) reported that the spreading of cytokinin isopentenyl adenosine produced by these plants into the culture medium leads to increase in the number of transformants with no vector backbone regenerated on the same medium. Thus, this method allows the selection of transformants free from selectable marker genes and vector backbone simultaneously.

A Dutch company (Avebe) in collaboration with the University of Wageningen produced an intragenic potato with high-amylopectin without using vector backbone and selectable marker sequences (Rommens 2004; de Vetten et al. 2003) due to the high transformation efficiency. Potato starch contains 80% amylopectin and 20% amylose. Both polysaccharides have different properties and industrial applications. Chemical separation of these two polysaccharides is environmentally unsustainable and costly. Available methods in conventional breeding are not able to provide potatoes without amylopectin. The intragenic alfalfa plants were produced through a highly competent and efficient transformation system that enables enough production of transformants without using selectable marker genes (Weeks et al. 2008).

The transformation efficiencies of vegetatively propagated crops like strawberry are low compared to potato. Therefore, selection cannot be excluded (Holme et al. 2013). For production of intragenic strawberries, the marker deletion method based on site-specific recombination was first developed (Schaart et al. 2004) and later applied to generate cisgenic and intragenic apple (Joshi et al. 2011; Vanblaere et al. 2011).

The site-specific recombinase strategy is based on the R/*Rs* system. Transformation and selection is initially performed with selectable marker genes that are flanked with *Rs* recombination target sites. The transformed plantlets are cloned to multiply a few plants before induction of the R recombination activity due to having low transformation efficiencies. In this way, the opportunity to obtain at least one clone from each transformant without any selectable marker is increased (Holme et al. 2013).

de Vetten et al. (2003) showed that transformation of marker-free plant in the vegetatively propagated species like potato is possible by using PCR subsequently for selection of the transformed plantlets. Transformation without selection in apple was carried out by Malnoy et al. (2007). Since only a few cells within a tissue are successfully transformed, most of the regenerated apple shoots are non-transformed plantlets. All the regenerated plantlets were analysed by PCR to identify the GM clones containing the target gene(s). However, the risk associated with regeneration of chimeric plants with only certain tissue harbouring the introduced genes is high due to the lack of selection pressure.

9.5.2 Co-transformation

Another efficient strategy to produce marker-free plants is co-transformation. There are three approaches for the co-transformation. The first approach involves one plasmid harbouring two T-DNAs in one strain of *Agrobacterium* (Breitler et al. 2004; Matheka et al. 2013; Ling et al. 2016), while the second approach contains two T-DNAs placed on two different plasmids in the same strain of *Agrobacterium* (Parkhi et al. 2005; Sripriya et al. 2008) and the third approach contains two T-DNAs in distinct strains of *Agrobacterium* (Dutt et al. 2012; Ling et al. 2016).

Disassembled integration might be the result of co-transformation of plants with both T-DNA vectors, the first harbouring the marker gene and the second having the target gene which can segregate in the progeny. The step of segregation could be avoided using co-transferring of two T-DNAs, one harbouring a selection gene and the other having the target gene, followed by a selection step. The objective is to integrate the T-DNA having the target gene only while another T-DNA which harvests the selectable marker gene will be expressed transiently and activated without being integrated into nuclear or plastid DNA (Rommens et al. 2004). Such an approach was successfully employed to develop cisgenic durum wheat (Gadaleta et al. 2008) and barley (Holme et al. 2012).

Agrobacterium-mediated plant transformation protocols can be modified to enable integration of the selectable marker gene and the transgene either into two different genomic loci or into two different chromosomes. This allows the two loci to be segregated through breeding, and cell line without marker gene but contains the target transgene can be produced (Woo et al. 2009). Selectable marker segregation should reach 25% of all co-transformed cell lines for the co-transformation to be regarded as a mature and efficient technology. However, screening of co-transformation lines becomes more costly and tedious because it increases the number of cell lines to be checked by fourfold. Functionally unrelated selectable marker gene sequences that are separated from the new transgenic trait will totally stop appearing in the mature plant (Konig 2003).

9.5.3 Site-Specific Recombination

Site-specific recombination is an alternative system to remove selectable marker genes and it was reported for the first time about 20 years ago (Dale and Ow 1991; Russell et al. 1992). These techniques use vectors that positioned the selectable marker and the recombinase genes in between two directly repeated recombinase recognition sites (Dalla Costa et al. 2016). Microbial site-specific recombinases cleave DNA at specific sites followed by ligation to the cleaved DNA at another target sequence. The removal of foreign DNA placed in between the recognition sites in a direct repeat orientation has served as the basis for eliminating unwanted transgenic DNA fragment from the nuclear genome of plants. Among the widely used recombination systems include Cre/*lox* from bacteriophage P1 (Hoess et al. 1982; Hoess and Abremski 1985), R/*RS* from *Zygosaccharomyces rouxii* (Araki et al. 1985) and FLP/*FRT* from *Saccharomyces cerevisiae* (Cox 1983; Senecoff et al. 1985).

The DNA sequence that encodes for the selectable marker gene is eliminated by the recombinase enzyme cutting two sequences for DNA recognition and ligating the free ends after the elimination of the DNA sequence placed in the middle. The site-directed nuclease system enabled targeted modification of a specific genomic sequence by replacing and integrating a different gene, which may be trans-, cis- or intragenic. This process is referred to as gene targeting that enables big regions of homologous DNA to be added. The area of the genome that needs to be modified is flanked by this homologous DNA. It does not exactly encircle the double strand breaks (DSBs) but may be adjacent within a minimum of a few hundred bases until a few kilobases, as processing of the DSBs may reach a long distance (Zhu et al. 2008; Hartung and Schiemann 2014).

The combination of the homologous DNA and a DSB increases the chance of homologous recombination. In homologous recombination, the middle sequence flanked by the homologous DNA is exchanged for the sequence encoded on the template DNA. Homologous recombination that is dependent on DSBs repair system has wide applications including for excision, alteration or integration of new genes (D'Halluin et al. 2008; Cai et al. 2009; Shukla et al. 2009; Hartung and Schiemann 2014). The strategy that relies on site-directed nuclease is quite similar to the crosses and wide crosses that occurred in traditional breeding. Removal of selectable marker gene by this technique should be further confirmed by other molecular biology techniques (Espinoza et al. 2013). Intragenic strawberries (Schaart 2004) as well as cisgenic and intragenic apples (Joshi et al. 2009; Vanblaere et al. 2014) have been produced by marker deletion methods based on site-specific recombination.

The bacteriophage P1 Cre/*lox* system that is mainly used in plants depends on three strategies: auto-excision, transient transfer and outbreeding with cell line harbouring Cre. These systems belong to the tyrosine recombinase family (Gidoni et al. 2008; Wang et al. 2011). After the reaction, a recombination site (*lox*, *FRT* or *RS*) stays in the genome and forms the site for integrative recombination. However, there have been no report on re-insertion of the eliminated fragment (Zuo and Chua 2000; Ebinuma et al. 2001), possibly due to the fact that excision is an intramolecular event, whereas integration requires interaction between unlinked sites, and since the excised circle cannot replicate autonomously, there is possibility for it to be rapidly lost in vivo (Hare and Chua 2002).

Application of the Cre/*loxP* approach has been reported in grapevine (Costa et al. 2010), apricot (Petri et al. 2012), rice (Nandy and Srivastava 2012), banana (Chong-Pérez et al. 2012), wheat (Blechl et al. 2012), citrus (Zou et al. 2013) and tobacco (García-Almodóvar et al. 2013). The R/*Rs* system has been used in citrus (Ballester et al. 2008), tomato (Khan et al. 2011), apple and pear (Righetti et al. 2014) and eggplant (Darwish et al. 2014). The FLP/*FRT* system has been successfully employed in tobacco (Woo et al. 2009), maize (Yang et al. 2009; Li et al. 2010), rice (Akbudak and Srivastava 2011; Nandy and Srivastava 2012), apple (Herzog et al. 2012) and grapevine (Dalla Costa et al. 2016).

9.5.4 pMF Marker-Free Technology Vectors

pMF vector was produced by Plant Research International, Wageningen, as a plant transformation vector for producing GM plants that do not contain undesirable DNA sequences. Its application in plant transformation allows effective removal of genes for antibiotic resistance from the transgenic plants and tissues of the transgenic plants. Elimination of selectable marker genes is important for the production of GM crops with better acceptance by consumers. Moreover, removal of selectable marker facilitates transgenes stacking by recurrent transformations.

9.5.4.1 The Principle of pMF Vector

The important feature of the pMF vector includes the presence of an inducible site-specific recombination system for the removal of unwanted DNA sequences. It allows negative selection using cytosine deaminase (*cod A*) gene to ensure the ultimate generation of completely selectable marker-free plants (Schaart et al. 2004). *Cod A* is a conditionally toxic and lethal dominant gene for the enzyme that converts non-toxic 5-fluorocytosine (5-FC) to cytotoxic 5fluorouracil (5-FU) (Schaart et al. 2004). Gleave et al. (1999) employed the negative selection capabilities of this gene to produce plants free from selectable marker gene. Kanamycin-resistant transgenic tobacco plants were able to be selected because the *lox* sequences flanked the *npt II* gene together with the *codA* gene.

9.5.4.2 Chemically Induced Marker Removal

The marker removal approach can be fitted easily in current transformation system using *Agrobacterium* (schemes in Figs. 9.1 and 9.2). Following positive selection of transgenic tissue or plants, chemical initiation of recombinase R activity will lead to

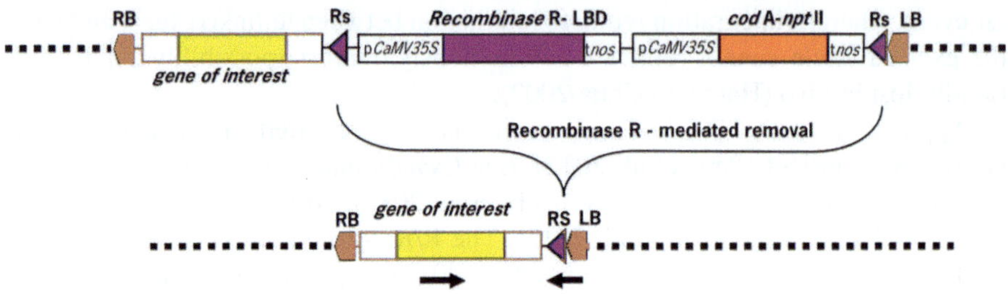

Fig. 9.2 Schematic representation describing the T-DNA of pMF1 vector before and after Recombinase R-mediated elimination of the DNA sequences (This entire system is present in a binary vector called pMF1 (Schaart et al. 2011), which is known as the marker-free system developed by Wageningen UR (https://www.wur.nl/en/Expertise-Services/Research-Institutes/plant-research/Marker-free-technology/Chemically-induced-marker-removal.htm))

the removal of DNA sequences flanked with the recombination sites (*Rs*). The recombinase R protein which is inactivated by ligand-binding domain (LBD) of the glucocorticoid receptor can be activated with overnight incubation of plant tissues in a solution containing 10 μM dexamethasone. Negative selection on 5-FC that is performed subsequently, stops the generation of plant tissues which still contain the *cod A*-gene, thus preventing the occurrence of chimeras due to excision of incomplete DNA (Joshi 2010).

9.5.4.3 The pMF1 Vector Map

The pMF1 vector (Fig. 9.3) harbours the *npt II* gene for kanamycin selection of transgenic plantlets. The two new versions, pMF2 and pMF3 contains *hpt* (hygromycin resistance) and *pat* (phosphinothricin resistance) as selectable marker gene, respectively, introduced into a unique Spe I-restriction site of pMF1. The *Spe* I-site can also be incorporated with reporter genes like *gus* or *gfp* for monitoring initial transformation efficiencies and validating the removal of marker visually. All selectable marker and reporter genes will be removed finally. So far, pMF vectors for marker-free technology has been used in a few plant species, including strawberry (Schaart et al. 2004), lily (Krens and Kronenburg-van de Ven 2011) and apple (Krens et al. 2015).

9.5.4.4 Transformation

The transformation efficiencies are comparable to that achieved using pBinplus (van Engelen et al. 1995), a pBin19 derivative which has normally been employed in transformation research. In several transgenic events, 10–50% of binary vector backbone sequences are transferred to the plant genome (De Buck et al. 2000; Lange et al. 2006). In *Agrobacterium*-mediated transformation systems transfer of backbone vector sequences is a common observation (Kuraya et al. 2004).

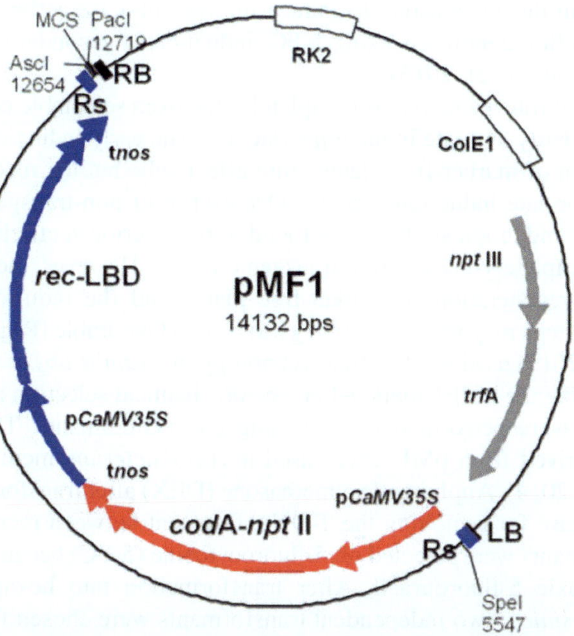

Fig. 9.3 Vector map of pMF1. *MCS* Multiple Cloning Site, *RB* Right Border of T-DNA, *Rs* Recombination Site, *rec-LBD* Fusion of Recombinase R-gene and LBD, *codA-npt II* Fusion of *codA* gene for negative selection on fluoro cytosine and *npt*II gene for positive selection on kanamycin, *SpeI* restriction site in which *hpt*II or *pat* genes have been inserted in pMF2 and pMF3, respectively, *LB* Left of T-DNA (This entire system is present in a binary vector called pMF1 (Schaart et al. 2011), which is known as the marker-free system developed by Wageningen UR (https://www.wur.nl/en/Expertise-Services/Research-Institutes/plant-research/Marker-free-technology/Chemically-induced-marker-removal.htm))

9.5.4.5 Induction of Recombinase and Negative Selection

For removal of selectable marker DNA application of negative selection and timing of induction of recombination activity have been studied (Fig. 9.1) (Schaart et al. 2004; Singh et al. 2015). Addition of dexamethasone induces the activity of recombinase (Schaart 2004). In pMF1 vector, the induction of recombinase activity together with the positive selection on kanamycin could be prevented and induction of regeneration continued under negative selection (addition of 5-FC). For example, the T-DNA of pMF1 vector containing the cisgene Rvi6 together with its native regulatory sequences and a cassette flanked by recombination sites was reported. The R recombinase gene was fused with the selectable marker genes *nptII* and *codA* to enable positive and negative selection on kanamycin- and 5-fluorocytosine selective medium, respectively (Vanblaere et al. 2014; Singh et al. 2015).

As an alternative, the late selectable marker removal strategy can be employed with continuous positive selection pressure during the initial regeneration stage of the transgenic plants. This was followed by regeneration of shoot tissue and treatment with dexamethasone to induce the recombinase activity. Finally, under negative selection conditions using 5-FC, induction of secondary regenerants was carried out (Krens et al. 2015).

Recovery of transgenic plants completely free from selectable could be achieved through both early and late induction strategies. The early induction strategy led to the production of marker-free plants quite effectively, but the recovery was not as efficient as the late induction strategy. Occurrence of non-transgenic escapes was high. The late induction strategy was found to be superior in efficiency for the generation of completely marker-free transgenic plants. However, the longer duration required for regeneration of marker-free plants and the requirement of double regeneration steps may make the strategy to be less favourable (Righetti et al. 2014).

Qi et al. (2015) used marker-free technology in *Crambe abyssinica.* After transformation using the pMF1 marker-free vector, chemical selection and meristematic regeneration were performed for screening the recombinants. The binary vector pJS-M14, derived from pMF1, was used in *Agrobacterium*-mediated transformation (Qi et al. 2014). Applying dexamethasone (DEX) after transformation activated the recombinase for removing the T-DNA fragment between recombination sites. The recombinants were selected on 5-fluorocytocine (5-FC) because *codA* converts 5-FC into toxic 5-fluorouracil. After transformation into hexaploid species of *Crambe abyssinica*, two independent transformants were chosen for DEX-induced recombination and subsequent 5-FC selection. This is different from previous pMF1 experiments because the strategy engaged stepwise selection based on meristematic regeneration. Recombinants were successfully obtained after a long period of 5-FC selection, but most of the survivors were wild type and non-recombinants.

When the pMF1 marker-free system was employed on *C. abyssinica*, increasing the DEX concentration did not enhance successful production of recombinants. Qi et al. (2015) suggested that both DEX-induced recombination and 5-FC negative selection appeared insufficient leading to the extremely high frequency occurrence of chimerism where both recombinant and non-recombinant cells exist together. The stepwise selection based on regeneration of meristem tissue was critical for the isolation of recombinant plant from the chimera. Often poor efficiencies of the DEX treatment and the subsequent 5-FC selection and the occurrence of a relatively high percentage of non-transgenic competing with the recombinant cells for growth and regeneration on 5-FC medium have been reported (Qi et al. 2015). The position effect of the T-DNA insertion might be important in determining accessibility of the *Rs* sites in the T-DNA to the recombinase (Matzke and Matzke 1998; Qi et al. 2015).

9.6 Use of Cis/Intragenesis in Improvement of Crops

Successful implementation of cisgenics technology in crop improvement strongly rely on the knowledge of the genes responsible for the desired trait. The presence of molecular markers for identification of these genes is strongly recommended as these are now established as essential tools to assist in traditional plant breeding (Collard and Mackill 2008).

Genes identification and isolation have been made much easier with the continuous achievements in plant genome sequencing. Databases that are progressive with updated information are indispensable in in silico research. There are now vast number of accessible sequences that can be explored for cisgenic or transgenic modification. Based on sequence similarities, some of these genes may be associated with specific functions. Such approach was used to identify sequences of putatively alike function through searches done on these databases for obtaining plant-derived DNAs (P-DNAs, used as alternative for conventional T-DNAs in *Agrobacterium*-mediated transformation) in selected plant genomes (Rommens et al. 2005). However, the exploitation of the recognized genes will involve comprehensive functional characterization that is costly and time-consuming. With the advent of genome editing approaches, their use in combination with other new plant breeding techniques including cisgenesis is the way forward for exploiting the plant genetic resources in crop improvement (Cardi 2016).

9.7 Advantages and Drawbacks of Cisgenesis

As the population of the world continues to increase, food remains a scare resource. Other challenges that should be given attention in crop improvement include global warming, (a)biotic stress, shortage of land resources and increasing demands for high food quality (Jacobsen and Nataraja 2008). Cisgenesis and intragenesis introduced as new breeding technologies exhibit a number of advantages over conventional plant breeding. The cisgenesis concept improves genetic modification through better accuracy, speed and sophistication of trait improvement by introducing genes from the conventional breeding pool (Jo 2013).

The first advantage in cisgenesis is the ability to achieve precise genetic modification at a faster rate in existing cultivar without any potential problem of linkage drag (Telem et al. 2013). The linkage drag refers to the problem associated with the gene of interest being tied to other undesirable ones which are sometimes deleterious genes that are being transferred together into the genome to be improved (Delwaide 2014).

Cisgenic breeding is useful for modifying the expression level of a trait that has limited allelic dissimilarity and variation within the sexually compatible gene pool (Schaart et al. 2016). Moreover, abiotic and biotic stress resistance genes from a crossable donor can be used to develop cultivars resistance to disease with obvious economic and environmental advantageous (Vanblaere et al. 2014).

Furthermore, cisgenes and intragenes basically involve exploitation of the breeders' gene pool. Cisgenes are natural genes and intragenes consist of functional components of natural genes from the crop plant itself or from sexual compatible species. Cisgenesis combines the applications of cisgenes with marker-free transformation, which basically represents a linkage drag free introgression breeding in a single step. Applications of cisgenesis and intragenesis strategies have been reported for a variety of crops (Jacobsen and Schouten 2009).

Food safety is another important privilege of cisgenesis. For example, wild species of *Solanum* was used as source of genetic variability to bring back various glycoalkaloids lost through breeding process (van Gelder 1989). Therefore, wild species can be a source of genetic variation for re-introduction of beneficial compounds that have been removed due to breeding and domestication (Jacobsen and Schouten 2008).

However, there are several drawbacks and risks associated with cisgenesis such as unknown insertion site and mutation at insertion site. A potential issue is the fact of having unforeseen risks because cisgenes are inserted into random places into the plant genome. The random insertion phenomenon of alien DNA pieces into the plant genome actually also occurs in conventional plant breeding. Induced translocation breeding in wheat (Friebe et al. 1996; Jacobsen and Schouten 2008) is one of the most prominent examples. In addition, this random insertion of one or more genes in the genome is also susceptible to the effect on expression by the genes located close to the insertion site or vice versa. de Cock et al. (2006) and Russell and Sparrow (2008) also highlighted that, contrary to what Schouten et al. (2006a) argued, cisgenic breeding may still result in adding novel traits to the cisgenic product and therefore give rise to novel hazards. The mutation made at the insertion site of the cisgene and the unexpected phenotypic changes resulting from this is another theoretical drawback of cisgenesis (Jacobsen and Schouten 2008).

9.8 Issues Associated with Genetic Modification

Precisely, cisgenic breeding is a particularly efficient method to improve plants with a long reproduction cycle for which the conventional breeding can be extremely long (Schaart and Visser 2009). Crop improvement could be accomplished by the introduction of the genes of interest or alleles only through genetic modification without co-insertion of undesirable genes (Jacobsen and Schouten 2007; Joshi et al. 2009).

Maintaining of the genetic makeup of proven cultivars is another advantage of genetic modification. For example, apple is a heterozygous and self-incompatible plant and conventional breeding can never fully be used to restore the genetic makeup of proven cultivars in the progeny. Introduction of the required genetic modifications remained a time- and labour-demanding process, as each individual alteration requires a cycle of breeding, selection and confirmation (Mans et al. 2015). Nevertheless, in the case of intragenesis and cisgenesis only one or a few genes of interest are attached to an existing cultivar.

9.9 Chimeras

9.9.1 Occurrence of Chimeras

Regeneration of multiple cells constituting both transformed and untransformed ones or after transformation of one cell in existing multicellular shoot meristem or embryogenic region can produce chimeras. A chimera consists of parts or tissues that differ in genetic composition (Joshi et al. 2009). Chimerism was observed in transformants of lime (Domínguez et al. 2004), sweet orange, pineapple, citrange (Domínguez et al. 2004; Ballester et al. 2007) and strawberry (Schaart et al. 2004) obtained from selectable marker-free transformation technologies. In plants that propagated vegetatively without fixation by a sexual phase, chimerism can be a problem (Ahuja 2013). Identifying of chimeras after genetic modification and regeneration is often difficult and generally not performed.

Recombination may also cause chimerism. The chimeras after the recombination consist of both cells free from selectable marker and non-recombined cells containing marker. These recombined cells and non-recombined cells can be identified through molecular characterization (Joshi 2010). Non-recombined cells may not be used further and may be discarded. This requires careful screening of the putative marker-free plants. Loss of function in mutation of the transgene, intragene or cisgene itself also may cause chimerism. This kind of chimerism may be found during marker-assisted selection to select for GM plants. In the cisgenic concept, all cells of a cisgenic plant have to be free from selectable marker genes (Joshi et al. 2009; Joshi 2010).

There are possible ways to minimize chimerism such as optimization of the primary regeneration process or subjecting the confirmed transformant to regeneration, grafting or budding and again testing for the presence and absence of the target gene and the selectable marker gene, respectively (Joshi et al. 2009). This method is good to identify chimeras in perennial crops such as apple which in natural situation may take years. The marker-free system pMf-1 was used by Schaart et al. (2004) but they did not experience any kind of somaclonal variation or chimerism. It is generally believed that an efficient binary vector system will not give rise to chimerism.

9.9.2 Identification of Chimeras

Reporter genes such as *β-glucuronidase (gus)* (Jefferson et al. 1987), *luciferase* (Koncz et al. 1990), green fluorescent protein (*GFP*) (Chalfie 1994), red fluorescent protein *DS-RED E-5* (Mirabella et al. 2004) or *MdMYB10* (Espley et al. 2007) can help identify chimeras. These reporter genes can place between recombination sites. Evaluation of transformants for chimerism can be conducted through positive selection for uniform production of red colour of *MdMYB10,* green or red-light emission of GFP and blue colour of *gus,* respectively, after transformation followed by uniform non-appearance of these colours in cisgenic plants after recombination (Schaart and Visser 2009). Faize et al. (2010) confirmed the identification and quantification of the chimeras in transformed apricots using quantitative real time PCR (qRT-PCR). Hence, qRT-PCR is also a reliable molecular technique to identify and quantify chimeras.

9.10 Stacking Genes or Alleles

The possibility to add more cisgenes into plants that have obtained cisgenes earlier can be considered as one of the important and significant privileges of cisgenesis and therefore marker-free technology (Joshi 2010). The absence of selectable marker genes enables the use of optimized selection system continually for cisgenic plant in secondary genetic modifications like stacking of genes. Marker-free plant production is essential for the successive stacking of genes in an elite cisgenic line (Righetti et al. 2014). Three ways to conduct gene stacking have been reported. The first method is by insertion of all the target genes at once to enable stacking of genes in a binary vector prior for the transformation event. The second approach is by insertion of the target genes successively through retransformation. In the third way two different transgenic lines are crossed and consequently the progeny harbours the desirable combination of introduced genes.

Through cisgenesis approach, Jo et al. (2014) introduced two broad spectrum potato late blight *R* genes, *Rpi-sto1* and *Rpi-vnt1.1* from two crossable species, *Solanum stoloniferum* and *Solanum venturii*, respectively, into different varieties of potato. The construct that they used for transformation contained both cisgenic late blight *R* genes, *Rpi-vnt1.1* and *Rpi-sto1*, without the bacterial kanamycin resistance (*nptII*) selectable marker gene, was introduced into three selected potato varieties using *Agrobacterium*-mediated transformation. In their experiment, they screen the gene transfer events via PCR among regenerated shoots. They pursued the screening and narrowing down the selection to eight independent events through subsequent analyses including greenhouse morphological observations, responsiveness to *Avr* genes and late blight resistance in detached leaf assays. Finally, these cisgenic events were chosen based on the broad spectrum late blight resistance due to the combined activities of both introduced *R* genes.

9.11 Conclusion and Future Prospects

This chapter describes cisgenesis, intragenesis and strategies for generating marker-free plants through these concepts as efficient alternatives to introduce and domesticate agriculturally important genes in improved existing varieties in one step. New parents can also be produced to be used for future crosses in producing new varieties. Future developments of generating and commercializing cisgenic and intragenic crops will be contingent on willingness to exert less severe regulation to these crops throughout the world (Holme et al. 2013). Breeders in small-scale breeding and seed companies are in need of helpful backings such as a less stringent regulation of cis/intragenic crops and reducing the costs for approval to gain an additional tool for crop improvement as well as to increase the number of cis/intragenic crops developed (Holme et al. 2013). Efforts towards modification of plants employing constructs assembled exclusively from DNA sequences derived from the same or sexually compatible plant species, cisgenics, have been going on in recent years (Schouten et al. 2006a). It is anticipated that the public acceptance of cisgenic plants will be without hesitation and reluctance in the future, and the technique will complement breeding programmes to pursue the introgression of desirable traits that were lost during domestication of wild crop species (Schouten et al. 2006a; Hou et al. 2014).

Several procedures can be employed to generate cisgenic or near-cisgenic plants such as using genome editing tools and homologous recombination of desirable genetic material (Hou et al. 2014) or *Agrobacterium*-mediated transformation of constructs containing cisgenic DNA (Schouten et al. 2006a). Among the genome editing technologies like chimeric DNA/RNA gene repair oligonucleotides (Gamper et al. 2000), zinc-finger nucleases (Townsend et al. 2009), homing endonucleases (Hafez and Hausner 2012), transcription activator-like effector nucleases (TALENs) (Bogdanove and Voytas 2011) and clustered regularly interspaced short palindromic repeats (CRISPR)/Cas9 (Shan et al. 2013), CRISPR/Cas9 holds many advantages and may be a more efficient choice in its application to generating cisgenic plants (Schaeffer and Nakata 2015).

Traditional *Agrobacterium*-mediated transformation procedures generate stable transgenic plants. Such procedures can develop plants harbouring bacterial-derived T-DNA borders which potentially disrupts a gene or regulatory DNA at the site of insertion. While, randomly insertion of a cisgene in the genome results potentially in having both native homologue in the same plant and a cisgene (Schouten et al. 2006a).

The CRISPR/Cas9 gene replacement (gene knock-in) technique has inherent privileges compared to *Agrobacterium*-mediated transformation systems as it allows the control of transgene copy number, introduction of only cisgenic material, has the ability to eliminate native genes to be replaced with a cisgene and perform site-specific genome integration events. This technique may perhaps be used to replace native regulatory regions for possible alteration of the spatial and temporal expression level, and/or the manner of expression of any given gene or set of genes. Moreover, specific genes, part of genes including the regulatory regions, introns, exons and targeting signals, or even SNPs can be introduced, modified or replaced through this approach.

References

Acquaah G (2015) Conventional plant breeding principles and techniques. In: Advances in plant breeding strategies: breeding, biotechnology and molecular tools. Springer International Publishing, Cham, pp 115–158

Ahuja MR (2013) Micropropagation of woody plants. Springer, Dordrecht

Akbudak MA, Srivastava V (2011) Improved FLP recombinase, FLPe, efficiently removes marker gene from transgene locus developed by Cre-*lox* mediated site-specific gene integration in rice. Mol Biotechnol 49:82–89

Akhond MAY, Machray GC (2009) Biotech crops: technologies, achievements and prospects. Euphytica 166:47–59

Ankita S, Joshi M, Devi EL (2015) Alternative to transgenesis: cisgenesis and intragenesis. In: Al-Khayri, JM, Jain SM, Johnson DV (eds) Advances in plant breeding strategies: breeding, biotechnology and molecular tools, pp 345–367

Araki H, Jearnpipatkul A, Tatsumi H, Sakurai T, Ushio K, Muta T, Oshima Y (1985) Molecular and functional organization of yeast plasmid pSR1. J Mol Biol 182:191–203

Bajaj S, Puthigae S, Templeton K, Bryant C, Gill G, Lomba P, Zhang H, Altpeter F, Hanley Z (2008) Towards engineering drought tolerance in perennial ryegrass using its own genome. In: 6th Canadian plant genomics workshop, p 62

Ballester A, Cervera M, Peña L (2007) Efficient production of transgenic citrus plants using isopentenyl transferase positive selection and removal of the marker gene by site-specific recombination. Plant Cell Rep 26:39–45

Ballester A, Cervera M, Peña L (2008) Evaluation of selection strategies alternative to *nptII* in genetic transformation of citrus. Plant Cell Rep 27:1005–1015

Barry GF, Rogers SG, Fraley RT, Brand L (1984) Identification of a cloned cytokinin biosynthetic gene. Proc Natl Acad Sci 81:4776–4780

Benjamin I, Kenigsbuch D, Galperin M, Abrameto JA, Cohen Y (2009) Cisgenic melons over expressing glyoxylate-aminotransferase are resistant to downy mildew. Eur J Plant Pathol 125:355–365

Blechl A, Lin J, Shao M, Thilmony R, Thomson J (2012) The *Bxb1* recombinase mediates site-specific deletion in transgenic wheat. Plant Mol Biol Report 30:1357–1366

Bogdanove AJ, Voytas DF (2011) TAL effectors: customizable proteins for DNA targeting. Science 333:1843–1846

Breitler JC, Meynard D, Boxtel JV, Royer M, Bonnot F, Cambillau L, Guiderdoni E (2004) A novel two T-DNA binary vector allows efficient generation of marker-free transgenic plants in three elite cultivars of rice (*Oryza sativa* L.) Transgenic Res 13:271–287

Breyer D, Kopertekh L, Reheul D (2014) Alternatives to antibiotic resistance marker genes for in vitro selection of genetically modified plants—scientific developments, current use, operational access and biosafety considerations. Crit Rev Plant Sci 33:286–330

Cai CQ, Doyon Y, Ainley WM, Miller JC, DeKelver RC, Moehle EA, Rock JM, Lee YL, Garrison R, Schulenberg L, Blue R, Worden A, Baker L, Faraji F, Zhang L, Holmes MC, Rebar EJ, Collingwood TN, Rubin-Wilson B, Gregory PD, Urnov FD, Petolino JF (2009) Targeted transgene integration in plant cells using designed zinc finger nucleases. Plant Mol Biol 69:699–709

Cardi T (2016) Cisgenesis and genome editing: combining concepts and efforts for a smarter use of genetic resources in crop breeding. Plant Breed 135:139–147

Chalfie M (1994) Green fluorescent protein as a marker for gene expression. Trends Genet 10:151

Chawla R, Shakya R, Rommens CM (2012) Tuber-specific silencing of *asparagine synthetase-1* reduces the acrylamide-forming potential of potatoes grown in the field without affecting tuber shape and yield. Plant Biotechnol J 10:913–924

Chizzali C, Gusberti M, Schouten HJ, Gessler C, Broggini GAL (2016) Cisgenic *Rvi6* scab-resistant apple lines show no differences in *Rvi6* transcription when compared with conventionally bred cultivars. Planta 243:635–644

Chong-Pérez B, Kosky RG, Reyes M, Rojas L, Ocaña B, Tejeda M, Pérez B, Angenon G (2012) Heat shock induced excision of selectable marker genes in transgenic banana by the Cre-*lox* site-specific recombination system. J Biotechnol 159:265–273

de Cock BT, van Bueren ETL, Haring MA, de Vriend HC, Struik PC (2006) 'Cisgenic' as a product designation. Nat Biotechnol 24:1329–1331

Collard BC, Mackill DJ (2008) Marker-assisted selection: an approach for precision plant breeding in the twenty-first century. Philos Trans R Soc B Biol Sci 363:557–572

Costa L, Mandolini M, Poletti V, Martinelli L (2010) Comparing 17-β-estradiol supply strategies for applying the XVE-Cre/*loxP* system in grape gene transfer (*Vitis vinifera* L.) VITIS J Grapevine Res 49:201–208

Cox MM (1983) The FLP protein of the yeast 2-microns plasmid: expression of a eukaryotic genetic recombination system in Escherichia Coli. Proc Natl Acad Sci 80:4223–4227

D'Halluin K, Ruiter R (2013) Directed genome engineering for genome optimization. Int J Dev Biol 57:621–627

D'Halluin K, Vanderstraeten C, Stals E, Cornelissen M, Ruiter R (2008) Homologous recombination: a basis for targeted genome optimization in crop species such as maize. Plant Biotechnol J 6:93–102

Dale EC, Ow DW (1991) Gene transfer with subsequent removal of the selection gene from the host genome. Proc Natl Acad Sci 88:10558–10562

Dalla Costa L, Piazza S, Campa M, Flachowsky H, Hanke MV, Malnoy M (2016) Efficient heat-shock removal of the selectable marker gene in genetically modified grapevine. Plant Cell Tissue Organ Cult 124:471–481

Darwish NA, Khan RS, Ntui VO, Nakamura I, Mii M (2014) Generation of selectable marker-free transgenic eggplant resistant to Alternaria Solani using the *R/RS* site-specific recombination system. Plant Cell Rep 33:411–421

De Buck S, De Wilde C, Van Montagu M, Depicker A (2000) Determination of the T-DNA transfer and the T-DNA integration frequencies upon cocultivation of *Arabidopsis thaliana* root explants. Mol Plant Microbe Interact 13:658–665

Delwaide AC (2014) European consumer's attitudes towards cisgenic rice. IMRD Consortium

Devi EL, Chongtham SK, Holeyachi P, Kousar N, Singh M, Behera C, Singh NB, Wani SH (2013) Cisgenesis and intragenesis: twin sisters for crop improvement. Res J Agric For Sci 1:22–26

Dhekney SA, Li ZT, Gray DJ (2011) Grapevines engineered to express cisgenic Vitis Vinifera thaumatin-like protein exhibit fungal disease resistance. In Vitro Cell Dev Biol Plant 47:458–466

Domínguez A, Cervera M, Pérez RM, Romero J, Fagoaga C, Cubero J, Lopez MM, Juarez JA, Navarro L, Pena L (2004) Characterisation of regenerants obtained under selective conditions after agrobacterium-mediated transformation of citrus explants reveals production of silenced and chimeric plants at unexpected high frequencies. Mol Breed 14:171–183

Dutt M, Li ZT, Dhekney SA, Gray DJ (2012) Co-transformation of grapevine somatic embryos to produce transgenic plants free of marker genes. Transgenic plants: methods and protocols, pp 201–203

Ebinuma H, Sugita K, Matsunaga E, Endo S, Yamada K, Komamine A (2001) Systems for the removal of a selection marker and their combination with a positive marker. Plant Cell Rep 20:383–392

van Engelen FA, Molthoff JW, Conner AJ, Nap JP, Pereira A, Stiekema WJ (1995) pBINPLUS: an improved plant transformation vector based on pBIN19. Transgenic Res 4(4):288–290

Espinoza C, Schlechter R, Herrera D, Torres E, Serrano A, Medina C, Arce-Johnson P (2013) Cisgenesis and intragenesis: new tools for improving crops. Biol Res 46:323–331

Espley RV, Hellens RP, Puterill J, Kutty-Amma S, Allan AC (2007) Red colouration in apple fruit is due to the activity of a MYB transcription factor, MdMYB10. Plant J 49:414–427

Faize M, Faize L, Burgos L (2010) Using quantitative real-time PCR to detect chimeras in transgenic tobacco and apricot and to monitor their dissociation. BMC Biotechnol 10:53

Friebe B, Jiang J, Raupp WJ, McIntosh RA, Gill BS (1996) Characterization of wheat-alien translocations conferring resistance to diseases and pests: current status. Euphytica 91:59–87

Gadaleta A, Giancaspro A, Blechl AE, Blanco A (2008) A transgenic durum wheat line that is free of marker genes and expresses 1Dy10. J Cereal Sci 48:439–445

Gamper HB, Parekh H, Rice MC, Bruner M, Youkey H, Kmiec EB (2000) The DNA strand of chimeric RNA/DNA oligonucleotides can direct gene repair/conversion activity in mammalian and plant cell-free extracts. Nucleic Acids Res 28:4332–4339

García-Almodóvar RC, Petri C, Padilla IMG, Burgos L (2013) Combination of site-specific recombination and a conditional selective marker gene allows for the production of marker-free tobacco plants. Plant Cell Tissue Organ Cult 116:205–215

Gidoni D, Srivastava V, Carmi N (2008) Site-specific excisional recombination strategies for elimination of undesirable transgenes from crop plants. In Vitro Cell Dev Biol Plant 44:457–467

Gleave AP, Mitra DS, Mudge SR, Morris BAM (1999) Selectable marker-free transgenic plants without sexual crossing: transient expression of Cre recombinase and use of a conditional lethal dominant gene. Plant Mol Biol 40:223–235

Hafez M, Hausner G (2012) Homing endonucleases: DNA scissors on a mission. Genome 55(8):553–569

Han KM, Dharmawardhana P, Arias RS (2011) Gibberellin-associated cisgenes modify growth, stature and wood properties in *Populus*. Plant Biotechnol J 9(2):162–178

Hansen M (2000) Genetic engineering is not an extension of conventional plant breeding [online]. Consumer Policy Institute

Hare PD, Chua NH (2002) Excision of selectable marker genes from transgenic plants. Nat Biotechnol 20:575

Hartung F, Schiemann J (2014) Precise plant breeding using new genome editing techniques: opportunities, safety and regulation in the EU. Plant J 78:742–752

Haverkort AJ, Struik PC, Visser RGF, Jacobsen E (2009) Applied biotechnology to combat late blight in potato caused by phytophthora infestans. Potato Res 52:249–264

Herzog K, Flachowsky H, Deising HB, Hanke MV (2012) Heat-shock-mediated elimination of the *nptII* marker gene in transgenic apple (*Malus x domestica Borkh*). Gene 498(1):41–49

Hoess RH, Abremski K (1985) Mechanism of strand cleavage and exchange in the Cre-*lox* site-specific recombination system. J Mol Biol 181:351–362

Hoess R, Ziese M, Sternberg N (1982) P1 site-specific recombination: nucleotide sequence of the recombining sites. Proc Natl Acad Sci 79:3398–3402

Holme IB, Dionisio G, Brinch-Pedersen H, Wendt T, Madsen CK, Vincze E, Holm PB (2012) Cisgenic barley with improved phytase activity. Plant Biotechnol J 10:237–247

Holme IB, Wendt T, Holm PB (2013) Intragenesis and cisgenesis as alternatives to transgenic crop development. Plant Biotechnol J 11:395–407

Hou H, Atlihan N, Lu ZX (2014) New biotechnology enhances the application of cisgenesis in plant breeding. Front Plant Sci 5:389

Hunter P (2014) "Genetically modified Lite" placates public but not activists. EMBO Rep 15(2): 138–141

Hunter CT, Suzuki M, Saunders J, Wu S (2014) Phenotype to genotype using forward-genetic mu-seq for identification and functional classification of maize mutants. Front Plant Sci 4:545

Jacobsen E, Nataraja KN (2008) Cisgenics-facilitating the second green revolution in India by improved traditional plant breeding. Curr Sci 94:1365–1366

Jacobsen E, Schouten HJ (2007) Cisgenesis strongly improves introgression breeding and induced translocation breeding of plants. Trends Biotechnol 25:219–223

Jacobsen E, Schouten HJ (2008) Cisgenesis, a new tool for traditional plant breeding, should be exempted from the regulation on genetically modified organisms in a step by step approach. Potato Res 51:75–88

Jacobsen E, Schouten HJ (2009) Cisgenesis: an important sub-invention for traditional plant breeding companies. Euphytica 170:235–247

Jefferson RA, Kavanagh TA, Bevan MW (1987) GUS fusions: beta-glucuronidase as a sensitive and versatile gene fusion marker in higher plants. EMBO J 6:3901–3907

Jo KR (2013) Unveiling and deploying durability of late blight resistance in potato from natural stacking to cisgenic stacking. Doctoral dissertation, Wageningen University

Jo KR, Kim CJ, Kim SJ, Kim TY, Bergervoet M, Jongsma MA, Visser RGF, Jacobsen E, Vossen JH (2014) Development of late blight resistant potatoes by cisgene stacking. BMC Biotechnol 14:50

Joshi SG (2010) Towards durable resistance to apple scab using cisgenes. Wageningen University, Wageningen

Joshi S, Soriano MJ, Schaart J, Broggini GAL, Szankowski I, Jacobsen E, Krens F, Schouten H (2009) Approaches for development of Cisgenic apples. Transgenic Plant J 3:40–46

Joshi SG, Schaart JG, Groenwold R, Jacobsen E, Schouten HJ, Krens FA (2011) Functional analysis and expression profiling of HcrVf1 and HcrVf2 for development of scab resistant cisgenic and intragenic apples. Plant Mol Biol 75:579–591

Khan RS, Nakamura I, Mii M (2011) Development of disease-resistant marker-free tomato by R/*RS* site-specific recombination. Plant Cell Rep 30:1041–1053

Koncz C, Langridge WHR, Olsson O, Schell J, Szalay AA (1990) Bacterial and firefly luciferase genes in transgenic plants: advantages and disadvantages of a reporter gene. Genesis 11:224–232

Konig A (2003) A framework for designing transgenic crops—science, safety and citizen's concerns. Nat Biotechnol 21:1274–1279

Krens F, van Kronenburg-van de Ven B (2011) Lily genetic modification at Wageningen UR. In: Helsley C, van Tuyl JM (eds) The Lily Yearbook of the North American Lily Society. The Society ©The North American Lily Society, Inc, pp 156–163

Krens FA, Pelgrom KTB, Schaart JG, Den Nijs APM, Rouwendal GJA (2003) Clean vector technology for marker-free transgenic fruit crops. In: XI Eucarpia symposium on fruit breeding and genetics, pp 431–436

Krens FA, Schaart JG, van der Burgh A, Tinnenbroek-Capel IEM, Groenwold R, Kodde LP, Broggini GAL, Gessler C, Schouten HJ (2015) Cisgenic apple trees; development, characterization, and performance. Front Plant Sci 6:286

Kuraya Y, Ohta S, Fukuda M, Hiei Y, Murai N, Hamada K, Ueki J, Imaseki H, Komari T (2004) Suppression of transfer of non-T-DNA 'vector backbone' sequences by multiple left border repeats in vectors for transformation of higher plants mediated by *Agrobacterium tumefaciens*. Mol Breed 14:309–320

Lange M, Vincze E, Møller MG, Holm PB (2006) Molecular analysis of transgene and vector backbone integration into the barley genome following *Agrobacterium*-mediated transformation. Plant Cell Rep 25:815–820

Li B, Li N, Duan X, Wei A, Yang A, Zhang J (2010) Generation of marker-free transgenic maize with improved salt tolerance using the FLP/*FRT* recombination system. J Biotechnol 145:206–213

Ling F, Zhou F, Chen H, Lin Y (2016) Development of marker-free insect-resistant indica rice by *Agrobacterium tumefaciens*-mediated co-transformation. Front Plant Sci 7:1–10

Lusser M, Parisi C, Plan D, Rodríguez-Cerezo E (2012) Deployment of new biotechnologies in plant breeding. Nat Biotechnol 30:231–239

Malnoy M, Jin Q, Borejsza-Wysocka EE, He SY, Aldwinckle HS (2007) Overexpression of the apple MpNPR1 gene confers increased disease resistance in *Malus* × *domestica*. Mol Plant Microbe Interact 20:1568–1580

Malnoy M, Boresjza-Wysocka EE, Norelli JL, Flaishman MA, Gidoni D, Aldwinckle HS (2010) Genetic transformation of apple (*Malus x domestica*) without use of a selectable marker gene. Tree Genet Genomes 6:423–433

Maltseva E, Ismagul A, Iskakova G, Chirkin A, Skiba Y, Ismagulova G, Eliby S, Aitkhozhina N (2014) Wheat cisgenic transformation with class I chitinase gene. J Biotechnol 185:S116–S117

Mans R, van Rossum HM, Wijsman M, Backx A, Kuijpers NGA, van den Broek M, Daran-Lapujade P, Pronk JT, van Maris AJA, Daran JMG (2015) CRISPR/Cas9: a molecular Swiss army knife for simultaneous introduction of multiple genetic modifications in Saccharomyces Cerevisiae. FEMS Yeast Res 15:1–15

Matheka JM, Anami S, Gethi J, Omer RA, Alakonya A, Machuka J, Runo S (2013) A new double right border binary vector for producing marker-free transgenic plants. BMC Res Notes 6:448

Matzke AJM, Matzke MA (1998) Position effects and epigenetic silencing of plant transgenes. Curr Opin Plant Biol 1:142–148

Mielby H, Sandøe P, Lassen J (2013) Multiple aspects of unnaturalness: are cisgenic crops perceived as being more natural and more acceptable than transgenic crops? Agric Hum Values 30:471–480

Mirabella R, Franken C, van der Krogt GNM, Bisseling T, Geurts R (2004) Use of the fluorescent timer *DsRED-E5* as reporter to monitor dynamics of gene activity in plants. Plant Physiol 135:1879–1887

Nandy S, Srivastava V (2012) Marker-free site-specific gene integration in rice based on the use of two recombination systems. Plant Biotechnol J 10:904–912

Nietsch J, Brügmann T, Becker D, Fladung M (2017) Old methods rediscovered: application and improvement of two direct transformation methods to hybrid poplar (*Populus tremula* × *P. alba*). Plant Cell Tiss Org Cult 130:183–196

Parkhi V, Rai M, Tan J, Oliva N, Rehana S, Bandyopadhyay A, Torrizo L, Ghole V, Datta K, Datta SK (2005) Molecular characterization of marker-free transgenic lines of indica rice that accumulate carotenoids in seed endosperm. Mol Gen Genomics 274:325–336

Parliament European (2013) Commission implementing regulation (EU) no 503/2013. Off J Eur Union L157:1–48
Perpiñá G, Esteras C, Gibon Y, Monforte AJ, Picó B (2016) A new genomic library of melon introgression lines in a cantaloupe genetic background for dissecting desirable agronomical traits. BMC Plant Biol 16:154
Petri C, Hily JM, Vann C, Dardick C, Scorza R (2011) A high-throughput transformation system allows the regeneration of marker-free plum plants (*Prunus domestica*). Ann Appl Biol 159:302–315
Petri C, López-Noguera S, Wang H, García-Almodóvar C, Alburquerque N, Burgos L (2012) A chemical-inducible Cre-*LoxP* system allows for elimination of selection marker genes in transgenic apricot. Plant Cell Tissue Organ Cult 110:337–346
Qi W, Tinnenbroek-Capel IEM, Schaart JG, Huang B, Cheng J, Visser RGF, Van Loo EN, Krens FA (2014) Regeneration and transformation of Crambe Abyssinica. BMC Plant Biol 14:235
Qi W, Tinnenbroek-Capel IEM, Salentijn EMJ, Schaart JG, Cheng J, Denneboom C, Zhang Z, Zhang X, Zhao H, Visser RGF, Huang B, Van Loo EN, Krens FA (2015) Screening for recombinants of Crambe Abyssinica after transformation by the pMF1 marker-free vector based on chemical selection and meristematic regeneration. Sci Rep 5:14033
Richael CM, Kalyaeva M, Chretien RC, Yan H, Adimulam S, Stivison A, Weeks JT, Rommens CM (2008) Cytokinin vectors mediate marker-free and backbone-free plant transformation. Transgenic Res 17:905–917
Righetti L, Djennane S, Berthelot P, Cournol R, Wilmot N, Loridon K, Vergne E, Chevreau E (2014) Elimination of the *nptII* marker gene in transgenic apple and pear with a chemically inducible R/*Rs* recombinase. Plant Cell Tissue Organ Cult 117:335–348
Romano A, Raemakers K, Bernardi J, Visser R (2003) Transgene organisation in potato after particle bombardment-mediated (co-) transformation using plasmids and gene cassettes. Transgenics 12:461–473
Rommens CM (2004) All-native DNA transformation: a new approach to plant genetic engineering. Trends Plant Sci 9:457–464
Rommens CM, Humara JM, Ye J, Yan H (2004) Crop improvement through modification of the plant's own genome. Plant Physiol 135:421–431
Rommens CM, Bougri O, Yan H, Humara J (2005) Plant-derived transfer DNAs. Plant Physiol 139:1338–1349
Rommens CM, Ye J, Richael C (2006) Improving potato storage and processing characteristics through all-native DNA transformation. J Agric Food Chem 54:9882–9887
Rommens CM, Haring MA, Swords K, Davies H (2007) The intragenic approach as a new extension to traditional plant breeding. Trends Plant Sci 12:397–403
Rommens CM, Yan H, Swords K (2008) Low-acrylamide French fries and potato chips. Plant Biotechnol J 6:843–853
Russell AW, Sparrow R (2008) The case for regulating intragenic GMOs. J Agric Environ Ethics 21:153–181
Russell SH, Hoopes JL, Odell JT (1992) Directed excision of a transgene from the plant genome. Mol Gen Genet MGG 234:49–59
Schaart JG (2004) Towards consumer-friendly cisgenic strawberries which are less susceptible to Botrytis cinerea. Ph.D. thesis, Wageningen University, Wageningen, The Netherlands
Schaart JG, Visser RGF (2009) Novel plant breeding techniques. Consequences of new genetic modification-based plant breeding techniques in comparison to conventional plant breeding. COGEM research report, The Netherlands Commission on Genetic Modification
Schaart JG, Krens FA, Pelgrom KTB, Mendes O, Rouwendal GJA (2004) Effective production of marker-free transgenic strawberry plants using inducible site-specific recombination and a bifunctional selectable marker gene. Plant Biotechnol J 2:233–240
Schaart JG, Krens FA, Wolters AM, Visser RGF (2011) Transformation methods for obtaining marker-free genetically modified plants. In: Steward CN Jr, Touraev A, Citovsky V, Tzfira T (eds) Plant Transformation Technologies. Wiley-Blackwell Publishing, Ames, pp 229–242

Schaart JG, van de Wiel CCM, Lotz LAP, Smulders MJM (2016) Opportunities for products of new plant breeding techniques. Trends Plant Sci 21:438–449
Schaeffer SM, Nakata PA (2015) CRISPR/Cas9-mediated genome editing and gene replacement in plants: transitioning from lab to field. Plant Sci 240:130–142
Schouten HJ, Krens FA, Jacobsen E (2006a) Cisgenic plants are similar to traditionally bred plants. EMBO Rep 7:750–753
Schouten HJ, Krens F, Jacobsen E (2006b) Do cisgenic plants warrant less stringent oversight? Nat Biotechnol 24:753
Senecoff JF, Bruckner RC, Cox MM (1985) The FLP recombinase of the yeast 2-micron plasmid: characterization of its recombination site. Proc Natl Acad Sci 82:7270–7274
Shan Q, Wang Y, Li J, Zhang Y, Chen K, Liang Z, Zhang K, Liu J, Xi JJ, Qiu JL, Gao C (2013) Targeted genome modification of crop plants using a CRISPR-Cas system. Nat Biotechnol 31(8):686–688
Shukla VK, Doyon Y, Miller JC, DeKelver RC, Moehle EA, Worden SE, Mitchell JC, Arnold NL, Gopalan S, Meng X, Choi VM, Rock JM, Wu YY, Katibah GE, Zhifang G, McCaskill D, Simpson MA, Blakeslee B, Greenwalt SA, Butler HJ, Hinkley SJ, Zhang L, Rebar EJ, Gregory PD, Urnov FD (2009) Precise genome modification in the crop species Zea mays using zinc-finger nucleases. Nature 459(7245):437–441
Singh A, Joshi M, Devi EL (2015) Alternative to transgenesis: cisgenesis and intragenesis. In: Advances in plant breeding strategies: breeding, biotechnology and molecular tools. Springer International Publishing, Cham, pp 345–367
Sripriya R, Raghupathy V, Veluthambi K (2008) Generation of selectable marker-free sheath blight resistant transgenic rice plants by efficient co-transformation of a cointegrate vector T-DNA and a binary vector T-DNA in one agrobacterium tumefaciens strain. Plant Cell Rep 27(10):1635–1644
Telem RS, Wani H, Singh NB, Nandini R, Sadhukhan R, Bhattacharya S, Mandal N (2013) Cisgenics—a sustainable approach for crop improvement. Curr Genomics 14(7):468–476
Townsend JA, Wright DA, Winfrey RJ, Fu F, Maeder ML (2009) High-frequency modification of plant genes using engineered zinc-finger nucleases. Nature 459(7245):442–445
Van Gelder WMJ (1989) Steroidal glycoalkaloids in Solanum species: consequences for potato breeding and food safety. PhD thesis, Wageningen University
Vanblaere T, Szankowski I, Schaart J, Schouten H, Flachowsky H, Broggini GAL, Gessler C (2011) The development of a cisgenic apple plant. J Biotechnol 154:304–311
Vanblaere T, Flachowsky H, Gessler C, Broggini GAL (2014) Molecular characterization of cisgenic lines of apple 'gala' carrying the Rvi6 scab resistance gene. Plant Biotechnol J 12:2–9
de Vetten N, Wolters A, Raemakers K, van der Meer I, ter Stege R, Heeres E, Heeres P, Visser R (2003) A transformation method for obtaining marker-free plants of a cross-pollinating and vegetatively propagated crop. Nat Biotechnol 21:439–442
Vidal JR, Kikkert JR, Donzelli BD, Wallace PG, Reisch BI (2006) Biolistic transformation of grapevine using minimal gene cassette technology. Plant Cell Rep 25:807–814
Wang Y, Yau YY, Perkins-Balding D, Thomson JG (2011) Recombinase technology: applications and possibilities. Plant Cell Rep 30:267–285
Weeks JT, Ye J, Rommens CM (2008) Development of an in-planta method for transformation of alfalfa (*Medicago sativa*). Transgenic Res 17:587–597
Woo HJ, Cho HS, Lim SH, Shin KS, Lee SM, Lee KJ, Kim DH, Cho YG (2009) Auto-excision of selectable marker genes from transgenic tobacco via a stress inducible FLP/*FRT* site-specific recombination system. Transgenic Res 18:455–465
Würdig J, Flachowsky H, Saß A, Peil A, Hanke MV (2015) Improving resistance of different apple cultivars using the Rvi6 scab resistance gene in a cisgenic approach based on the FLP/*FRT* recombinase system. Mol Breed 35:95
Yang M, Djukanovic V, Stagg J, Lenderts B, Bidney D, Carl Falco S, Alexander Lyznik L (2009) Targeted mutagenesis in the progeny of maize transgenic plants. Plant Mol Biol 70:669–679

Zhu Z, Chung WH, Shim EY, Lee SE, Ira G (2008) Sgs1 helicase and two nucleases Dna2 and Exo1 resect DNA double-strand break ends. Cell 134:981–994

Zou X, Peng A, Xu L, Liu X, Lei T, Yao L, He Y, Chen S (2013) Efficient auto-excision of a selectable marker gene from transgenic citrus by combining the Cre/*loxP* system and *ipt* selection. Plant Cell Rep 32:1601–1613

Zuo J, Chua NH (2000) Chemical-inducible systems for regulated expression of plant genes. Curr Opin Biotechnol 11:146–151

Chapter 10
Use of dsRNA in Crop Improvement

Lau Su Ee, Purabi Mazumdar, Pooja Singh, and Jennifer Ann Harikrishna

10.1 Introduction and Discovery of RNA Silencing

RNA silencing, also referred to as RNA interference (RNAi), is a gene silencing phenomenon that results in inhibition of gene expression or protein synthesis induced by the presence of double-stranded RNA (dsRNA). Before the discovery of RNA silencing pathways, RNA-mediated gene silencing was reported as co-suppression in *Petunia hybrida* (Napoli et al. 1990), post-transcriptional gene silencing (PTGS) in tobacco (Lindbo et al. 1993; Béclin et al. 1998) and quelling in the fungus *Neurospora crassa* (Romano and Macino 1992). The phenomenon of RNA silencing was probably first reported as early as 1928, where tobacco plants infected with *tobacco ringspot virus* showed symptoms only on the initially infected leaves (Wingard 1928). Newly emerging upper leaves of the plants had somehow became immune to the virus, were asymptomatic and resistant to secondary infection. At the time the apparent recovery of the plants was a mystery, as there was no obvious way to explain the specificity of the resistance to secondary infection, however the practical application of infecting plants with a mild strain of virus to protect them from a more virulent strain, known as cross-protection, has been widely used in agriculture (Ogwok et al. 2012; reviewed in Syller 2016).

Many years later and after the establishment of reliable plant transformation systems, researchers introducing transgene copies of the flower pigmentation gene,

L.S. Ee • P. Mazumdar • P. Singh
Centre for Research in Biotechnology for Agriculture, University of Malaya, 50603 Kuala Lumpur, Malaysia

J.A. Harikrishna (✉)
Centre for Research in Biotechnology for Agriculture, University of Malaya, 50603 Kuala Lumpur, Malaysia

Institute of Biological Sciences, Faculty of Science, University of Malaya, 50603 Kuala Lumpur, Malaysia
e-mail: jennihari@um.edu.my

S.N.A. Abdullah et al. (eds.), *Crop Improvement*,
DOI 10.1007/978-3-319-65079-1_10

chalcone synthase (*CHS*) into *Petunia hybrida* flowers with an aim to produce a more intense purple colouration, found instead some variegated and some completely white flowers rather than the hypothetical intense purple flower (Napoli et al. 1990). The mechanism was still not understood, but was reproducible and was termed "co-suppression". A phenomenon similar to the co-suppression observed in petunia was described in the fungus *Neurospora crassa* and termed "quelling" (Romano and Macino 1992). In 1993, Lindbo et al. transformed tobacco plants with a gene sequence with homology to the *Tobacco etch virus coat protein* (TEV CP), to confer TEV resistance. The transformed tobacco plants managed to outgrow TEV infection at approximately 3–5 weeks after inoculation and the leaves were reported to be resistant to TEV, while still susceptible to a control virus, *Potato virus X* (PVX). Further molecular analyses of the recovered leaf tissue revealed that TEV mRNA failed to accumulate even though the introduced TEV sequences were actively transcribed. This study explained the PTGS mechanism in a sequence-specific context (Lindbo et al. 1993). A few years later, Ratcliff, working with David Baulcombe, showed that a cytoplasmic RNA trigger was responsible for both virus resistance and homology-dependent gene silencing in plants (Ratcliff et al. 1997). In 1998, Fire et al. initiated the term "RNAi" on discovery that dsRNA was more effective than antisense single-stranded RNA (ssRNA) at inducing gene silencing in the nematode, *Caenorhabditis elegans* and confirming that a similar mechanism was likely responsible in nematodes and plants. This was quickly followed by further reports from the Baulcombe group that showed presence and sequence specificity of 21–25 nt small RNA (sRNA) species in plants during co-suppression of an endogenous gene following homologous transgene insertion; during PTGS; during systemic silencing of a transgene and in plants infected with virus PVX (Hamilton and Baulcombe 1999). The authors also showed correlation between the level of sRNA accumulation and the silencing efficiency conferred by each of these systems (Hamilton and Baulcombe 1999). Subsequently, endogenous small RNA species [small interfering RNA (siRNA) and microRNA (miRNA)] were discovered in plants as a natural mechanism of negative gene regulation (Llave et al. 2002; Reinhart et al. 2002) and as suppressors of mobile genetic elements (reviewed in Borges and Martienssen 2015). Following this, the major frameworks related to the biogenesis of genome-derived siRNAs and miRNAs as genetic regulators and in response to virus infections have been well established and reviewed (reviewed in Borges and Martienssen 2015; Dalakouras et al. 2015).

Since the discovery of the RNA silencing mechanism and pathways, the technology to employ RNA silencing using exogenous constructs to regulate endogenous nucleic acids has been developed and applied via the genetic engineering of plants to alter secondary metabolite profiles, alter nutritional content and to provide resistance or tolerance to various biotic and abiotic stresses. Crop improvement through RNA silencing can be established through the steps of target gene(s) identification, followed by development of suitable RNA silencing constructs, introduction to whole plants or plant tissue and finally phenotypic evaluation. This chapter provides a review of techniques used for inducing dsRNA-mediated silencing (RNAi) in plants and the application for crop improvement.

10.1.1 Mechanism of dsRNA-Mediated Silencing in Plants

RNA silencing is catalytically triggered by long dsRNAs or short-hairpin RNA (shRNA) precursors, which are homologous in sequence to the targeted gene. These arise naturally in plants from endogenous miRNA and siRNA trans- and cis-loci, but the potential to exploit RNA silencing by deliberate introduction of dsRNA was soon realised (Baulcombe and Hamilton 2012). RNA silencing mediated by dsRNA has been widely explored in plants as a process of homology-dependent inhibition of gene expression, at the transcriptional and post-transcriptional levels (Tables 10.1 and 10.2).

The presence of endogenous dsRNAs (from miRNA genes and siRNA generating genetic elements including transposable and repetitive elements) or the introduction of exogenous dsRNAs (either from infection by RNA viruses or laboratory manipulated genes) act as an initiator for gene silencing. RNAse III enzymes belonging to the DICER-LIKE (DCL) family process dsRNA molecules into typically 21–25 nt length sRNAs called siRNAs (Bernstein et al. 2001; reviewed in Chen 2009). siRNA can also arise by the action of RNA-dependent RNA polymerases. Together with proteins from the ARGONAUTE (AGO) family, siRNA form various versions of an RNA-induced silencing complex (RISC) which (depending on the siRNA and the other specific components of the RISC) will effect gene silencing either as PTGS (Fig. 10.1a) or transcriptional gene silencing (TGS) (Fig. 10.1b). In PTGS, an mRNA with sequence complementarity to the siRNA is either cleaved (Baumberger and Baulcombe 2005) or prevented from binding with a ribosome so that translation is inhibited (Brodersen et al. 2008). The dsRNA that triggers PTGS is derived from nuclear gene expression by RNA polymerase II (Pol II) and synthesis by RNA-DEPENDENT RNA POLYMERASE 6 (RDR6), or by exogenous dsRNA or hairpin RNA (hpRNA) treatments (Fig. 10.1a). The dsRNA is processed by RNAse III enzymes belonging to the DICER-LIKE family (DCL1/2/3/4) into 21–25 nt siRNAs with two-base overhangs on the hydroxylated 3′ ends. The short siRNAs duplexes are protected from degradation by 2′-O-methylation at both 3′ termini by methyltransferase HUA ENHANCER 1 (HEN1) after which endonucleases from the AGO family (AGO1/2/7/10) catalyse the unwinding of the siRNA duplex. Once the siRNAs are unwound, the passenger (sense) strand is degraded by an RNA-induced silencing complex (RISC) while the guide (antisense) strand is incorporated into RISC in association with AGO and effector proteins to target mRNAs with sequence complementarity to the guide siRNA (reviewed in Borges and Martienssen 2015).

TGS (Fig. 10.1b) involves the methylation of cytosine bases in nuclear DNA by RNA-directed DNA methylation (RdDM), which prevents transcription of the targeted gene (Jones et al. 1998; Mette et al. 2000; Wassenegger et al. 1994; Bond and Baulcombe 2015). In plants, TGS initiates from dsRNA transcription by RNA polymerase IV (Pol IV) and synthesis by RNA-DEPENDENT RNA POLYMERASE 2 (RDR2). The dsRNA is processed by DICER-LIKE family RNAse III, DCL3 into 21–25 nt siRNA. The short siRNAs duplexes are 2′-*O*-methylated at both 3′ termini by methyltransferase HUA ENHANCER 1 (HEN1). Endonucleases from the AGO

Table 10.1 Examples of dsRNA-mediated interference in plants (stable transformation)

Technique used to induce silencing	Plant	Silencing target	Promoter or coding region targeted	Percentage of silencing	Trait	Reference
PTGS by agrobacterium-mediated transformation	Barley	Barley yellow dwarf virus-PAV (*BYDV-PAV*)	Coding region	Not reported	Resistance against barley yellow dwarf virus (BYDV)	Wang et al. (2000)
PTGS by agrobacterium-mediated transformation	Cotton, Arabidopsis	*Δ12 desaturase, Δ9 desaturase*	Coding region	80%, 60%	High-stearic and high-oleic acid content in seed oil	Wesley et al. (2001)
PTGS by agrobacterium-mediated transformation	Cotton	*Δ12 desaturase, Δ9 desaturase*	Coding region	53%, 60%	High-stearic and high-oleic acid content in seed oil	Liu et al. (2002)
PTGS by agrobacterium-mediated transformation	Rice	*Glutelin B gene GluB*	Coding region	60–100%	Low glutenin content	Kusaba et al. (2003)
PTGS by agrobacterium-mediated transformation	Opium poppy	*Codeine reductase*	Coding region	100% (but unstable and not heritable)	High-yield accumulation of nonnarcotic alkaloid reticuline	Allen et al. (2004)
PTGS by agrobacterium-mediated transformation	Oilseed rape, Arabidopsis	*B-type MADS box*	Coding region	22.6%	High sepal content	Byzova et al. (2004)

PTGS by agrobacterium-mediated transformation	Coffee bean	*Theobromine synthase*	Coding region	Not reported	Decaffeinated varieties	Ogita et al. (2004)
TGS via agrobacterium-mediated transformation	Maize	Male sterility factor 45, dihydrol flavenol reductase, cytochrome P450	Promoter	78–90%	Male-sterile plants for studying anther-specific gene function	Cigan et al. (2005)
PTGS by agrobacterium-mediated transformation	Tomato	*DET1*	Coding region	Not reported	Carotenoid and flavonoid	Davuluri et al. (2005)
PTGS by particle bombardment	Wheat	*Vernalization gene VRN-1*	Coding region	Not reported	Early flowering	Loukoianov et al. (2005)
PTGS by agrobacterium-mediated transformation	Tobacco	*Chalcone isomerase*	Coding region	100% (level of silencing was reported to be variable with different construct)	Flavonoid and flower colour	Nishihara et al. (2005)
PTGS by agrobacterium-mediated transformation	Tomato	*ACC oxidase*	Coding region	27–87%	Decreased ethylene	Xiong et al. (2005)
PTGS by agrobacterium-mediated transformation	Potato	*Granule bound starch synthase I*	Coding region	48–87%	Inhibition in amylose production	Heilersig et al. (2006a)

(continued)

Table 10.1 (continued)

Technique used to induce silencing	Plant	Silencing target	Promoter or coding region targeted	Percentage of silencing	Trait	Reference
PTGS and TGS via agrobacterium-mediated transformation	Potato	*Granule bound starch synthase I*	Promoter and coding region	50%	Inhibition in amylose production	Heilersig et al. (2006b)
PTGS by agrobacterium-mediated transformation	Wheat	Starch branching enzyme IIa, transmembrane protein	Coding region	92%, 33%	High-amylose	Regina et al. (2006)
PTGS by agrobacterium-mediated transformation	Cotton	*Delta-cadinene synthase*	Coding region	99%	Reduction of toxic terpenoid gossypol	Sunilkumar et al. (2006)
PTGS by particle bombardment	Wheat	*Phytoene desaturase* (*PDS*)	Coding region	78%	Inhibition of the carotenoid biosynthesis and photobleaching	Travella et al. (2006)
PTGS by particle bombardment	Wheat	*NAM/NAC transcription factor*	Coding region	29%	Grain protein, zinc, and iron content	Uauy et al. (2006)
PTGS by agrobacterium-mediated transformation	Tobacco	*Splicing factor and integrase*	Coding region	Not reported	Resistance against PPN (*Meloidogyne incognita*)	Yadav et al. (2006)
PTGS by agrobacterium-mediated transformation	Maize	*Vacuolar ATPase* (*V-ATPaseA*)	Coding region	Not reported	Resistance against corn root worm (*Diabrotica virgifera virgifera*)	Baum et al. (2007)

PTGS by particle bombardment	Bean	*Replication-associated protein (AC1) coding region*	Coding region	Not reported	Resistance against bean golden mosaic virus (BGMV)	Bonfim et al. (2007)
PTGS by agrobacterium-mediated transformation	Cotton	*Cytochrome P450 gene (CYPAE14)*	Coding region	Not reported	Resistance against cotton bollworm *Helicoverpa armigera*	Mao et al. (2007)
PTGS by agrobacterium-mediated transformation	Rice	*Gibberellin20-oxidase (OsGA20ox2)*	Coding region	35–53%	Improved grain yield	Qiao et al. (2007)
PTGS by agrobacterium-mediated transformation	Tomato	*Chalcone synthase*	Coding region	99%	Seedless fruit	Schijlen et al. (2007)
PTGS by agrobacterium-mediated transformation	Rice	*Fatty acid desaturase OsFAD7* and *OsFAD8*	Coding region	Not reported	Resistance against rice blast fungus *Magnaporthe grisea*	Yara et al. (2007)
PTGS by agrobacterium-mediated transformation	Rapeseed	*Lycopene epsilon cyclase*	Coding region	Not reported	Increased carotenoid content	Yu et al. (2007)
PTGS by particle bombardment	Wheat	Seed storage protein	Coding region	100%	Elasticity and viscosity of the gluten complex	Yue et al. (2007)
PTGS by agrobacterium-mediated transformation	Onion	*Lachrymatory factor Synthase (LFS)*	Coding region	Not reported	Tearless onion	Eady et al. (2008)

(continued)

Table 10.1 (continued)

Technique used to induce silencing	Plant	Silencing target	Promoter or coding region targeted	Percentage of silencing	Trait	Reference
TGS via agrobacterium-mediated transformation	Rice	Heme oxygenase, RAC GTPase 1, RAC GTPase 3, RAC GTPase 4 Putative PolyA binding protein, putative CBS domain protein, putative ribosomal protein L5	Promoter	Not reported	siRNA targets endogenous promoters induce DNA methylation	Okano et al. (2008)
PTGS by agrobacterium-mediated transformation	Soybean	*Oleosin A*	Coding region	Not reported	Modified oil content	Schmidt and Herman (2008)
PTGS by agrobacterium-mediated transformation	Rice	*Fatty acid desaturase OsSSI2*	Coding region	Not reported	Resistance against rice blast fungus *Magnaporthe grisea* and the bacterial blight pathogen *Xanthomonas oryzae*	Jiang et al. (2009)
PTGS by agrobacterium-mediated transformation	Rice	*C-kinase1(RACK1)*	Coding region	Not reported	Enhanced drought tolerance	Li et al. (2009)
PTGS by agrobacterium-mediated transformation	Cotton	*Myb transcription factor*	Coding region	Not reported	Early fibre and trichome development	Machado et al. (2009)

PTGS by agrobacterium-mediated transformation	Rice	Viroplasm matrix protein (PNS12)	Coding region	Not reported	Resistance against Rice dwarf virus (RDV)	Shimizu et al. (2009)
PTGS by agrobacterium-mediated transformation (floral dip)	Canola	*Farnesyltransferase*	Coding region	Not reported	Protection against drought	Wang et al. (2009)
PTGS by agrobacterium-mediated transformation	Tomato	*α-man/β- hex*	Coding region	99%	Increased shelf life	Meli et al. (2010)
PTGS by agrobacterium-mediated transformation	Rice	*Oryza sativa Delayed seed germination 1* (*OsDSG1*)	Coding region	Not reported	Enhanced drought tolerance	Park et al. (2010)
PTGS by agrobacterium-mediated transformation	Potato	*SYNTAXIN-related 1* (*SYR1*)	Coding region	Not reported	Resistance against late blight *Phytophthora infestans*	Eschen-Lippold et al. (2012)
PTGS by agrobacterium-mediated transformation	Tomato	*9-cis-epoxycarotenoid dioxygenase* (*SlNCED1*)	Coding region	20–50%	Increased β-carotene and lycopene content	Sun et al. (2012)
PTGS by agrobacterium-mediated transformation	Tomato	*ACC synthase* (*ACS*)	Coding region	30–50%	Decreased ethylene	Gupta et al. (2013)

Table 10.2 Examples of dsRNA-mediated interference in plants (transient transformation)

Technique used to induce silencing	Plant	Silencing target	Promoter or coding region targeted	Percentage of silencing	Trait	Reference
PTGS by particle bombardment	Barley	GAMyb transcription factor, slender protein, ABA-inducible kinase	Coding region	Not reported	Seed dormancy	Zentella et al. (2002)
Crude extracts bacterially expressed dsRNA	*Nicotiana. Benthamiana*	Pepper mild mottle virus RNA	Coding region	Not reported	Resistance against pepper mild mottle virus RNA	Tenllado et al. (2003)
PTGS by Agroinfiltration	Wheat	*Vernalization gene VRN2*	Coding region	Not reported	Early flowering	Yan et al. (2004)
VIGS (based on brome mosaic virus) via mechanical inoculation	Barley, rice and maize	*Phytoene desaturase* (*PDS*) and *Green fluorescent protein* (*GFP*)	Coding region	Not reported	Photobleaching (amenability of a brome mosaic virus strain as a gene silencing vector in rice and maize)	Ding et al. (2006)
PTGS by particle bombardment	Bean	*Replication-associated protein* (*AC1*) *coding region*	Coding region	Not reported	Resistance against bean golden mosaic virus (BGMV)	Bonfim et al. (2007)
PTGS by agrobacterium-mediated transformation	Wheat	*Mildew locus O* (*MLO*)	Coding region	Not reported	Resistance against powdery mildew *Blumeria graminis* f. Sp. tritici	Riechen, (2007)
PTGS by agrobacterium-mediated transformation	Potato	*Rar1*	Coding region	Not reported	Late blight resistance	Bhaskar et al. (2009)

(continued)

VIGS (based on cucumber mosaic virus) via sap inoculation	Soybean	*Flavonoid 3′-hydroxylase*	Coding region	97.6%	Reduced pigmentation	Nagamatsu et al. (2009)
VIGS (based on tobacco mosaic virus) via Agroinfiltration	Tobacco and tomato	*Phytoene desaturase (PDS) gene*	Coding region	Not reported	Photobleaching (optimization of system)	Velásquez et al. (2009)
Crude extracts bacterially expressed dsRNA	Maize	Sugarcane mosaic virus coat protein	Coding region	Not reported	Resistance against sugarcane mosaic virus	Gan et al. (2010)
VIGS (based on brome mosaic virus) via mechanical inoculation	Barley	*Phytoene desaturase* (PDS) and *Ubiquitin* gene	Coding region	Not reported	Photo bleaching (amenability of *Barley stripe mosaic virus* as a gene silencing vector in barley (*Hordeum vulgare*) roots, *Brachypodium distachyon* and Avena sativa (oat)	Pacak et al. (2010b)
PTGS by agrobacterium-mediated transformation	Lemon	*Citrus phytoene desaturase (PDS)* and *callose synthase (CalS1) genes*	Coding region	Not reported	Resistance against citrus canker (*Xanthomonas citri* subsp. citri (Xcc))	Enrique et al. (2011)
VIGS (based on tobacco rattle virus) via agrobacterium infiltration	Cotton	*KATANIN* and *WRINKLED1*	Coding region	99%	Increased fibre length but reduced oil seed content	Qu et al. (2012)

Table 10.2 (continued)

Technique used to induce silencing	Plant	Silencing target	Promoter or coding region targeted	Percentage of silencing	Trait	Reference
Crude extracts bacterially expressed dsRNA	Cattleya	Cymbidium mosaic virus coat protein	Coding region	Not reported	Resistance against cymbidium mosaic virus	Lau et al. (2014)
VIGS (based on apple latent spherical virus) via particle bombardment	Sweet cherry	*Phytoene desaturase* (*PDS*)	Coding region	Not reported	Photobleaching (amenability of apple latent spherical virus as a gene silencing vector in Prunus species)	Kawai et al. 2016
VIGS (based on foxtail mosaic virus) via Agroinfiltration	Barley, wheat and foxtail millet	*Phytoene desaturase* (*PDS*), magnesium chelatase in barley and PDS and Cloroplastos alterados1 in wheat	Coding region	Not reported	Photobleaching (amenability of a foxtail mosaic virus as a gene silencing vector in barley and wheat)	Liu et al. (2016a)
VIGS (based on tobacco rattle virus) via Agroinfiltration	Tomato	*Solanum lycopersicum Stress-related NAC1*	Coding region	70%	Resistance against grey mould *Botrytis cinerea*	Liu et al. (2016b)
VIGS (based on cucumber mosaic virus) via vascular puncture inoculation	Maize	*Phytoene desaturase* (*PDS*) and sphingosine 1-phosphate	Coding region	Not reported	Photobleaching (amenability of a cucumber mosaic virus (CMV) strain as a gene silencing vector in rice and maize)	Wang et al. (2016)

wfamily (AGO4/AGO6/AGO9) catalyse the unwinding of the siRNA duplex and the guide (antisense) strand of the siRNA activates de novo methylation at the targeted promoter region, resulting in transcriptional inhibition of the target gene (reviewed in Borges and Martienssen 2015).

As gene silencing is harnessed as a tool for crop improvement, the distribution of the silencing effect throughout a plant is of interest. This is especially important if the dsRNA is only expressed, or introduced into a single site or tissue. Gene silencing by sRNAs has been shown to spread from cell to cell through plasmodesmata (Lucas et al. 1995; Kim et al. 2003) and to spread systemically from the site of initiation throughout the plant by movement of mobile silencing signals between neighbouring plant cells through the vascular phloem tissue of the plant (Yoo et al. 2004; Buhtz et al. 2008). These mobile silencing signal components include sRNA molecules operating in a nucleotide-sequence-specific manner (Palauqui et al. 1997; Herr and Baulcombe 2004). While the exact details of the mobile signal molecules are still under debate, studies have shown that systemic silencing is mediated by both PTGS and TGS (reviewed in Mermigka et al. 2015; Pumplin and Vionnet 2013), the latter involving DNA methylation (Bai et al. 2011; Bond and Baulcombe 2015).

10.2 Techniques Used for Inducing dsRNA-Mediated Silencing in Plants

RNA-mediated silencing can be induced in plants at the transcriptional and post-transcriptional levels by a variety of dsRNA forms, including sense/antisense RNA, small/long hairpin RNA (hpRNA) and artificial miRNA precursors (reviewed in Simon Mateo and Garcia 2011). Induction of gene silencing by TGS involves the design of a construct which produces dsRNA that shares homology with the gene promoter region of the silencing target (Mette et al. 2000). This type of promoter-directed RNA silencing is heritable and persists even in the absence of the inducing transgene (reviewed in Verdel et al. 2009). On the other hand, PTGS involves silencing of a target mRNA by introduction of construct producing an inverted repeat with homology to an mRNA encoding sequence of the targeted gene (Fig. 10.1a). The techniques reported for induction of targeted TGS and PTGS using dsRNA in plants are described below.

10.2.1 Induction of RNA Silencing Using Agrobacterium-Mediated Transformation

The elucidation of RNA silencing mechanisms coincided with the period of expansion of stable genetic transformation techniques in plants in the 1990s. This provided the means for researchers to exploit RNA silencing via *Agrobacterium tumefaciens*-mediated transformation, and to silence target genes in a

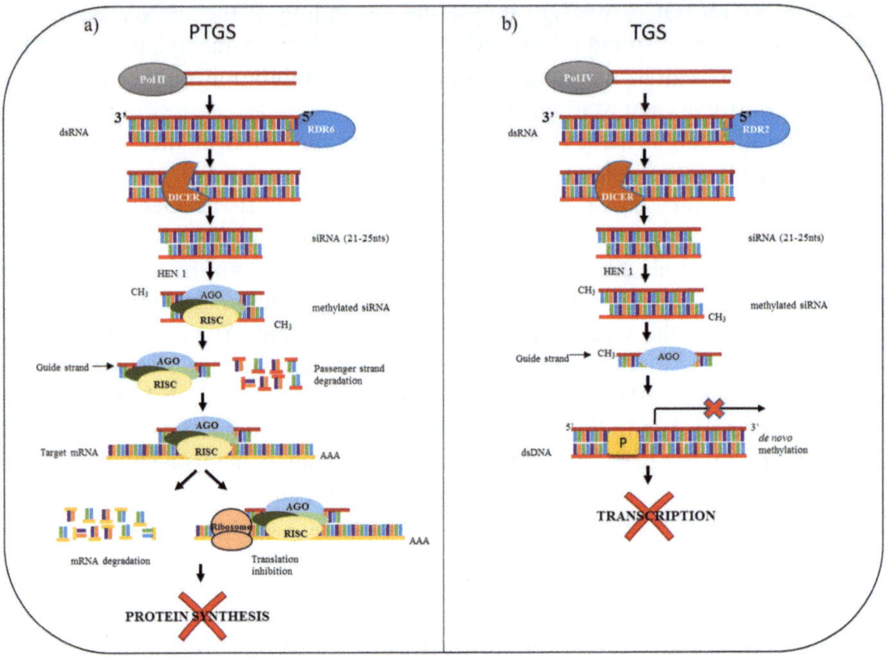

Fig. 10.1 Representative pathways for RNA silencing in plants. (**a**) Post-transcriptional gene silencing (PTGS) in plants. dsRNA forms from mRNA transcribed by RNA polymerase II (Pol II) with complementary strand synthesis by RNA-DEPENDENT RNA POLYMERASE 6 (RDR6). DICER-LIKE family enzymes (DICER) produce siRNAs which are 2′-*O*-methylated by methyltransferase HUA ENHANCER 1 (HEN1). AGO family enzymes unwind the siRNA duplex, then the guide (antisense) strand is incorporated into RNA-induced silencing complex (RISC) in association with AGO and effector proteins. Target mRNAs with sequence complementarity to the guide siRNA are either cleaved by AGO family RNases and degraded, or their transcription is prevented by blocking of ribosomal binding, inhibiting translation into proteins. (**b**) Transcriptional gene silencing (TGS) in plants. dsRNA forms from RNA transcribed by RNA polymerase IV (Pol IV) with complementary strand synthesis by RNA-DEPENDENT RNA POLYMERASE 2 (RDR2). DICER-LIKE family RNAse III enzymes (DICER) produce siRNA which are 2′-*O*-methylated by methyltransferase HUA ENHANCER 1 (HEN1). AGO family endonucleases (AGO) unwind the siRNA duplex and the guide (antisense) strand of the siRNA activates de novo methylation at the targeted promoter region, resulting in transcriptional inhibition of the target gene

sequence-specific manner for crop improvement and for functional genomic research in crop plants. RNA silencing in plants was first described when transgenic plants were developed to express sense RNA (earlier termed as co-suppression) (Napoli et al. 1990), antisense RNA (Van der Krol et al. 1990; Van Blokland et al. 1994; Waterhouse et al. 1998; Davuluri et al. 2005) and hpRNA (Smith et al. 2000) with a matching targeted gene sequence. Since then there have been numerous reports of RNA silencing to produce plants with modified phenotypes, including modified metabolite composition; improved stress tolerance; virus resistance; modified oil content; altered seed storage capacity and improved nutritional value

(Tables 10.1 and 10.2). RNA silencing has been shown to be effective for producing stable and heritable silencing of targeted traits in various crop plants. In addition to *Agrobacterium*-mediated transformation, particle bombardment (Yue et al. 2007) and electroporation (Akashi and Miyagishi 2004) have been used to introduce RNA silencing constructs into plants (Tables 10.1 and 10.2). However, the introduction of RNA silencing traits into plants via plant transformation is time consuming, laborious, costly and faces biosafety regulatory hurdles for field testing and commercial application, thus other approaches have been developed and are particularly useful for research.

10.2.2 Virus-Induced Gene Silencing (VIGS)

A popular alternative for inducing RNA silencing in plants is the virus-mediated silencing method commonly known as VIGS. In VIGS, recombinant phytopathogenic viruses containing target gene silencing sequences are used to inoculate host plants (Gammon and Mello 2015). The inoculated plant silences the target genes as part of the defence mechanism of the plant perceiving a virus infection. The main advantage of the VIGS technique is that the response of the plant is rapid, with results obtained within 3–4 weeks after inoculation. This technique offers an alternative for plant species which are recalcitrant to *Agrobacterium*-mediated transformation, as VIGS also can be induced by a sap inoculation method (Lu et al. 2003) or via particle bombardment (Krenz et al. 2010). The first VIGS vector reported was based on the RNA virus, *Tobacco mosaic virus* (TMV), and was used to silence *Phytoene desaturase* (PDS) mRNA in *Nicotiana benthamiana* (Kumagai et al. 1995). Since then several RNA and DNA viruses have been modified to design VIGS vectors (Liu et al. 2016a) and applied for gene silencing in several crop species including tomato, soybean, barley, rice and maize (Tables 10.1 and 10.2). A major limitation of the VIGS approach is the restriction of the host range of the viruses for which VIGS vectors are available, for example, there are limited options for monocot crops, however VIGS constructs have been reported based on some viruses that can be used in some monocot species, these include *Barley stripe mosaic virus* (Pacak et al. 2010a), *Brome mosaic virus* (Ding et al. 2006), *Bamboo mosaic virus* (Liou et al. 2014), *Foxtail mosaic virus* (Liu et al. 2016a) and *Cucumber mosaic virus* (Wang et al. 2016) .

10.2.3 RNA Silencing via Bacterially Expressed dsRNA

Bacterially expressed dsRNA is perhaps the most simple and cost-effective approach for inducing RNA silencing in plants. In this approach, crude extracts from bacteria expressing a specific dsRNA are directly applied or sprayed on plant tissues to induce silencing of a targeted gene. Similar to gene silencing methods used in nematodes, the direct application of dsRNA by mechanical inoculation of in vitro synthesized dsRNA on tobacco leaves resulted in sequence-specific gene silencing (Tenllado and Díaz-Ruíz 2001). The treated plants showed effective resistance against *Tobamovirus*, *Potyvirus* and *Alfamovirus*. However, the researchers noted that the technique would be costly and laborious to apply on a large scale, so was unsuitable for commercial application (Tenllado and Díaz-Ruíz 2001). To overcome this problem, they developed a cost-effective technique using bacteria to biosynthesize dsRNA in vivo, and then used crude extracts of the bacteria to directly inoculate plant tissue. This was able to produce resistance to *Tobamovirus* and *Potyvirus* (Tenllado et al. 2003). Several studies successfully induced RNA silencing in crop plants via bacterially expressed dsRNA (Tables 10.1 and 10.2). However, in contrast to the stable and heritable resistance that can be achieved in a transgenic plant, direct application of dsRNA is unable to confer long-term protection for the maintenance of RNA silencing, so continuous or repetitive treatments with dsRNA are required, if the desired trait requires long-term silencing.

10.3 Application of dsRNA-Mediated Silencing for Crop Improvement

RNA-induced gene silencing has been found to be a very effective approach for the development of plants with loss-of-function phenotypes that result in improved crop traits and also to target plant pest and disease agents. The following sections and tables provide examples of dsRNA for crop improvement. Tables 10.1 and 10.2 list representative examples of RNA silencing methods that have been explored in crops and Table 10.3 lists some key RNA silencing events in crops that have been approved for field release.

10.3.1 dsRNA-Mediated Silencing for Crop Protection

After the discovery of the involvement of RNA silencing as a natural mechanism of plant host protection against viruses, there was rapid development of RNA silencing based methods against not only viruses, but also against other major plant pests and disease causing organisms. Considering that many crops are intended for food or animal feed, most RNA silencing approaches are designed to target mRNA that are distinct in sequence from, or have no equivalent gene, in humans and animals (Casacuberta et al. 2015). The next section describes some of the examples where RNA silencing has been developed towards crop protection.

10.3.1.1 Virus Resistance

Plant viruses generally cause undesirable effects on plant phenotypes by generating a wide range of symptoms, including discolouration and distortion of flowers and leaves, and loss of vigour and yield. Pathogen-derived resistance (PDR), as the name suggests, derives resistance from sequences of the genome of the pathogen being targeted (Sanford and Johnston 1985; reviewed in Yeam 2016). PDR was validated in 1986 when tobacco plants transformed to contain a TMV CP showed delayed symptom development, with 10–60% of the plants resistant to TMV infection (Abel et al. 1986). A few years later, transformation of tobacco plants with *TEV CP* conferred resistance to TEV (Lindbo and Dougherty 1991, 1992). The discovery of the involvement of virus-derived siRNA as triggers of RNA silencing (Ratcliff et al. 1997) paved the way for more precise use of PDR via RNA silencing in plants (Tenllado and Llave 2004; reviewed in Baulcombe 2004). PDR via RNA silencing had a major impact in rescue of the papaya crop in Hawaii from *Papaya ringspot virus* in 1992 (Ferreira et al. 2002) and has been successfully demonstrated to produce virus resistance in several other plants by targeting viral genes in a sequence-specific manner using in vivo or in vitro transcribed siRNAs; through plasmid-encoded shRNA and by direct application of dsRNA (Lau et al. 2014). Among all RNA silencing-mediated traits in crops that have been approved for field release, virus resistance is the most prominent (Table 10.3).

The direct application of crude extracts of bacteria expressing dsRNA onto plant surfaces, without plant transformation showed high efficacy against *Tobamovirus* and *Potyvirus* infection and was shown to have effects in cells far distant from the initial site of dsRNA delivery (Tenllado et al. 2003). The in vivo or in vitro production of virus CP dsRNA has been applied for resistance in a number of crops, including sugarcane (*Sugarcane mosaic virus*; Gan et al. 2010), potato (*Potato virus Y*; Sun et al. 2010), orchid (*Cymbidium mosaic virus*; Lau et al. 2014) and pea (*Pea seed borne mosaic virus*; Šafářová et al. 2014).

10.3.1.2 Insect and Other Pest Resistance

Insects and other pests, such as nematodes, cause huge losses in crop production and genetic modification strategies can help address this issue, with the advantage of reducing the exposure of farm workers and the environment to pesticides. The ideal targets for an RNA silencing approach are mRNA from genes for which expression is essential in the insect or pest and with no similar target in the host plant or in beneficial herbivores or humans. Since widespread and repeated spraying of dsRNA on crops is not economically feasible, there is an obvious advantage to expressing dsRNA in the plant parts on which the targeted pest feeds, generally the leaf, root or plant sap.

Several studies on insect feeding showed effectiveness of dsRNA in controlling insect infestation by silencing insect regulatory genes such as acetylcholinesterase gene (*ace*), aminopeptidase P-like gene (*APP*-like), Cathedrin receptor gene

Table 10.3 Selected approved events in crops modified for RNA silencing phenotypes

Crop	Modified trait	Developer	Country and year of approval	Type of approval
Tomato (*Lycopersicon esculentum*)	Delayed ripening/senescence	Monsanto Company	USA (1994)	Food, feed and cultivation
Tomato (*Lycopersicon esculentum*)	Delayed fruit softening	Zeneca Plant Science and Petoseed Company	USA (1994) and Mexico (1996)	Approved as food, feed and cultivation in USA but in Canada it approved as food only
Squash (*Cucurbita pepo*)	Resistance against Cucumber Mosaic Cucumovirus (CMV), Zucchini Yellow Mosaic Potyvirus and Watermelon Mosaic Potyvirus 2	Seminis Vegetable Seeds (Canada) and Monsanto Company (Asgrow)	USA (1994) and Canada (1998)	Approved as food, feed and cultivation in USA but in Canada it approved as food only
Tomato (*Lycopersicon esculentum*)	Delayed ripening/senescence	DNA Plant Technology Corporation (USA)	Canada (1995), USA (1995) and Mexico (1998)	Approved as food, feed and cultivation in USA but in Canada and Mexico it approved as food only
Tomato (*Lycopersicon esculentum*)	Delayed ripening/senescence	Huazhong Agricultural University (China)	China (1997)	Food, feed and cultivation
Papaya (*Carica papaya*)	Resistance against papaya ringspot virus	Cornell University and University of Hawaii	USA (1997), Canada (2003) and Japan (2011)	Approved as food, feed and cultivation in USA but in Japan and Canada it approved as food only
Sweet pepper (*Capsicum annuum*)	Resistance against Cucumber Mosaic virus	Beijing University	China (1998)	Food and cultivation
Potato (*Solanum tuberosum*)	Resistance against Potato Virus Y	Monsanto Company	USA (1998)	Food and feed
Potato (*Solanum tuberosum*)	Resistance against Potato Leaf Roll Virus	Monsanto Company	USA (1998), Canada (1999), Australia (2001), Japan (2001), Mexico (2001), New Zealand (2001), Philippines (2004) and South Korea (2004)	Approved as food in all countries and also as feed in Canada, Philippines and USA but cultivation is allowed only in Canada and USA

Tomato (*Lycopersicon esculentum*)	Delayed ripening	Beijing University	China (1999)	Food, feed and cultivation
Plum (*Prunus domestica*)	Resistance against Plum pox virus	United States Department of Agriculture—Agricultural Research Service	USA (2009)	Food, feed and cultivation
Common bean (*Phaseolus vulgaris*)	Resistance against Bean Golden Mosaic Virus	EMBRAPA (Brazil)	Brazil (2011)	Food, feed and cultivation
Soybean (*Glycine max*)	Increase in oleic acid content in seed oil	Monsanto Company	Australia (2011), Canada (2011), Mexico (2011), New Zealand (2011), USA (2011), Japan (2012), South Korea (2013), Taiwan (2013), Colombia (2014), Philippines (2014), European Union (2015), Indonesia (2015), Singapore (2015), Vietnam (2015)	Approved as food in all countries and as feed in Canada, Colombia, European Union, Japan, Philippines, South Korea, USA and Vietnam but cultivation is allowed only in Canada and USA
Potato (*Solanum tuberosum*)	Modified starch	BASF	USA (2014)	Food and feed
Potato (*Solanum tuberosum*)	Reduced acrylamide potential, black spot bruise tolerance	J.R. Simplot Co.	USA (2014)	Food, feed and cultivation
Potato (*Solanum tuberosum*)	Modified starch/carbohydrate, reduced acrylamide potential, black spot bruise tolerance	J.R. Simplot Co.	USA (2014), Canada (2016) and South Korea (2016)	Food, feed and cultivation in Canada and USA. In south Korea approved only as feed
Golden delicious apple (*Malus domestica*)	Non-Browning Phenotype	Okanagan Specialty Fruits Incorporated	Canada and USA (2015)	Food, feed and cultivation

Data is for crops with approval for field release and is sourced from http://www.isaaa.org/gmapprovaldatabase/

(*Cad1b*), Ryanodine receptor genes (*RyR*) (reviewed in Kim et al. 2015) and osmoregulatory genes expressed in the gut including *systemic RNA interference deficient-1* (*SID1*), sucrase, aquaporin and sugar transporter genes (Tzin et al. 2015). Once it had been established that insect and other pests could be inhibited by feeding on dsRNA targeting the expression of gut enzymes and other key metabolic targets. Several research groups began to develop transgenic plants expressing such dsRNA sequences (reviewed in Kim et al. 2015).

Baum et al. (2007) developed transgenic corn producing a Vacuolar ATPase gene (*v-ATP*) dsRNA from the Western corn rootworm, *Diabrotica virgifera virgifera* LeConte. Rootworm feeding on the transgenic plants showed delayed larval development and increased mortality. Mao et al. (2007) targeted a cytochrome P450, *CYP6AE14*, a midgut-associated gene found to detoxify gossypol (secondary metabolite in cotton plant) in the cotton bollworm, *Helicoverpa armigera*. A cytochrome P450, *CYP6AE14* dsRNA was expressed in *Nicotiana tabacum* and *Arabidopsis thaliana*, and larvae fed on either of the transgenic plants showed retarded growth. In 2012, Bhatia et al. showed that feeding on transgenic *A. thaliana* expressing serine protease dsRNA decreased the protease activity (31%) and reduced numbers of progeny of the green peach aphid, *Myzus persicae*.

More recently, Zhang et al. (2015) demonstrated the concept of expressing long dsRNA in plastids for conferring crop protection. This addressed the concern that dsRNA-mediated silencing can be hampered by the endogenous plant RNA silencing machinery in which dsRNA is processed as siRNA, resulting in reduction in siRNA levels and also availability (as the siRNA will be bound to RISC in the plant cells). Chloroplasts lack RNA silencing machinery and hence expressing long dsRNA in chloroplasts protect against dicer activity and other nucleases in the cell. Transplastomic potato plants producing dsRNA against the *β-actin* of Colorado potato beetle, *Leptinotarsa decemlineata* showed dsRNA accumulation up to 0.4% of the total cellular RNA in the chloroplast. Larvae fed on the plants showed retarded growth, supporting the technique to have potential for crop protection.

Xiu-Ming et al. (2016) developed transgenic cotton expressing NADH dehydrogenase (ubiquinone) flavoprotein 2 (NDUFV2) dsRNA. NDUFV2 is a subunit of mitochondrial complex I and catalyses NADH dehydrogenation in the respiratory chain in insects. Cotton bollworm, *Helicoverpa armigera* fed on the transgenic plants showed mortality up to 80% within 5 days and complete elimination of the insect occurred within 7 days post feeding because of altered mitochondrial structure and activity. In a similar study, transgenic expression of hpRNA was applied to develop tobacco and tomato plants targeting *chitinase* mRNA in cotton bollworm. Continuous feeding on the leaves of transgenic lines affected the overall growth and development of the insects (Mamta et al. 2016).

In another recent study, Murphy et al. (2016) engineered genetically modified yeast, *Saccharomyces cerevisiae* expressing dsRNA of a *Drosophila suzukii*, γ-tubulin. Yeast naturally grows on the surface of fruit crops and attracts Drosophila; the study relied on symbiotic interaction between the Drosophila, fruit crop and yeast. The naturally attractive yeast biopesticide expressing dsRNA was observed to be effective to reduce the numbers and locomotor skill of larvae and reproduction in adults.

In addition to insect and pests, plant parasitic nematodes (PPN) are major causes of crop yield loss (Sinha 2010). Yadav et al. (2006) first demonstrated the successful use of dsRNA-mediated silencing against the PPN, *Meloidogyne incognita* in tobacco by targeting integrase and splicing factor mRNA, which was lethal to the nematode. Klink and Matthews (2009) reported expression of four embryonic lethal genes in *Glycine max* (soya bean) roots. The expression of dsRNA targeting ribosomal protein 3 and 4, synaptobrevin and a spliceosome SR protein reduced infection by *Heterodera glycines*. In 2013, Iberkleid et al. developed tomato hairy roots expressing hpRNA of *Meloidogyne javanica* fatty acid and retinol binding protein *mj-far-1* which led to an 80% reduction of the nematodes in the transgenic tomato plants. Likewise several studies in plants have demonstrated the delivery of dsRNA and target potential genes for inhibiting crop destruction caused by PPN (reviewed in Dutta et al. 2015).

10.3.1.3 Bacterial and Fungal Resistance

Bacterial and fungal diseases are major threats to crop production. Many of these diseases spread rapidly through crop plants, often via insect vectors, and are difficult to control. Hence, in addition to other crop protection strategies, RNA silencing has been explored towards reducing the severity of bacterial and fungal infections in plants or to develop crops resistant to these diseases. In 2009, Jiang et al. reported an RNA silencing approach to reduce expression of fatty acid desaturase gene *OsSS12* in rice to confer resistance against the bacterial pathogen *Xanthomonas oryzae* pv. Oryzae and the rice blast fungus, *Magnaporthe grisea*. Another example of dsRNA-mediated silencing against bacterial infection was the silencing of two tumour formation bacterial genes (*iaaM* and *ipt*) which was able to reduce crown gall tumours induced by *Agrobacterium tumefaciens* by 98.5–100% in *A. thaliana* and 75.8–100% in *Lycopersicon esculentum* (Escobar et al. 2001). Examples that have shown promising results for RNA silencing-mediated control of fungal pathogens have been demonstrated in plants including transgenic *Arabidopsis* lines that silenced fungal pathogenicity related *Fusarium oxysporum* Wilt 2 (FOW2), F-box protein Required for Pathogenicity 1 (*FRP1*) and 12-oxophytodienoate-10, 11-reductase gene (*OPR*) mRNA in *Fusarium oxysporum*, resulting in a significant reduction in the level of infection in transgenic plants (Hu et al. 2015); and transgenic tobacco plants expressing a *chitin synthase* (*chs*) hpRNA construct to inhibit cell wall formation and reduce infection by *Sclerotinia sclerotiorum* (Andrade et al. 2015).

10.3.2 dsRNA-Mediated Interference for Improved Crop Quality

As RNA silencing is a negative regulator, this limits application in crop improvement to traits where a reduction in gene expression can provide a desirable change in phenotype. Good examples of this have been demonstrated for several quality traits in various crops including the manipulation of nutritional or chemical content and the alternation of colour in flowers and fruits, as described in the following sections.

10.3.2.1 Crop Improvement for Shelf Life

The visual appearance, flavour, texture, and the nutritional value of fruits and vegetables are factors critical to consumer acceptance. For optimal flavour, fruits and vegetables must be harvested at or close to maturity; however, rapid ripening after harvesting is responsible for short shelf lives and product spoilage, which are major causes of economic loss for farmers and the fruit industry. Ripening of climacteric fruits requires ethylene, so delaying ethylene activity can enhance shelf life. The first genetically modified plant product to be commercialized, the FlavorSavr® tomato was based on "antisense" technology, prior to full understanding of RNA silencing mechanisms. Tomato engineered to produce an antisense (reverse compliment) polygalacturonidase mRNA showed delayed ripening of fruit (Sheehy et al. 1988). Later, dsRNA-mediated silencing of 1-Aminocyclopropane-1-carboxylate (ACC) oxidase (responsible for the oxidation of ACC to ethylene) was reported to suppress the ethylene production and extend the post-harvest storage of tomato (Xiong et al. 2005) and papaya (Sekeli et al. 2014). Shelf life in tomato fruits (Meli et al. 2010) was also observed to be enhanced by 30 days via silencing of two ripening-specific enzymes, α-mannosidase and β-acetyl hexosaminidase. The fruits obtained were approximately 2.5 fold firmer from the α-mannosidase (α-Man) silenced lines and twofold firmer from the β-acetyl hexosaminidase (β-Hex) silenced lines. Similarly, in capsicum, RNA-mediated gene silencing of *α-Man* and *β-Hex* yielded fruits that were two times firmer compared with the control fruit and shelf life was extended by approximately 7 days (Ghosh et al. 2011). In a study on banana fruit ripening, Elitzur et al. (2016) functionally characterized two banana E class SEPALLATA3 MADS box genes, *MaMADS1* and *MaMADS2*, homologous to the tomato *RIN-MADS* ripening gene. Transgenic banana plants generated to silence either gene via RNA-mediated silencing showed delay in fruit ripening and enhanced shelf life phenotypes, including delayed colour development and softening.

10.3.2.2 Crop Improvement for Nutrition and Health

Genetic modification for crop improvement related to nutritional value as food or feed relies on two key points, the first is enhancement of desirable nutrient components and the second is a reduction in the anti-nutrient/toxic components. Of these, the more obvious for an RNA silencing approach is the down-regulation of gene expression of undesired proteins (anti-nutrients). As the most common targets for modified nutritional content are the fruit and seeds of plants, the use of tissue-specific promoters can restrict the knock-down effect to these specific tissues avoiding "off-target" silencing in other parts of the plant that might impact normal development or defence mechanisms (Sunilkumar et al. 2006; Tang et al. 2007).

Starch, a plant-derived carbohydrate constituting amylopectin (75–80%) and amylose (20–25%) polysaccharides, is the major source for human and animal calories from "staple" food and feed crops (Maskarinec et al. 2006). Non-starch polysaccharides and resistant starch are the chief components of dietary fibre and help in faecal bulking as well as laxative processes. Amylose is known to reduce the gel strength of starch and therefore amylose-free starch is preferred for industrial and food processing purposes as a thickener, water binder, emulsion stabilizer and gelling agent. Regina et al. (2006) used RNA silencing to reduce the expression of two isoforms of the starch-branching enzyme (SBE) II (SEB2a and SEB2b) in wheat endosperm (*Triticum aestivum*) to increase the amylose content. The silencing of isoform SEB2a alone had no significant effect on the amylose content, while silencing of both SEB2a and SEB2b resulted in a 70% increase in amylose content. Feeding of the RNA silenced high amylose wheat grain to rats improved their digestion and bowel movement, supporting the potential of high amylose wheat grain to improve human digestive health. In a similar study, Anderson et al. (2006) used RNA silencing to inhibit starch-branching enzymes encoded by *Sbe1* and *Sbe2* in potato. Constructs were made with either 100 bp or 200 bp segments of both branching enzyme genes, cloned as inverted repeats and controlled by a granule-bound starch synthase promoter (GBSSI). The construct with the two, 200 bp inverted repeat segments produced higher yielding amylose tubers in more than 50% of the transgenic potato lines.

Apart from being a good source of carbohydrates, rice is known to be one of the richest sources of dietary protein. Endosperm protein, glutelin the major component of rice seeds, is encoded by *GluA* and *GluB* gene subfamilies (Takaiwa et al. 1991). Kusaba et al. (2003) reported a rice mutant line LGC-1 (Low Glutelin Content), which was the first report of a commercially useful cultivar developed by using hpRNA-mediated silencing. This low glutelin rice was reported to be useful for the patients with diseases requiring a protein-restricted diet (like diabetes and kidney diseases). It was observed that this dominant mutant produced hpRNA from an inverted repeat for two *glutelin* genes. The resulting RNA-mediated silencing suppressed expression of the chief seed storage protein glutelin to lower glutelin content in the rice grains. This mutant trait was observed to be stable for 20 generations.

Plant seed oils are another key source of calories and an important component of the human and animal diet. Triacylglycerol (TAG), a main component in seed oils, consists of glycerol backbone with three esterified fatty acid groups, the most common of which are palmitic acid (C16:0), oleic acid (C18:1) and linoleic acid (C18:2) (Cagliari et al. 2011). While a high polyunsaturated fatty acid content is considered good for health, this is less desirable in processed foods as they have a higher oxidation rate than saturated and monounsaturated fatty acid, which reduces shelf life (Ebenezer et al. 2015). RNA silencing of omega-3 fatty acid desaturase led to decreased linolenic acid content in transgenic soybean and silencing was observed to be stable across multiple generations (Flores et al. 2008). Similarly in flax, RNA silencing of fatty acid desaturase 2 was observed to increase the oleic acid up to 80% of TAG in 69 RNA silenced lines and was found to be stable across successive generations with oleic acid up to 77% of TAG in homozygous T3 progeny (Chen et al. 2015). In a similar study, cotton seed rich in palmitic acid was developed by RNA silencing of cotton β-ketoacyl-ACP synthase II (ghKAS2), the enzyme responsible for converting palmitoyl-ACP to stearoyl-ACP and oleoyl ACP (Liu et al. 2017).

Other than the macronutrient-rich components, including carbohydrates, oils, fats and proteins, other biochemical compounds that are found in lower levels are also important components in various food crops. One of the essential components in both coffee (*Coffea arabica*) and tea (*Camellia senesis*) is caffeine. However, due to the sometimes undesired metabolic effects of consuming coffee, there is a sizeable demand for caffeine-free coffee, mostly produced using chemical decaffeination, a process which adds costs and may use strong solvents. In 2004, Ogita et al. reported an RNA silencing method for producing decaffeinated coffee plants. Their study of the biosynthetic pathway in coffee suggested that three different N-methyltransferases, namely xanthosine methyltransferase (XMT), 3,7-dimethylxanthine methyltransferase (DXMT; caffeine synthase) and 7-N-methylxanthine methyltransferase (MXMT; theobromine synthase), are key components for caffeine synthesis. RNA silencing by dsRNA was able to suppress *theobromine synthase* (*CaMXMT1*) expression with transformed tissues showing reduced transcript levels for all the three N-methyltransferases and reducing the caffeine content by 100% in embryogenic tissues and 70% in the plantlets. This report showed the feasibility of dsRNA for the development of decaffeinated varieties of plants such as coffee and tea.

Phytonutrients such as carotenoids and flavonoids are known for their various health benefits. Tomato fruit is a good source of carotenoids and flavonoids. Genetic manipulation by overexpression of genes encoding biosynthetic enzymes, or regulation of the related transcription factors, was reported to be suitable techniques for enhancing the levels of phytonutrients independently (Niggeweg et al. 2004). In 2005, Davuluri et al. used an RNA silencing approach to elevate the nutritional value in tomato, using a fruit-specific promoter to suppress an endogenous photomorphogenesis regulatory gene (*DET1*). The transgenic fruits had low transcript levels of *DET1*, which led to significant increase in the carotenoid and flavonoid contents, while other qualities of the fruit remained unaffected.

Another interesting example of loss-of-function by RNA silencing in plants addressed the problem of eye irritation and tearing when chopping fresh onions (*Allium cepa*). When onions cells are damaged during cutting, an enzyme allinase is released. Allinase cleaves sulphurous secondary metabolites in the onion into sulphenic acids and volatile sulphur compounds, which are responsible for providing flavour to onion. However, this group of volatile compounds also include lachrymatory factor (LF), which causes the eye irritation and tearing experienced when cutting onions. Eady et al. (2008) used RNA silencing to prevent the conversion of 1-propenyl sulphenic acid to LF by targeting the mRNA encoding the enzyme, lachrymatory factor synthase (LFS). The levels of the enzyme were reduced up to 1544-fold and produced onion with lower levels of LF (Eady et al. 2008).

Many crops are used for multiple downstream products. In addition to the valuable fibres of the plant, cotton seeds are rich in oil (21%) and protein (23%), which can serve as a nutrient rich source of food and feed. However, wild-type cotton plants often contain a toxic terpenoid metabolite gossypol, which is induced as a defence response to biotic stress related to microbial infection or infestation by insects (Lusas and Jividen 1987; Hedin et al. 1992). Gossypol-free cotton glandless mutants were produced and commercialized (McMichael 1960) and the seed of such varieties are considered safe for consumption for monogastric animals and humans (Hedin et al. 1992). However, under field conditions, the glandless cottonseed varieties were very susceptible to pathogen attacks, as they lacked protective terpenoids (Jenkins 1995). Gossypol is a sesquiterpene derived from (+)-δ-cadinene via δ-cadinene synthase. In 2006, Sunilkumar et al. used seed tissue-specific RNA silencing of δ-cadinene synthase mRNA to disrupt terpenoid biosynthesis and produce gossypol-free cottonseeds. As the dsRNA expression was limited to seed, gossypol and related plant protective terpenoids were able to be produced in other parts of the plants for maintaining the defensive barrier against pathogens. Recently, Zhongping et al. (2016) used RNA silencing to increase the oil accumulation in cottonseeds via down-regulation of phosphoenolpyruvate carboxylase 1 (GhPEPC1). The cottonseed oil content in the transgenic lines increased by up to 16.7% without exhibiting any observed undesired phenotypic changes to the plants. Use of RNA silencing for functional genomic studies and trait improvement in cotton *Gossypium* sp. has been recently reviewed by Abdurakhmonov et al. (2016).

Tobacco is consumed primarily in the form of cigarettes. People become addicted to tobacco because it contains the chemical nicotine, which is a para-sympathomimetic alkaloid and is a stimulant drug (Chakraborty et al. 2014). Tobacco leaves (*N. tabacum*) contain nicotin, nornicotine, anabasine and anatabine alkaloids, where nicotine represents ~90% of the total alkaloid pool (Cai et al. 2012). During the tobacco curing processes, nitrosamines (TSNAs), a class of carcinogenic toxicants, are formed by nitrosation of these tobacco alkaloids (Cai et al. 2016). RNA silencing has been applied to produce tobacco with low nicotine via suppression of nicotine demethylase, which produced a sixfold decrease in nicotine content relative to control plants (Lewis et al. 2010). In another study, down-regulation of a tobacco plasma membrane-localized nicotine uptake permease (NUP1) was reported to reduce total alkaloid levels in tobacco plants (Hildreth et al. 2008).

10.3.2.3 Modification of Flower and Fruit Pigmentation

The colours of flowers and fruits are important quality traits and have long been manipulated by plant breeders in the horticultural field. Conventional breeding using genetically related species and natural or induced mutants have been widely applied and successful, however some desirable traits, for instance blue roses and carnations, could not be achieved, due to the lack of genes encoding delphinidin-based anthocyanins (reviewed in Nishihara and Nakatsuka 2011). The development of genetic engineering using transgenic approaches was able to address this gap, and eventually both blue roses (Katsumoto et al. 2007) and blue carnations (Fukui et al. 2003) were developed and commercialized.

The colour of plant organs such as flowers and fruits arise from three main groups of secondary metabolites, the flavanoids, betalains and carotenoids (reviewed in Tanaka et al. 2010). Of these, flavonoids are the most straightforward targets for genetic manipulation, as their role is not critical for plant survival. In contrast, carotenoids not only have important light harvesting and photoprotective roles, but are also precursors for the synthesis of plant hormones including abscisic acid, strigolactone and gibberellins (reviewed in Grotewold 2006; Tanaka et al. 2010; Nishihara and Nakatsuka 2011). Due to this, selection of tissue-specific promoters is again important to avoid "off-target" effects. RNA silencing (then known as antisense suppression) was reported by van der Krol et al. (1988) producing white flowers in petunia and tobacco by constitutive expression of an antisense *CHS* gene. White and pale-coloured shades of flowers were similarly produced in other plants such as *Gentiana* sp. (Nishihara et al. 2006) and *Fragaria ananassa* (Lunkenbein et al. 2006). After the discovery of the RNA silencing mechanism, comparison of the effectiveness of antisense (ssRNA with the complementary mRNA strand) and of RNA silencing technology (dsRNA) in *Torenia* plants showed that dsRNA-mediated silencing of the *anthocyanidin synthase* (*ANS*) gene resulted in higher frequency of white flowers and better stability than the expression of sense and antisense *ANS* gene (Nakamura et al. 2006). Other reports using dsRNA-mediated silencing of the *ANS* include production of a paler blue *Gentian* sp. flower (Nakatsuka et al. 2008) and white *Torenia* flower (Nakamura et al. 2006). In other reports, dsRNA-mediated silencing of *flavonoid 3′,5′-hydroxylase* caused the modification of flowers of *Gentian* sp. from blue to magenta (Nakatsuka et al. 2008), while dsRNA-mediated silencing of *chalcone synthase* (*CHS*) successfully produced flowers with colour modified from the wild-type reddish-purple, to white and pale blue in *Tricyrtis* sp. (Kamiishi et al. 2012) and from the wild-type blue to white in *Gentian* sp. (Nakatsuka et al. 2008).

Improved knowledge on the genetic regulation of plant pigment biosynthesis has provided a greater number of potential targets for colour manipulation using RNA silencing. A complex formed from three plant transcription factors, *R2R3-MYB*, basic helix loop helix (*bHLH*) and WD4-repeat protein (MBW complex), coordinates control of the metabolic flux and activates the transcription of flavonoid and anthocyanin biosynthesis genes (Gonzalez et al. 2008; reviewed by Xu et al. 2015). Examples of studies using RNA silencing to alter plant phenotypes by reducing expression of transcription factors include the dsRNA-mediated silencing of *NtAn2* (R2R3-MYB) in tobacco which resulted in white flower colour and the inhibition of the late pathway genes (Pattanaik et al. 2010). Further investigation showed that silencing of *NtAn2* prevented the activation of the promoters of *CHS* and *dihydroflavonol reductase*, which are two main flavonoid pathway genes, thus by silencing this activator, the downstream genes were also effectively silenced (Pattanaik et al. 2010). An example where a repressor (rather than activator) was a target of gene silencing is the transient dsRNA-mediated silencing of *FcMYB1* in white Chilean strawberry producing fruit with a more intense fruit colour than the wild-type fruit (Salvatierra et al. 2013). The silencing of the negative regulator, *FcMYB1*, resulted in higher expression of the *ANS* (compared to controls) and a strong down-regulation of *anthocyanidin reductase* and *leucoanthocyanidin reductase* (Salvatierra et al. 2013). In addition to the presence of various pigments, flower colour can be influenced by the shape of epidermal cells (Kay et al. 1981; Baumann et al. 2007). A recent study by our group targeted *DhMYB1* in *Dendrobium hybrida* by the direct application of dsRNA (to wild-type plants) to observe the effect on cell shape and flower phenotype. While there was no percievable change in flower colour, the direct application of dsRNA successfully changed the phenotype of floral epidermal cells with the epidermal cells on the flower lip becoming much flatter compared to those of untreated control flowers (Lau et al. 2015).

10.4 Conclusions and Future Prospects

Over the last century, the global population has dramatically increased. According to Census (2012) we may reach a human population of 9.1 billion by 2050. The resulting increase in demand for food, feed, fibre and renewable energy, while competing for arable land and water against growing urbanization and climate change, brings many challenges, which must be addressed at least in part by crop improvement. The development and application of dsRNA to crops to date, indicate that this is a promising technology, not only to assist in discovery research, such as function validation of novel gene sequences, but also to develop genetically modified crops and for the transient inhibition of mRNAs that may have various and useful field applications. While it is certain that any single type of technology alone will not be able to address every need, among the currently available techniques, RNA silencing approaches have shown their potential as cost-effective and sustainable technologies for crop improvement, as elaborated in this chapter. The mechanism of dsRNA-mediated silencing has been shown to offer promising solutions for combating crop pests (including insects and nematodes); important crop diseases (viral, bacterial and fungal); to improve nutritional and heath related traits and to manipulate useful commercial traits such as flower colour. There remain a number of challenges for the widespread development and deployment of dsRNA crops and crop protective products based on dsRNA. These include stability of the RNA and stability of transgene expression, optimization of the delivery of dsRNA from plant host to external pest, and the development and harmonization of regulatory procedures governing release of dsRNA-based crops and products across different legislative districts (at state, country and international levels). In each area, there is active research and collaboration in academia and industry towards producing the data and knowledge needed to address the issues and, together as integrated approaches to crop improvement, it seems certain that many more dsRNA crops will be seen in the future.

References

Abdurakhmonov IY, Ayubov MS, Ubaydullaeva KA, Buriev ZT, Shermatov SE, Ruziboev HS, Shapulatov UM, Saha S, Ulloa M, Yu JZ, Percy RG, Devor EJ, Sharma GC, Sripathi VR, Kumpatla SP, van der Krol A, Kater HD, Khamidov K, Salikhov SI, Jenkins JN, Abdukarimov A, Pepper AE (2016) RNA interference for functional genomics and improvement of cotton (*Gossypium* sp.) Front Plant Sci 7:202

Abel PP, Nelson RS, De B, Hoffmann N, Rogers SG, Fraley RT, Beachy RN (1986) Delay of disease development in transgenic plants that express the tobacco mosaic virus coat protein gene. Science 232:738–743

Akashi H, Miyagishi TK (2004) RNAi expression vectors in plant cells. Methods Mol Biol 252:533–543

Allen RS, Millgate AG, Chitty JA, Thisleton J, Miller JA, Fist AJ, Gerlach WL, Larkin PJ (2004) RNAi-mediated replacement of morphine with the nonnarcotic alkaloid reticuline in opium poppy. Nat Biotechnol 22:1559–1566

Anderson MM, Melander PP, Larsson H, Bulow L, Hofvander P (2006) Targeted gene suppression by RNA interference: an efficient method for production of high-amylose potato lines. J Biotechnol 123:137–148

Andrade CM, Tinoco MLP, Rieth AF, Maia FCO, Aragão FJL (2015) Host-induced gene silencing in the necrotrophic fungal pathogen *Sclerotinia sclerotiorum*. Plant Pathol 65:626–632

Bai S, Kasai A, Yamada K, Li T, Harada T (2011) A mobile signal transported over a long distance induces systemic transcriptional gene silencing in a grafted partner. J Exp Bot 62:4561–4570

Baulcombe D (2004) RNA silencing in plants. Nature 431:356–363

Baulcombe DC, Hamilton AJ (2012) U.S. Patent No. 8349607. U.S. Patent and Trademark Office, Washington, DC

Baum JA, Bogaert T, Clinton W, Heck GR, Feldmann P, Ilagan O, Johnson S, Plaetinck G, Munyikwa T, Pleau M, Vaughn T, Roberts J (2007) Control of coleopteran insect pests through RNA interference. Nat Biotechnol 25:1322–1326

Baumann K, Perez-Rodriguez M, Bradley D, Venail J, Bailey P, Jin HL, Koes R, Roberts K, Martin C (2007) Control of cell and petal morphogenesis by R2R3 MYB transcription factors. Development 134:1691–1701

Baumberger N, Baulcombe DC (2005) Arabidopsis ARGONAUTE1 is an RNA slicer that selectively recruits microRNAs and short interfering RNAs. Proc Natl Acad Sci U S A 102:11928–11933

Béclin C, Berthomé R, Palauqui JC, Tepfer M, Vaucheret H (1998) Infection of tobacco or Arabidopsis plants by CMV counteracts systemic post-transcriptional silencing of nonviral (trans) genes. Virology 252:313–317

Bernstein E, Caudy AA, Hammond SM, Hannon GJ (2001) Role for a bidentate ribonuclease in the initiation step of RNA interference. Nature 409:363–366

Bhaskar PB, Venkateshwaran V, Wu L, Ané JM, Jiang J (2009) *Agrobacterium*-mediated transient gene expression and silencing: a rapid tool for functional gene assay in potato. PLoS One 4:e5812

Bhatia V, Bhattacharya R, Uniyal PL, Singh R, Niranjan RS (2012) Host generated siRNA attenuate expression of serine protease gene in *Myzus persicae*. PLos One 7:e46343

Bond DM, Baulcombe DC (2015) Epigenetic transitions leading to heritable, RNA-mediated de novo silencing in *Arabidopsis thaliana*. Proc Natl Acad Sci U S A 112:917–922

Bonfim K, Faria JC, Nogueira EO, Mendes EA, Aragao FJ (2007) RNAi-mediated resistance to be an golden mosaic virus in genetically engineered common bean (*Phaseolus vulgaris*). Mol Plant-Microbe Interact 20:717–726

Borges F, Martienssen RA (2015) The expanding world of small RNAs in plants. Nat Rev Mol Cell Biol 16(12):727–741

Brodersen P, Sakvarelidze-Achard L, Bruun-Rasmussen M, Dunoyer P, Yamamoto YY, Sieburth L, Voinnet O (2008) Widespread translational inhibition by plant miRNAs and siRNAs. Science (New York, NY) 320:1185–1190

Buhtz A, Springer F, Chappell L, Baulcombe DC, Kehr J (2008) Identification and characterization of small RNAs from the phloem of *Brassica napus*. Plant J 53:739–749

Byzova M, Verduyn C, De Brouwer D, De Block M (2004) Transforming petals into sepaloid organs in Arabidopsis and oilseed rape: implementation of the hairpin RNA-mediated gene silencing technology in an organ-specific manner. Planta 218:379–387

Cagliari L, Margis R, Maraschin FS, Turchetto-Zolet AC, Loss G, Margis-Pinheiro M (2011) Biosynthesis of Triacylglycerols (TAGs) in plants and algae. Int J Plant Biol 2:1

Cai B, Siminszky B, Chappell J, Dewey RE, Bush LP (2012) Enantioselective demethylation of nicotine as a mechanism for variable Nornicotine composition in tobacco leaf. J Biol Chem 287:42804–42811

Cai B, Huihua J, Franklin FF, Bush LP (2016) Contribution of nicotine and Nornicotine toward the production of N-Nitrosonornicotine in air-cured tobacco (Nicotiana tabacum). J Nat Prod 79:754–759

Casacuberta JM, Devos Y, Du Jardin P, Ramon M, Vaucheret H, Nogue F (2015) Biotechnological uses of RNAi in plants: risk assessment considerations. Trends Biotechnol 33:145–147

Census (2012) The official website of U.S. Department of Commerce, U.S. Census Bureau-World POPClock Projection. http://www.census.gov. Accessed 25 May 2011

Chakraborty A, Apoorv G, Abhinay KS, Pranav P (2014) Effect of oxidative phytochemicals on nicotine-stressed UMNSAH/DF-1 cell line. J Tradit Complement Med 4:126–131

Chen XM (2009) Small RNAs and their roles in plant development. Ann Rev Cell Dev Biol 25:21–44

Chen Y, Zhou XR, Zhang ZJ (2015) Development of high oleic oil crop platform in flax through RNAi-mediated multiple FAD2 gene silencing. Plant Cell 34:643–653

Cigan AM, Unger-Wallace E, Haug-Collet K (2005) Transcriptional gene silencing as a tool for uncovering gene function in maize. Plant J 43:929–940

Dalakouras A, Dadami E, Wassenegger M (2015) Engineering viroid resistance. Virus 7:634–646

Davuluri GR, Van Tuinen A, Fraser PD, Manfredonia A, Newman R, Burgess D, Brummell DA, King SR, Palys J, Uhlig J, Bramley PM, Pennings HM, Bowler C (2005) Fruit-specific RNAi-mediated suppression of DET1 enhances carotenoid and flavonoid content in tomatoes. Nat Biotechnol 23:890–895

Ding XS, Schneider WL, Chaluvadi SR, Mian MA, Nelson RS (2006) Characterization of a brome mosaic virus strain and its use as a vector for gene silencing in monocotyledonous hosts. Mol Plant-Microbe Interact 19:1229–1239

Dutta TK, Banakar P, Rao U (2015) The status of RNAi-based transgenic research in plant nematology. Front Microbiol 5:1–7

Eady CC, Kamoi T, Kato M, Porter NG, Davis S, Shaw M, Kamoi A, Imai S (2008) Silencing onion lachrymatory factor synthase causes a significant change in the sulphur secondary metabolite profile. Plant Physiol 147:2096–2106

Ebenezer AV, Arulazhagan P, Kumar SA, Yeom IT, Banu JR (2015) Effect of deflocculation on the efficiency of lowenergy microwave pretreatment and anaerobic biodegradation of waste activated sludge. Appl Energy 145:104–110

Elitzur T, Yakir E, Quansah L, Zhangjun F, Vrebalov J, Khayat E, Giovannoni JJ, Haya Friedman H (2016) Banana MaMADS transcription factors are necessary for fruit ripening and molecular tools to promote shelf-life and food security. Plant Physiol 171:380–391

Enrique R, Siciliano F, Favaro MA, Gerhardt N, Roeschlin R, Rigano L, Sendin L, Castagnaro A, Vojnov A, Marano MR (2011) Novel demonstration of RNAi in citrus reveals importance of citrus callose synthase in defence against *Xanthomonas citri* subsp. *citri*. Plant Biotechnol J 9:394–407

Eschen-Lippold L, Landgraf R, Smolka U, Schulze S, Heilmann M, Heilmann I, Hause G, Rosahl S (2012) Activation of defense against *Phytophthora infestans* in potato by down-regulation of syntaxin gene expression. New Phytol 193:985–996

Escobar MA, Edwin LC, Summerfelt KR, Dandekar AM (2001) RNAi-mediated oncogene silencing confers resistance to crown gall tumorigenesis. Proc Natl Acad Sci U S A 98:13437–13442

Ferreira SA, Pitz KY, Manshardt R, Zee F, Fitch M, Gonsalves D (2002) Virus coat protein transgenic papaya provides practical control of papaya ringspot virus in Hawaii. Plant Dis 86:101–105

Fire A, Xu S, Montgomery MK, Kostas SA, Driver SE, Mello CC (1998) Potent and specific genetic interference by double-stranded RNA in *Caenorhabditis elegans*. Nature 391:806–811

Flores T, Karpova O, Su X, Zeng P, Bilyeu K, Sleper DA, Nguyen HT, Zhang ZJ (2008) Silencing of *GmFAD3* gene by siRNA leads to low alpha-linolenic acids (18:3) of fad3-mutant phenotype in soybean [Glycine max (Merr.)]. Transgenic Res 17:839–850

Fukui Y, Tanaka Y, Kusumi T, Iwashita T, Nomoto K (2003) A rationale for the shift in colour towards blue in transgenic carnation flowers expressing the flavonoid 3′,5′-hydroxylase gene. Phytochemistry 63:15–23

Gammon DB, Mello CC (2015) RNA interference-mediated antiviral defence in insects. Curr Opin Insect Sci 8:111–120

Gan D, Zhang J, Jiang H, Jiang T, Zhu S, Cheng B (2010) Bacterially expressed dsRNA protects maize against SCMV infection. Plant Cell Rep 29:1261–1268

Ghosh S, Meli VS, Kumar A, Thakur A, Chakraborty N, Chakraborty S, Datta A (2011) The N-glycan processing enzymes a-mannosidase and b-α-N-acetylhexosaminidase are involved in ripening-associated softening in the non-climacteric fruits of capsicum. J Exp Bot 62:571–582

Gonzalez A, Zhao M, Leavitt JM, Lloyd AM (2008) Regulation of the anthocyanin biosynthetic pathway by the TTG1/bHLH/Myb transcriptional complex in Arabidopsis seedlings. Plant J 53:814–827

Grotewold E (2006) The genetics and biochemistry of floral pigments. Annu Rev Plant Biol 57:761–780

Gupta A, Pal RK, Rajama MV (2013) Delayed ripening and improved fruit processing quality in tomato by RNAi-mediated silencing of three homologs of 1-aminopropane-1-carboxylate synthase gene. J Plant Physiol 170:987–995

Hamilton AJ, Baulcombe DC (1999) A species of small antisense RNA in posttranscriptional gene silencing in plants. Science 286:950–952

Hedin PA, Parrott WL, Jenkins JN (1992) Relationship of glands, cotton square terpenoid aldehydes and other allelochemicals to larvae of Heliothis Virescens (Lepidoptera: Noctuidae). J Econ Entomol 85:359–364

Heilersig HJ, Loonen A, Bergervoet M, Wolters AM, Visser RG (2006a) Post-transcriptional gene silencing of GBSSI in potato: effects of size and sequence of the inverted repeats. Plant Mol Biol 60:647–662

Heilersig BH, Loonen AE, Janssen EM, Wolters AM, Visser RG (2006b) Efficiency transcriptional gene silencing of GBSSI in potato depends on the promoter region that is used in an inverted repeat. Mol Gen Genomics 275:437–449

Herr AJ, Baulcombe DC (2004) RNA silencing pathways in plants. Cold Spring Harb Symp Quant Biol 69:363–370

Hildreth SB, Gehmana EA, Yang H, Lua R-He KCR, Hariche KC, Yua S, Lind J, Sandoea JL, Okumotoa S, Murphy SA, Jeleskoa John G (2008) Tobacco nicotine uptake permease (NUP1) affects alkaloid metabolism. PNAS 44:18179–18184

Hu Z, Parekh U, Maruta N, Trusov Y, Botella JR (2015) Down-regulation of fusarium oxysporum endogenous genes by Host-Delivered RNA interference enhances disease resistance. Front Chem 3:1

Iberkleid I, Vieira P, de Almeida EJ, Firester K, Spiegel Y, Horowitz SB (2013) Fatty acid-and retinol-binding protein, Mj-FAR-1 induces tomato host susceptibility to root-knot nematodes. PLoS One 8:e64586

Jenkins JN (1995) Host resistance to insects in cotton. In: Constable GA, Forrester NW (eds) Challenging the future, proceedings of the world cotton research conference-1. Commonwealth Scientific and Industrial Research Organization, Melbourne, pp 359–372

Jiang CJ, Shimono M, Maeda S, Inoue H, Mori M, Hasegaw M, Sugano S, Takatsuji H (2009) Suppression of the rice fatty-acid desaturase gene *OsSSI2* enhances resistance to blast and leaf blight diseases in rice. Mol Plant Microbe Int 22:820–829

Jones AL, Thomas CL, Maule AJ (1998) De novo methylation and cosuppression induced by a cytoplasmically replicating plant RNA virus. EMBO J 17:6385–6393

Kamiishi Y, Otani M, Takagi H, Han DS, Mori S, Tatsuzawa F, Okuhara H, Kobayashi H, Nakano M (2012) Flower color alteration in the liliaceous ornamental *Tricyrtis* sp by RNA interference-mediated suppression of the *chalcone synthase* gene. Mol Breed 30:671–680

Katsumoto Y, Fukuchi-Mizutani M, Fukui Y, Brugliera F, Holton TA, Karan M, Nakamura N, Yonekura-Sakakibara K, Togami J, Pigeaire A, Tao GQ, Nehra NS, Lu CY, Dyson BK, Tsuda S, Ashikari T, Kusumi T, Mason JG, Tanaka Y (2007) Engineering of the rose flavonoid biosynthetic pathway successfully generated blue-hued flowers accumulating delphinidin. Plant Cell Physiol 48:1589–1600

Kawai T, Gonoi A, Nitta M, Yamagishi N, Yoshikawa N, Tao R (2016) Virus-induced gene silencing in various Prunus species with the apple latentspherical virus vector. Sci Hortic 199:103–113

Kay QON, Daoud HS, Stirton CH (1981) Pigment distribution, light-reflection and cell structure in petals. Bot J Linn Soc 83:57–83

Kim JY, Yuan Z, Jackson D (2003) Developmental regulation and significance of KNOX protein trafficking in Arabidopsis. Development 130:4351–4362

Kim YH, Moustapha SI, Anastasia MWC, Kun YZ (2015) RNA interference: applications and advances in insect toxicology and insect pest management. Pest Biochem Physiol 120:109–117

Klink VP, Matthews BF (2009) Emerging approaches to broaden resistance of soybean to soybean cyst nematode as supported by gene expression studies. Plant Physiol 151:1017–1022

Krenz B, Wege C, Jeske H (2010) Cell-free construction of disarmed abutilon mosaic virus-based gene silencing vectors. J Virol Methods 169:129–137

Kumagai MH, Donson J, Della-Cioppa G, Harvey D, Hanley K, Grill LK (1995) Cytoplasmic inhibition of carotenoid biosynthesis with virus-derived RNA. Proc Natl Acad Sci U S A 92:1679–1683

Kusaba M, Miyahara K, Iida S, Fukuoka H, Takano T, Sassa H, Nishimura M, Nishio T (2003) Low glutelin content1: a dominant mutation that suppresses the glutelin multigene family via RNA silencing in rice. Plant Cell 15:1455–1467

Lau SE, Mazumdar P, Hee TW, Song ALA, Othman RY, Harikrishna JA (2014) Crude extracts of bacterially-expressed dsRNA protect orchid plants against *Cymbidium mosaic virus* during transplantation from in vitro culture. J Hortic Sci Biotechnol 89:569–576

Lau SE, Schwarzacher T, Othman RY, Harikrishna JA (2015) dsRNA silencing of an R2R3-MYB transcription factor affects flower cell shape in a *Dendrobium* hybrid. BMC Plant Biol 15:194

Lewis RS, Bowen SW, Keogh MR, Dewey RE (2010) Three nicotine demethylase genes mediate nornicotine biosynthesis in *Nicotiana tabacum* L.: functional characterization of the CYP82E10 gene. Phytochemistry 71:1988–1998

Li DH, Liu H, Yang Y, Zhen PP, Liang JS (2009) Down-regulated expression of RACK1 gene by RNA interference enhances drought tolerance in rice. Rice Sci 16:14–20

Lindbo JA, Dougherty WG (1991) Pathogen-derived resistance to a potyvirus: immune and resistant phenotypes in transgenic tobacco expressing altered forms of a potyvirus coat protein nucleotide sequence. Mol Plant Microbe Interact 5:144–153

Lindbo JA, Dougherty WG (1992) Untranslatable transcripts of the *tobacco etch virus coat protein* gene sequence can interfere with *tobacco etch virus* replication in transgenic plants and protoplasts. Virology 189:725–733

Lindbo JA, Silvarosales L, Proebsting WM, Dougherty WG (1993) Induction of a highly specific antiviral state in transgenic plants: implications for regulation of gene expression and virus resistance. Plant Cell 5:1749–1759

Liou MR, Huang YW, Hu CC, Lin NS, Hsu YH (2014) A dual gene-silencing vector system for monocot and dicot plants. Plant Biotechnol J 12:330–343

Liu Q, Singh SP, Green AG (2002) High-stearic and high-oleic cottonseed oils produced by hairpin RNA-mediated posttranscriptional gene silencing. Plant Physiol 129:1732–1743

Liu N, Xie K, Jia Q, Zhao J, Chen T, Li H, Wei X, Diao X, Hong Y, Liu Y (2016a) Foxtail mosaic virus-induced gene silencing in monocot plants. Plant Physiol 171:1801–1807

Liu B, Ouyang Z, Zhang Y, Li X, Hong Y, Huang L, Liu S, Zhang H, Li D, Song F (2016b) Tomato NAC transcription factor SlSRN1 positively regulates Defense response against biotic stress but negatively regulates abiotic stress response. PLoS One 9:102067

Liu Q, Wu M, Zhang B, Shrestha P, Petrie J, Green AG, Singh SP (2017) Genetic enhancement of palmitic acid accumulation in cotton seed oil through RNAi down-regulation of ghKAS2 encoding b-ketoacyl-ACP synthase II (KASII). Plant Biotechnol J 15(1):132–143

Llave C, Kasschau KD, Rector MA, Carrington JC (2002) Endogenous and silencing-associated small RNAs in plants. Plant Cell 14:1605–1619

Loukoianov A, Yan L, Blech A, Sanchez A, Dubcovsky J (2005) Regulation of VRN-1 vernalization genes in normal and transgenic polyploid wheat. Plant Physiol 138:2364–2373

Lu R, Martin-Hernandez AM, Peart JR, Malcuit I, Baulcombe DC (2003) Virus-induced gene silencing in plants. Methods 30:296–303

Lucas WJ, Bouché-Pillon S, Jackson DP, Nguyen L (1995) Selective trafficking of KNOTTED1 homeodomain protein and its mRNA through plasmodesmata. Science 270:1980–1983

Lunkenbein S, Coiner H, de Vos CHR, Schaart JG, Boone MJ, Krens FA, Schwab W, Salentijn EMJ (2006) Molecular characterization of a stable antisense *chalcone synthase* phenotype in strawberry (*Fragaria* x *ananassa*). J Agric Food Chem 54:2145–2153

Lusas EW, Jividen GM (1987) Glandless cottonseed: a review of the first 25 years of processing and utilization research. J Am Oil Chem Soc 64:839–854

Machado A, Wu Y, Yang Y, Llewellyn DJ, Dennis ES (2009) The MYB transcription factor GhMYB25 regulates early fibre and trichome development. Plant J 59:52–62

Mamta K, Reddy RK, Rajam MV (2016) Targeting chitinase gene of *Helicoverpa armigera* by host-induced RNA interference confers insect resistance in tobacco and tomato. Plant Mol Biol 90:281–292

Mao YB, Cai WJ, Wang JW, Hong GJ, Tao XY, Wang LJ, Huang YP, Chen XY (2007) Silencing a cotton bollworm P450 mono oxygenase gene by plant-mediated RNAi impairs larval tolerance of gossypol. Nat Biotechnol 25:1307–1313

Maskarinec G, Takata Y, Pagano I, Carlin L, Goodman MT, Le Marchand L, Nomura AM, Wilkens LR, Kolonel LN (2006) Trends and dietary determinants of overweight and obesity in a multiethnic population. Obesity (Silver Spring) 14:717–726

McMichael SC (1960) Combined effects of the glandless genes gl2 and gl3 on pigment glands in the cotton plant. Agron J 52:385–386

Meli VS, Ghosh S, Prabha TN, Chakraborty N, Chakraborty S, Datta A (2010) Enhancement of fruit shelf life by suppressing N-glycan processing enzymes. Proc Natl Acad Sci U S A 107:2413–2418

Mermigka G, Verret F, Kalantidis K (2015) RNA silencing movement in plants. J Integr Plant Biol 58(4):328–342

Mette MF, Aufsatz W, Van der Winden J, Matzke MA, Matzke AJM (2000) Transcriptional silencing and promoter methylation triggered by double-stranded RNA. EMBO J 19:5194–5201

Murphy KA, Tabuloc CA, Cervantes KR, Chiu JC (2016) Ingestion of genetically modified yeast symbiont reduces fitness of an insect pest via RNA interference. Sci Rep 6:22587

Nagamatsu A, Masuta C, Matsuura H, Kitamura K, Abe J, Kanazawa A (2009) Down-regulation of flavonoid 3′-hydroxylase gene expression by virus-induced silencing in soybean reveals the presence of a threshold mRNA level associated with pigmentation in pubescence. J Plant Physiol 166:32–39

Nakamura N, Fukuchi-Mizutani M, Miyazaki K, Suzuki K, Tanaka Y (2006) RNAi suppression of the *anthocyanidin synthase* gene in *Torenia hybrida* yields white flowers with higher frequency and better stability than antisense and sense suppression. Plant Biotechnol 23:13–17

Nakatsuka T, Mishiba KI, Abe Y, Kubota A, Kakizaki Y, Yamamura S, Nishihara M (2008) Flower color modification of gentian plants by RNAi-mediated gene silencing. Plant Biotechnol 25:61–68

Napoli C, Lemieux C, Jorgensen R (1990) Introduction of a chimeric *chalcone synthase* gene into petunia results in reversible co-suppression of homologous genes in trans. Plant Cell 4:279–289

Niggeweg R, Michael AJ, Martin C (2004) Engineering plants with increased levels of the antioxidant chlorogenic acid. Nat Biotechnol 22:746–754

Nishihara M, Nakatsuka T (2011) Genetic engineering of flavonoid pigments to modify flower color in floricultural plants. Biotechnol Lett 33:433–441

Nishihara M, Nakatsuka T, Yamamura S (2005) Flavonoid components and flower color change in transgenic tobacco plants by suppression of chalcone isomerase gene. FEBS Lett 579:6074–6078

Nishihara M, Nakatsuka T, Hosokawa K, Yokoi T, Abe Y, Mishiba K, Yamamura S (2006) Dominant inheritance of white-flowered and herbicide-resistant traits in transgenic gentian plants. Plant Biotechnol 23:25–31

Ogita S, Uefuji H, Morimoto M, Sano H (2004) Application of RNAi to confirm theobromine as the major intermediate for caffeine biosynthesis in coffee plants with potential for construction of decaffeinated varieties. Plant Mol Biol 54:931–941

Ogwok E, Odipio J, Halsey M, Gaitán-Solís E, Bua A, Taylor NJ, Fauquet CM, Alicai T (2012) Transgenic RNA interference (RNAi)-derived field resistance to cassava brown streak disease. Mol Plant Pathol 13:1019–1031

Okano Y, Miki D, Shimamoto K (2008) Small interfering RNA (siRNA) targeting of endogenous promoters induces DNA methylation, but not necessarily gene silencing, in rice. Plant J 53:65–77

Pacak A, Geisler K, Jørgensen B, Barciszewska-Pacak M, Nilsson L, Nielsen TH, Johansen E, Grønlund M, Jakobsen I, Albrechtsen M (2010a) Investigations of barley stripe mosaic virus as a gene silencing vector in barley roots and in *Brachypodium distachyon* and oat. Plant Methods 6:26

Pacak A, Strozycki PM, Barciszewska-Pacak M, Alejska M, Lacomme C, Jarmołowski A, Szweykowska-Kulińska Z, Figlerowicz M (2010b) The brome mosaic virus-based recombination vector triggers a limited gene silencing response depending on the orientation of the inserted sequence. Arch Virol 155:169–179

Palauqui JC, Elmayan T, Pollien JM, Vaucheret H (1997) Systemic acquired silencing: transgene-specific post-transcriptional silencing is transmitted by grafting from silenced stocks to non-silenced scions. EMBO J 16:4738–4745

Park GG, Park JJ, Yoon J, Yu SN, An G (2010) A ring finger E3 ligase gene, *Oryza sativa* delayed seed germination 1 (*OsDSG1*), controls seed germination and stress responses in rice. Plant Mol Biol 74:467–478

Pattanaik S, Kong Q, Zaitlin D, Werkman JR, Xie CH, Patra B, Yuan L (2010) Isolation and functional characterization of a floral tissue-specific R2R3 MYB regulator from tobacco. Planta 231:1061–1076

Pumplin N, Voinnet O (2013) RNA silencing suppression by plant pathogens: defence, counter-defence and counter-counter-defence. Nat Rev Microbiol 11:745–760

Qiao F, Yang Q, Wang CL, Fan YL, Wu XF, Zhao KJ (2007) Modification of plant height via RNAi suppression of *OsGA20ox2* gene in rice. Euphytica 158:35–45

Qu J, Ye J, Geng YF, Sun YW, Gao SQ, Zhang BP, Chen W, Chua NH (2012) Dissecting functions of KATANIN and WRINKLED1 in cotton fiber development by virus-induced gene silencing. Plant Physiol 160:738–748

Ratcliff F, Harrison BD, Baulcombe DC (1997) A similarity between viral defense and gene silencing in plants. Science 276:1558–1560

Regina A, Bird A, Topping D, Bowden S, Freeman J, Barsby T, Hashemi BK, Li Z, Rahman S, Matthew M (2006) High-amylose wheat generated by RNA interference improves indices of large bowel health in rats. Proc Natl Acad Sci U S A 103(10):3546–3551

Reinhart BJ, Weinstein EG, Rhoades MW, Bartel B, Bartel DP (2002) MicroRNAs in plants. Genes Dev 16:1616–1626

Riechen J (2007) Establishment of broad-spectrum resistance against *Blumeria graminis* f. Sp. triticiin *Triticum aestivum* by RNAi-mediated knock-down of MLO. J Verbrauch Lebensm 2:120

Romano N, Macino G (1992) Quelling: transient inactivation of gene expression in *Neurospora crassa* by transformation with homologous sequences. Mol Microbiol 6:3343–3353

Šafářová D, Brazda P, Navratil M (2014) Effect of artificial dsRNA on infection of pea plants. Czech J Genet Plant Breed 50:105–108

Salvatierra A, Pimentel P, Moya-León MA, Herrera R (2013) Increased accumulation of anthocyanins in Fragaria chiloensis fruits by transient suppression of *FcMYB1* gene. Phytochemistry 90:25–36

Sanford JC, Johnston SA (1985) The concept of parasite-derived resistance-deriving resistance genes from the parasite's own genome. J Theor Biol 113:395–405

Schijlen EGWM, deVos RCH, Martens S, Jonker HH, Rosin FM, Molthoff JWM, Tikunov YM, Angenent GC, Tunen AJV, Bovy AG (2007) RNA interference silencing of chalcone synthase, the first step in the flavonoid biosynthesis pathway, leads to parthenocarpic tomato fruits. Plant Physiol 144:1520–1530

Schmidt MA, Herman EM (2008) Suppression of soybean oleosin produces micro-oil bodies that aggregate into oil body/ER complexes. Mol Plant 1:910–924

Sekeli R, Abdullah JO, Namasivayam P, Muda P, Abu Bakar UK, Yeong WC, Pillai V (2014) RNA interference of 1-Aminocyclopropane-1-carboxylic acid oxidase (ACO1 and ACO2) genes expression prolongs the shelf life of Eksotika (Carica papaya L.) papaya fruit. Molecules 19:8350–8362

Sheehy RE, Kramer M, Hiatt WR (1988) Reduction of polygalacturonase activity in tomato fruit by antisense RNA. Proc Natl Acad Sci U S A 85:8805–8809

Shimizu T, Yoshii M, Wei T, Hirochika H, Omura T (2009) Silencing by RNAi of the gene for Pns12, a viroplasm matrix protein of Rice dwarf virus, results in strong resistance of transgenic rice plants to the virus. Plant Biotechnol J 7:24–32

Simon-Mateo C, Garcia JA (2011) Antiviral strategies in plants based on RNA silencing. Biochim Biophys Acta 809:722–731

Sinha SK (2010) RNAi induced gene silencing in crop improvement. Physiol Mol Biol Plants 16:321–332

Smith NA, Singh SP, Wang MB, Stoutjesdijk PA, Green AG, Waterhouse PM (2000) Total silencing by intron-spliced hairpin RNAs. Nature 407:319–320

Sun ZN, Yin GH, Song YZ, An HL, Zhu CX, Wen FJ (2010) Bacterially expressed double-stranded RNAs against hot-spot sequences of *tobacco mosaic virus* or *potato virus Y* genome have different ability to protect tobacco from viral infection. Appl Biochem Biotechnol 162:1901–1914

Sun L, Yuan B, Zhang M, Wang L, Cui M, Wang Q, Leng P (2012) Fruit-specific RNAi mediated suppression of SlNCED1 increases both lycopene and β-carotene contents in tomato fruit. J Exp Bot 63:3097–3108

Sunilkumar G, Campbell LM, Puckhaber L, Stipanovic RD, Rathore KS (2006) Engineering cotton seed for use in human nutrition by tissue-specific reduction of toxic gossypol. Proc Natl Acad Sci U S A 103:18054–18059

Syller J (2016) Antagonistic within-host interactions between plant viruses: molecular basis and impact on viral and host fitness. Mol Plant Pathol 17:769–782

Takaiwa F, Oono K, Wing D, Kato A (1991) Sequences of three members and expression of a new major subfamily of glutelin genes from rice. Plant Mol Biol 17:875–885

Tanaka Y, Brugliera F, Kalc G, Senior M, Dyson B, Nakamura N, Katsumoto Y, Chandler S (2010) Flower color modification by engineering of the flavonoid biosynthetic pathway: practical perspectives. Biosci Biotechnol Biochem 74:1760–1769

Tang G, Galili G, Zhuang X (2007) RNAi and microRNA: breakthrough technologies for the improvement of plant nutritional value and metabolic engineering. Metabolomics 3:357–369

Tenllado F, Díaz-Ruíz JR (2001) Double-stranded RNA mediated interference with plant virus infection. J Virol 75:12288–12297

Tenllado F, Llave C (2004) RNA interference as a new biotechnological tool for the control of virus diseases in plants. Virus Res 102:85–96

Tenllado F, Martínez-García B, Vargas M, Díaz-Ruíz JR (2003) Crude extracts of bacterially expressed dsRNA can be used to protect plants against virus infections. BMC Biotechnol 3:3

Travella S, Klimm TE, Keller B (2006) RNA interference based gene silencing as an efficient tool for functional genomics in hexaploid bread wheat. Plant Physiol 142:6–20

Tzin V, Xiaowei Y, Xiangfeng J, Kai GJ, Angela ED (2015) RNA interference against gut osmoregulatory genes in phloem-feeding insects. J Insect Physiol 79:105–112

Uauy C, Distelfeld A, Fahima T, Blechl A, Dubcovsky J (2006) A NAC gene regulating senescence improves grain protein, zinc, and iron content in wheat. Science 314:1298–1301

Van Blokland R, Van der Geest N, Mol JMN, Kooter JM (1994) Transgene-mediated suppression of chalcone synthase expression in *Petunia hybrida* results from an increase in RNA turnover. Plant J 6:861–877

Van der Krol AR, Lenting PE, Veenstra J, van der Meer IM, Koes RE, Gerats AG, Mol JNM, Stuitje AR (1988) An antisense *chalcone synthase* gene in transgenic plants inhibits flower pigmentation. Nature 333:866–869

Van der Krol AR, Mur LA, Beld M, Mol JMN, Stuitje AR (1990) Flavonoid genes in petunia: addition of a limited number of gene copies may lead to suppression of gene expression. Plant Cell 2:291–299

Velásquez AC, Chakravarthy S, Martin GB (2009) Virus-induced gene silencing (VIGS) in *Nicotiana benthamiana* and tomato. J Vis Exp 28:1292. doi:10.3791/1292

Verdel A, Vavasseur A, Le Gorrec M, Touat-Todeschini L (2009) Common themes in siRNA mediated epigenetic silencing pathways. Int J Dev Biol 53:245–257

Wang M, Abbott DC, Waterhouse PM (2000) A single copy of a virus-derived transgene encoding hairpin RNA gives immunity to barley yellow dwarf virus. Mol Plant Pathol 1:347–356

Wang Y, Beaith M, Chalifoux M, Ying J, Uchacz T, Sarvas C, Griffiths R, Kuzma M, Wan J, Huang Y (2009) Shoot-specific down-regulation of protein farnesyltransferase (alpha-subunit) for yield protection against drought in canola. Mol Plant 2:191–200

Wang R, Yang X, Wang N, Liu X, Nelson RS, Li W, Fan Z, Zhou T (2016) An efficient virus-induced gene silencing vector for maize functional genomics research. Plant J 86:102–115

Wassenegger M, Heimes S, Riedel L, Sänger HL (1994) RNA-directed de novo methylation of genomic sequences in plants. Cell 76:567–576

Waterhouse PM, Graham MW, Wang MB (1998) Virus resistance and gene silencing in plants can be induced by simultaneous expression of sense and antisense RNA. Proc Natl Acad Sci U S A 95:13959–13964

Wesley SV, Helliwell CA, Smith NA (2001) Construct design for efficient, effective and high throughput gene silencing in plants. Plant J 27:581–590

Wingard SA (1928) Hosts and symptoms of ring spot, a virus disease of plants. J Agric Res 37:127–153

Xiong AS, Yao QH, Peng RH, Li X, Han PL, Fan HQ (2005) Different effects on ACC oxidase gene silencing triggered by RNA interference in transgenic tomato. Plant Cell Rep 23:639–646

Xiu-Ming W, Chang-Qing Y, Ying-Bo M, Ling-Jian W, Xiao-Xia S, Xiao-Ya C (2016) Targeting insect mitochondrial complex I for plant protection. Plant Biotechnol J 14:1925–1935

Xu W, Dubos C, Lepiniec L (2015) Transcriptional control of flavonoid biosynthesis by MYB–bHLH–WDR complexes. Trends Plant Sci 20:176–185

Yadav BC, Veluthambi K, Subramaniam K (2006) Host-generated double stranded RNA induces RNAi in plant-parasitic nematodes and protects the host from infection. Mol Biochem Parasitol 148:219–222

Yan L, Loukoianov A, Blechl A, Tranquilli G, Ramakrishna W, SanMiguel P, Bennetzen JL, Echenique V, Dubcovsky J (2004) The wheat VRN2 gene is a flowering repressor down-regulated by vernalization. Science 303:1640–1644

Yara A, Yaeno T, Hasegawa M, Seto H, Montillet JL, Kusumi K, Seo S, Iba K (2007) Disease resistance against *Magnaporthe grisea* is enhanced in transgenic rice with suppression of o-3fatty acid desaturases. Plant Cell Physiol 48:1263–1274

Yeam I (2016) Current advances and prospectus of viral resistance in horticultural crops. Hortic Environ Biotechnol 57:113–122

Yoo BC, Kragler F, Varkonyi-Gasic E, Haywood V, Archer-Evans S, Lee YM, Lough TJ, Lucas WJ (2004) A systemic small RNA signaling system in plants. Plant Cell 16:1979–2000

Yu B, Lydiate DJ, Young LW, Schafer UA, Hannoufa A (2007) Enhancing the carotenoid content of Brassica napus seeds by down regulating *Lycopene epsilon* cyclase. Transgenic Res 17:573–585

Yue SJ, Li H, Li YW, Zhu YF, Guo JK, Liu YJ, Chen Y, Jia X (2007) Generation of transgenic wheat lines with altered expression levels of 1Dx5 high-molecular weight glutenin subunit by RNA interference. J Cereal Sci 47:153–161

Zentella R, Yamauchi D, Ho TH (2002) Molecular dissection of the gibberellin/abscisic acid signaling pathways by transiently expressed RNA interference in barley aleurone cells. Plant Cell 14:2289–2301

Zhang J, Khan SA, Hasse C, Ruf S, Heckel DG, Bock R (2015) Full crop protection from an insect pest by expression of long double-stranded RNAs in plastids. Science 347:991–994

Zhongping X, Li J, Guo X, Jin S, Zhang X (2016) Metabolic engineering of cottonseed oil biosynthesis pathway via RNA interference. Sci Rep 6:33342

Chapter 11
Alternative Strategy in Crop Protection: Protease Inhibitors from Turmeric

Chan Seow-Neng, Norliza Abu Bakar, Maziah Mahmood, Ho Chai-Ling, and Noor Azmi Shaharuddin

11.1 Introduction

Agriculture is one of the essential components for economy development of a country. In order to achieve high production yields, most of the crop plantations are conducted in a gigantic scale. However, one of the major challenges in maintaining and increasing the production yield is the spoilage of the crops due to pest and pathogen infestations. If left unattended, the situation could cause a major loss of profits to the planters and eventually the producing country. Instead of using chemical pesticides which could be harmful to the environment and consumers, the production of crop plants with enhanced traits through biotechnology technique is one of the alternatives to overcome these problems.

Advancement in technology has enabled the production of genetically engineered crop plants where the crops acquired beneficial defensive genes from other plants (Dunaevsky et al. 2005). One example of these beneficial genes responsible in natural plant protection is genes coding for protease inhibitors (PIs). PI genes have been identified from plant genome and some are being utilized in the produc-

C. Seow-Neng • M. Mahmood • N.A. Shaharuddin (✉)
Faculty of Biotechnology and Biomolecular Sciences, Department of Biochemistry, Universiti Putra Malaysia (UPM), 43400 Serdang, Malaysia
e-mail: noorazmi@upm.edu.my

N. Abu Bakar
Biotechnology and Nanotechnology Research Center, Malaysian Agricultural Research and Development Institute (MARDI), 43400 Serdang, Malaysia

H. Chai-Ling
Institute of Plantation Studies, Universiti Putra Malaysia, UPM, 43400 Serdang, Selangor, Malaysia

Department of Cell and Molecular Biology, Faculty of Biotechnology and Biomolecular Sciences, Universiti Putra Malaysia, UPM, 43400 Serdang, Selangor, Malaysia

S.N.A. Abdullah et al. (eds.), *Crop Improvement*,
DOI 10.1007/978-3-319-65079-1_11

tion of genetically engineered plants that express phytoprotection characteristic towards certain pest infestation (Christou et al. 2006; Dunaevsky et al. 2005; Lawrence and Koundal 2002; Schlüter et al. 2010). In addition, PIs or the recombinant protein of PIs could also be utilized as active compounds in biopesticides. Their anti-pathogenic properties were proven in direct test assays whereby pest insects showed a profound reduction in body size when the PIs were ingested (Cruz et al. 2013; Haq et al. 2004; Macedo et al. 2010).

Nonetheless, over the course of time, pest insects and pathogens are constantly changing their pathogenesis strategy and adapting to the hosts' defensive response (Habib and Fazili 2007; van Wyk et al. 2016). Haq et al. (2004) proposed that one of the necessary preparations to overcome the adaptability and resistance towards crop protections was by developing a second generation of PIs from novel sources, such as non-host plants. Therefore, it is postulated that pest or pathogens that have not been exposed to novel PIs are not likely to establish resistance towards the PIs.

In line with the idea, turmeric plant could serve as a suitable source for the discovery of novel PIs. Turmeric (*Curcuma longa*) is a well-known traditional medicine and is commonly used as spices in culinary arts in country like India, Malaysia, Thailand and other Asian countries. In this chapter, plant pathogenesis, the involvement of pathogenic protease and general plant defence mechanism are briefly discussed. Current findings on plant protease inhibitors were reviewed including their classification, regulations, characteristics and applications in the industry.

11.2 Plant Pathogenesis

Pathogenesis is a mechanism where diseases are developed and transmitted in susceptible hosts for the mediation and sustenance of diseases (Newton et al. 2010). Plants are in a constant battle to defend themselves against the attacks of their natural predators that include insects and microorganisms, notably fungi, bacteria, viruses and nematodes (Habib and Fazili 2007; Odjakova and Hadjiivanova 2001). Pathogenesis and predation are common occurrences in plants mainly due to the fact that plants are the primary food producer. Plants also serve as a shelter and host for disease mediation. During pathogenesis, a host–pathogen interaction begins with pathogens attacking the host to extract nutrients from the host. The targeted host usually react by developing resistance responses to prevent damages or being killed by the pathogens and at the same time restrict the availability of their nutrients to the pathogens (Schaible and Kaufmann 2005).

In general, bacteria that are involved in plant pathogenesis are mainly from Gram-negative bacteria which include *Agrobacterium, Erwinia, Pseudomonas, Xanthomonas, Ralstonia* and *Xylella species*, while the common Gram-positive bacteria includes *Clavibacter* and *Streptomyces* (Alfano and Collmer 2001; Vranova et al. 2013). A report by Mansfield et al. (2012) has ranked *Pseudomonas syringae* as the most economically important plant diseases causing bacteria, followed by

Ralstonia solanacearum and *Agrobacterium tumefaciens*. These bacteria reside in the intercellular spaces between cells and are usually facultative parasites.

Similarly, fungal pathologists have ranked *Magnaporthe oryzae* as the most scientifically and economically important fungal pathogens, followed by *Botrytis cinerea, Puccinia* spp. and *Fusarium graminearum* (Dean et al. 2012). As for pest insects, orders classification that affect the most economically to the crops plants are identified as Lepidoptera, Diptera and Coleoptera such as *Tribolium confusum, Manduca sexta* and *Sesamia inferen* (Habib and Fazili 2007; van Wyk et al. 2016).

11.2.1 Protease in Plant Pathogenesis

Proteases, also identified as peptidase or proteinases, are proteolytic enzyme that can be found in all living organisms including in the natural predators of plant (Dunaevskii et al. 2008; Ingmer and Brøndsted 2009; Shen and Chou 2009). Proteases are divided into two groups: exo- and endo-proteases. However a more detailed classification based on their reaction mechanisms further divides proteases into six different types. They are aspartic, cysteine, glutamic, metalloprotease, serine and threonine (Rawlings et al. 2015). Being a proteolytic enzyme, the fundamental activity of protease is to break down proteins by cleaving peptide bonds. Therefore proteases play major biological processes in the synthesis, modification and destruction of proteins (Seife 1997; Shen and Chou 2009). While proteases are beneficial to the living organisms, they also can be harmful to the host if excessive amount of proteases are accumulated and left unregulated in the body.

There have been numerous studies reported that one common feature of pathogenesis reaction, including plant pathogenesis, is the utilization of digestive proteases (Dunaevskii et al. 2008; Ingmer and Brøndsted 2009; Li et al. 2012; Polya 2003). Among the digestive proteases involved in pathogenesis is serine proteases, cysteine proteases and aspartic proteases (Habib and Fazili 2007; Vranova et al. 2013). During plant pathogenesis, proteases released by pathogen are used to degrade plant proteins into small compounds (Li et al. 2012; Seife 1997; Shen and Chou 2009), which facilitate nutrients absorption. The pathogens utilize the nutrients for growth and proliferation inside the host cell. In addition, the secreted proteases are used to disrupt the activation of the cascade reactions by protein activator and inactivation of protease inhibitors inside the host plant (Travis et al. 1995; Lantz 1997). The attacks would lead to serious damages and even death to the host plant if it is left uncontrolled.

Plants are usually resistant to most of the pathogen attacks as some plants are considered non-host for the pathogens. However for some host plants, they contain resistance genes that recognize specific pathogens elicitors (Scheel 1998; Schulze-Lefert and Panstruga 2011; Sels et al. 2008). When the plant is being attacked, it will elicit several defensive responses which can be broadly classified into pre-formed defences and inducible defences. Pre-formed defences are defensive measurement present in plant even before the attack, such as physical and chemical barriers. Inducible defences are resistance activated after pathogen attack and can be further divided into two categories: qualitative and quantitative resistance (Chen 2008; Zhang et al. 2013). Qualitative resistance involved activation of defence response which only combat against a specific pathogen race while quantitative resistance is defence response contributed and controlled by multiple genes, which combat against a wide range of pathogens (Chen 2008; Zhang et al. 2013). It has been reported that quantitative resistance is a more durable defensive response as compared to qualitative resistance (Parlevliet 2002). Due to the important role played by digestive proteases in pathogenesis, plant produce a diversity of protease inhibitors as a type of quantitative resistance response and it is one of the major defence mechanism elicited by plant against pathogen attacks (Koiwa et al. 1997; Habib and Fazili 2007; Chen 2008; Sels et al. 2008; Zhang et al. 2013).

11.3 Protease Inhibitors

The existence of protease inhibitors (PIs) were firstly reported at the end of the nineteenth century by Fermi and Pernossi where they discovered the presence of 'anti-trypsin activity' in serum. During the 1930s and 1940s, M. Kunitz pioneered the work of isolation, identification and characterization of these PIs and he had introduced numerous fundamental concepts regarding proteases and their inhibitors interactions (Birk 2010). Since then, research on PIs has attracted the attention of researchers from a variety of disciplines mainly due to their clinical applications.

In general, PIs are termed as molecules that are responsible in inhibiting the proteolytic activity of proteases, either intracellular or extracellular proteases. In plant, PIs generally play an important role as regulator of intracellular proteases (Brady 2003; Koiwa et al. 1997; Schlüter et al. 2010) as the overexpression of proteases can be detrimental to the plant. Thus, digestive proteases such as chymotrypsin, pepsin and trypsin are often produced in the form of inactive zymogen and are activated only after secretion and proteolytic processing.

Most of the time, a wide range of PIs with different inhibition activity can be found inside a single source (organism). A single PI could have inhibitory activities specifically on one enzyme or on several different specific enzymes (Benchabane et al. 2008; Liener 1970; Lomate and Hivrale 2012). PI that is having strong inhibition effect towards a particular enzyme found in one source might not have the same inhibition effect on the same enzyme found in other sources (Huang et al. 2010; Mallory and Travis 1975). The varied characteristic of different PIs from different sources has made it difficult for grouping and classification. Previously, one of the common criteria used to group PIs was based on their average weight. For example, PIs between 3 and 25 kDa are grouped under the same types of protease that they inhibit (Birk 2010). However since then, there have been few systematically methods being introduced for PIs classification.

11.3.1 Classification of Protease Inhibitors

In the early 1980s, Laskowski and Kato classified PIs based on their activity towards certain protease and the PIs were named after their sources. For example, PIs identified from pancreas and that were capable to inhibit trypsin were grouped as 'pancreatic trypsin inhibitor'. However, this nomenclature system is sometimes confusing and unfit as some of the PIs contain multiple active inhibitory domains towards different types of protease. Therefore, Rawlings et al. (2015) suggested a new nomenclature system where PIs are classified into 48 families based on the similarity of amino acid sequences on the inhibitory domain or 'inhibitory unit'. Inhibitory unit is defined as a segment of amino acid sequences contained in the single reactive site after the removal of other unrelated amino acid sequences which do not contribute to the inhibitory activity of the PI (Rawlings et al. 2015; Rawlings and Barrett 1999). Therefore PIs with different inhibitory unit are grouped under a different family although they are functionally similar. According to this nomenclature system, PIs in 31 families are further grouped into 26 clans based on the similarity of their three-dimensional structures. This nomenclature system has been implemented in the MEROPS peptidase database (database containing information on proteinase or protease and their respective inhibitors) although the common families name is still widely used (Rawlings et al. 2015).

Throughout the study of plant PIs, there are PIs families that receive more attention and in-depth studies from researchers as compared to the other families. This is due to their applicable potentials in the agricultural and pharmacological industries (Habib and Fazili 2007; Haq et al. 2004; Shamsi et al. 2016). These PIs are from the families commonly known as serpin, Bowman Birk inhibitors, Kunitz family, Squash inhibitors, Cereal trypsin/α-amylase inhibitors, Mustard trypsin inhibitor (MSI), Potato type I and II, Cystatin superfamily, Aspartyl and metallocarboxypeptidase inhibitors, as listed in Table 11.1 (Habib and Fazili 2007; Shamsi et al. 2016).

11.3.2 Protease Inhibitors in Plant Defence Mechanism

The PIs in plant contribute to about 10–15% of the total protein content and are the major constituents found in seeds and storage organs (Jongsma and Bolter 1997; Lawrence and Koundal 2002; Shamsi et al. 2016). The amount of PIs found in plant is far more excessive than it is required to regulate intracellular proteolysis. Besides being responsible for the regulation of key plant biological processes such as seed maturation, programmed cell death and responses against stresses, the PIs are also suggested to take part in plant defence mechanism (Dutt et al. 2012; Habib and Fazili 2007; Jongsma and Bolter 1997; Li et al. 2008; Shamsi et al. 2016).

A few observations of PIs in plant have led to this suggestion. Firstly, during pathogen infections, proteases released by pathogens and the digestion products induce the expression of various types of PIs along with other resistance responses.

Table 11.1 Examples of protease inhibitors' source and their target proteases based on their common family name and MEROPS family, as suggested by Rawlings et al. (2004)

MEROPS family	Common name	Source	Target protease
I12	Bowman-Birk	*Glycine max* *Arachis hypogaea* *Helianthus annuus*	Trypsin, Chymotrypsin Trypsin, Chymotrypsin Trypsin, Cathepsin G Elastase, Chymotrypsin and thrombin
I3A	Kunitz Family	*Glycine max* *Hordeum vulgare* *Psophocarpus tetragonolobus* *Solanum tuberosum*	Trypsin, Chymotrypsin Subtilisin, α-amylase α-Chymotrypsin Cysteine proteases
I3B		*Sagittaria sagittifolia* *Canavalia lineate* *Solanum tuberosum* *Acacia confusa*	Trypsin, Chymotrypsin, Kallikerin, Subtilisin-type Microbial serine protease Cathepsin D, Trypsin Trypsin, α-chymotrypsin
I7	Squash inhibitors	*Momordica charantia* *Momordica charantia* *Momordica cochinchinensis* *Cucumis sativus*	Pancreatic elastase Trypsin Trypsin Trypsin
I6	α-Amylase Inhibitors	*Eleusine coracana* *Hordeum vulgare* *Triticum aestivum* *Zea mays*	α-Amylase α-Amylase, Trypsin α-Amylase, Trypsin Mammalian trypsin, activate hageman factor
I18	Mustard Trypsin Inhibitors (MSI)	*Sinapis alba* *Brassica hirta* *Brassica napus*	β-Trypsin Bovine β-trypsin, α-chymotrypsin Trypsin, Chymotrypsin
I13	Potato Type I	*Solanum tuberosum* *Momordica charantia* *Hordeum vulgare* *Triticum aestivum*	Chymotrypsin, Trypsin *Glu S. griseus* protease Subtilisin Subtilisin, Chymotrypsin *B. lichenoformis* subtilisin, α-chymotrypsin

Adapted from Habib and Fazili (2007)

In fact, the concentration of PIs in plant are reported to increase tremendously at the site of attack and later on spread systemically to other parts of the plant when there are presences of extracellular proteases, attacks or mechanical injuries detected on the organs of the plant (Bangrak and Chotigeat 2011; Koiwa et al. 1997; Ryan 1990).

Secondly, the accumulated plant PIs after pathogenesis are reported to be capable of controlling and inhibiting the activity of a wide range of extracellular digestive protease secreted by the pathogens (Bergey et al. 1996; Dunaevskii et al. 2008; Hildmann et al. 1992; Schlüter et al. 2010). The inhibition activities of PIs are antimicrobial and anti-inflammatory in nature as the effects from the inhibition of proteases showed reduction in microbial growth and activities (Kim et al. 2005, 2009). In the case of insects attack, plant PIs function inside the gut of insects as antimetabolic compound. In the beginning, wounds made on the plant during the attack induce the accumulation of PIs around the wound area (Chen 2008; Cruz et al. 2013; Koiwa et al. 1997). As feeding continues, PIs produced by the plant are taken into the gut of the insects where digestion of proteins and nutrient absorption takes place. From there, inhibition of digestive proteases by plant PIs occurs and thus reduces the efficiency of protein digestion of insect in the gut (Jongsma and Bolter 1997). Furthermore, a healthy mature plant has a very low amount of PIs (Bergey et al. 1996; Jongsma and Bolter 1997; Oliva et al. 2010). However, the concentration of PIs increases rapidly during wounding.

11.3.3 Overproduction of Protease Inhibitors in Plants

The production of plant PIs is dependent on the efficacy of effectors involved in the resistance, while the efficiency of plant defence mechanism is dependent on an effective production of PIs together with the types of PIs being produced. In other words, the concentration of PIs is paramount for an effective inhibition. Overproduction of PIs in plant is a tightly controlled process and their activation undergoes a series of biochemical reactions involving several intermediates and signalling compounds. Once a wound or insect attack has been made, the accumulation of PIs can be found not only at the injury area (locally) but subsequently at distal locations away from the site of injury of the plant (Bangrak and Chotigeat 2011; Koiwa et al. 1997). It is believed that overexpression of PIs is strategized to be induced in the whole plant to prevent further spreading of the digestive proteases and provide an overall defence system. The PIs overproduction is activated through signal transduction pathways where signal molecules from predation and insects attack are transduced as wound response in plant cells (Koiwa et al. 1997; Lomate and Hivrale 2012). The response would then initiate the PIs synthesis through two types of signalling pathways, comprising local and systemic signal pathways (Fig. 11.1).

In local signal pathway, wounds inflicted on plant elicit the release of pectin oligosaccharide fragments by endopolygalacturonases from the cell wall. This compound, together with chitosan oligomers of the fungal cell wall during pathogenesis, is presumed to be the elicitor of the pathway which leads to the expression of genes coding for PI (Bergey et al. 1996; Lawrence and Koundal 2002). As for the systemic signal pathways, different types of systemic signals have been proposed in transducing the wound signal but only a few of them have been of interest. They are systemin, abscisic acid (ABA), hydraulic potentials and electrical signals.

Systemin are polypeptides that are found to be an inducer to PIs gene expression and was first isolated from a wounded tomato (Bergey et al. 1996; Chen 2008; Moura and Ryan 2001). It has been proposed that vascular tissues such as phloem

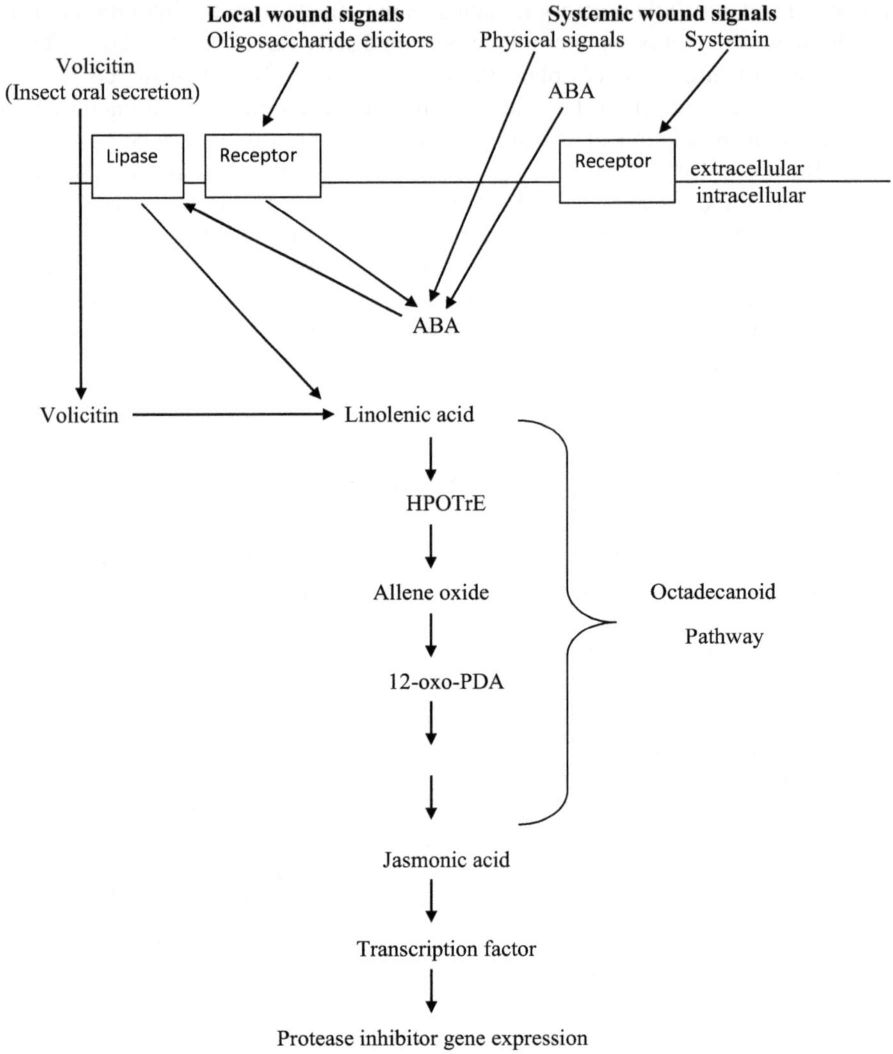

Fig. 11.1 PI gene expression pathway in plants. The pathway includes local and systemic signals as the elicitor of abscisic acid (ABA) inside the plant cell. ABA stimulates the breakdown membrane lipases into linolenic acid and activates the octadecanoid pathway. Volicitin, an oral secretion which are having a similar structure with linolenic acid also activates the octadecanoid pathway. The pathway produces jasmonic acid and leads to the mRNA expression of PIs (Adapted from Koiwa et al. 1997)

and xylem are responsible as the medium for the transmission of the signals. Research has shown that once a wound is made, the activities of the four systemic signals can be found on the area surrounding the wound, in phloem and xylem and later at other parts of the plant (Habib and Fazili 2007; Koiwa et al. 1997; Lawrence and Koundal 2002). It is also suggested that multiple alternative pathways are present in order to allow redundancy in the systemic signals transduction and providing a comprehensive defence mechanism.

Both of the signal pathways then activate another pathway known as octadecanoid pathway by using ABA as an intermediate compound (Kim et al. 2009; Koiwa et al. 1997; Lomate and Hivrale 2012). The activated octadecanoid pathway was reported to produce a large amount of jasmonic acids, starting with the breakdown of membrane lipases into linolenic acid (Fig. 11.1). Volicitin, a structurally similar compound to linolenic acid, which originates from the oral secretion of insects, may also breakdown to increase the concentration of jasmonate through the activation of octadecanoid pathway. It is believed that jasmonate would lead to mRNA expression and accumulation of PIs transcripts. Furthermore, research also showed that ethylene may play similar function as jasmonate in producing PIs as inhibited ethylene action also led to inhibition of PIs mRNA accumulation (Dutt et al. 2012; Kang et al. 2002; Koiwa et al. 1997). However to-date, the pathway downstream from jasmonate until to the PI mRNA expression is not well defined yet.

11.3.4 Modes of Protease Inhibition

The inhibition and regulation of proteases by PIs is achieved through the formation of a protein complex. The complex formed between the PIs and the target protease resulted in a less active or a fully inactive protease (Habib and Fazili 2007; Haq et al. 2004). There are a few different inhibition modes used by the PIs for the complex formation (Habib and Fazili 2007; Shamsi et al. 2016). The magnitude of increment in the PI concentrations produced by the plant for complete inhibition is dependent on the K_m values of the complex. In some cases, a low concentration of PIs is sufficient while for others, 100 folds of PIs concentration against the concentration of the protease are needed (Jongsma and Bolter 1997).

The most common mode of inhibition of the PI complex is through direct blockage to the active centre by mimicking the structures of the substrates of the target protease. It is a competitive inhibition reaction and is the canonical inhibition for trypsin, serine, cysteine and aspartyl proteinase (Habib and Fazili 2007; Haq et al. 2004). Other mode of inhibition used by PIs is through indirect blockage where the PIs cover the active site of the protease without having to mimic the structure of the substrates. The third inhibition mode is through the binding of PIs on the adjacent site or exocite binding site, which is located next to the active site, and this binding would cause the target protease to be inactivated. Lastly is the allosteric interaction between PIs and the protease. The binding of PIs on the allosteric site changes the conformation of the active site of protease and leads to inactivity (Habib and Fazili 2007).

11.4 Applications of Plant Protease Inhibitors in Industry

To date, numerous plant defence-related proteins with anti-pathogenic properties (including PIs) have been identified and applied into various industrial uses especially in the field of agriculture (Johnson and Pellecchia 2006; Shamsi et al. 2016). In fact, the applications of plant defence-related proteins (such as *Bt* toxins and PIs) and the genes coding for them in the respective applied field have proven to be effective in defending against pathogens. While the application of defence-related proteins is highly feasible, it is also expected to face challenges such as the development of resistance and insensitivity to the pathogens after a period of application (Benchabane et al. 2010; Habib and Fazili 2007; Haq et al. 2004; Rawlings and Barrett 1999).

11.4.1 Candidate for Genetic Engineering in Phytoprotection

Different classes of plant PIs have been characterized from a variety of plants and commonly used to produce genetically engineered crop plants. Generally, genes coding for the plant PIs that have inhibitory activity against pathogens are identified and transformed into the genome of important crop plants (Benchabane et al. 2008). During pathogen attack, the transgene PIs will be activated and provide protection to the plant. This biotechnology technique helps to improve the traits and the genetic make-up of the crops and enhance their survivability against the deadly attacks of their predators. This is important, as plant pathogens such as fungi and bacteria are deleterious and often are able to cause massive damage to the crop plant.

Transgenic plant that confers enhanced protection towards specific pathogens is not new and have been proven to be effective as shown by *Bacillus thuringiensis* (Bt) entomotoxic protein which was genetically engineered into plants for protection against a wide range of insect pests (Haq et al. 2004; Schlüter et al. 2010). Similarly, plant PIs genes with anti-pathogenic activities can be used as an alternative or can be expressed together with the currently available phytoprotection genes (Habib and Fazili 2007; Haq et al. 2004; Schlüter et al. 2010; Shamsi et al. 2016). Previous researches had proved that transgenic plants that expressed serine PIs or cysteine PIs exhibited an enhanced resistance to predation by insects as most of the insects contained digestive protease such as serine and cysteine (Johnson et al. 1989; Urwin et al. 1995). Other researchers have reported that tobacco that carried tomato's serine protease inhibitor II gene successfully showed insecticidal activity (Zhang et al. 2004). Other examples of genetically engineered crop plants with PIs are summarized in Table 11.2 (Falco and Silva-Filho 2003; Morton et al. 2000; Schlüter et al. 2010).

Genetically engineered plants with PIs increase yields of the crops indirectly as the plants gained better survival advantages and at the same time reduce the usage of chemical pesticides. With fewer chemicals usage or uses not only this would help in improving health of the farmers and consumers, but at the same time it also leads

Table 11.2 Examples of successfully genetically engineered plants expression plant PIs

Genetically engineered plant	Plant protease inhibitor	Target (insects, nematodes, pathogens)
		Insects
Alfalfa	Oryzacystatin II	*Phytodecta formicata*
Apple	*Nicotiana alata* PI	*Epiphyas postvittana*
Arabidopsis	Mustard Trypsin Inhibitor 2	*Plutella xylostella*
Oilseed rape	Mustard Trypsin Inhibitor 2	*Plutella xylostella*
Peas	Bean α-amylase Inhibitor 1	*Bruchus pisorum*
Potato	Multidomain cystatin fusions	*Frankliniella occidentalis*
	Barley cystatin, HvCPI	*Leptinotarsa decemlineata*
Rice	Barley trypsin inhibitor	*Sitophilus oryzae*
	Soybean trypsin inhibitor	*Nilaparvata lugens*
		Cnaphalocrocis medinalis
	Maize protease inhibitor	*Chilo suppresssalis*
	Cowpea trypsin inhibitor	*Cnaphalocrocis medinalis*
Sugarcane	Soybean Kunitz trypsin inhibitor	*Diatraea saccharalis*
	Soybean Bowman-Birk inhibitor	*Diatraea saccharalis*
Tobacco	*Brassica juncea* Trypsin inhibitor	*Spodoptera litura*
	Mustard Trypsin Inhibitor 2	*Spodoptera littoralis*
	Sporamin and Taro cystatin	*Helicoverpa armigera*
	Buckwheat serine protease inhibitor	*Trialeurodes vaporariorum*
		Root parasitic nematodes
Alfalfa	Oryzacystatin I	*Pratylenchus penetrans*
Potato	Oryzacystatin I	*Meloidogyne incognita*
		Globodera pallida
Tomato	Taro cystatin	*Meloidogyne incognita*
Wheat	Potato proteinase inhibitor 2	*Heterodera avenae*
		Pathogens
Potato	Buckwheat serine proteinase inhibitor	*Pseudomonas syringae*
		Clavibacter michiganensis
Rice	Potato carboxypeptidase inhibitor	*Magnaporthe orzyae*
		Fusarium verticilliodes
Tobacco	Buckwheat serine proteinase inhibitor	*Pseudomonas syringae*
		Clavibacter michiganensis
	Sporamin and Taro cystatin	*Pythium aphanidermatum*

Adapted from Schlüter et al. (2010)

to cost reduction in the plantation of the important crops (Shamsi et al. 2016; van Wyk et al. 2016). So far there is no report on the negative side of PIs genes that were transformed into crops and ingested by mammals. In fact it has even been suggested that the introduction of PIs help to improve the nutritional values of the crops (Haq et al. 2004; Christou et al. 2006).

11.4.2 Food Preservation Industry

PIs have been applied in the field of food preservative especially for seafood industry. The usage of PIs is an effective way to preserve many types of seafood like salted fish as it inhibits the deteriorative processes of protease during food processing and preservation (Bijina et al. 2011; Reppond and Babbitt 1993). It has been suggested that PIs can become an alternative and natural method in preserving food. The application of PIs in preserving food is of advantage and beneficial because approximately, as much as 25% of food spoilage worldwide is reported to be due to microbial activity (Bijina et al. 2011; Lund et al. 2000) and the usage of PIs could replace the use of chemical preservations in food. Bijina et al. (2011) reported that PIs isolated from matured leaves and seeds of *Moringa oleifera* possessed potential to be used as seafood preservative as it could prevent proteolysis from occurring in a commercially valuable shrimp, *Penaus monodon,* during its storage. The activity of PIs from *Moringa oleifera* was shown to be effective and had high inhibitory activity towards serine protease thrombin, elastase, chymotrypsin and cysteine protease cathepsin B and papain. Other than seafood, PIs have also been used in preserving foods for export.

11.4.3 Resistance Development Against Engineered Protection in Industry

In a plant–pathogen relationship, pest insects and pathogens are constantly changing its pathogenesis strategy and adapt to the hosts' defensive response in order to achieve a successful invasion (Habib and Fazili 2007; Haq et al. 2004). For instance in agriculture, pest insects are reported to acquire immunity against genetically engineered crops which expressed PI genes and they adapted by secreting new digestive proteases which are insensitive to the expressed PIs or by increasing the secretion of other types of digestive proteases than the ones which are being inhibited (Habib and Fazili 2007).

Haq et al. (2004) stated in their review that one of the necessary preparations to overcome the adaptability and resistance towards crop protection (expression of PIs) portrayed by pest insects and pathogens are by developing a second generation of PIs (novel PIs) from novel sources. Owing to the novelty, PIs with inhibitory activity from novel sources (such as non-host plant) have not been exposed to the target pest and pathogens before, and hence there are none or less resistance being established by the target pest and pathogens. The incorporation of novel PIs from novel sources in addition to the currently available insecticidal protein did not just overcomes the resistance development by the pest insects, but also enhances the overall protection efficacy (Shamsi et al. 2016; Schlüter et al. 2010; Habib and Fazili 2007; Haq et al. 2004). The discovery of novel PIs is crucial to ensure a continuous effective protection against the pathogens.

11.5 Turmeric as Novel Source of PIs

Turmeric (*Curcuma longa*) is a type of perennial plant which belongs to the family of Zingiberaceae and in the genus of *Curcuma.* The plants have an average height of 1 m tall with a short, underground stems and rhizomes. The shapes of the rhizomes are usually ovate, oblong, pyriform and often short-branched (Eigner and Scholz 1999) (Fig. 11.2). The distribution of turmeric is mainly in India, China, Malaysia and other Asian countries (Amalraj et al. 2017; Govindarajan 1980). Commonly found used in food preparations, rhizomes of turmeric are made into powder form and it is added to food as aromatic spices, flavouring and preservatives (Govindarajan 1980; Krup et al. 2013).

As a member of the Zingiberaceae family, turmeric is a type of herb and medicinal plant. It is widely used as folk medicines, especially in India's Ayurvedic medicine, to treat various infectious diseases. Medicinal properties such as antibacterial, antifungal and especially antiviral properties have been identified from turmeric crude extracts (Mukhtar et al. 2008).

Traditionally, turmeric is applied internally as stomachic, tonic and blood purifier and externally for the treatment of skin diseases (The Wealth of India 2001). Turmeric is also reported to be effective against biliary disorders, anorexia, coryza, cough, diabetic wounds, hepatic disorders, rheumatism and sinusitis (Ammon et al. 1992; Krup et al. 2013). As time progresses, more research have been carried out on turmeric to expose its medicinal potential. The major medicinal values reported to date includes antioxidant activity, anti-protozoal activity, antimicrobial activity, antivenom activity, anti-HIV, antitumour activity and anti-inflammatory activity (Amalraj et al. 2017; Araújo and Leon 2001; Bhat et al. 2016; Jayaprakasha et al. 2005). Turmeric also helps in lowering cholesterol, suppressing diabetes, enhancing wound healing, modulating multiple sclerosis and Alzheimer's disease and blocking HIV replication (Amalraj et al. 2017; Okada et al. 2001).

With all these reported medicinal properties, turmeric is suitable to be developed to replace or to be integrated into current conventional medicines. In addition, besides posing as a cheaper alternative of medication, there are strong supports from developed countries to integrate plant-based products into the current healthcare systems (Mukhtar et al. 2008).

11.5.1 Protease Inhibition by Zingiberaceae Family

In year 2005, a study conducted by Sookkongwaree et al. showed that methanol and aqueous extracts of eight members of Zingiberaceae family, including *Curcuma longa*, are able to inhibit viral proteases from human immunodeficiency virus type 1 (HIV-1), hepatitis C virus (HCV) and human cytomegalovirus (HCMV). This study also showed that extracts of turmeric possessed potent inhibitory effects on the replication of viruses (Sookkongwaree et al. 2006). However identification of

Fig. 11.2 Rhizome of turmeric, *Curcuma longa*. The shape of the turmeric's rhizome is oblong with root-like branches

the active compound responsible for the protease inhibition in *Curcuma longa* was not conducted in the study.

One of the major active ingredients that are responsible for the pharmaceutical and medical properties of turmeric is curcuminoids. Curcuminoid is a compound that gives the yellow pigment to turmeric (Bhat et al. 2016; Chiou et al. 1983; Moken et al. 1984). Curcuminoids consisted of diferuloylmethane (curcumin I), demethoxycurcumin (curcumin II), bisdemethoxycurcumin (curcumin III) and recently-discovered member, cyclocurcumin (Goel et al. 2008). Curcumin I is a phenylpropanoid compound and it might be a non-protein type of plant protease inhibitors, grouped under metallopeptidase inhibitors. The specific proteases inhibited by this curcumin are those grouped under aminopeptidase (Polya 2003).

Although curcuminoids have been widely reported to provide various medicinal properties in turmeric, a few have suggested that curcumin is responsible only for anti-inflammatory properties and not the antimicrobial properties (Apisariyakul et al. 1995; Lantz et al. 2005). Apisariyakul et al. (1995) proved that curcumin actually had no antifungal activity on fungi and molds whereas compound that showed antifungal activity was turmeric oil rather than the curcumin itself. Besides, crude and unfractionated chemical extracts in turmeric has shown a better biological activity as compared to fractionated and sub-fractionated curcuminoids (Lantz et al. 2005). However, the exact active compound which gives the anti-pathogenic properties in turmeric is still unknown.

11.6 Conclusion and Future Prospects

Currently we have identified and characterized two novel PI genes from turmeric belonging to the phytocystatin superfamily (*CypCl*) and Kunitz trypsin inhibitor family (*ClKTI*), respectively. Both the genes sequence has been deposited in the GenBank of National Center for Biotechnology Information (NCBI) with the accession number KF545954.1 for *CypCl* and KF889322.1 for *ClKTI*. Based on our results, it is suggested that *ClKTI* involved in the plant defence mechanism as its expression was induced by the treatment of methyl jasmonate which are known to mimic and elicit plant defence genes expression (Chan et al. 2017). PIs found in turmeric might have played pivotal role in providing anti-protease and anti-pathogenic properties as it is one of the secondary metabolites released to fight against pathogens.

References

Alfano JR, Collmer A (2001). Mechanisms of bacterial pathogenesis in plants: familiar foes in a foreign kingdom. In: Principles of bacterial pathogenesis, pp 179–226

Amalraj A, Pius A, Gopi S, Gopi S (2017) Biological activities of curcuminoids, other biomolecules from turmeric and their derivatives—a review. J Tradit Complement Med 7(2):205–233

Ammon HP, Anazodo MI, Safayhi H, Dhawan BN, Srimal RC (1992) Curcumin: a potent inhibitor of leukotriene B4 formation in rat peritoneal polymorphonuclear neutrophils (PMNL). Planta Med 58:226

Apisariyakul A, Vanittanakom N, Buddhasukh D (1995) Antifungal activity of turmeric oil extracted from *Curcuma longa* (Zingiberaceae). J Ethnopharmacol 49:163–169

Araújo CC, Leon LL (2001) Biological activities of *Curcuma longa* L. Mem Inst Oswaldo Cruz 96:723–728

Bangrak P, Chotigeat W (2011) Molecular cloning and biochemical characterization of a novel cystatin from Hevea rubber latex. Plant Physiol Biochem 49:244–250

Benchabane M, Goulet MC, Dallaire C, Côté PL, Michaud D (2008) Hybrid protease inhibitors for pest and pathogen control—a functional cost for the fusion partners? Plant Physiol Biochem 46:701–708

Benchabane M, Schluter U, Vorster J, Goulet MC, Michaud D (2010) Plant cystatins. Biochimie 92:1657–1666

Bergey DR, Howe GA, Ryan CA (1996) Polypeptide signaling for plant defensive genes exhibits analogies to defense signaling in animals. Proc Natl Acad Sci U S A 93:12053–12058

Bhat SV, Amin T, Nazir S (2016) Biological activities of turmeric (*Curcuma longa* Linn.)—an overview. BMR Microbiol 2(1):1–5

Bijina B, Chellappan S, Krishna JG, Basheer SM, Elyas KK, Bahkali AH, Chandrasekaran M (2011) Protease inhibitor from Moringa oleifera with potential for use as therapeutic drug and as seafood preservative. Saudi J Biol Sci 18:273–281

Birk Y (2010) Plant protease inhibitors: significance in nutrition, plant protection, cancer prevention and genetic engineering. Springer, Berlin

Brady RL (2003) Plant protease inhibitors: significance in nutrition, plant protection, cancer prevention and genetic engineering. Phytochemistry, vol 64. Springer, Berlin

Chan SN, Abu Bakar N, Mahmood M, Ho CL, Dzaki NM, Shaharuddin NA (2017) Identification and expression profiling of a novel Kunitz trypsin inhibitor (KTI) gene from turmeric, *Curcuma longa* by real-time quantitative-PCR (RT-qPCR). Acta Physiol Plant 39:12. doi:10.1007/s11738-016-2311-7

Chen M (2008) Inducible direct plant defense against insect herbivores: a review. Insect Sci 15:101–114

Chiou J, Wei H, Chunk-Kuo (1983) Preliminary study on the anti-oxidative components of some species grown in Taiwan. Nung Yeh Hua Hsueh Hui Chih 1–2:97–103

Christou P, Capell T, Kohli A, Gatehouse JA, Gatehouse AMR (2006) Recent developments and future prospects in insect pest control in transgenic crops. Trends Plant Sci 11:302–308

Cruz ACB, Massena FS, Migliolo L, Macedo LLP, Monteiro NKV, Oliveira AS, Santos EA (2013) Bioinsecticidal activity of a novel Kunitz trypsin inhibitor from Catanduva (*Piptadenia moniliformis*) seeds. Plant Physiol Biochem 70:61–68

Dean R, Van Kan JAL, Pretorius ZA, Hammond-Kosack KE, Di Pietro A, Spanu PD, Foster GD (2012) The top 10 fungal pathogens in molecular plant pathology. Molecular plant pathology. Blackwell, Hoboken

Dunaevskii YE, Matveeva AR, Fatkhullina GN, Belyakova GA, Kolomiets TM, Kovalenko ED, Belozersky MA (2008) Extracellular proteases of mycelial fungi as participants of pathogenic process. Russ J Bioorg Chem 34:286–289

Dunaevsky YE, Elpidina EN, Vinokurov KS, Belozersky MA (2005) Protease inhibitors in improvement of plant resistance to pathogens and insects. Mol Biol 39:608–613

Dutt S, Gaur VS, Taj G, Kumar A (2012) Differential induction of two different cystatin genes during pathogenesis of Karnal bunt (Tilletia indica) in wheat under the influence of jasmonic acid. Gene 506:253–260

Eigner D, Scholz D (1999) Ferula asa-foetida and Curcuma longa in traditional medical treatment and diet in Nepal. J Ethnopharmacol 67:1–6

Falco MC, Silva-Filho MC (2003) Expression of soybean proteinase inhibitors in transgenic sugarcane plants: effects on natural defense against *Diatraea saccharalis*. Plant Physiol Biochem 41:761–766

Goel A, Kunnumakkara AB, Aggarwal BB (2008) Curcumin as "Curecumin": from kitchen to clinic. Biochem Pharmacol 75:787–809

Govindarajan VS (1980) Turmeric—chemistry, technology, and quality. Crit Rev Food Sci Nutr 12:199–301

Habib H, Fazili KM (2007) Plant protease inhibitors: a defense strategy in plants. Biotechnol Mol Biol Rev 2:68–85

Haq SK, Atif SM, Khan RH´(2004) Protein proteinase inhibitor genes in combat against insects, pests, and pathogens: natural and engineered phytoprotection. Arch Biochem Biophys 431:145–159

Hildmann T, Ebneth M, Peña-Cortés H, Sánchez-Serrano JJ, Willmitzer L, Prat S (1992) General roles of abscisic and jasmonic acids in gene activation as a result of mechanical wounding. Plant Cell 4:1157–1170

Huang H, Qi SD, Qi F, Wu CA, Yang GD, Zheng CC (2010) NtKTI1, a Kunitz trypsin inhibitor with antifungal activity from *Nicotiana tabacum*, plays an important role in tobacco's defense response. FEBS J 277:4076–4088

Ingmer H, Brøndsted L (2009) Proteases in bacterial pathogenesis. Res Microbiol 160:704–710

Jayaprakasha GK, Jagan Mohan Rao L, Sakariah KK (2005) Chemistry and biological activities of C. longa. Trends Food Sci Technol 16:533–548

Johnson SL, Pellecchia M (2006) Structure- and fragment-based approaches to protease inhibition. Curr Top Med Chem 6:317–329

Johnson R, Narvaez J, An G, Ryan C (1989) Expression of proteinase inhibitors I and II in transgenic tobacco plants: effects on natural defense against Manduca sexta larvae. Proc Natl Acad Sci U S A 86:9871–9875

Jongsma MA, Bolter C (1997) The adaptation of insects to plant protease inhibitors. J Insect Physiol 43:885–895

Kang SG, Choi JH, Suh SG (2002) A leaf-specific 27 kDa protein of potato Kunitz-type proteinase inhibitor is induced in response to abscisic acid, ethylene, methyl jasmonate, and water deficit. Mol Cells 13:144–147

Kim JY, Park SC, Kim MH, Lim HT, Park Y, Hahm KS (2005) Antimicrobial activity studies on a trypsin-chymotrypsin protease inhibitor obtained from potato. Biochem Biophys Res Commun 330:921–927

Kim JY, Park SC, Hwang I, Cheong H, Nah JW, Hahm KS, Park Y (2009) Protease inhibitors from plants with antimicrobial activity. Int J Mol Sci 10:2860–2872

Koiwa H, Bressan RA, Hasegawa PM (1997) Regulation of protease inhibitors. Trends Plant Sci 2:379–384

Krup V, Prakash LH, Harini A (2013) Pharmacological activities of turmeric (*Curcuma longa* linn): a review. J Homeopath Ayu Med 02:133. doi:10.4172/2167-1206.1000133

Lantz MS (1997) Are bacterial proteases important virulence factors? J Periodontal Res 32(1):126–132

Lantz R, Chen G, Solyom A, Jolad S, Timmermann B (2005) The effect of turmeric extracts on inflammatory mediator production. Phytomed Int J Phytother Phytopharmacol 12:445–452

Lawrence PK, Koundal KR (2002) Plant protease inhibitors in control of phytophagous insects. Electron J Biotechnol 5:93–109

Li J, Brader G, Palva ET (2008) Kunitz trypsin inhibitor: an antagonist of cell death triggered by phytopathogens and fumonisin B1 in Arabidopsis. Mol Plant 1:482–495

Li H, Child MA, Bogyo M (2012) Proteases as regulators of pathogenesis: examples from the Apicomplexa. Biochim Biophys Acta 1824:177–185

Liener IE (1970) Toxic constituents of plant foodstuffs. The proceedings of the nutrition society, vol 29. Academic, New York

Lomate PR, Hivrale VK (2012) Wound and methyl jasmonate induced pigeon pea defensive proteinase inhibitor has potency to inhibit insect digestive proteinases. Plant Physiol Biochem 57:193–199

Lund BM, Baird-Parker TC, Gould GW, Grahame W (2000) The microbiological safety and quality of food. Aspen, New York

Macedo MLR, Durigan RA, da Silva DS, Marangoni S, Freire MDGM, Parra JRP (2010) Adenanthera pavonina trypsin inhibitor retard growth of *Anagasta kuehniella* (Lepidoptera: Pyralidae). Arch Insect Biochem Physiol 73:213–231

Mallory PA, Travis J (1975) Inhibition spectra of the human pancreatic endopeptidases. Am J Clin Nutr 28:823–830

Mansfield J, Genin S, Magori S, Citovsky V, Sriariyanum M, Ronald P, Foster GD (2012) Top 10 plant pathogenic bacteria in molecular plant pathology. Mol Plant Pathol 13:614–629

Moken Y, Xianping D, Yaoshu T (1984) Studies on the chemical constituents of common turmeric (*Curcuma longa*). Zhongcoayoa 5:197–198

Morton RL, Schroeder HE, Bateman KS, Chrispeels MJ, Armstrong E, Higgins TJ (2000) Bean alpha-amylase inhibitor 1 in transgenic peas (*Pisum sativum*) provides complete protection from pea weevil (*Bruchus pisorum*) under field conditions. Proc Natl Acad Sci U S A 97:3820–3825

Moura DS, Ryan CA (2001) Wound-inducible proteinase inhibitors in pepper. Differential regulation upon wounding, systemin, and methyl jasmonate. Plant Physiol 126:289–298

Mukhtar M, Arshad M, Ahmad M, Pomerantz RJ, Wigdahl B, Parveen Z (2008) Antiviral potentials of medicinal plants. Virus Res 131:111–120

Newton AC, Fitt BDL, Atkins SD, Walters DR, Daniell TJ (2010) Pathogenesis, parasitism and mutualism in the trophic space of microbe-plant interactions. Trends Microbiol 18:365–373

Odjakova M, Hadjiivanova C (2001) The complexity of pathogen defense in plants. Bulg J Plant Physiol 27:101–109

Okada K, Wangpoengtrakul C, Tanaka T, Toyokuni S, Uchida K, Osawa T (2001) Curcumin and especially tetrahydrocurcumin ameliorate oxidative stress-induced renal injury in mice. J Nutr 131:2090–2095

Oliva MLV, Silva MCC, Sallai RC, Brito MV, Sampaio MU (2010) A novel subclassification for Kunitz proteinase inhibitors from leguminous seeds. Biochimie 92:1667–1673
Parlevliet JE (2002) Durability of resistance against fungal, bacterial and viral pathogens; present situation. Euphytica 124:147–156
Polya MG (2003) Protein and non-protein protease inhibitors from plants. Stud Nat Prod Chem 29(Part J):567–641
Rawlings ND, Barrett AJ (1999) MEROPS: the peptidase database. Nucleic Acids Res 27:325–331
Rawlings ND, Tolle DP, Barrett AJ (2004) MEROPS: the peptidase database. Nucleic Acids Res 32(90001):160D–164D
Rawlings ND, Barrett AJ, Finn R (2015) Twenty years of the *MEROPS* database of proteolytic enzymes, their substrates and inhibitors. Nucleic Acids Res 44(D1):D343–D350
Reppond KC, Babbitt JK (1993) Protease inhibitors affect physical properties of arrowtooth flounder and walleye pollock surimi. J Food Sci 58:96–98
Ryan CA (1990) Protease inhibitors in plants: genes for improving defenses against insects and pathogens. Annu Rev Phytopathol 28:425–449
Schaible UE, Kaufmann SHE (2005) A nutritive view on the host-pathogen interplay. Trends Microbiol 13:373–380
Scheel D (1998) Resistance response physiology and signal transduction. Curr Opin Plant Biol 1:305–310
Schlüter U, Benchabane M, Munger A, Kiggundu A, Vorster J, Goulet MC, Michaud D (2010) Recombinant protease inhibitors for herbivore pest control: a multitrophic perspective. J Exp Bot 61:4169–4183
Schulze-Lefert P, Panstruga R (2011) A molecular evolutionary concept connecting nonhost resistance, pathogen host range, and pathogen speciation. Trends Plant Sci 16:117–125
Seife C (1997) Blunting nature's swiss army knife. Science 277:1602–1603
Sels J, Mathys J, De Coninck BMA, Cammue BPA, De Bolle MFC (2008) Plant pathogenesis-related (PR) proteins: a focus on PR peptides. Plant Physiol Biochem 46:941–950
Shamsi TN, Parveen R, Fatima S (2016) Characterization, biomedical and agricultural applications of protease inhibitors: a review. Int J Biol Macromol 91:1120–1133
Shen HB, Chou KC (2009) Identification of proteases and their types. Anal Biochem 385:153–160
Sookkongwaree K, Geitmann M, Roengsumran S, Petsom A, Danielson UH (2006) Inhibition of viral proteases by Zingiberaceae extracts and flavones isolated from Kaempferia parviflora. Pharmazie 61:717–721
The Wealth of India (2001) An encyclopedia of India's Raw Material Resources. National Institute of Science Communication, CSIR, New Delhi
Travis J, Potempa J, Maeda H (1995) Are bacterial proteinases pathogenic factors? Trends Microbiol 3:405–407
Urwin PE, Atkinson HJ, Waller DA, McPherson MJ (1995) Engineered oryzacystatin-I expressed in transgenic hairy roots confers resistance to Globodera pallida. Plant J Cell Mol Biol 8:121–131
Vranova V, Rejsek K, Formanek P (2013) Proteolytic activity in soil: a review. Appl Soil Ecol 70:23–32
van Wyk SG, Kunert KJ, Cullis CA, Pillay P, Makgopa ME, Schlüter U, Vorster BJ (2016) Review: the future of cystatin engineering. Plant Sci 246:119–127
Zhang H, Xie X, Xu Y, Wu N (2004) Isolation and functional assessment of a tomato proteinase inhibitor II gene. Plant Physiol Biochem 42:437–444
Zhang Y, Lubberstedt T, Xu M (2013) The genetic and molecular basis of plant resistance to pathogens. J Genet Genomics 40:23–35

Chapter 12
Sustainable Supply Chains: Utilisation of Whole Crops for New Product Development

Rebecca Smith

12.1 Introduction

Food security is well defined by the FAO food security brief (FAO 2006) as 'when all people, at all times, have physical and economic access to sufficient, safe and nutritious food that meets their dietary needs and food preferences for an active and healthy life'. In order for this to occur, food supply chains need to be well balanced and tailored to the requirements of individual regions and countries. In 2014, there were one billion people undernourished, one billion who were hungry and two billion who were overweight or obese (United Nations 2014b), showing how unbalanced the current food supply chain is and how far away from food security we currently are. As the population is set to rise from seven billion in 2016 to nine billion in 2050 (United Nations 2014b), this problem is only going to get worse if we make no changes to the way our food is produced, processed, supplied and consumed. Reducing avoidable waste, closing the loop on supply chains, and making use of all the food and by-products that are produced is one of the ways we can contribute towards making the food supply chain more sustainable.

R. Smith (✉)
Department of Food and Nutritional Sciences, University of Reading, Whiteknights, Reading RG6 6AP, UK

Valley Produce, Springalls Farm, Swallowfield, Reading RG7 1RN, UK
e-mail: rebecca@valleyproduce.co.uk

S.N.A. Abdullah et al. (eds.), *Crop Improvement*,
DOI 10.1007/978-3-319-65079-1_12

12.2 The Scale of Food Waste

In 2015, the United Nations Sustainable Development Goals came into action, setting targets for countries to achieved by 2030 to 'end poverty, protect the planet and ensure prosperity for all' (United Nations 2014a). Goal 12, to ensure sustainable consumption and production patterns, includes a target to halve per capita food waste by 2030 along the entire supply chain. There is, however, no universal definition for food wastage. The Food Agricultural Organization (FAO 2011) splits the definition of food wastage into two categories: food loss and food waste. Food loss is any decrease in food mass in the part of the supply chain that specifically leads to edible food for human consumption. Food losses, therefore, take place in the production, post-harvest and processing stages of the supply chain. Food waste takes place later in the supply chain at retail or consumption where a behavioural decision leads to food being discarded. The FAO definition of food wastage generally disregards unavoidable or inedible waste in this definition. The EU includes both edible and inedible parts of food in its definition of food wastage, but like the FAO, only includes that which the current user intended or reasonably expected to be part of the food supply chain (FUSIONS 2014). Although the EU definition does include unavoidable food waste such as crop left in the field due to lower than expected demand, or of a quality lower than that required upstream in the supply chain, it does not include crop residues such as straw or the outer husks of corn or nuts. Whichever definition of food waste is used, it is generally agreed that loss from the supply chain should be avoided, or where unavoidable be recovered or utilised back into the food supply chain or into other industries.

12.2.1 Volumes of Waste

With no agreed definition on food wastage and with countries differing in the way they collect and record this information, if at all, it is difficult to get a figure for global food waste. There are also hidden areas of the supply chain, within field crop residues and unharvested crops generally not recorded. The most widely quoted figures, and those adopted by large organisation such as the FAO and UN, are that the total world food wastage is estimated to be 1.6 billion tonnes, of which 1.3 billion tonnes is from edible parts (FAO 2013). This volume of edible food waste is 21% of the total world domestic agricultural production, inclusive of non-food commodities, and is generated along the whole supply chain (FAO 2013). The levels of food wastage differ between high-income industrialised and low-income regions. Total per capita waste volumes for Europe and North America are 95–115 kg per year, whilst in sub-Saharan Africa and South and Southeast Asia this is significantly lower at 6–11 kg per year (FAO 2011). The differences in the volumes of food waste are related to the general availability of food, with higher waste in countries where food is abundant. The relationship consumers and stakeholders in the supply chain have with food and the types of food in the supply chain will affect the volume and format of the waste produced, with higher-income regions producing more processing waste during the manufacturing of luxury food items and fresh, ready-to-eat foods.

As previously mentioned, the FAO do not include crop residues in their definition of food wastage. Although crop residues are not intended to enter the human food supply chain, they are a by-product of the industry which is not entirely utilised and a potential source of materials and compounds for both inside and out of the food supply chain. According to calculations by Vaclav Smil (1999) for the mid-1990s, the global annual output of crop residues was 1.4 times the annual aggregate harvest at 3.75 billion tonnes. This calculation does not include crop processing residues which would substantially increase the total. A proportion of crop residues are harvested and sold as animal feed, bedding and building materials, and in recent years more and more crop residues are being used to generate energy in anaerobic digesters. In 2014, around 150 anaerobic digester plants were sited on UK farms, growing from just under 25 in 2005 (ABDA 2015). Although sustainable energy production is beneficial, the potential ingredients and products that could be developed from these residues should also be investigated.

12.2.2 Origin and Causes of Wastage in the Supply Chain

The causes of wastage in the supply chain are diverse, Table 12.1 lists each stage of the supply chain and some of the main waste causes. Some causes are specific to that stage of the supply chain, for example, processing produces large quantities of its waste through extracting the desired product and discarding biomass or waste water, whilst a portion of agricultural waste is crop residues in the field. Other drivers of waste span larger portions of the supply chain; quality and specifications are usually set by the end user or retailer, with the processors and in turn primary producers having to work to those specifications and deal with the resulting waste themselves. Finding answers to the problem of food wastage is therefore not going to be a one-size-fits-all answer, the cause of the waste needs to be understood and each stakeholder involved in the chain engaged in the solution.

The food supply chain can be split into ‘upstream’; agricultural production, post-harvest handling and storage, and ‘downstream’; processing, distribution and consumption. Currently, 54% of global food waste is produced upstream in the supply chain, with 46% produced downstream (FAO 2013). However, the type, reason for and weighting of waste volumes vary depending upon the region of the world producing the waste. As Fig. 12.1 shows, industrialised regions produce more of their waste downstream in the supply chain and lower-income regions produce more waste upstream in the supply chain. The reason for this difference can be explained by both behavioural relationships of consumers with food, but also the financial and structural investment in all areas of the supply chain. In low-income regions populations tend to be more rural, with the consumer closer to the site of production

Table 12.1 Causes of plant-based food waste at each stage of the supply chain

Stage of supply chain	Cause of waste
Agricultural production	Not meeting market specifications
	– Adverse weather damaging crop
	– Faster than expected growth
	– Pest/disease damage
	– Weeds in harvest
	– Lower nutritional/functional quality
	Residues (stubble, husks, stems)
	Fluctuation in demand—no outlet for sale
Post-harvest handling and storage	Pest/disease damage—unsuitable for sale Dehydration Storage facility breakdown Quality/functional deterioration Grading out Trimming Mishandling
Processing	Grading out Rejections (not to specification) Fluctuations in demand Peeling Cutting/slicing Discarded biomass Wash or processing water
Distribution	Infrastructure limitations Storage facility breakdown Break in the chill chain Fluctuations in demand Past best-before/use-by date Change in pack format/marketing
Consumption	Over purchasing Low value given to food Quality deterioration Food preparation waste Leftovers Past best-before/use-by date

resulting in shorter supply chains. Food losses are largely due to structural, technical and financial limitations in agricultural techniques, transport infrastructure and storage in often difficult climate conditions. Downstream in the supply chain wastage is low due to the high value placed on food by consumers. In industrialised regions, large portions of the population live in urbanised towns and cities away from agricultural production. To support this type of society the infrastructure and distribution chain are robust and tailored to the requirements of perishable food items, and agricultural techniques are advanced which limits losses downstream in the supply chain. Upstream in the supply chain, the consumer and, to some extent, the retailers are removed from food production. The ease of access to relatively cheap, fresh food throughout the year often results in a low value being assigned to

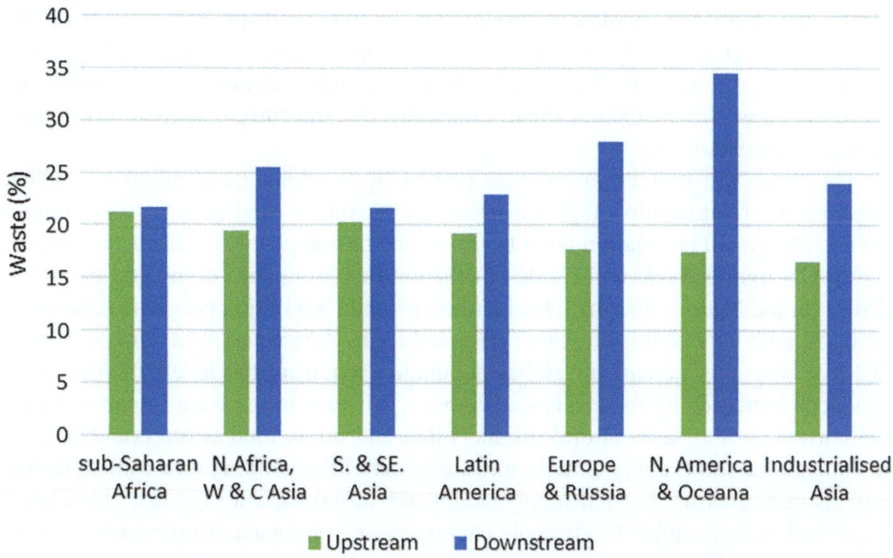

Fig. 12.1 Estimated waste for plant-based commodity groups upstream and downstream in the supply chain (in the percentage of the total entering each stream) (Data taken from FAO 2011)

food and consequently high volumes of waste. As previously mentioned the volume of waste across the whole supply chain is high in industrialised regions; populations are able to afford more luxury and perishable food items with high quality expectations. This leads to high waste levels across the supply chain, with more processing waste and losses at production due to crops not meeting specifications for often aesthetic reasons.

12.3 Pressures on Supply Chains

For the majority of the industrialised world, it is difficult to see food wastage as a problem. Unless your local supermarket has a broken-down refrigerator, there is rarely a time when you cannot get the food you want, when you want. For less affluent regions of the world, the relationship to food is very different, eating what the land can produce and when it can produce it. With one billion people undernourished, one billion who are hungry, two billion who are overweight or obese (United Nations 2014b) and 21% of agricultural production ending up as waste, the food supply chain is already unbalanced. The World Resources Institute estimated that if all the food calories available in 2009 were distributed amongst the expected population in 2050 without any losses along the supply chain, each person would fall short of the daily recommended calorie intake by 200 kcal (Searchinger et al. 2013). If the same level of waste was to be present in the supply chain in 2050 as in 2009 this would increase to 900 kcal; to close this gap with current levels of waste the global agricultural production would need to increase by 103% (Searchinger et al. 2013). This gap in food supplied and food required by future populations coupled with the global shifts in population dynamics, population size increasing and climate change clearly indicates that the current state of food supply chains is unsustainable.

The term 'the perfect storm' was coined by Sir John Beddington in 2009 to describe the relationship of global pressures on future food security (Beddington 2009). The global population is set to rise to one billion people by 2050, and according to UN figures, 66% will reside in urbanised environments compared to 54% in 2014 (United Nations 2014b). This increase will add 2.5 billion people to urban populations, with 90% of this increase accounted for by Africa and Asia (United Nations 2014b). With the increase in global population requiring the levels of food to be produced increased by the levels stated above, it is easy to see what impact reducing the current 21% of agricultural products that end up as food waste could have on making global food supply chains more sustainable. The shift in population dynamics will also add pressure to food supply chains; urbanised societies typically have much larger and more complex food supply chains, relying on robust infrastructure to move food from the site of production in rural areas to the urban environment where the population resides. Where growth in urban environments is fast and unplanned without the investment in transport, road or market infrastructure, this will have a negative impact on the ability to bring food to these areas and on the amount of waste generated in the supply chain (Parfitt et al. 2010). As the world population increases, so will the economic status of those regions of growth which in turn impacts upon the types of food consumed. The consumption of basic commodities such as grain tends to be higher is less affluent regions (Bennett 1941), with increasing levels of meat, dairy and fruit and vegetables as populations shift to greater levels of middle-income people. This shift to shorter-life food products again requires a robust food supply chain to be able to move products quickly and under controlled storage conditions.

12.4 Food Waste Utilisation

The problem of food waste needs to be tackled at all points in the supply chain. The first option to explore should always be prevention. Investment in infrastructure, agricultural techniques, novel control for pest and disease and process engineering will lower the levels of preventable waste upstream in the supply chain. Education and engagement with consumers on the practical ways food waste can be prevented will also help alleviate the waste at the final point in the supply chain. Even with these prevention strategies in place, there will always be by-products of processing and waste on the farm which are rich sources of potential functional and nutritional ingredients, as well as fibres, enzymes and pigments for use both in food, but also outside of the food industry. Food waste at the farm and processing level is typically ploughed back into the land or sent for animal feed or fuel production, often at a cost to the waste producer. With increasing financial pressures to bring down the cost of food for consumers and the cost of labour and inputs increasing, primary producers and processors are having to find ways to make their operations more efficient and generate 'added-value' to by-products of production which are currently treated as waste.

12.4.1 Product Development from Plant Food Waste

In recent years there has been an increase in fundamental research and product development using by-products and waste from the food industry. All sectors of the food supply chain are facing increasing pressures on the need to produce high quality goods for low prices, at the same time as the cost of raw materials and labour costs are putting financial strain on the businesses involved. This financial sustainability, alongside legislative pressures, is a driving force in the increased focus the industry has seen in closing the loop in supply chains. One of the major hurdles to utilisation of plant waste is the high water content and perishability of the raw material. Water content of fresh vegetables and fruits ranges from 70 to 100% of the total mass (Russ and Meyer-Pittroff 2004), transporting this water significantly increases costs, whilst removal of the water presents the problem of disposal of the waste water which will have a high level of biological material and therefore a high biological oxygen demand. This water will require treatment before it can be returned to the normal water supply. Depending upon the processing steps the plant waste originates from, there will be high levels of bacteria and other microorganisms present which will act on the material to reduce the functionality, quality and safety. Enzymatic activity in the plant material will also impact upon the functionality and quality of the plant waste. Waste utilisation therefore needs to take these factors into account, by either processing the waste into its final form or stabilising the waste without compromising on quality or functionality at source.

12.4.1.1 Format NPD

A large proportion of fruit and vegetable waste consists of unharvested or graded out produce that is edible and fit for human consumption, but is either surplus to market demands or does not meet the aesthetic or size quality specifications required for sale. This produce usually retains the flavour, colour and nutritional value, with the format of the product becoming the barrier to use and sale. In field grown herbs, an additional 20% of crop is grown in order to meet programmed sales due to expected shortfalls in specification of the final crop (Valley Produce 2016). In addition to this, around 30% of potential yield is not utilised. This potential yield consists of older plant growth left in the field when the top 15 cm of young growth is harvested, as well as second growth of plants after harvest which is tougher than the first growth and therefore not utilised for fresh sales (Valley Produce 2016). Figure 12.2 displays the levels of volatile flavour and aroma compounds in coriander at three stages of

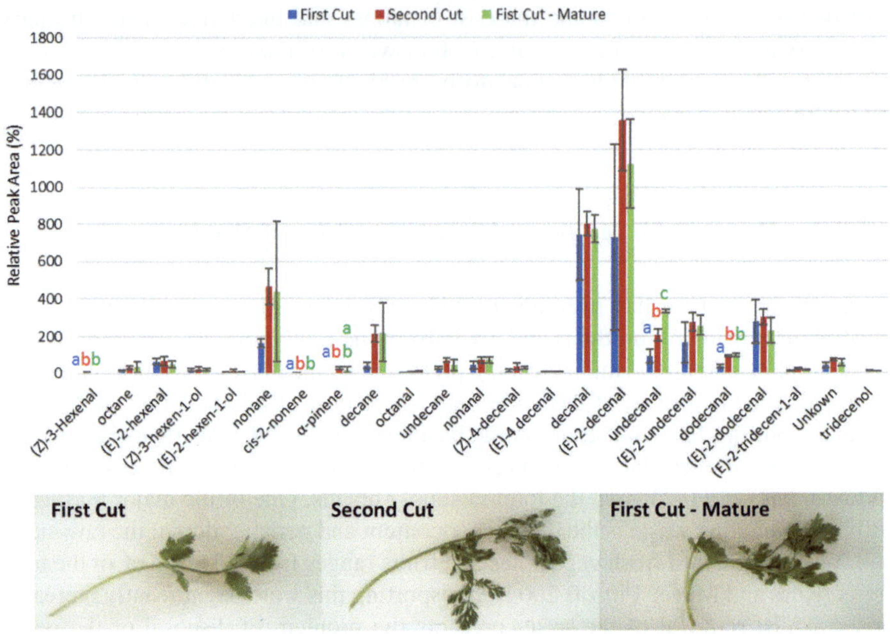

Fig. 12.2 Flavour and aroma volatiles in coriander at different stages of maturity as measured by SPME-GCMS. Coriander of the same variety was harvested at commercial maturity (first cut), beyond commercial maturity (first cut—mature) and the second cut after harvest (second cut). Letters indicate significant differences in volatile levels between maturities ($P < 0.05$). Volatile levels are expressed as relative peak areas compared to an internal standard. Average values taken from three replicate samples

maturity. Although there are significant differences in some compounds, this is actually lower than the seasonal differences for the same coriander plant (data not shown). Therefore, if the colour and/or flavour of herbs is required, a change in format of the out of specification crop would enable utilisation of the entire grown crop and a more sustainable production practice for the primary producer.

A successful example of a waste problem being solved by changing the format of the raw material into juice is ‘B Fresh’ Cold Pressed juices. Phillip Maddock, the owner of B-Fresh, had an entire spinach crop damaged by a hailstorm resulting in the crop being unsuitable for sale and therefore wasted (Briggs 2015). Two years of research later the ‘B-Fresh’ juices were born, utilising lower specification crop grown on the farm in the UK and through partner producers and ensuring sustainable production in the future. In addition to unharvested waste at the farm gate, vegetable and fruit trimmings and grade-outs from further up the supply chain in processing and manufacturing represent a potential source of raw material which retains the flavour, colour and nutrient value of the fresh whole produce. Processing out of specification fruit and vegetable waste and trimmings into juices is a relatively simple way to remove the boundary to sale of aesthetic imperfections. The company

Provalor started trading in vegetable juices made from waste vegetable trimmings in 2005 (Waldron 2009; Provalor 2016). The process of producing the juices itself generates waste in the form of de-watered pulp with a high dry matter content. Provalor have developed processes to extract further ingredients and high-value products from this pulp and in 2015 sold the juicing arm of the business in order to continue development in this field. The company is actively involved in EU research funding on the valorisation of waste under project titled RESFOOD and NOSHAN with the aim of extracting maximum value from vegetable waste streams and making vegetable supply chains more sustainable. Research into the valorisation of fruit and vegetable waste into added-value products and ingredients is high on the agenda of most national and continental funding bodies agendas. Successful commercialisation of technologies and processes to utilise this waste, and potential waste streams to be exploited will be outlined in the following sections.

12.4.1.2 Fibre and Cell Wall Material

The majority of plant-based food waste contains fibre from the cell wall material (CWM). Dependent upon the source of the fibre there will be varying levels of lignin, cellulose, hemicelluloses and polysaccharides such as pectin. These fibres have uses as functional ingredients in foods. A review by Foster (2011) bought together the emerging science on the use and application of CWM as structuring agents within the manufacture of soft solid composites such as ice cream and sauces. This review, along with articles by Appelqvist et al. (2010) and Redgwell (2010), reports that as well as providing a sustainable structuring material, CWM can also add to the nutritional value of the foods as added fibre or essential vitamins and minerals. Research into improved raw CWM and the properties that can be gained from them has spurred studies into how they could be used in food products to replace current structuring agents or high energy processing techniques. Sánchez et al. (2003) studied the 'adding back' of tomato slurry to tomato sauces, which is usually removed during the processing of tomato pastes. This addition of particulate tomato matter resulted in improved rheological properties. Another study by McCann et al. (2011) used carrot CWM to stabilise yoghurts and found that the CWM occupied the void space within casein particle network, aiding in the gelation and rheological properties of the network. As well as a structuring functional role, fibre can also be used for nutritional purposes in foods. Orange pomace left over from the juicing industry added back into orange juice increased the satiating level in healthy adults to levels comparable of whole oranges (Dong et al. 2016). Carrot pomace and cauliflower trimmings have also been utilised in a novel extruded snack, resulting in a product with a higher than average protein (10.25 g/100 g) and fibre (0.84 g/100 g) content than standard extruded snacks and high acceptability in consumer sensory analysis (Mohammed et al. 2016). Carrot pomace has also been successfully commercialised in the form of KAROPRO by the company FoodSoutionsTeam (FST 2016). FST developed a process to produce a powder from dried carrot pomace which can act as a binding agent in many foods. The same company are researching product developments from other sources of food waste.

Fibres from plant waste streams also have a potential functional use outside of the food industry. The company Cellucomp based in the UK has successfully bought to market the product 'Curran' which utilises the properties of cellulose nanofibres extracted from vegetable residues such as sugar beets and carrots. The fibres bring structuring properties to bio-composites, improve the rheological properties of liquid materials and reinforce dried paints and resins (Cellucomp 2016). Cellulose nanofibres can also be utilised by the packaging industry to produce sustainable and biodegradable food packaging. Sugarcane bagasse, the residue left over from sugar processing, contains around 40–50% cellulose and was investigated by Ghaderia et al. (2014) as a sustainable source of cellulose for packaging, the resulting films produced were of comparable tensile strength to other biodegradable and non-biodegradable films. The company LondonBioPackaging makes packaging from sugar cane bagasse alongside more traditional bio-packaging formats from plant starch and recycled paper and wood, demonstrating the commercial viability of this concept. The EU-BADANA project is another example of product development success from waste. The consortium successfully developed a process to extract fibres from banana residues for use as a sustainable reinforcement polymer composite in moulded products (Pickering 2015). In a truly closed loop approach to business operations, the company Solidus solutions has developed a procedure to produce solid board from tomato plants mixed with waste paper, which is now used to package tomatoes (Solidus Solutions 2016). This innovative solution utilises 85,000 tonnes of plant crop residues which becomes available every year after a full cycle of tomato harvest has been completed from one grower in the Netherlands.

12.4.1.3 Polyphenols and Bioactive Compounds

Annually around 5.5 million tonnes of solid waste are produced by the juicing industry (Panouillé et al. 2007). The amount of waste varies depending upon the fruit or vegetable being processed but averages at around 30–50% of the raw material (Panouillé et al. 2007). This type of food waste is an unavoidable part of processing which if not utilised represents a disposal problem to the processor. Fruit and vegetables are a rich source of phenolic compounds; there is increasing evidence for their role in the prevention of coronary heart disease and cancer (Crozier et al. 2007; Tomás-Barberán and Espín 2001), and they are widely accepted as having health-promoting properties. The pomace left over from processing into beverages which consists of seeds, peels, pulp and stones represent a sustainable source of these compounds. In fruit and vegetable pomace antioxidants are often found with high levels of dietary fibre; this led to the concept of Antioxidant Dietary Fibre (ADF) being used as an ingredient in food products as a nutrient rich source of dietary fibre (Saura-Calixto 1998). The benefit of utilising fruit and vegetable wastes in this way is that there is relatively little preparation and processing required to produce a value-added product. To stabilise the pomace for storage and inclusion in a wide variety of foods, the water needs to be removed by drying before grinding to a fine powder. ADF from a variety of waste sources have been investigated, including grape wine pomace in yoghurts and dressings (Tseng and Zhao 2013), apple pomace in cake (Sudha et al. 2007), carrot pomace in pasta (Gull et al. 2015) and extruded snacks (Alam et al. 2016), to improve the functional properties and nutritive value of the food.

Polyphenols, carotenoids and antioxidant compounds can also be extracted from fruit and vegetable wastes as ingredients in their own right. Vesna et al. (2016) added microencapsulated sour cherry extract, a by-product from the juicing industry, to cookies. The resulting product had improved nutritional value both in polyphenolic content and antioxidant activity which remained stable during storage, due to the encapsulation technique. A similar microencapsulation technique has also been employed on beetroot pomace extract (Šaponjac et al. 2016), a source of significant levels of polyphenols and flavonoids, to be used as a potential food additive in the food and pharmaceutical industries. Pomegranates are another high waste producing fruit when processed into juice, with around 50% of the weight resulting in waste. Goula et al. (2017) used ultrasound to extract the carotenoid compounds directly into vegetable oils to produce an enriched oil as an ingredient for the food industry. The method resulted in 85–93% extraction efficiency of carotenoids but in a much greener and clearer method than traditional solvent methods of extraction. The kernel and peel of mango fruits are also a significant source of polyphenols and carotenoids, with the seed making up 20–60% of the mango fruit and the kernel 45–75% of the whole seed this is a good potential source of bioactive compounds (Asif et al. 2016). A comprehensive review of the therapeutic potential of these compounds has been conducted by Asif et al. (2016), highlighting the potential therapeutic, antibacterial and antifungal properties of bioactive compounds that have been isolated from this plant waste source. The author highlighted the need to optimise the extraction techniques before they can be fully utilised. The EU funded project BERRYPOM is a consortium of European researchers aiming to analyse the value of and optimise the extraction of bioactive compounds from berry pomace, with the aim of incorporating the resulting processed berry pomace into cereal-based foods. A recent review of the potential processing techniques and characterisation of the polyphenols was conducted by one of the consortia members, again highlighting the requirement for pilot scale testing and optimisation of processing techniques (Struck et al. 2016) before the technology and ingredients can be commercialised.

12.4.1.4 Pigments

The trend for natural colourants in food is increasing, with plant wastes as a potential source of these compounds. Carotenoids can be sourced from carrot pomace, orange peel and tomato pomace (Oreopoulou and Tzia 2007). Grape pomace left over from the wine making industry is a good source of anthocyanins for use as pigments (Mazza and Francis 1995), with these compounds also being extracted from banana bracts, a by-product of the banana growing industry (Pazmiño-Durán et al. 2001). Red pigments, for example, betacyanin, have also been extracted from prickly pear pulps and seeds, with the extraction efficiency enhanced through using 'green' pulsed electric field and ultrasound technology (Koubaa et al. 2016). Betalain, another red pigment with higher pH stability than anthocyanins, is found in high quantities in red beet stalks, a by-product of beetroot farming and processing. Destro dos Santos et al. (2016) investigated the use of filtration techniques to improve the stability of this pigment extracted from beet stalks and found that this process improved the intensity, clarity and stability; demonstrating that added-value could be gained from this by-product of red beet processing. Pigments can also be obtained from microbial fermentation, and with the medium for fermentation representing 38–73% of the total production costs, the utilisation of plant wastes is a potentially cost-effective and sustainable alternative (Panesar et al. 2014). A recent review by Panesar outlined successful pigment mediums from plant wastes such as corn steep liquor, a by-product of the corn milling industry, jackfruit and Durian seeds, grape pomace, tomato pomace and various crop residues including husks and pea pods.

12.4.1.5 Proteins

Plant proteins are becoming ever more popular as a more sustainable and animal free source for use as a functional ingredient in foods. Plant residues, waste and by-products are being increasingly investigated as a raw material for protein manufacture. The EU project BIORICE, a consortium of 7 SMEs and research institutes and companies, aims to investigate rice protein by-product pre-treatment, peptide isolation and bioactivity and safety testing, with the aim of producing peptides and other bio-actives for use in nutraceuticals, cosmetics and functional foods (BIORICE 2016). The raw material being utilised is the waste stream from rice-starch processing. Rice starch is itself extracted from broken rice grains which are a by-product of the rice production industry; however, the process to extract the starch results in wastewater with a high protein content. This waste water costs businesses around 100,000 Euros a year to discard (Tassoni 2015). Producing added value ingredients from this waste water would therefore make the businesses more financially sustainable, but also reduce the environmental pressures of waste-water disposal and generate a sustainable and cost-effective set of ingredients for the food, cosmetic and pharmaceutical industries. Another EU project APROPOS was successful in extracting bioactive peptides and proteins from rapeseed cake, a by-product of the rapeseed oil processing industry, on a lab and pilot scale. The technology to scale this up to industrial processing is, however, not yet available but the fact the concept and pilot processes have now been proven should mean that the step to production of these ingredients is now possible (APROPOS 2014).

12.4.1.6 Substrates

The annual global volume of spent coffee grounds (SCG) left over from the brewing process is approximately 6 million tonnes (Asmita et al. 2016). Like many waste materials, spent coffee grounds (SCG) have been investigated for their use as biomass for fuel production. Vardon et al. (2013) used a three-step process to not only produce biodiesel and biofuel to recognised standards, but also to produce 'biochar' for use as a fertiliser. When used as a fertiliser on sorghum-sudangrass seedlings the biomass increased by twofold compared to controls. The antioxidant capacity and carotenoid levels in lettuce have also been increased when cultivated with fresh SCGs which the author attributed to increased stress levels of the plants (Cruz et al. 2013). SCGs have also been used to remove heavy metal ions from waste water (Tokimoto et al. 2005), a study by Asmita et al. (2016) used bioelastomeric foams containing 60 wt% of SCG and 40 wt% of a silicone elastomer to remove Pb^{2+} and Hg^{2+} ions from wastewater with an overall removal efficiency of 80%.

12.5 Conclusions and Future Prospects

This chapter has highlighted some of the ways by-products from the current supply chains can be utilised through new product development. With food sustainability targets on the United Nations agenda, there is now a focus on food waste and sustainable food production in national and regional funding programmes, which are available for both business and academia. Stand-alone research into new methods of waste valorisation, identification of waste streams and examination of the what exactly various plant wastes and by-products are made of is a useful mine of information for businesses and organisations who have a waste problem. However, to be able to make production of food a truly closed loop system, there needs to be a systems-based approach. The reasons for food waste are often complex and affected by decisions and changes further up or down the chain. As well as valorisation of waste, the whole supply chain needs to come together to ensure there is no avoidable waste produced, and unavoidable waste has a way to be utilised.

References

ABDA (2015) Anaerobic digestion market report July 2015. http://adbioresources.org/docs/July_2015_Market_Report.pdf. Accessed 18 Sept 2016

Alam MS, Kaur J, Khaira H, Gupta K (2016) Extrusion and extruded products: changes in quality attributes as affected by extrusion process parameters: a review. Crit Rev Food Sci Nutr 56(3):445–473

Appelqvist I, Day L, Lundin L (2010) Natural plant cell wall materials—value to food processors and consumers. Food Sci Technol J 24:21–23

APROPOS (2014) Final project report: added value from high protein and high oil industrial costreams. http://cordis.europa.eu/docs/results/289/289170/final1-apropos-final-report-excluding-confidential-part-2015-02-27.pdf. Accessed 18 Sept 2016

Asif A, Farooq U, Akram K, Hayat Z, Shafi A, Sarfraz F, Sidhu MAI, Rehman H, Aftab S (2016) Therapeutic potentials of bioactive compounds from mango fruit wastes. Trends Food Sci Technol 53:102–112

Asmita AC, Javier P, Ioannis L, Ilker SB, Simone L, Athanassia A, Despina F (2016) Spent coffee bioelastomeric composite foams for the removal of Pb2+ and Hg2+ from water. ACS Sustain Chem Eng 4(10):5495–5502. doi:10.1021/acssuschemeng.6b01098

Beddington J (2009) Food, energy, water and the climate: a perfect storm of global events? In: Conference presentation given to the sustainable development UK annual conference, QEII Conference Centre, London, 19 March 2009. http://webarchive.nationalarchives.gov.uk/20121212135622/http://www.bis.gov.uk/assets/goscience/docs/p/perfect-storm-paper.pdf

Bennett MK (1941) Wheat in national diets. Wheat Studies of the Food Research Institute. 18(2)

BIORICE (2016) http://www.biorice.eu/default.aspx. Accessed 18 Sept 2016

Briggs F (2015) British cold pressed fruit juices, B.Fresh, launch onto UK market. Retail times. http://www.retailtimes.co.uk/british-cold-pressed-fruit-juices-b-fresh-launch-onto-uk-market/. Accessed 18 Sept 2016

Cellucomp (2016) http://cellucomp.com/. Accessed 18 Sept 2016

Crozier A, Jaganath IB, Clifford MN (2007) Phenols, polyphenols and tannins: an overview. In: Plant secondary metabolites. Blackwell, Hoboken, pp 1–24

Cruz R, Gomes T, Ferreira A, Mendes E, Baptista P, Cunha S, Pereira JA, Ramalhosa E, Casal S (2013) Antioxidant activity and bioactive compounds of lettuce improved by espresso coffee residues. Food Chem 145:95–101

Dong H, Sargent LJ, Chatzidiakou Y, Saunders C, Harkness L, Bordenave N, Rowland I, Spencer J, Lovegrove JA (2016) Orange pomace fibre increases a composite scoring of subjective ratings of hunger and fullness in healthy adults. Appetite 107:478–485

Dos Santos CD, Scherer RK, Cassini AS, Marczak LDF, Tessaro IC (2016) Clarification of red beet stalks extract by microfiltration combined with ultrafiltration. J Food Eng 185:35–41

FAO (2006) Policy brief—food security. http://www.fao.org/forestry/13128-0e6f36f27e0091055bec28ebe830f46b3.pdf. Accessed 8 June 2017

FAO (2011) Global food losses and food waste: extent, causes and prevention. Rome. http://www.fao.org/docrep/014/mb060e/mb060e.pdf. Accessed 18 Sept 2016

FAO (2013) Food wastage footprint: impacts on natural resources—summary report. http://www.fao.org/docrep/018/i3347e/i3347e.pdf. Accessed 18 Sept 2016

Foster TJ (2011) Natural structuring with cell wall materials. Food Hydrocoll 25:1828–1832

FST (2016) http://www.foodsolutionsteam.ch/en. Accessed 18 Sept 2016

Fusions (2014) FUSIONS definitional framework for food waste—full report. ISBN 978-91-7290-331-9

Ghaderia M, Mousavia M, Yousefib H, Labbafia M (2014) All-cellulose nanocomposite film made from bagasse cellulose nanofibers for food packaging application. Carbohydr Polym 104:59–65

Goula AM, Ververi M, Adamopoulou A, Kaderides K (2017) Green ultrasound-assisted extraction of carotenoids from pomegranate wastes using vegetable oils. Ultrason Sonochem 34:821–830

Gull A, Prasad K, Kumar P (2015) Effect of millet flours and carrot pomace on cooking qualities, color and texture of developed pasta. LWT- Food Sci Technol 63(1):470–474

Koubaa M, Barba FJ, Grimi N, Mhemdi H, Koubaa W, Boussetta N, Vorobiev E (2016) Recovery of colorants from red prickly pear peels and pulps enhanced by pulsed electric field and ultrasound. Innov Food Sci Emerg Technol 37(Part C):336–344

Mazza G, Francis FJ (1995) Anthocyanins in grapes and grape products. Crit Rev Food Sci Nutr 35(4):341–371

McCann TH, Fabre F, Day L (2011) Microstructure, rheology and storage stability of low-fat yoghurt structured by carrot cell wall particles. Food Res Int 44:884–892

Mohammed SA, Shivani P, Anjali S (2016) Optimization of the extrusion process for development of high fibre soybean-rice ready-to-eat snacks using carrot pomace and cauliflower trimmings. LWT- Food Sci Technol 74:135–144

Oreopoulou V, Tzia C (2007) Utilization of plant by-products for the recovery of proteins, dietary fibers, antioxidants, and colorants. In: Uitilisation of bi-products and treatment of waste in the food industry. Springer, New York, pp 209–232

Panesar R, Kaur R, Panesar PS (2014) Production of microbial pigments utilizing agro-industrial waste: a review. Curr Opin Food Sci 1:70–76

Panouillé M, Ralet M.-C, Bonnin E, Thibault J-F (2007) Recovery and reuse of trimmings and pulps from fruit and vegetable processing. In: Waldron KW (ed) Handbook of waste management and co-product recovery in food processing. Woodhead Publishing, Cambridge, pp 417–439

Parfitt J, Barthel M, Macnaughton S (2010) Food waste within food supply chains: quantification and potential for change to 2050. Philos Trans R Soc B 365:3065–3081

Pazmiño-Durán EA, Giusti MM, Wrolstad RE, Glória MBA (2001) Anthocyanins from banana bracts (Musa × paradisiaca) as potential food colorants. Food Chem 73(3):327–332

Pickering J (2015) Final report summary—BADANA (Development of an automated process to extract fibres from the waste of banana food production for exploitation as a sustainable reinforcement in injection- and rotomoulded products). Project reference: 232287. http://cordis.europa.eu/result/rcn/149409_en.html. Accessed 18 Sept 2016

Provalor (2016) http://www.provalor.nl/EN/. Accessed 18 Sept 2016

Redgwell RJ (2010) Dietary fibre in food fabrication: a changing landscape for consumer and industry. Food Sci Technol J 24:18–20

Russ W, Meyer-Pittroff R (2004) Utilizing waste products from the food production and processing industries. Crit Rev Food Sci Nutr 44(1):57–62

Sánchez MC, Valencia C, Ciruelos A, Latorre A, Gallegos C (2003) Rheological properties of tomato paste: influence of the addition of tomato slurry. J Food Sci 68:551–554

Šaponjac VT, Čanadanović-Brunet J, Ćetković G, Jakišić M, Djilas S, Vulić J, Stajčić S (2016) Encapsulation of beetroot pomace extract: RSM optimization, storage and gastrointestinal stability. Molecules 21(5):584–600

Saura-Calixto F (1998) Antioxidant dietary Fiber product: a new concept and a potential food ingredient. J Agric Food Chem 46:4303–4306

Searchinger T, Hanson C, Ranganathan J, Lipinski B, Waite R, Winterbottom R, Dinshaw A and Heimlich R (2013) The great balancing act: working paper, instalment 1 of creating a sustainable food future. World Resources Institute, Washington, DC. http://www.worldresourcesreport.org. Accessed 18 Sept 2016

Smil V (1999) Crop residues: agricultures largest harvest. Bioscience 49(4):299–308

Solidus Solutions (2016) http://solidus-solutions.com/en/home/. Accessed 18 Sept 2016

Struck S, Plaza M, Turner C, Rohm H (2016) Berry pomace—a review of processing and chemical analysis of its polyphenols. Int J Food Sci Technol 51:1305–1318

Sudha ML, Baskaran V, Leelavathi K (2007) Apple pomace as a source of dietary fiber and polyphenols and its effect on the rheological characteristics and cake making. Food Chem 104(2):686–692

Tassoni A (2015) The bi-product is improved efficiency. EU Res 2015(1):49–51

Tokimoto T, Kawasaki N, Nakamura T, Akutagawa J, Tanada S (2005) Removal of lead ions in drinking water by coffee grounds as vegetable biomass. J Colloid Interface Sci 281(1):56–61

Tomás-Barberán FA, Espín JC (2001) Phenolic compounds and related enzymes as determinants of quality in fruits and vegetables. J Sci Food Agric 81(9):853–876

Tseng A, Zhao Y (2013) Wine grape pomace as antioxidant dietary fibre for enhancing nutritional value and improving storability of yogurt and salad dressing. Food Chem 138:356–365

United Nations (2014a) Report of the Open Working Group of the General Assembly on Sustainable Development Goals. Document number A/68/970. http://undocs.org/A/68/970. Accessed 18 Sept 2016

United Nations (2014b). Department of Economic and Social Affairs. Population Division World Urbanization Prospects: The 2014 Revision, Highlights (ST/ESA/SER.A/352). https://esa.un.org/unpd/wup/Publications/Files/WUP2014-Highlights.pdf. Accessed 18 Sept 2016

Valley Produce (2016) Calculations of annual herb crop waste—personal communication

Vardon DR, Moser BR, Zheng W, Witkin K, Evangelista RL, Strathmann TJ, Rajagopalan K, Sharma BK (2013) Complete utilization of spent coffee grounds to produce biodiesel, bio-oil, and biochar. Am Chem Soc 1:1286–1294

Vesna TŠ, Gordana C, Jasna CB, Biljana P, Sonja D, Jovana P, Ivana L, Sladana S, Jelena V (2016) Sour cherry pomace extract encapsulated in whey and soy proteins: incorporation in cookies. Food Chem 207:27–33

Waldron KW (2009) Production of high-value functional vegetable juices from food co-products. In: Handbook of waste management and co-product recovery in food processing. Woodhead Publishing, Cambridge, pp 376–390

Chapter 13
Food Losses in Rice Milling

Ye Aung

13.1 Introduction

Our world is approaching the inevitable food scarcity problem as food insufficiency is becoming eminent in certain parts of the world, mainly in Africa, Asia, and South America (Buerkle 2006). Furthermore, for major cereal crops of the world, there has been a steady decline in the rate of growth in yield (FAO 2015). This may lead to food chaos that could possibly occur within the next 30 years if food security-related issues are not properly addressed. Among the many reports on this issue, the report from the United Nations' Food and Agriculture Organization is more precise. According to the report, world's population is expected to exceed 9 billion people by 2050 and it was affirmed that agricultural output needs to increase by 70% in 40 years to feed the global population, therefore countries of the world were urged to reverse a 20-year decline in investment and put back into agricultural section (FAO 2009).

Rice is one of the main staple foods for the people of the world. More than half of the global population consumes it. Demand for it is steadily increasing along with the population growth. It is interesting to note that rice demand in the European countries has risen up recently (Ferrero and Tinarelli 2007). Fresh data are showing that internationally traded volume of the rice has increased more than 10 million tons compared to the time when our world was turning into the twenty-first century (The World Bank 2014). This rising trend seems to be continuing at least for some time.

The surge of demand also highlights the loss that occurs presently in the postharvest sector of rice. The loss has already reached to an alarming level if carefully calculated. In 1975, the United Nations in its 7th Special Session stressed for efforts to reduce postharvest food losses in developing countries to be given priority.

Y. Aung (✉)
Buhler Asia Pacific Ltd, Yangon, Myanmar
e-mail: ye.aung@drye-rice.com

S.N.A. Abdullah et al. (eds.), *Crop Improvement*,
DOI 10.1007/978-3-319-65079-1_13

While the world body is keeping caution on the quantity, more losses came from the corner of quality. The interest in milling quality rice has increased since then. Instead of focusing on waste-control as a response to the world body, the industries had shifted its focus to the beauty of rice which leads to more rice losses. With an awareness of this trend, one food monitoring organization has announced in 1978 the importance to focus on recommendation for the efficiency of rice production system and to reduce losses (National Academy of Sciences, Washington DC, 1978). However, the thirst for beauty still goes on.

13.2 State of the Paddy at Harvest

Rice originated from a wild grass called *Oryza sativa.* It is a seasonal crop and usually grown in tropical countries which receive sufficient sunshine. The plant can be grown on muddy soil with little water dwelling on it. The plants are of various heights producing numerous leaves and branches. Paddy grains are developed on a long and thin panicle hanging out from the branch. A panicle can carry many grains and there are a number of panicles on a plant.

Most paddy varieties mature just over hundred days and ready for harvesting. Harvesting is carried out by cutting upper part of the plant where the panicles are hanging out. Manual harvesting uses sickle or a special knife cutter that cuts a handful amount of the branch batch by batch. Harvested branches are bound into bundles for threshing. Modern combined harvester can deliver threshed paddy simultaneously in the time of harvesting (Liang et al. 2017). Deciding the right time to harvest is very important to avoid immature and cracked grains. Early harvest may collect more immature grains while late harvest will collect cracks.

Threshing of harvested branches is required to receive paddy seeds separately. Traditional threshing method is carried out by beating the bundles across a grate or stamped over by a team of buffalos. Paddy seeds are collected always with plant parts like stems, leaves, panicles, and also with earth pieces. Hence, cleaning of threshed paddy is required. Farmers themselves clean it to a certain extent by manual or natural winnowing. Along the steps, losses occur and the amount is from 5 to 15%. Manual harvesting leads to more losses than mechanical harvesting.

13.3 Deconstructing a Rice Grain

Outer piece of the rice is the husk formed by two leaves covering the entire rice (caryopsis) loosely. The two has joined at opposite sides by overlapping folds extending the entire length (Figs. 13.1 and 13.2). The leaves are mostly fiber and supported by hard spines. Short or long tiny hairs are always found on the surface. Hairs are somehow hard, straight, and pointed. Therefore rice is called spikelet in the botanical term.

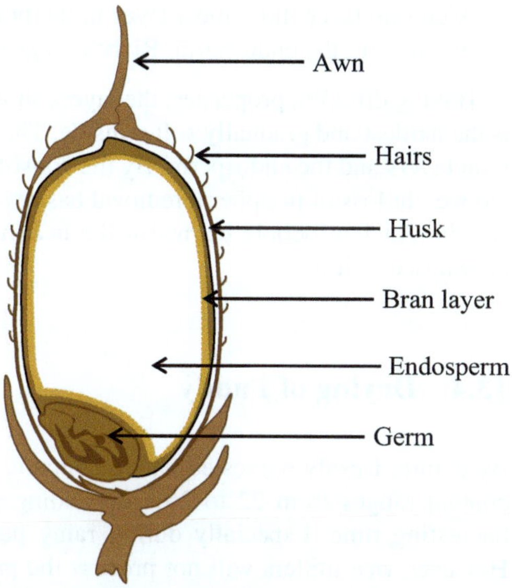

Fig. 13.1 Paddy rice

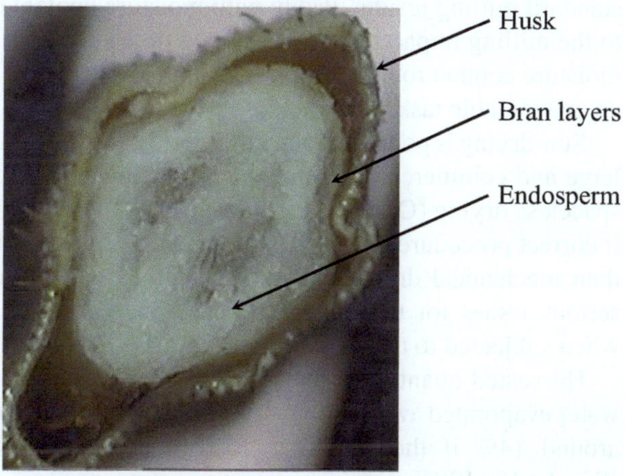

Fig. 13.2 Cross section of a paddy seed

Inside the husk lies a seed called rice containing a reproductive organ called germ. Based on the color, the rice is called brown rice. Germ is the most nutritious part of rice and people prefer to consume it. The rice is traded under the name of cargo rice. However, the germ is crushed and thrown out as by-product during the milling. The bran layers are brown in color which is second to germ in terms of nutrition. Again, these nutritious parts are removed during whitening. Based on the properties, bran layers are divided into three groups:

1. Pericarp layer: It is the outermost layer of the brown rice; a thin fibrous hard cell.
2. Tegmen layer: Just beneath the pericarp layer and less fibrous than pericarp; but relatively richer in oil. A small amount of protein also exists.

3. Aleurone layer: Innermost layer in the bran groups and genuine bran layer bordering with the endosperm. Richest in protein, vitamins, minerals, and fat.

Having different properties, the layers show off the different hardness. The outer is the hardest and gradually softer inside. There is no definite boundary between the bran layers and the endosperm. By the constituents, bran is considered empty when the weight loss of peripheral removal has reached to 8% from the initial brown rice weight. Ancient people learned to be healthier than today's rice eaters by eating brown rice as it is.

13.4 Drying of Paddy

By nature, freshly harvested paddy contains high moisture content. Generally, the content ranges from 22 to 26% depending on the regional condition around the harvesting time. Especially during rainy period, the moisture content is higher. However, rice millers will not process the paddy with this moisture range. Paddy must be dried down to 14% which is the norm and proven level of safe storage and standard milling results. Paddy with moisture content over 14% is weak to respond to the milling impacts and deteriorated in the storage. Long-term storage needs the moisture content to be a bit lower than this. Therefore, drying of harvested paddy is an unavoidable task.

Sun-drying is popular around the world and still being practiced widely. However, large and commercial rice mills prefer artificial (mechanical) dryers for quick and spaceless drying (Okeke and Oluka 2017). Any drying method can crack the paddy if correct procedures are not followed. Sun-drying comparatively offers less control than mechanical drying and as such infamous for sun-cracks. Drying cracks are serious issues for the milling because once cracked, rice is readily broken even when subjected to a light milling process.

Harvested quantity of the rice will drastically reduce by weight after drying as water evaporated away during drying. The drying loss was observed and found to be around 14% if the paddy of 26% moisture content was dried down to 14% (Wimberley 1983).

13.5 De-husking or Hulling

13.5.1 Cleaning

Paddy is always mixed with a certain quantity of impurities whenever the paddy arrived at the rice mills. The sort of impurities includes both organic and inorganic items. At the rice mills, the impurity contents range between 2 and 6%, requiring a series of cleaning processes. These include removal of seedless and very light immature grains. Impurities can block the inlet and outlet of the machine, quite often, the impurities block the screens. Even if they do not damage the machines, they can jeopardize the performance of the machines. In the worst case, they shorten the lifespan of the machinery parts.

As the impurities are of different kinds, different technologies are required in eliminating them. The standard cleaning practice is to carry out the impurity elimination process in two stages. Large and coarse impurities are first removed and the fines are removed in the second stage. The first cleaning work is called pre-cleaning, usually carried out just after receiving the paddy. The second time is in the milling line before any major operations and is called secondary cleaning. Secondary cleaning uses more précised and complicated technologies. Some people called the process fine cleaning.

13.5.2 Hulling Machine (Huller)

De-hulling or hulling is the first operation in which rice is broken for the first time. In earlier days, de-hulling is carried out by disc huller made of iron with partial stone coating (Bhattacharya and Ali 2015). Stones are emery grains firmly coated around circumferential areas of the discs. There are two discs in one machine assembled in parallel by keeping stone coated surfaces inside. Lower disc is attached to a shaft and rotates during operation while the upper one is fixed and remained stationary. Paddy is fed through the hole made at the center of the upper disc and let spread onto the lower disc. Fed paddies spread in all directions by the influence of the rotating surface and reach between the stone coated areas. By the rough surface, the stones crush the paddy and husks are split into pieces. The admixture of brown rice, husks, and unhusked paddies instantly reach out of the discs after hulling. The products are first dropped through an aspiration box to blow the husks out and the rests are sent to another machine called paddy separator.

Later, rubber-rolls are used to replace the iron discs (Bhattacharya and Ali 2015). Hulling breakage significantly reduced since then. Rolls are cylindrical-shaped rubber shoes lying in parallel with different rating speeds. One roll is movable forward and backward enabling adjustment of the clearance between the two. Paddy is fed into the clearance and the husk is sheared off by the different speeds. Hulled rice are pushed out simultaneously along with the husks.

Working principle of the rubber roll was invented in view of exploiting loosened wrapping and joining pattern of the husk leaves. Overlap-folding of the husks indicates the way of sliding them (Fig. 13.2). Elasticity of the rubber, clearance between the husk and rice, and the folding pattern, all contribute in the removal of the husks with little damage to the rice.

The machine was modified subsequently by adding a feeding system in which paddies are guided to dive into the rolls vertically (Bhattacharya and Ali 2015). The invention further improves the performance of the machine both in efficiency and capacity.

Whatever the model and type, performance of a huller is always judged by the two fractions in its output, the unhusked paddies and the broken ones.

Lesser unhusked paddies translates into a high degree of hulling requirement while small amount of broken rice means good performance as well as good rice quality. A huller is considered the best if the percentages of both unhusked paddies and broken rice it produced is the lowest. The weight contributed by the husks to the paddy at 14% moisture stands from 18 to 23% depending on the variety. Most long grains show 21% husk content at the said moisture.

13.5.3 *Separating Unhusked Paddies*

It is a common understanding that not all of the paddies fed to a huller will have the husk removed. A certain quantity will always be left unhusked. Unhusked paddies are in fact thin grains that escaped out of the clearance set for the majority of the grains. The amount of unhusked paddy is high when hulling is done on early-harvested paddy. Timely harvested paddy will have the minimum amount. This is because not all paddies matured at the time of harvest. In the real situation, paddies are not uniformed even on the same panicle as they do not flower at the same time. There are always early and late bloomers on a paddy plant. Hence, the stock contains immature grains. Therefore, hulling is always set for the majority of the grains and the thin grains are let to go free. When it is set for thin grains, the majority of the grains are likely to be broken in a massive quantity.

Unhusked paddies are not allowed to proceed for further processing, especially to the whitening machines for fear of wear and tear. Whitening machine works with compressive force in which paddy with its abrasive nature could erode machine parts. Paddy separator segregates the paddies and brown rice by exploiting different surface properties between the two. Paddy is rough and brown rice is smooth. Separated paddies are discharged separately and sent back to the hulling section while brown rice including broken rice is being forwarded to the whitening section. Purity of the products is adjustable by changing machine inclination, number of frequency, and length of the stroke. Determination is done by examining the products especially in brown rice fraction.

13.5.4 *Recovery of Brown Rice*

Keeping record of brown rice recovery is uncommon in small and conventional rice mills. However, the practice is necessary for large and commercial-scale mills because the data highlights the impurity content after hulling and also the losses during whitening. Based on the husk weight from the standard content and brown rice weight, content of the impurity can be easily discovered by deducting brown rice recovery from the input paddy quantity. Modern rice mills include a weighing machine before whitening for this purpose.

In most major rice producing countries, 75% of brown rice recovery is generally accepted (Baradi and Elepaño 2012). This leaves impurity content at 5% including dusts, chaffs, seedless grains, and very light immature grains. However, the data should be considered as a bottom line as some rice mills that count starting from cleaned paddy take 78% as the acceptable brown rice recovery.

Actual weight of the husk is occasionally checked in some rice mills where frequent variety changes take place. Standard procedure is comparing 1000-kernel weight between the paddy and brown rice, whole grain. The same method is sometimes applied for determining the loss of moisture content.

Brown rice is usually checked for head rice and broken rice contents, so that necessary adjustments can be done to salvage the broken rice.

13.6 Whitening

13.6.1 Brown Rice and Digestibility

Whitening became a principal part in the rice milling since people ate white rice. Whitening is not part of the primitive milling method but the primitive way includes hulling. Rice milling in the ancient time involved pounding the paddies in a mortar by a pestle. The primary objective then was only to remove the husk. Pounding was stopped because only a small amount of paddy was able to be unhusked. A lot of broken rice was produced after pounding. If the amount was deemed too much, the stock was screened out in a bamboo tray before consumption, otherwise, taken as it is.

Nowadays, a rice mill without a whitening machine is not regarded as a rice mill. Some countries even issued regulations for the inclusion of this machine. One rationale is that eating quality of the brown rice is inferior to white rice. However new studies showed that brown rice contain more essential nutrients compared to milled rice (Pan et al. 2017).

Whitening operation changes the brown color into white. The degree of whiteness increases with the whitening time as more bran is being removed. Exposure of the endosperm indicates that the whiteness has reached the highest level.

13.6.2 Structure of a Whitening Machine

Whitening of rice is performed by the abrasion produced on the rice surface using the stone specially fabricated for this purpose. The rice is rubbed with the stone by which abrasion is created that is just sufficient for dismantling the rice surface. The rubbing speed supports the effectiveness of the abrasion. The stone is made of emery grains with crystalized multiple corners forming sharp edges. Stones made of larger emery grains produced rough abrasion whereas smaller grains are for fine abrasions. Whatever the degree of abrasion that is made, uniformity of the grains is important in order to prevent the rice from breaking.

Stone made for conventional whitening machine is conical in shape and made in one solid piece. Modern machines use smaller discs and a number of discs are used in one machine. The discs are assembled on a single shaft that rotates all the stones simultaneously. A cylindrical screen surrounds the stones with a distance from the stone surface providing the space for rice to come into. The clearance of the space was predetermined citing whitening effects. Then, circular hollow area is formed into segments by inserting vertical bars (brakes). The brakes are in equal distances and attached to the screen leaving a narrow gap towards the stones. The configuration in cone-type machine is the same. Both models have openings at the top and bottom for rice feeding and discharging, respectively. However, discharging rate is controllable based on the desired period for the rice to be maintained inside the machine.

While the machine is running, the rice is fed from the top and guided to fall equally into each chamber. Disc type uses mechanical feeding whereas cone type uses gravity feeding. The rotation process enables the running stones to rub the rice while slowly moving downwards and being blocked by brakes from running around. Surface of the rice is scratched and bran layers are removed. The brans that are disintegrated into pieces are screened under the influence of the rotation of the stone. Residual period is regulated by the discharge apparatus. Rice whitening cannot be carried out without breaking the rice because the brans are integral parts of the rice (caryopsis) and bound organically to the endosperm.

13.6.3 Heat and Breakage

Rice is always warm during whitening and if the warmness is high, self-breaking of the rice occurs leading to high whitening breakage. Rice temperature increases along with the time of whitening and the pressure applied on it. Once warmed, lateral pressure increased and creates thermal stress inside the rice causing cracking. Cracked rice could not withstand the milling stresses and breaks easily. Thermally stressed rice breaks easily though cracks are yet to be visible. In order to minimize breakage, the whitening operation is carried out in multiple passes to minimize heat generation.

The most popular concept is the 3-pass whitening system corresponding to the number of bran groups. Stones are set as coarse, medium, and fine starting from the first pass. The same way is applied for the peripheral speeds, which are fast, medium, and slow. By having intervals between the passes, it is possible to limit the level of warmness leading to reduction in breakage.

13.7 Polishing

Polishing is the follow-up process to the whitening as a complementary work to make the rice attractive. Rice looks coarse after abrasive whitening due to the scratches inflicted by the stone. Polishing sealed it by filling with loose bran particles on the rice. The rice becomes neat and tidy after the scratches have disappeared.

Friction polisher gives better surface finish to the milled rice (Koga 1969). Further passes of polishing hardens the fillets and smoothens the surface.

This kind of facial improvement misled the rice millers to apply more passes of polishing with the hope of having a better look to fulfill the consumers' demand. In fact, polishing is simply a beauty-making process and no evidence has been found that rice gives better taste by intensive polishing.

13.7.1 Structure of a Polishing Machine

Unlike the whitening machine, polishing machine works horizontally. The rice is fed from one end and discharged at the opposite end. Being a horizontally working machine, rice is fed mechanically and moved forward by the feeding force. There is a counter force from the discharge side to hold the rice tightly. The working chamber is formed by screen-baskets keeping cam-rotor in the middle. Unlike the whitening machine, the screens are folded in design to create resistance to the rice which is supposed to be moving around with the cam-rotor. While rotating, the tips of the cam beat the rice enabling rice kernels to be displaced repeatedly. The moment of displacement together with the tightly packed conditions prompts the rice to rub each other. This is called self-frictioning and the polishing machine is called frictioning machine. Rice is warmer than when it was in the whitening machine.

Subsequently, the polishing method was further developed to include water addition. Water mist is blown in during the operation via a central shaft and openings of the cam. The rice is wet and the surfaces are softened by the water. Polishing heat cooks the rice immediately resulting in the surface layer to form a coating. The coating is very thin and looks like a silk layer. The rice from wet polisher is called silky rice. The rice kernels are brighter than that produced by dry-polishing.

13.8 Losses

13.8.1 Physical Losses

White rice that has been subjected to a high degree of polishing produces slippery surface and better reflective quality that stands better in today's market. These attributes are only determined by personal judgments, and as such there are frequent disputes. For fear of disputes, millers often carry out the whitening and polishing processes to a greater extent. Unfortunately, the practice intensifies further the losses that occur.

Food loss is particularly boosted during the smoothening process in which rice parts are cut off. If we look at the cross section, it can be seen that the rice body is uneven and undulating. There are six numbers of ridges and furrows (Fig. 13.3a). The undulating body is due to the grooves on the rice surface that formed during

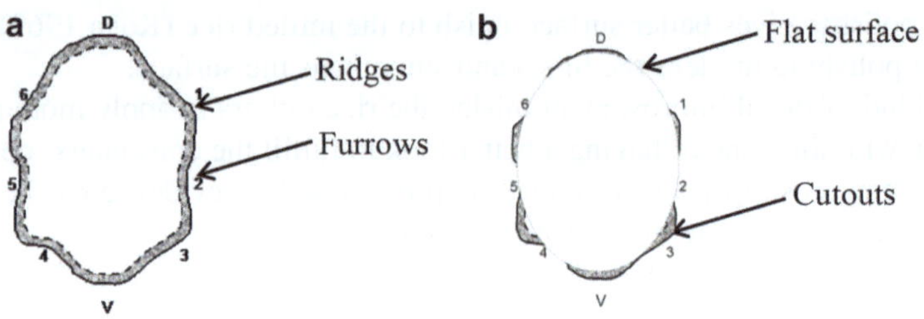

Fig. 13.3 (**a**) Cross section of rice. (**b**) Cross section of grounded rice

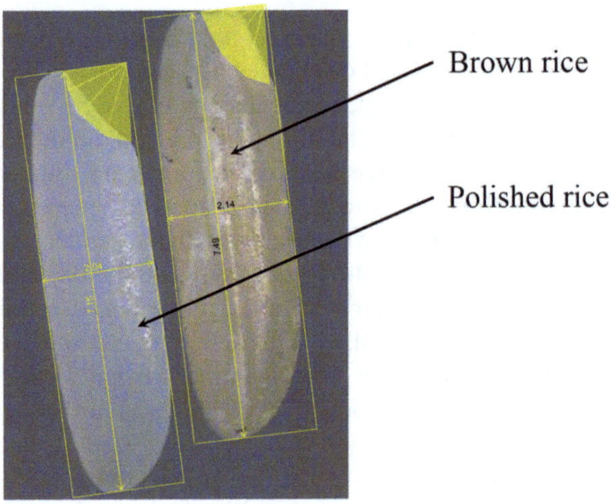

Fig. 13.4 Brown rice and the smaller highly polished rice

growth of the seeds (Bhattacharya 1980). The depth of the grooves varies from 1 to 88 μm (Bhashyam and Srinivas 1984) depending on the growing location.

Smoothening rice surface requires flattening the ridges to the level of the furrows (Fig. 13.3b). In this procedure, rice with deep grooves requires more grinding; however, this leads to additional losses.

In order to get rid of the ridges, whitening must be aggressive, prompting rice to be broken extraordinarily. In this scenario, whitening operation eventually becomes a size-reduction process (Fig. 13.4). Massive rice parts are found in the discarded bran in this kind of whitening condition (Fig. 13.5).

An analysis of the size reductions on a Thai premium rice that has been milled until 11% of weight loss from brown rice.

Reduction in length: 4.5%,
Reduction in breadth: 5.1%, and
Reduction in thickness: 6.9%

Higher reduction on the thickness is observed because of the fact that rice mostly has kernel whose shape is somewhat flat. Among them, the length reduction is the most risky finding as rice could be broken when its tips are trimmed. Besides,

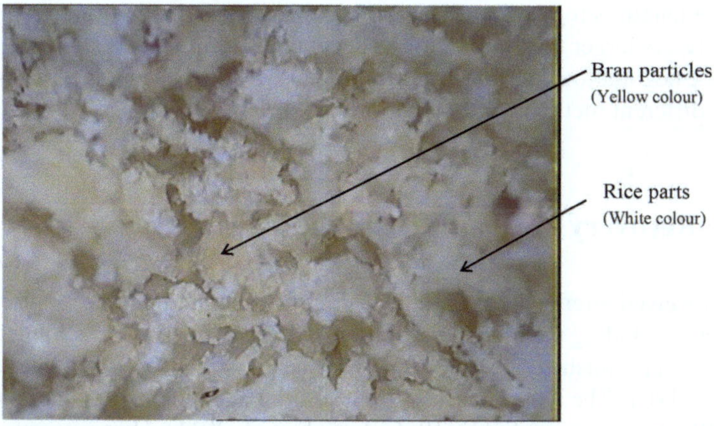

Fig. 13.5 Rice cells in the bran after aggressive whitening

shorter length could bring down the price it usually fetches. Overall volume loss in such case amounted to 15% which could lead the people to consume more rice for satisfaction.

13.8.2 Nutritional Losses

Apart from physical losses, overly milled rice lost nutrition too. Proteins, fats, vitamins, and minerals are present in greater quantity in the germs and outer layers than in the starchy endosperm (Heinemann et al. 2005). The rates of nutritional reduction are further examined on the white rice. White rice that has been milled at 7–8% bran removal showed nutrient losses as 29% proteins, 79% fat, 84% lime, and 67% iron. Towards these findings, FAO published this notification in its publication, *"Highly polished rice produced in modern mills further increases the incidence of beriberi and other deficiency diseases such as keratomalacia, stomatitis, glossitis, cheilosis, and hepatic cirrhosis"* (FAO 1948). Rates of the losses seem more critical now with respect to the higher degree of milling (Abbas et al. 2011).

13.9 Grading

The responsibility of the grading operation is to fix the broken rice in the finished rice as per specifications. The operation selects the correct sizes of broken rice that are present. Sizes of the broken rice in a rice quality measurement are generally identified by decimal numbers on tenth basis of the rice length. For example, 0.8, 0.75, 0.65, 0.5, 0.4, 0.3, 0.2, and 0.1. Some use fractional whole numbers by eighth basis such as 10/10, 9/10, 8/10, 7/8, 6/8, 5/8, 4/8, 3/8, 2/8, and 1/8. In general, 3/4 of the rice length is regarded as head rice. However, top quality long grain types of rice are always larger than this ratio. As accuracy of content and size are very critical in price negotiation, grading is carried out in several steps in a specified sequence by using different methods of segregation.

13.10 Recovery of Head Rice in the Milled Rice

The rice received after whitening and polishing is called milled rice representing the whole mass including the broken rice. Of the two, head rice is the deciding factor for the final yield of finished rice. The ratio of the head rice must be the highest possible in the milled rice. The type and design of the milling machinery influence the milling results (Bhattacharya and Subba 1966; Appiah et al. 2011). However, other factors such as plant design, processing method, and ambient conditions also play a role and affect the rate of broken rice. From the material side, immature grains and cracked kernels are serious factors and effectively bring down the head rice recovery.

13.10.1 Double Losses in High Quality Rice

Every time when high-quality rice is produced, the process involves a certain rate of bran removal followed by consequential damages leading to production of broken rice. In the past, so-called high-quality rice are made up of 7–8% bran removal (7–8% weight loss from input paddy). Nowadays at 11% weight loss, the head rice recovery in the milled rice has reduced substantially, much lower than that of the past (Fig. 13.6). In today's milled rice, head rice is hardly found at 65% while in the past it was 75% or more (Table 13.1). The disappearance of 10% head rice in the milled rice is the clear reason for the yield loss for the same quality of milling in the current scenario.

13.11 Rice Specifications

Rice specification describes details what a certain rice quality wants and what are not part of its specification. The contents of head rice and broken rice are the main criteria. A rice quality is formed based on these two components in which the former is the principal fraction and stands as the higher portion. The higher the quality grade, the higher is the content of the head rice. Given below are important quality descriptions which affect the price and consumers' perception.

1. Content of head rice
2. Content of whole kernel (part of head rice)
3. Content of broken rice
4. Size of broken rice permitted

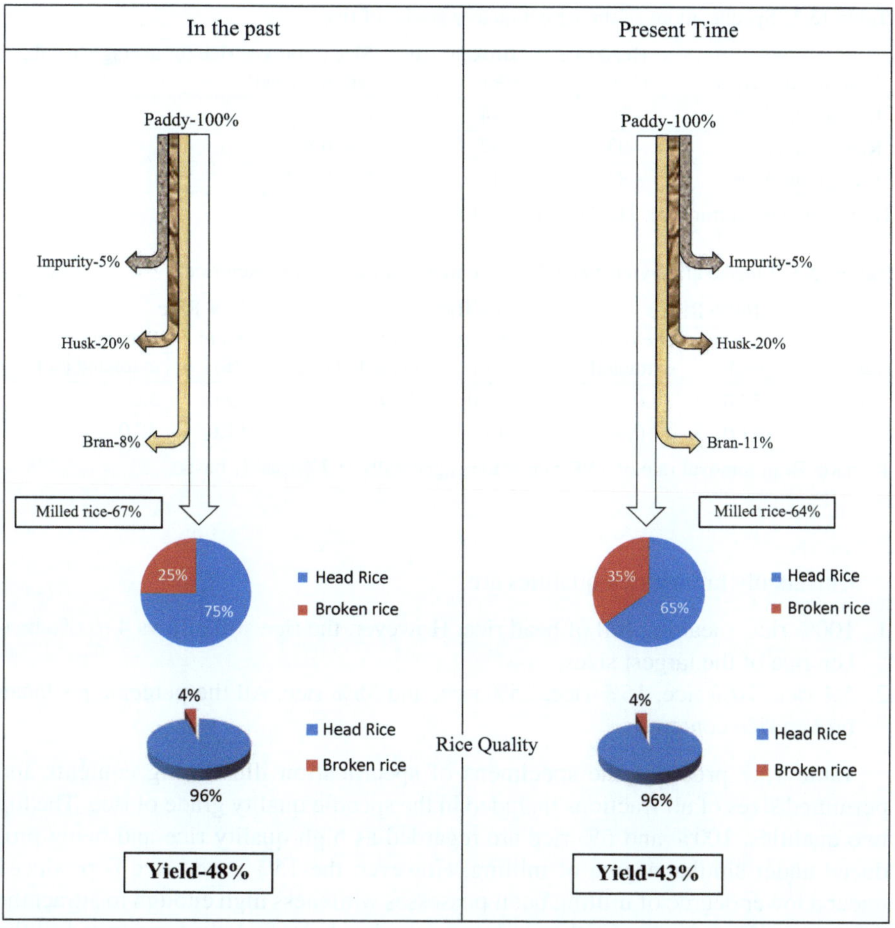

Fig. 13.6 Yield comparison between milled rice from the past and present

Table 13.1 Head rice recoveries for the same high-quality rice in the past and at present

Input paddy (%)	Impurity (%)	Husk (%)	Bran (%)	Milled rice (%)	Broken rice (%) in milled rice
Past 100 % Rice Quality	4.0	21.0	6.5	68.5	*75 (25)*[a]
Present 100% Rice Quality	-Same-	-Same-	+11.0	65.0	*65 (35)*[a]

[a]Ratio of broken rice in the milled rice the two figures make 100%.

Qualities of the rice are named based on the content of the broken rice. The system enables immediate determination to compare the price and quantity of broken rice in finished rice. This is because the broken rice is a cheaper item compared to the head rice.

Table 13.2 Specifications of three high-quality grades of rice

High-quality grade	Head rice (%)	Broken rice (%)	Size of broken rice (of average length, whole kernel)
Rice grade, 100%	+96	−4	0.5–0.8
Rice grade, 5%	+93	−7	0.4–0.8
Rice grade, 15%	+83	−17	0.3–0.7

Notification of Commerce, Thailand B. E 2540 (1997)

Table 13.3 Finished rice yields and the percentages of extracted broken rice

Bran removal	100% Rice		5% Rice		15% Rice	
	Yield (%)	Broke rice extracted (%)	Yield (%)	Broken rice extracted (%)	Yield (%)	Broken rice extracted (%)
8.0	50.0	*14.0*	52.0	*12.0*	55.0	*5.0*
11.0	41.0	*22.0*	43.0	*19.0*	51.0	*10.0*

Remark: Bran removal rate of 15% rice grade is generally at 7% (paddy basis)

Universally known rice qualities are:

1. 100% rice, meaning full of head rice. However, the rice still allows 4 to 6% broken rice of the largest sizes.
2. 5% rice, 10% rice, 15% rice, 25% rice, and 35% rice. All the percentages mean broken rice contents.

Table 13.2 provides the specimens of specification illustrating contents and permitted sizes of all fractions included in the specific quality grade of rice. The top two qualities, 100% and 5% rice are regarded as high-quality rice and being produced under similar degree of milling. However, the 15% grade rice is produced under a lower degree of milling but it possesses whiteness high enough to attract the consumers. Comparatively, the content of the head rice is lower compared to the former two higher quality grades of rice and it allows higher content of broken rice. Therefore, the yield of the rice is higher (Table 13.3).

Although the contents of head rice and broken rice are more important for the pricing, the criteria to be considered first and of the utmost importance is the milling degree. Milling degree descriptions are given below for the respective quality grade but some degrees are entitled for two or three quality grades which are very close together.

1. "Ordinarily well-milled," for the three lowest qualities including 25% lower grade. There are 25% and 45% rice lower than 25% rice.
2. "Reasonably well-milled," from super grade of 25% rice to 15% quality.
3. "Well-milled," for the rice of 10% and 5% grades.
4. "Extra well-milled," for three grades of 100% rice such as grades, A, B, and C.

The definitions are vague and not specified by any type of measurements, therefore, over-milling usually occurs.

13.12 Yield of Finished Rice

The term yield in the rice milling refers to the quantity of finished rice in relation to the input paddy usually shown in percentages. Some countries have their own local descriptions but still convertible to percentages.

Yield can be fluctuated by the following factors regardless of the quality grade.

1. Recovery of milled rice after whitening and polishing
2. Content of head rice in the milled rice
3. Content of head rice in the finished rice

13.13 Summary

Harvested paddy should not be estimated as actual food available for the people's consumption because the quantity slides down even before milling through drying and handling losses. The amount is reduced by around 20% when it is ready for conversion into edible rice.

Further irrelevance is the ready-reckoners used for converting paddy into rice. The data 60 or 65% are in fact replica of milled rice recovery, not finished rice. Consumers' trend that tend to not eat broken rice should be drawn into consideration, as well as the actual practice of removing broken rice from milled rice. About 10–25% of input paddy quantity is being thrown out in the milling as broken rice resulting in the average yield to be less than the existing converting figure of 60%. One cause of reduction is today's milling degree which is higher than in the past. Rice eaters should know that increased whitening and polishing do not necessarily give superior eating quality (Roya et al. 2008).

The illustration below shows how much of rice disappeared from human consumption from the same quality grade comparing past and at present time. This is the present scenario. Out of this, the rice volume loss and the feedable number of population can be estimated as given below.

World paddy production	650 million per year
Standard conversion rate	0.60
Milled rice availability	390 million tons per year
Estimated high-quality rice (assumed 20% of milled rice)	78 million tons
By the average yield of 43%, paddy consumed for 84 million ton rice would be	181 million tons
If 181 million tons of paddy was milled to the same quality by the past bran removal rate of 8% and the yield, 48%, white rice would be	87 million tons
So, extra rice is (87–78)	9 million tons
Number of people that can be fed by extra rice at the rate of per capita consumption of 150 kg	60 million people for 1 year

13.14 Conclusion and Future Prospects

Practically, it is impossible to change the consumer's preference for white rice. However, the craving for rice-beauty seems to be changeable if they are convinced by the losses and the incoming food shortage. Nutritional losses and the diseases associated with highly polished rice could also be helpful. Possible reduction of the price by the lower power consumption if qualities are brought down to standard level can be considered.

On the other hand, technical solution should be sought by focusing on the breeding of thin-husked paddies and shallow undulating varieties so that weight and material losses would be less even though the present thirst for rice-beauty persists. At the same time, regulatory bodies concerning the rice quality standards should also include eating values in its regulations and standardization programs by highlighting the benefits. Therefore, optimum degree of milling could be determined on the basis of eating quality which will in turn reduce the losses.

On the other hand, eating of low-quality rice is not uncommon practice, because some countries had exercised the practice when they were facing food shortages. Government of India had once recommended 4 to 5% bran removal only (Bhole 1976).

Citing these examples, suggested efforts to save the food in the time of eminent food insufficiency would not be vain attempts. Otherwise, governments would face a time when they have to punish their own people whose behavior seems to be wasting food, even for a tiny amount. However, this kind of regulatory actions would create social instabilities and eventually diminish the peace and harmony of the world.

References

Abbas A, Murtaza S, Aslam F, Khawar A, Rafique S, Naheed S (2011) Effect of processing on nutritional value of rice (Oryza sativa). World J Med Sci 6(2):68–73

Appiah F, Guisse R, Dartey PK (2011) Post-harvest losses of rice from harvesting to milling in Ghana. J Stored Prod Postharvest Res 2(4):64–71

Baradi MAU, Elepaño AR (2012) Aroma loss in RIce as affected by various conditions during postharvest operations. Philipp Agric Scientist 95:260–266

Bhashyam MK, Srinivas T (1984) Varietal difference in the topography of rice grain and its influence on milling quality. J Food Sci 49(2):393–395

Bhattacharya KR (1980) Simplified determination of water-insoluble amylose content of rice. Starch 32:409–411

Bhattacharya KR, Ali SZ (2015) Introduction to rice-grain technology. Woodhead Publishing India Pvt Ltd, New Delhi

Bhattacharya KR, Subba RPV (1966) Processing conditions and milling yield in parboiling of rice. J Agric Food Chem 14(5):473–475

Bhole NG (1976) Study of rice milling phenomenon in abrasive rice polishers. PhD Thesis (unpublished) submitted to Agric. Engg. Dept. IIT Kharagpur

Buerkle T (2006) 40 countries face food shortages worldwide. FAONewsroom. http://www.fao.org/Newsroom/en/news/2006/1000416/index.html

FAO (1948) The state of food and agriculture-1948, p 50. http://www.fao.org/docrep/016/ap636e/ap636e.pdf

FAO (2009) How to feed the world in 2050, p 2. http://www.fao.org/fileadmin/templates/wsfs/docs/expert_paper/How_to_Feed_the_World_in_2050.pdf
FAO (2015) World food and agriculture. Statistical yearbook 2015. The Food and Agriculture Organization of the United Nations, p 369
Ferrero A, Tinarelli A (2007) Rice cultivation in the E.U. ecological conditions and agronomical practices. In: Pesticide risk assessment in rice paddies, pp 1–24
Heinemann RJB, Fagundes PL, Pinto EA, Penteado MVC, Lanfer-Marquez UM (2005) Comparative study of nutrient composition of commercial brown, parboiled and milled rice from Brazil. J Food Compos Anal 18:287–296
Koga Y (1969) Drying, husking and milling in Japan IV and V. Farming, Japan
Liang Z, Yaoming L, Xu L, Zhao Z, Tang Z (2017) Optimum design of an array structure for the grain loss sensor to upgrade its resolution for harvesting rice in a combine harvester. Biosyst Eng 157:24–34
Okeke CG, Oluka SI (2017) A survey of rice production and processing in South East Nigeria. Niger J Technol (NIJOTECH) 36:227–234
Pan T, Zhao L, Lin L, Wang J, Liu Q, Wei C (2017) Changes in kernel morphology and starch properties of high-amylose brown rice during the cooking process. Food Hydrocoll 66:227–236
Roya P, Ijiri T, Okadome H, Nei D, Orikasa T, Nakamura N, Shiina T (2008) Effect of processing conditions on overall energy consumption and quality of rice (Oryza sativa L.) J Food Eng 89(3):343–348
The World Bank (2014) Myanmar: capitalizing on rice export opportunities. Southeast Asia Sustainable Development Unit East Asia and Pacific Region, p 9
Wimberley JE (1983) Paddy rice, postharvest industry in developing countries. The International Rice Research Institute, Los Banos

Chapter 14
Elicitors and Their Roles in Plant Defence Against Pathogens Particularly Basidiomycetes

Sathyapriya Hamid and Mui-Yun Wong

14.1 Introduction

Microbes in nature establish mutualistic or harmful interactions with plants depending on their lifestyles. While beneficial microbes enhance defence responses of host plants towards pathogens, pathogens affect the plant physiology in detrimental ways (Zeilinger et al. 2015). These interactions often involve recognition of microbial signals by plant receptors. These signals are generally termed as elicitors and have become a major area of interest within the field of plant–microbe interactions.

Typically, bioactive signals such as pathogen-associated molecular patterns (PAMPs), microbe-associated molecular patterns (MAMPs) or danger-associated molecular patterns (DAMPs) are perceived by surface-localised pattern recognition receptors (PRR) of plants (Zipfel 2014). PRRs are often coupled with other receptors particularly receptor-like kinases (RLKs) or receptor-like proteins (RLPs) to form molecular complexes to facilitate ligand recognition, signal transduction or regulation PRR-triggered immunity against pathogens (Monaghan and Zipfel 2012). Within minutes, PAMP-PRR perception triggers the first level of inducible defences including accumulation of reactive oxygen species (ROS) in plasma membrane (Boller and Felix 2009). This plant defence response is known as PAMP-triggered immunity (PTI). Failure to overcome these defence mechanisms results in

S. Hamid
Laboratory of Plantation Science and Technology, Institute of Plantation Studies, Universiti Putra Malaysia, 43400 Serdang, Selangor, Malaysia

M.-Y. Wong (✉)
Laboratory of Plantation Science and Technology, Institute of Plantation Studies, Universiti Putra Malaysia, 43400 Serdang, Selangor, Malaysia

Department of Plant Protection, Faculty of Agriculture, Universiti Putra Malaysia, 43400 Serdang, Selangor, Malaysia
e-mail: muiyun@upm.edu.my

S.N.A. Abdullah et al. (eds.), *Crop Improvement*,
DOI 10.1007/978-3-319-65079-1_14

activation of effective PTI and prevention of pathogen proliferation in the apoplast (Gohre and Robatzek 2008). In general, successful pathogenic fungi deliver various effectors (race-specific elicitors) into host cells to suppress the PTI components. Upon perception of these effectors by resistance (R) proteins of plants, a second layer of defence known as effector-triggered immunity (ETI) is actuated (Wiesel et al. 2014). Consequently, localised cell death (hypersensitive response) takes place to kill both the invading pathogen and infected plant cells (Jones and Dangl 2006).

Elicitor compounds are remarkably important in the development of local and systemic resistance in plants. In general, plants utilise systemic acquired resistance (SAR) or induced systemic resistance (ISR) to initiate systemic immunity during their interactions with elicitors. SAR which relies on salicylic acid (SA) signalling pathway (Gao et al. 2015) provides robust and long-term protection against an array of pathogens. It is typically associated with programmed cell death that is confined to the infected area and accumulation of pathogenesis-related (PR) proteins (Durrant and Dong 2004). In contrast, ISR depends on jasmonic acid (JA) and ethylene (ET) pathways and is not associated with programmed cell death (Wiesel et al. 2014).

14.2 Plant Defence Elicitors

Elicitors are fast becoming the key instruments in plant disease control due to their competence in activating or inducing resistance in plants. The term "elicitor" was initially used only for molecules capable of inducing the phytoalexin production in plant cells (Keen et al. 1972; Ebel and Cosio 1994). However, it is now often used for any signal-inducing compounds recognised by the innate immune system involved in priming and/or induction of defence responses regardless of its origin (Walters et al. 2013; Wiesel et al. 2014).

In general, elicitors comprise both microbe- or plant-derived compounds and synthetically made substances such as acibenzolar-S-methyl (ASM) and beta-amino-butyric acid (BABA) mimicking plant-derived compounds. Microbe-derived compounds, the exogenous elicitors are often regarded as pathogen- or microbe-associated molecular patterns (PAMPs or MAMPs) depending on the nature of the microbes (Maffei et al. 2012). On the other hand, plant-derived compounds, the endogenous elicitors are released within plant cells after attacks from herbivores or microbes, and thus termed as damage- or danger-associated molecular patterns (DAMPs) (Mazzota and Kemmerling 2011; Henry et al. 2012).

Typically, these compounds can be divided into two groups, namely general elicitors or race-specific elicitors based on the specificity of elicitation of defence responses in plants. While general elicitors induce defence responses both in host and non-host plants, race-specific elicitors which are better known as effectors trigger defence responses only in specific host plant species (Montesano et al. 2003). Avirulence (*avr*) genes encoding effectors from certain race of pathogens and their corresponding resistance (*R*) genes in specific host cultivars have been identified as major contributing factors for gene-for-gene resistance whereby the absence of products produced by any of these genes can cause disease (Cohn et al. 2001).

Elicitors are structurally distinct and can be grouped into diverse classes of compounds mainly oligosaccharides, proteins, glycoproteins and lipids. Montesano et al. (2003) have reviewed various classes of plant defence and defence-like responses-eliciting compounds that are present in plant pathogenic microbes. Here, we have listed oomycete-, fungal- and bacteria-derived elicitors which have been extensively studied for their defence-eliciting capability on various plant species (Table 14.1). According to Wiesel et al. (2014), elicitors provide a signal of potential pathogen attack to the host plant irrespective to their nature and origin. In many cases, elicitor compounds exist as structural components of the microbes. These include chitin, glucan and mannoprotein fragments from fungi (Wiesel et al. 2014), flagellin from the Gram-negative bacteria (Felix et al. 1999) and viral coat protein from tobacco mosaic virus (Culver and Dawson 1989; Wahlroos and Susi 2015). On the other hand, pathogen-derived elicitors such as harpins secreted through type III secretion system in certain Gram-negative bacteria and pectic enzymes which degrade plant cell walls including endopolygalacturonases and pectate lyases from several phytopathogenic bacteria have been reported as crucial virulence factors in some plant–pathogen interactions (Montesano et al. 2003; Choi et al. 2013; Benali et al. 2014). The latter examples are able to release plant cell wall fragments particularly oligogalacturonides (OGs) which act as DAMPs within plant cells upon infections (Boller and Felix 2009; Galletti et al. 2009; Vorhölter et al. 2012; Ferrari et al. 2013). Interestingly, plants are able to recognise PAMPs and MAMPs as "non-self" or DAMPs as "damaged or infectious self"-signals using specialised receptors (Monaghan and Zipfel 2012; Heil and Land 2014; Zipfel 2014). As a result, defence signalling pathways leading to stimulation of plant basal immunity are triggered.

14.3 Elicitor Perception Systems in Plants

Bioactive signal molecules (PAMPs, MAMPs or DAMPs) produced or released during plant–microbe interactions are usually recognised by PRRs located in the plasma membrane of plants. Among these PRRs, the role of RLKs and RLPs in PTI has been well studied (Monaghan and Zipfel 2012). RLKs comprise an ectodomain of leucine-rich repeats (LRRs) which is involved in the perception of conserved microbial molecules, and an intracellular kinase domain which is involved in signal transduction through mitogen activated protein kinase (MAPK) cascades (Jones and Dangl 2006; Coll et al. 2011). Meanwhile, RLPs are lacked of a cytoplasmic signalling domain, thus may require a protein kinase as their signalling partner to initiate downstream signalling in plants (Liebrand et al. 2014). This was exemplified in the studies undertaken by Liebrand et al. (2013) and Zhang et al. (2013a, b).

The role of LRR-RLK as PRR in Arabidopsis was first demonstrated using forward genetic screen by Gómez-Gómez and Boller (2000). In their ground-breaking study, they identified Flagellin Sensing 2 (FLS2) of *Arabidopsis thaliana* as the receptor involved in the recognition of flg22, a bacteria-derived elicitor flagellin. Since then, many studies have concentrated on identifying LRR-RLK receptors in

Table 14.1 Summary of oomycete-, fungal- and bacteria-derived elicitors which have been extensively studied for its defence-eliciting capability on various plant species

Elicitor producing organism	Elicitor[a]	Chemical class of elicitor	Effects shown in	Mode of action[b]	Reference
Oomycetes					
Aphanomyces euteiches	Glucan-chitosaccharides	Oligosaccharides	*Medicago truncatula*	Induction of PR genes and nuclear calcium oscillations	Nars et al. (2013)
Hyaloperonospora arabidopsidis	HaNLP3 (nlp24)	Protein (peptide)	Arabidopsis	Activation of immunity-related gene expression, and resistance to downy mildew	Oome et al. (2014)
Phytophthora colocasiae	No description	Glycoprotein	*Colocasia esculenta*	Formation of HR lesions, induction of lipoxygenase activity and SAR	Mishra et al. (2009)
Phytophthora cryptogea	Cryptogein	Protein	Tobacco	Induction of signal pathway leading to the oxidative burst and pH changes	Kasparovsky et al. (2004)
Phytophthora infestans	Eicosapentaenoic acid	Fatty acid	Potato	Elicitation of defences and reduced symptoms of late blight	Henriquez et al. (2012)
	β-Glucans	Oligosaccharides	Potato	Suppression of host defence responses	Ozeretskovskaya et al. (2001)
	Elicitin	Protein	Potato	Activation of PTI	Du et al. (2015)
	INF1 (elicitin)	Protein	Tomato	Activation of JA and ET signalling pathways and enhanced resistance to bacterial wilt pathogen	Kawamura et al. (2009)
	EPIC1, EPIC2B	Protein	Tomato	Inhibition of the tomato protease activities	Tian et al. (2007), Kaschani et al. (2010)
Pythium oligandrum	Oli-D1, Oli-D2	Protein	Tobacco, tomato	Induction of HR, accumulation of ROS and autofluorescence production and activation of immune responses through the JA/ET-mediated signalling pathway	Ouyang et al. (2015)

Phytophthora parasitica	CBEL	Glycoprotein	Arabidopsis	Induction of necrosis and immune responses	Larroque et al. (2013)
	OPEL	Protein	Tobacco	Cell death, callose deposition, production of ROS and induced expression of PTI markers and SA-responsive defence genes	Chang et al. (2015)
Phytophthora sojae	Pep-13	Oligopeptide	Potato	Induction of salicylic acid-dependent and -independent defence reactions	Halim et al. (2004)
	XEG1	Protein	Soybean	Virulence factor; elicitation of PTI	Ma et al. (2015)
Fungi					
Alternaria tenuissima	Hrip1	Protein	Tobacco	Induction of early defence responses and activation of SAR	Kulye et al. (2012)
	PeaT1	Protein	Tobacco	Enhancement of SAR against TMV	Zhang et al. (2011)
Botrytis cinerea	PebC1	Protein	Tomato	Induction of PAL, POD, PPO and induced resistance against gray mold fungus	Zhang et al. (2010)
	BcSpl1	Protein	Tomato, tobacco, Arabidopsis	Induction of HR and defence genes	Frías et al. (2011)
	BcGs1	Glycoprotein	Tomato, tobacco	Induction of necrosis on leaves	Zhang et al. (2015)
Ceratocystis platani	CP	Protein	*Platanus acerifolia*	Induction of cell death through programmed cell death mechanism	Lombardi et al. (2013)
Ceratocystis populicola	Cerato-populin (Pop1)	Protein	*Platanus acerifolia*	Induction of cell death through programmed cell death mechanism	Lombardi et al. (2013)
Cladosporium fulvum	Ergosterol	Lipid	Tomato, tobacco	Induction of ROS, changes of ion fluxes and production of the phytoalexin	Granado et al. (1995), Kasparovsky et al. (2003), Vatsa et al. (2011)
	Avr2	Protein	Tomato	Inhibition of apoplastic proteases to avoid activation of host immunity	Rooney et al. (2005)
	Avr4, Ecp6	Protein	Tomato	Suppression of host resistance by protecting fungal cell walls against plant chitinases	Van der Burg et al. (2006), Sánchez-Vallet et al. (2013)

(continued)

Table 14.1 (continued)

Elicitor producing organism	Elicitor[a]	Chemical class of elicitor	Effects shown in	Mode of action[b]	Reference
Colletotrichum lindemuthianum	CLPG1 (endoPG)	Polypeptide	Tobacco	Induction of AOS and β-1,3-glucanase activity followed by leaf necrosis	Boudart et al. (2003)
Fusarium oxysporum	Nep1	Protein	Arabidopsis	Induction of several classes of genes involved in ROS production, signal transduction, ET biosynthesis, membrane modification, apoptosis and stress	Bae et al. (2006)
Fusarium oxysporum f. sp. *lycopersici*	Cerebroside C	Glycosphingolipid	Chilli	Stimulation of early hydrogen peroxide accumulation, production of plant defence-related enzymes, accumulation of capsidiol and enhanced resistance to *Colletotrichum capsici*.	Naveen et al. (2013)
Fusarium oxysporum strain CS-20	CS20EP	Polypeptide	Tomato	Induction of reversible alteration of extracellular pH, increased chitinase activity, enhancement of PR1 gene expression	Shcherbakova et al. (2015)
Leptosphaeria maculans	Cell wall components	Oligosaccharides	Oilseed rape	Induction of PR1 expression, hydrogen peroxide accumulation and enhanced resistance	Kim et al. (2013)
Magnaporthe grisea	β-1,3-Glucan fragments	Oligosaccharides	Rice	Production of phytoalexins	Yamaguchi et al. (2002)
	PemG1	Protein	Rice	Induction of transient expression of PR genes, and increased accumulation of hydrogen peroxide, and activation of SAR	Peng et al. (2011)
	MgSM1	Protein	Arabidopsis	Induction of HR, enhancement of broad-spectrum resistance with up-regulation of defence-related genes	Yang et al. (2009)

Magnaporthe oryzae	MoHrip1, MoHrip2	Protein	Tobacco, rice	Enhancement of systemic resistance to the rice blast pathogen	Chen et al. (2012, 2014)
	MSP1	Protein	Rice	Induction of hydrogen peroxide, cell death and expression of defence-related genes	Wang et al. (2016)
	Slp1	Protein	Rice	Suppression of chitin-induced plant immune responses and sequesters chitin oligosaccharides to prevent PTI	Mentlak et al. (2012)
	Chitin oligosaccharides	Oligosaccharides	Rice	Induction of cell death and increased production of ROS and nitrogen species	Kishimoto et al. (2010)
Mycosphaerella graminicola	Mg3LysM	Protein	Wheat	Suppression of chitin-induced plant immune responses	Marshall et al. (2011)
Trichoderma asperellum	EplT4	Protein	Soybean	Induction of glucanase, chitinase III-A, cysteine proteinase inhibitor, and peroxidase genes, activation of early cellular events and increased resistance to *Cercosporidium sofinum*	Wang et al. (2013)
Trichoderma longibrachiatum	Hydrophobin	Protein	Tomato	Induction of systemic resistance and activation of ROS accumulation, phytoalexin production and PR protein activity	Ruocco et al. (2015)
Trichoderma virens	Sm1, Sm2	Protein	Cotton, maize	Activation of ROS production, phenolic compounds accumulation and systemic expression of defence-related genes	Djonovic et al. (2006), Gaderer et al. (2015)
Trichoderma virens strain Gv29-8	TvBI, TvBII	Peptides	Cucumber	Induction of systemic protection to *Pseudomonas syringae* pv. *lachrymans*, induction of antimicrobial compounds and up-regulation of hydroxyperoxide lyase, PAL and PO	Viterbo et al. (2007)
Trichoderma atroviride	Epl1	Protein	Tomato	Induction of peroxidase expression and systemic disease resistance	Salas-Marina et al. (2015)

(continued)

Table 14.1 (continued)

Elicitor producing organism	Elicitor[a]	Chemical class of elicitor	Effects shown in	Mode of action[b]	Reference
Trichoderma viride	Xylanase	Protein	Tomato, tobacco, rice	Induction of HR and regulation of phytoalexin biosynthesis	Ron and Avni (2004), Kurusu et al. (2010), Hamada et al. (2012)
Ustilago maydis	Pit2	Protein	Maize	Inhibition of maize protease activities	Mueller et al. (2013)
	Pep1	Protein	Maize	Inhibition of the PO-driven oxidative burst and suppression of early immune responses	Hemetsberger et al. (2012)
Verticillium dahliae	PevD1	Protein	Tobacco	Induction of hydrogen peroxide accumulation, extracellular-medium alkalisation, callose deposition, phenolics metabolism, and lignin synthesis	Wang et al. (2012)
Bacteria					
Bacillus cereus C1L	Dimethyl disulphide	Disulphide	Tobacco, corn	Activation of ISR	Huang et al. (2012)
Bacillus subtilis/ amyloliquefaciens complex	Surfactin	Cyclic lipopeptide	Tobacco, tomato	Induction of oxidative burst and activation of ISR	Cawoy et al. (2014)
Bacillus thuringiensis	CspD	Peptide	Tobacco	Enhancement of resistance against TMV and *Alternaria longipes*	Kromina and Dzhavakhiia (2006)
Burkholderia gladioli	EPS	Polysaccharide	Cucumber	Activation of ISR	Park et al. (2008)
Escherichia coli GI826	EF-Tu (efl18)	Peptide	Arabidopsis	Induction of an oxidative burst and ethylene biosynthesis	Kunze et al. (2004)
Pseudomonas aeruginosa	Rhamnolipid	Glycolipid	Grape	Activation of early defence responses, HR-like response generation, expression of defence genes and enhancement of resistance	Varnier et al. (2009)
Pseudomonas fluorescens WCS374r	Pseudobactin	No description	Rice	Induction of systemic resistance against *M. oryzae*	De Vleesschauwer et al. (2008)

Pseudomonas syringae pv tabaci	Flagellin	Peptide	Tomato	Induction of medium alkalinisation, oxidative burst and ethylene biosynthesis	Felix et al. (1999)
Staphylococcus aureus	PGN	Sugar-amino acid polymer	Tobacco, tomato	Activation of PTI	Nguyen et al. (2010)
Xanthomonas campestris pv. *campestris* strain 8004	LPS	Glycolipid	Pepper	Induction of acidic β-1,3-glucanase expression	Newman et al. (2002)
Xanthomonas oryzae pv. *oryzae*	$Hpa1_{Xoo}$ (harpin)	Polypeptide	Cotton	Induction of hydrogen peroxide production and activation of SAR	Miao et al. (2010)
	Ax21	Peptide	Rice	Activation of XA21-mediated immunity	Lee et al. (2009)

[a]*Avr2* Avirulence protein 2, *Avr4* Avirulence protein 4, *BcSpl1 Botrytis cinerea* Secreted LysM protein, *CBEL* Cellulose-binding elicitor lectin, *CP* Cerato-platanin, *CspD* Cold shock protein D, *Ecp6* Extracellular protein 6, *EF-Tu* Elongation factor Tu, *endoPG* Endopolygalacturonase, *EPS* Exopolysaccharide, *HaNLP3 Hyaloperonospora arabidopsidis* necrosis and ethylene-inducing peptide 1 (Nep1)-like protein, *Hrip1* Hypersensitive response-inducing protein1, *LPS* Lipopolysaccharide, *LysM* Lysin motif, *MoHrip Magnaporthe oryzae* hypersensitive response-inducing protein, *MSP1 Magnaporthe oryzae* snodprot1, *Nep1* Necrosis- and ethylene- inducing protein1, *Oli-D1* Oligandrin1, *Oli-D2* Oligandrin2, *PGN* Peptidoglycan, *Slp1* Secreted LysM Protein1, and *XEG1* xyloglucan-specific endo-β-1,4-glucanase

[b]*AOS* Active oxygen species, *ET* Ethylene, *HR* Hypersensitive response, *ISR* Induced systemic resistance, *JA* Jasmonic acid, *PAL* Phenylalanine ammonia-lyase, *PO* Peroxides, *PPO* Polyphenol oxidase, *PR* Pathogenesis-related, *PR1* Pathogenesis-related protein 1, *PTI* Pattern-triggered immunity, *ROS* Reactive oxygen species, *SA* Salicylic acid, *SAR* Systemic acquired resistance, and *TMV* Tobacco Mosaic Virus

various plants to better understand the mechanisms of PRRs and their roles in PTI. Besides, flg22/FLS pair, Ax21/XA21 (Lee et al. 2008, 2009) and elf18/26 peptides/ EFR1 (Zipfel et al. 2006) also have been well studied. To date, several putative plant PRRs have been recognised and their molecular-recognition mechanisms have been described (Zipfel 2014; Trdá et al. 2015). Here, we have listed the well-characterised PRR proteins involved in perception of PAMPs/MAMPs/DAMPs from selected model plants, namely rice, Arabidopsis and tomato (Table 14.2). PRRs containing extracellular lysin (LysM) motifs including chitin oligosaccharide elicitor-binding protein (CEBip) and chitin elicitor receptor-like kinase (CERK1) are highly sensitive to chitin and have been reported to involve in the perception of fungal-derived chitin in rice (Miya et al. 2007; Shimizu et al. 2010). However, Zipfel (2014) has highlighted the differences in chitin recognition mechanism of rice and Arabidopsis. Interestingly, OsCERK1 has been found to be important for both peptidoglycan (PGN) and chitin signalling initiated by two chitin- and PGN-sensing proteins, namely OsLYP4 and OsLYP6 in rice (Ao et al. 2014).

With regard to LRR-RLPs, tomato Cf proteins which correspond to the avirulence proteins encoded by *Avr* genes of tomato leaf mold disease-causing pathogen, *Cladosporium fulvum* have been vastly reported (Jones et al. 1994; Wulff et al. 2001; Rooney et al. 2005; van den Burg et al. 2006). More recently, BRI1-ASSOCIATED KINASE 1/SOMATIC EMBRYOGENESIS RECEPTOR KINASE 3 (BAK1/ SERK3) was shown to be essential for Cf-4-mediated resistance of tomato against *C. fulvum* (Postma et al. 2016). Besides, ethylene-inducing xylanase (EIX)/ LeEix1 and LeEix2, and Ve1/ Ave1 pairs have been reported in tomato–fungus interaction (Ron and Avni 2004; de Jonge et al. 2012). In addition, receptors involved in detection of a proteinaceous fungal PAMPs as well as fungal endopolygalacturonase have been reported in Arabidopsis (Jehle et al. 2013; Zhang et al. 2013a, b, 2014).

Perception of DAMPs by plant PRRs has been well-reviewed (Heil and Land 2014; Zipfel 2014). Oligogalacturonides (OGs) released from plant cell walls during infection or wounding are perceived by wall-associated kinase 1 (WAK1) which contains epidermal growth factor (EGF) motif (Brutus et al. 2010; Zipfel 2014). In addition, AtPep peptides (DAMPs) have been found to be perceived by AtPEPR1 and AtPEPR2 in Arabidopsis and this perception is crucial for stimulation of systemic immunity in plants (Yamaguchi et al. 2006, 2010). More recently, Choi et al. (2014) have identified DORN1/LecRK-I.9, a plant receptor for extracellular ATP (eATP), a DAMP released on cell rupture during infection or wounding (Zipfel 2014). Expression of DORN1/LecRK-I.9 has improved resistance to *Phytophthora infestans* in tobacco and potato (Bouwmeester et al. 2014).

14.4 Activation of Plant Immunity by Elicitors

Upon perception of a bioactive signal molecule (PAMP, MAMP or DAMP) by its cognate receptor, PTI is activated (Jones and Dangl 2006). PTI is typically associated with calcium influx, accumulation of reactive oxygen and nitrogen species,

Table 14.2 Overview of pattern recognition receptors (PRRs) present in rice, Arabidopsis and tomato

Plant	Receptor[a]	Receptor type[b]	Ligand[c]	Ligand type[d]	Reference
Rice	CEBiP, OsCERK1	LysM-RLP	Chitin	Fungal PAMP	Kaku et al. (2006), Shimizu et al. (2010)
	OsLYP4, OsLYP6	LysM-RLP	Chitin, PGN	Fungal and bacterial PAMP/MAMP	Liu et al. (2013a, b), Ao et al. (2014)
	XA21	LRR-RLK	Rax proteins; Ax21	Bacterial PAMP	da Silva et al. (2004), Lee et al. (2008, 2009)
Arabidopsis	AtCERK1	LysM-RLK	Chitin	Fungal PAMP	Miya et al. (2007), Wan et al. (2008), Petutschnig et al. (2010)
	LYK4	LysM-RK	Chitin	Fungal PAMP	Wan et al. (2012)
	AtLYM1, AtLYM3	LysM-RK	PGN	Bacterial PAMP	Willmann et al. (2011)
	FLS2	LRR-RLK	Flagellin	Bacterial PAMP	Gómez-Gómez and Boller (2000)
	EFR1	LRR-RLK	Elf18/26	Bacterial PAMP	Zipfel et al. (2006)
	AtRLP1/ReMAX	LRR-RLP	Proteinaceous elicitor	Bacterial PAMP	Jehle et al. (2013)
	AtRLP30	LRR-RLP	SCFE1	Fungal PAMP	Zhang et al. (2013a, b)
	RBPG1/AtRLP42	LRR-RLP	Endopolygalacturonase	Fungal PAMP	Zhang et al. (2014)
	AtPEPR1, AtPEPR2	LRR-RLK	Atpep	DAMP	Yamaguchi et al. (2006, 2010)
	WAK1	EGF-RLK	Oligogalacturonides	DAMP	Brutus et al. (2010)

(continued)

Table 14.2 (continued)

Plant	Receptor[a]	Receptor type[b]	Ligand[c]	Ligand type[d]	Reference
Tomato	LeEix1, LeEix2	LRR-RLP	EIX	Fungal PAMP/MAMP	Ron and Avni, (2004)
	Ve1	LRR-RLP	Ave1	Fungal PAMP	De Jonge et al. (2012)
	Cf-4/Cf-9	LRR-RLP	Avr4/avr9	PAMP	Jones et al. (1994), Wulff et al. (2001), Rooney et al. (2005), van den Burg et al. (2006)

[a]*AtCERK1 Arabidopsis thaliana* chitin elicitor receptor kinase 1, *AtLYM1 Arabidopsis thaliana* LysM domain protein 1, *AtLYM3 Arabidopsis thaliana* LysM domain protein 3, *AtRLP30 Arabidopsis thaliana* receptor-like protein 30, *AtRLP1/ReMAX Arabidopsis thaliana* receptor-like protein 1 ReMAX, *CEBiP* Chitin elicitor-binding protein, *Cf-4 Cladosporium fulvum*-4 protein, *Cf-9 Cladosporium fulvum*-9 protein, *EFR1* EF-Tu receptor 1, *FLS2* Flagellin-sensitive 2, *LYK4* LysM-containing receptor-like kinase 4, *OsCERK1 Oryza sativa* chitin elicitor receptor kinase 1, *OsLYP4 Oryza sativa* LysM-containing protein 4, *OsLYP6 Oryza sativa* LysM-containing protein 6, *RBPG1/AtRLP42* RESPONSIVENESS TO BOTRYTIS POLYGALACTURONASE 1/*Arabidopsis thaliana* receptor-like protein 42, and *WAK1* Wall-associated kinase 1

[b]*LysM-RLP* Lysin motif-receptor-like protein, *LysM-RLK* Lysin motif-receptor-like kinase, *LysM-RK* Lysin motif-receptor kinase, *LRR-RLK* Leucine-rich repeat-receptor-like kinase, *LRR-RLP* Leucine-rich repeat-receptor-like protein, and *EGF-RLK* Epidermal growth factor-receptor-like kinase

[c]*Avr4* Avirulence 4, *Avr9* Avirulence 9, *Ave1* Avirulence on Ve1, *EIX* Ethylene-inducing xylanase, *PGN* Peptidoglycan, and *SCFE1* SCLEROTINIA CULTURE FILTRATE ELICITOR1

[d]*DAMP* Damage-associated molecular pattern, *MAMP* Microbe-associated molecular pattern, and *PAMP* Pathogen-associated molecular pattern

synthesis of phytoalexins, expression of PR proteins, activation of MAPK pathways and cell wall modifications. These basal defence responses are produced to confine the growth of pathogens (Wiesel et al. 2014) while activating systemic resistance in plants that provides boosted protection against pathogens in subsequent attacks (Henry et al. 2012).

In most cases, adapted pathogens counteract the plant's basal defence responses by secreting effector proteins into the plant cells (O'Leary et al. 2016). Successful suppression of PTI often results in effector-triggered susceptibility (ETS) as illustrated in the "zigzag" model by Jones and Dangl (2006) where the plants become susceptible to diseases. Nevertheless, effector-triggered immunity (ETI) is activated when a particular effector protein is recognised either directly or indirectly by its cognate resistance (R) protein in plants (Jones and Dangl 2006; Liu et al. 2013). ETI is an accelerated and augmented response (Coll et al. 2011) which typically results in HR, a localised cell death at the site of infection that kills both the invading pathogen and infected plant cells (Newman et al. 2013). Consequently, SAR in the host is activated (Jones and Dangl 2006) where the host plant becomes resistant to subsequent attacks from obligate biotrophic or hemibiotrophic pathogens. ETI is ineffective against necrotrophic pathogens (Glazebrook 2005). Several studies have shown that necrotrophic pathogens secrete a specialised group of necrotrophic effectors which are known as host-specific toxins (HSTs) that confer susceptibility to the pathogen upon recognition by host PRRs and leads to HST-induced programmed cell death (Friesen et al. 2008a, b; Oliver and Solomon 2010).

14.5 Roles of Fungal- and Oomycete-Derived Elicitors in Plant Defence Against Pathogens

14.5.1 Saccharides

Chitin, chitosan and chitooligosaccharides are among the well-studied fungal-derived oligosaccharide elicitors which trigger defence responses both in monocots and dicots. However, the molecular mechanisms by which plants recognise and transduce the signals of oligosaccharide elicitors remain elusive (Trouvelot et al. 2014). Chitin, a linear β-1,4 linked homopolymer of *N*-acetyl-D-glucosamine (GlcNAc), and chitosan, a deacetylated form of chitin, occur in fungal cell walls (Bowman and Free 2006; Pawlowski et al. 2016). Many chitin-binding proteins designated as CEBiP or CERK1 and its analogues have been reported in both monocotyledons and dicotyledons (Yin et al. 2016). In rice, chitin has been reported to induce the expression of defence-related genes encoding PR proteins, WRKY transcription factors, and R proteins (Wan et al. 2008). However, some pathogenic fungi have developed some strategies to suppress chitin-induced PTI in plants. For instance, *Colletotrichum graminicola*, *Puccinia graminis* f. sp. *tritici* and *Uromyces fabae* convert surface-exposed chitin to chitosan, a poor elicitor, by de-N-acetylation to reduce PAMP and protect themselves against fungal cell wall degradation by host chitinases (El Gueddari et al. 2002). On the other hand, chitin-binding lectin Avr4 in *C. fulvum* specifically binds to the fungal cell wall and protects the chitin against hydrolysis by host endochitinases during infection (Gohre and Robatzek 2008). Besides, extracellular protein 6 (Ecp6), a conserved LysM effector of *C. fulvum* sequesters chitin oligosaccharides released from invading fungal hyphae to avoid activation of host immunity through recognition of chitin by CEBip (de Jonge et al. 2010). Likewise, Slp 1, a chitin-binding protein with two LysM domains from *Magnaporthe oryzae* competitively inhibited CEBiP and suppressed chitin-induced immune responses in rice (Mentlak et al. 2012).

On the other hand, chitooligosaccharides (COSs) are released from fungal cell walls when plant chitinases act on chitins upon fungal infection (Wan et al. 2008). It has been reported that long-chain COSs (degree of polymerisation = 6–8) are strong inducers of plant innate immunity (Liang et al. 2014). Besides, COSs larger than hexose have also been reported to trigger expression of several defence-related genes, extracellular alkalinisation and ion efflux, changes in protein phosphorylation, transient membranes depolarisation and generation of ROS in suspension-cultured rice cells (Mishra et al. 2012). In addition, a carbohydrate moiety elicitor from *Blumeria graminis* f. sp. *tritici* has been reported to exhibit elicitor activity by inducing thaumatin-like proteins in a broad range of cereals (Schweizer et al. 2000).

Apart from that, the roles of oligosaccharide elicitors derived from the β-glucans of pathogenic oomycetes, such as *P. sojae* in plant defence, have been well reviewed (Shibuya and Minami 2001; Montesano et al. 2003; Fesel and Zuccaro 2016). However, least is known about fungal β-glucans cognate PRRs, the recognition as well as signal transduction in plants. The β-1,3-glucan fragments from *Magnaporthe grisea*, the rice blast pathogen induced production of phytoalexins in rice suspension-cultured cells (Yamaguchi et al. 2002). In general, several fungal pathogens and root endophytic fungi employ several strategies to avoid or attenuate β-1,3-glucan-activated resistance in plants during invasion. For example, Oliveira-Garcia and Deising (2013) reported that the expression of β-1,3-glucan synthetase, *GLS1*, in hemibiotrophic fungus, *Cl. graminicola* was tightly regulated in a temporal and spatial manner during infection of maize plants to prevent β-1,3-glucan-triggered immunity. Furthermore, a study done on two β-1,3-glucan synthesis genes, namely, *KRE5* and *KRE6* in biotrophic hyphae of *Cl. graminicola* also demonstrated that downregulation of production and prevention of exposure of branched β-1,3-β-1,6-glucan are essential to attenuate PTI in maize (Oliveira-Garcia and Deising 2016). Apart from that, Fujikawa et al. (2009, 2012) have reported that modulation of cellular surface by α-glucan apposition in *M. oryzae* prevented the fungal cell wall degradation and delayed PTI during invasion in rice. On the other hand, fungal-specific β-glucan-binding lectin, FGB1 secreted by root endophyte, *Piriformospora indica* was reported to alter fungal cell wall composition and suppress glucan-triggered reactive oxygen species (ROS) production in Arabidopsis, barley and tobacco, resulting in effective fungal colonisation (Wawra et al. 2016). Besides, laminarin, a non-defined β-1,3-glucan with β-1,6-glycosidic side branches from a marine brown alga, *Laminaria digitata*, was reported to be able to elicit an array of defence responses such as medium alkalisation, H_2O_2 production, salicylic acid accumulation and PR genes induction in several plants as reviewed by Fesel and Zuccaro (2016). However, sulphated laminarin did not induce SA-signalling but specifically activated expression of *PR1* in *A. thaliana* ecotype Columbia-0 (Col-0) (Ménard et al. 2004) and the reason for this remains elusive.

14.5.2 Xylanases

Endo-β-1,4-xylanases which are present both in pathogenic and non-pathogenic fungi show elicitor activities in plants. For instance, the ethylene-inducing xylanase (EIX) from *Trichoderma viride,* which binds to two putative membrane receptors, LeEix1 and LeEix2 can trigger defence in specific cultivars of tomato and tobacco (Ron and Avni 2004). Besides, EIX was reported to induce ROS accumulation and activation of OsMPK6, a MAPK in rice cells along with hypersensitive cell death (Kurusu et al. 2005). Kurusu et al. (2010) have reported that xylanase from *T. viride* (TvX)/EIX triggers mitochondrial dysfunction and hypersensitive cell death in rice. They also found that two Calcineurin B-like protein-interacting protein kinases (OsCIPK14/15) are essential for regulation of defence signalling pathway in rice cell cultures. Besides, endo-β-1,4-xylanase from *Botrytis cinerea*, Xyn11A has been found to induce necrosis that facilitates its infection process in tomato plant tissues (Noda et al. 2010). Interestingly, several studies have indicated that the defence elicitation and xylan-degrading activity of EIX protein are independent properties. For example, site-directed mutation of glutamic acid residue in xylanase II from *T. reesei* did not alter its elicitor activity in leaves of tobacco and tomato (Enkerli et al. 1999). Likewise, substitution of the pentapeptide TKLGE to VKGT on β-strand of the EIX protein inhibited the elicitation activity but not the β-1-4-endoxylanase activity in tobacco (Rotblat et al. 2002).

14.5.3 Sterols

Ergosterol, a 5,7-diene oxysterol is the main sterol present in most fungal cell membranes. It is often perceived as a general defence signal followed by activation of lipid-based signalling pathway in plants (Klemptner et al. 2014). To date, the exact mechanism by which ergosterol is recognised by plants and signal transduction pathway induced upon ergosterol recognition have not been reported yet. However, Granado et al. (1995) have proposed that plants have specialised receptors for ergosterol recognition based on elicitation of extracellular alkalinisation by ergosterol purified from spores of *C. fulvum* as compared to endogenous plant sterols, namely, campesterol, stigmasterol and sitosterol in tomato cells. This hypothesis was also supported by Amborabé et al. (2003) who reported the induction of different early events at the plasma membrane level by ergosterol as compared to chitosan in *Mimosa pudica*. Ergosterol has been reported to induce changes in membrane potential, generation of active oxygen species in addition to production of phytoalexins (Rossard et al. 2006; Klemptner et al. 2014) as well as modifications of H^+ fluxes in plants (Rossard et al. 2010). Interestingly, sterol was reported to be important in elicitation of defence responses including Ca^{2+}, extracellular alkalinisation and SAR by cryptogein, the proteinaceous elicitor from *Phytophthora* spp. in tobacco (Osman et al. 2001). On the other hand, involvement of signalling molecules, namely, nitric oxide and calmodulin-dependent protein kinases and SA were reported to activate defence-related genes following ergosterol treatment in tobacco (Dadakova et al. 2013).

14.5.4 Pectolytic Enzymes

Pectolytic enzymes degrade pectin, the constituent of plant cell wall releasing endogenous elicitors within plant cells upon infections. Endopolygalacturonases (EndoPGs) from *Colletotrichum lindemuthianum* have been shown to hydrolyse homogalacturonan domains of pectic polysaccharides resulting in cell wall disintegration and tissue maceration followed by induction of active oxygen species and β-1,3-glucanase activity in tobacco (Boudart et al. 2003). Besides, a functional study done on two endoPG-coding genes (*Tvg1* and *Tvg2*) from *T. virens* revealed the potential of endoPGs in eliciting ISR while establishing mutualistic interaction with tomato plants (Baroncelli et al. 2016). In addition, Herron et al. (2000) reported that PelC, a pectate lyase secreted by *Erwinia chrysanthemi* which cleaves the pectate component of plant cell walls is recognized as a virulence factor.

14.5.5 Peptaibols

Peptaibols are antibiotics and regarded as a novel class of fungal proteinaceous elicitors. For instance, chrysospermin, a 19-mer peptaibol from *Apiocrea chrosospermin* has been reported to reduce tobacco mosaic virus infection in tobacco plants through activation of plant resistance (Kim et al. 2000). Besides, alamenticin from *T. viride* was reported to trigger defence responses by synthesising SA and volatile compounds in lima bean (Engelberth et al. 2001) and *A. thaliana* (Viterbo et al. 2007) when applied exogenously. Viterbo et al. (2007) also reported the action of an 18-residue peptaibol from *T. virens* on cucumber plants against *Pseudomonas syringae* pv. *lachrymans*. This peptide was found to trigger plant defence through up-regulation of genes encoding peroxidase, hydroxyperoxide lyase and phenylalanine ammonia lyase.

14.5.6 Cerato-platanins

Cerato-platanins (CPs) which are exclusively present in fungi are also potent elicitors of primary defence responses. Cerato-platanin was first identified in *Ceratocystis fimbriata* f. sp. *platani*, an aggressive fungal pathogen causing canker stain on plane trees and was reported as a new phytotoxic protein (Pazzagli et al. 1999). To date, many studies have been carried out to elucidate the roles of CPs in plant defence and several of them are presented in Table 14.1. It includes Sm1 from *T. virens* (Djonovic et al. 2006; Salas-Marina et al. 2015), Pop1 from *Cr. populicola* and CP from *Cr. platani* (Lombardi et al. 2013), Epl1 from *T. atroviride* (Salas-Marina et al. 2015), MgSM1-CP-like protein from *M. grisea* (Yang et al. 2009), necrosis- and ethylene-inducing peptide (Nep1) from *Fusarium oxysporum* (Bae et al. 2006), *MoHrip* protein from *M. oryzae* (Chen et al. 2014), BcGs1 from *B. cinerea* (Zhang et al. 2015) as well as Snodprot1 from *Phaeosphaeria nodorum,* MpCP1 from *Moniliophthora perniciosa* and Sp1 from *Leptosphaeria maculans* (de Oliveira et al. 2011).

14.5.7 Avirulence Proteins

Activation of ETI by race-specific protein products of avirulence (*Avr*) genes from a number of phytopathogenic fungi has also been reported. Among Avr2, Avr4, Avr4E and Avr9 from *C. fulvum* which are recognised by the corresponding Cf R proteins in tomato, only Avr2, a cysteine protease inhibitor, and Avr4, a chitin armour, exhibited virulence functions through suppression and avoidance of host defences (Rooney et al. 2005; van den Burg et al. 2006). Interestingly, functional ortholog of *C. fulvum* Avr4 has been reported in *Mycosphaerella fijiensis*, the causal agent of black Sigatoka disease of banana (Stergiopoulos et al. 2010). Six1 and Six3 effectors were required for full virulence of *F. oxysporum* f. sp. *lycopersici* (*Fol*) during infection in tomato plants (Houterman et al. 2007). Meanwhile, *Fol* Six4 (Avr1) was involved in suppression of I-2- and I-3-mediated resistance to establish infection in tomato (Houterman et al. 2007). Several studies have also proposed the virulence functions of AvrLm1d and AvrLm4-7 from *L. maculans* in plants (Huang et al. 2006, 2010). The mechanisms by which oomycete Avr effectors, the RXLRs suppress host defence have been critically reviewed by Stassen and Van den Ackerveken (2011). Other than that, Avr-Pii from rice blast fungus was found to target OsExo70-F3 and exocytosis pathway is crucial for ETI in rice (Fujisaki et al. 2015). Furthermore, the interactions between five blast Avr effectors, namely AvrPita, Avr-Pik, AvrPiz-t, Avr-Pia and AVr1-Co39 and their cognate rice R proteins have been reviewed in Liu et al. (Liu et al. 2013a, b) and Liu and Wang (2016).

14.6 Roles of Elicitors in Plant Defence Against Economically Important Basidiomycetes

During the past 20 years, many reviews concentrated on the potential of using biotic and synthetic elicitors in activating immune responses in various plants against a wide range of plant pathogens. However, most literatures primarily focused on the efficiency of elicitors against an array of bacterial pathogens as well as fungal pathogens from the phylum Ascomycota. This section focuses on the studies which have been made to elucidate the effects of several elicitors in inducing resistance in plants against several economically important basidiomycetes which affect the productivity of food crops and plantation crops globally.

14.6.1 Benzothiadiazole

Since 1996, a great number of studies have reported the potential of the synthetic elicitor, benzothiadiazole (BTH; trade name, BION®), synonymous ASM, a functional analogue of SA which is a secondary signalling molecule of plant origin, in inducing SAR in various plants against a wide range of pathogens including basidiomycetes. Typically, the triggered immune responses in plants include synthesis of PR protein mostly PR-1 proteins through activation of MPPK3 (Beckers and Conrath 2007), accumulation of phytoalexins and strengthening of cell walls. To date, BTH has attracted attention of many researchers due to its high competence in disease control against a wide array of pathogens (as reviewed in Iriti and Faoro 2003) besides posing low or zero toxicity to plants, animals and the environment (Tomlin 2001).

In a study on Eucalyptus plants, BION® and autoclaved preparation made from *Saccharomyces cerevisiae* were found to increase the activity of chitinase, peroxidase and phenylalanine ammonium lyase (PAL) enzymes together with the induction of a HR in both susceptible and resistant cultivars that resulted in an enhanced resistance against *Puccinia psidii* (Boava et al. 2009). In a different study, Boava et al. (2010) reported higher expression of gene encoding chitinase in the leaves of Eucalyptus pre-treated with ASM and challenged with *Pc. psidii* suggesting the preconditioning effect of ASM in plants. Besides, Han et al. (2012) demonstrated that disease resistance to *Pc. striiformis* f. sp. *tritici* (*Pst*) which lasted for at least 60 days was induced in mature wheat plants pre-treated with BTH. Furthermore, alterations in peroxidase and PAL activities as well as in total phenol content revealed the ability of BTH in inducing resistance against *Pst* in Tamuz-2 (moderately susceptible) and AL-8/70 (susceptible) wheat genotypes, mimicking the function of β-aminobutyric acid (BABA), indole acetic acid (IAA) and SA (Al-Maaroof et al. 2014).

The findings of Iriti and Faoro (2003) highlighted the efficiency of a single application of BTH in preventing rust infection caused by *U. appendiculatus* on French bean plants. Consistent with this finding, Maffi et al. (2011) found that 0.3 mM BTH was able to induce oxidative burst in 24 h and altered PAL expression for several days in bean plants providing protection against rust disease. In the case of *Coffea arabica*, which is prone to coffee leaf rust disease caused by *Hemileia vastatrix* Berk, a study has been conducted to understand the effects of the resistance inducing formulas, namely, "Greenforce CuCa" (formulation based on a by-product of coffee and citrus industries, developed in UFLA) and Bion® on its resistance (Possa et al. 2014). Initial reports found that these elicitor formulas affected the types of protein being accumulated in the coffee leaves at cellular level (Possa et al.

2014). More recently, the findings of Barilli et al. (2015) highlighted the role of BTH in reducing pre- and post-penetration of rust pathogen, *U. pisi* through priming of phytoalexins accumulation (scopoletin, pisatin and medicarpin) in pea leaves. However, they have also suggested that the response mediated by BTH was genotype-dependent as compared to DL-BABA.

In vitro assessment demonstrated that 0.025 and 0.05 g/L of ASM (BION® 50 WG) were neither fungicidal nor fungistatic against *Ganoderma boninense*, a notorious fungal pathogen that causes basal stem rot (BSR) disease of oil palm (Table 14.3, unpublished data). On the other hand, in a preliminary in planta study, both spraying (50 mg/L) and drenching (25 mg/L) of ASM on *G. boninense* infected oil palm seedlings were effective in reducing BSR disease incidence (DI) through the activation of PR genes (chitinase and glucanase) on the seedlings compared to the untreated oil palms (Wong and Sariah 2007) suggesting that ASM could be used to induce resistance of oil palm for BSR disease management.

14.6.2 Silicon

Silicon (Si), an inorganic elicitor has been gaining popularity as a plant elicitor due to its prophylactic role in increasing plant defence mechanisms against plant pathogens. Si possibly induces defence responses similar to SAR (Cai et al. 2009). The dual actions of Si as a physical barrier and resistance activator cause an increased production of phenolic compounds, phytoalexins and lignin, increased activity of defence enzymes as well as transcription of genes associated with defence in various plants (Pozza et al. 2015; Rodrigues et al. 2015). Si application has been used in many different plants, including soybean, to protect against soybean rust (SBR). Foliar application of Si was found to reduce the intensity of SBR in soybean caused by the fungus *Phakopsora pachyrhizi* (Rodrigues et al. 2009). Delayed onset of SBR and reduction in area under disease progress curve (AUDPC) for treated soybean plants with Si under glasshouse and field conditions have also been reported (Lemes et al. 2011). Moreover, Arsenault-Labrecque et al. (2012) have highlighted the potential of Si in protecting soybean plants particularly cultivar Hikmok sorip against SBR through mediated resistance which was associated with HR. Furthermore, Cruz et al. (2013) have reported an increased activity of chitinase

Table 14.3 Effect of different concentrations of BION® on in vitro mycelial growth inhibition of *G. boninense* based on each day after inoculation (DAI)

	DAI 3	DAI 7	DAI 9	DAI 10	DAI 13
BION® Treatment	PIRG (%)				
T1 (ddH_2O)	0	0	0	0	0[a]
T2 (0.05 g/L)	19.26	17.65	4.86	4.65	0[a]
T3 (0.025 g/L)	0.88	0.74	0	0	0[a]

[a]*PIRG* percent inhibition of radial growth. Values show the means of seven replications for each treatment. Means with the same letter is not significant at LSD at $P \leq 0.05$

in the onset of Asian soybean rust (ASR) symptoms on plants grown in soil amended with calcium silicate, a source of Si, compared to ASM and JA suggesting the important role of Si in ameliorating SBR severity.

Apart from soybean, several studies have been carried out to determine the potential of Si in enhancing rice resistance against *Rhizoctonia solani*, the causal agent of sheath blight. Zhang et al. 2013a, b reported that Si application improved resistance against *R. solani* particularly in Ningjing 1, a susceptible cultivar and the resistance was associated with increased production of phenolics, flavonoids and lignin, and increased polyphenoloxidase (PPO) and PAL activities. Similarly, increased activities of PAL, PPO, chitinase and peroxidases (PO) were observed in the leaf sheaths of susceptible rice plants of BR-Irga 409 resulting in reduced sheath blight development (Schurt et al. 2014). Najihah et al. (2015) have tested the efficacy of silicon oxide, potassium silicate, calcium silicate, sodium silicate and sodium meta-silicate at different concentrations against BSR disease in oil palm and found that treatment with 1200 mg/L of sodium silicate was most effective in suppressing BSR by 53%. Moreover, foliar spray of potassium silicate was reported to be able to reduce the severity of coffee leaf rust caused by *H. vastatrix* (Carré-Missio et al. 2012). These studies have highlighted the potential of Si in controlling diseases caused by basidiomycetes; however, it is essential to provide information on biochemical and molecular aspects of resistance induced by Si and its signalling pathways in treated plants.

14.6.3 Saccharin

In recent years, there has been an increasing interest in using saccharin, a metabolite of probenazole (PBZ), as defence activator in several plants. For example, Srivastava et al. (2011) have highlighted the potential of this compound in activating SAR of soybean in response to its pathogen *Ph. pachyrhizi.* An augmented level of resistance was recorded 1 day after application of saccharin as a root drench and increased systemic protection against rust infection until 15 days after treatment. Saccharin also reduced the number and diameter of pustules, sporulation of bean rust and rust severity in the susceptible bean plants cultivar Guará inoculated with *U. appendiculatus* (Delgado et al. 2013).

In another study, saccharin was shown to induce systemic resistance when applied as a soil drench to *Vicia faba* L. cv. Aquadulce at three-leaf stage, against *U. viciae-fabae,* the rust fungus (Boyle and Walters 2005). Ahmed et al. (2013) have proposed that rust disease control in broad beans applied with saccharin was due to wyerone acid, the phytoalexin which was formed by host tissues and it increased with gradual increasing of saccharin concentration. However, understanding the biochemical or molecular mechanisms employed by saccharin in triggering resistance in plants is necessary to explore the potential of this elicitor in controlling diseases caused by basidiomycetes.

14.7 Conclusions and Perspectives

Elicitors of biological and non-biological origins play crucial roles in establishing beneficial or pathogenic interactions with plants. To date, elicitors which are able to control plant pathogens by activating plant innate immunity are being extensively studied due to their great potential for crop protection and environmental compatibility. However, it is crucial to note that the mechanism of several elicitors is highly dependent on their origins and plant species. Thus, more detailed studies on the plant defence elicitation mechanisms by elicitors and practical applications in the field are essentially required for the development of biopesticides against pathogens, particularly basidiomycetes, in the near future. Taking into account the advantages of elicitors to the farmers, consumers and environment, both biotic and abiotic elicitors have the potential to be used as a component of integrated pest management in agriculture.

References

Ahmed BM, Helal IM, Mohamed GA (2013) Control of rust disease in irradiated *Vicia faba* by using safe alternative methods. Isot Radiat Res 45:629–637

Al-Maaroof E, Fayadh HA, Fattah FA (2014) Use of some chemical inducers to improve wheat resistance to *Puccinia striiformis* f. sp. *tritici*. Ratar Povrt 51:83–90

Amborabé BE, Rossard S, Pérault JM, Roblin G (2003) Specific perception of ergosterol by plant cells. C R Biol 326:363–370

Ao Y, Li Z, Feng D, Xiong F, Liu J, Li JF, Wang HB (2014) OsCERK1 and OsRLCK176 play important roles in peptidoglycan and chitin signaling in rice innate immunity. Plant J 80:1072–1084

Arsenault-Labrecque G, Menzies JG, Bélanger RR (2012) Effect of silicon absorption on soybean resistance to *Phakopsora pachyrhizi* in different cultivars. Plant Dis 96:37–42

Bae H, Kim MS, Sicher RC, Bae HJ, Bailey BA (2006) Necrosis-and ethylene-inducing peptide from *Fusarium oxysporum* induces a complex cascade of transcripts associated with signal transduction and cell death in Arabidopsis. Plant Physiol 141:1056–1067

Barilli E, Rubiales D, Amalfitano C, Evidente A, Prats E (2015) BTH and BABA induce resistance in pea against rust (*Uromyces pisi*) involving differential phytoalexin accumulation. Planta 242:1095–1106

Baroncelli R, Matarese F, Moncini L, Vannacci G, Vergara M (2016) Two endopolygalacturonase genes in *Trichoderma virens*: In silico characterization and expression during interaction with plants. J Phytopathol 164:18–28

Beckers GJ, Conrath U (2007) Priming for stress resistance: from the lab to the field. Curr Opin Plant Biol 10:425–431

Benali S, Mohamed B, Eddine HJ (2014) Virulence strategies of phytopathogenic bacteria and their role in plant disease pathogenesis. Afr J Microbiol Res 8:2809–2815

Boava LP, Kuhn OJ, Pascholati SF, Di Piero RM, Furtado EL (2009) Effect of acibenzolar-S-methyl and *Saccharomyces cerevisiae* on the activation of Eucalyptus defences against rust. Australas Plant Pathol 38:594–602

Boava LP, Laia ML, Jacob TR, Dabbas KM, Gonçalves JF, Ferro JA, Ferro MI, Furtado EL (2010) Selection of endogenous genes for gene expression studies in Eucalyptus under biotic (*Puccinia psidii*) and abiotic (acibenzolar-S-methyl) stresses using RT-qPCR. BMC Res Notes 3:43

Boller T, Felix G (2009) A renaissance of elicitors: perception of microbe-associated molecular patterns and danger signals by pattern-recognition receptors. Annu Rev Plant Biol 60:379–406

Boudart G, Charpentier M, Lafitte C, Martinez Y, Jauneau A, Gaulin E, Dumas B (2003) Elicitor activity of a fungal endopolygalacturonase in tobacco requires a functional catalytic site and cell wall localization. Plant Physiol 131:93–101

Bouwmeester K, Han M, Blanco-Portales R, Song W, Weide R, Guo LY, van der Vossen EA, Govers F (2014) The Arabidopsis lectin receptor kinase LecRK-I. 9 enhances resistance to *Phytophthora infestans* in Solanaceous plants. Plant Biotechnol J 12:10–16

Bowman SM, Free SJ (2006) The structure and synthesis of the fungal cell wall. Bioessays 28:799–808

Boyle C, Walters D (2005) Induction of systemic protection against rust infection in broad bean by saccharin: effects on plant growth and development. New Phytol 167:607–612

Brutus A, Sicilia F, Macone A, Cervone F, De Lorenzo G (2010) A domain swap approach reveals a role of the plant wall-associated kinase 1 (WAK1) as a receptor of oligogalacturonides. Proc Natl Acad Sci U S A 107:9452–9457

van den Burg HA, Harrison SJ, Joosten MH, Vervoort J, de Wit PJ (2006) *Cladosporium fulvum* Avr4 protects fungal cell walls against hydrolysis by plant chitinases accumulating during infection. Mol Plant Microbe Interact 19:1420–1430

Cai K, Gao D, Chen J, Luo S (2009) Probing the mechanisms of silicon-mediated pathogen resistance. Plant Signal Behav 4:1–3

Carré-Missio V, Rodrigues FA, Schurt DA, Moreira WR, Rezende DC, Korndorfer GH, Zambolim L (2012) Proteção local, não sistêmica, do silicato de potássio reduz os sintomas da ferrugem do cafeeiro. Trop Plant Pathol 37:275–280

Cawoy H, Mariutto M, Henry G, Fisher C, Vasilyeva N, Thonart P, Ongena M (2014) Plant defence stimulation by natural isolates of *Bacillus* depends on efficient surfactin production. Mol Plant Microbe Interact 27:87–100

Chang YH, Yan HZ, Liou RF (2015) A novel elicitor protein from *Phytophthora parasitica* induces plant basal immunity and systemic acquired resistance. Mol Plant Pathol 16:123–136

Chen M, Zeng H, Qiu D, Guo L, Yang X, Shi H, Zhao J (2012) Purification and characterization of a novel hypersensitive response-inducing elicitor from *Magnaporthe oryzae* that triggers defence response in rice. PLoS One 7:e37654

Chen M, Zhang C, Zi Q, Qiu D, Liu W, Zeng H (2014) A novel elicitor identified from *Magnaporthe oryzae* triggers defence responses in tobacco and rice. Plant Cell Rep 33:1865–1879

Choi MS, Kim W, Lee C, Oh CS (2013) Harpins, multifunctional proteins secreted by Gram-negative plant-pathogenic bacteria. Mol Plant Microbe Interact 26:1115–1122

Choi J, Tanaka K, Cao Y, Qi Y, Qiu J, Liang Y, Lee SY, Stacey G (2014) Identification of a plant receptor for extracellular ATP. Science 343:290–294

Cohn J, Sessa G, Martin GB (2001) Innate immunity in plants. Curr Opin Immunol 13:55–62

Coll NS, Epple P, Dangl JL (2011) Programmed cell death in the plant immune system. Cell Death Differ 18:1247–1256

Cruz MFAD, Rodrigues FÁ, Polanco LR, Curvêlo CRDS, Nascimento KJT, Moreira MA, Barros EG (2013) Inducers of resistance and silicon on the activity of defence enzymes in the soybean-*Phakopsora pachyrhizi* interaction. Bragantia 72:162–172

Culver JN, Dawson WO (1989) Tobacco mosaic virus coat protein: an elicitor of the hypersensitive reaction but not required for the development of mosaic symptoms in *Nicotiana sylvestris*. Virology 173:755–758

Dadakova K, Klempova J, Jendrisakova T, Lochman J, Kasparovsky T (2013) Elucidation of signaling molecules involved in ergosterol perception in tobacco. Plant Physiol Biochem 73:121–127

De Vleesschauwer D, Djavaheri M, Bakker PA, Höfte M (2008) *Pseudomonas fluorescens* WCS374r-induced systemic resistance in rice against *Magnaporthe oryzae* is based on pseudobactin-mediated priming for a salicylic acid-repressible multifaceted defence response. Plant Physiol 148:1996–2012

Delgado DZ, de Freitas MB, Stadnik MJ (2013) Effectiveness of saccharin and ulvan as resistance inducers against rust and angular leaf spot in bean plants (*Phaseolus vulgaris*). Crop Prot 47:67–73

Djonovic S, Pozo MJ, Dangott LJ, Howell CR, Kenerley CM (2006) Sm1, a proteinaceous elicitor secreted by the biocontrol fungus *Trichoderma virens* induces plant defence responses and systemic resistance. Mol Plant Microbe Interact 19:838–853

Du J, Verzaux E, Chaparro-Garcia A, Bijsterbosch G, Keizer LP, Zhou J, van der Vossen EA (2015) Elicitin recognition confers enhanced resistance to *Phytophthora infestans* in potato. Nat Plants 1:15034

Durrant WE, Dong X (2004) Systemic acquired resistance. Annu Rev Phytopathol 42:185–209

Eber J, Cosio EG (1994) Elicitors of plant defence responses. Int Rev Cytol 148:1–36

El Gueddari NE, Rauchhaus U, Moerschbacher BM, Deising HB (2002) Developmentally regulated conversion of surface-exposed chitin to chitosan in cell walls of plant pathogenic fungi. New Phytol 156:103–112

Engelberth J, Koch T, SchülerG BN, Rechtenbach J, Boland W (2001) Ion channel-forming alamethicin is a potent elicitor of volatile biosynthesis and tendril coiling. Cross talk between jasmonate and salicylate signaling in lima bean. Plant Physiol 125:369–377

Enkerli J, Felix G, Boller T (1999) The enzymatic activity of fungal xylanase is not necessary for its elicitor activity. Plant Physiol 121:391–398

Felix G, Duran JD, Volko S, Boller T (1999) Plants have a sensitive perception system for the most conserved domain of bacterial flagellin. Planta J18:265–276

Ferrari S, Savatin DV, Sicilia F, Gramegna G, Cervone F, De Lorenzo G (2013) Oligogalacturonides: plant damage-associated molecular patterns and regulators of growth and development. Front Plant Sci 4:49

Fesel PH, Zuccaro A (2016) β-Glucan: crucial component of the fungal cell wall and elusive MAMP in plants. Fungal Genet Biol 90:53–60

Frías M, González C, Brito N (2011) BcSpl1, a cerato-platanin family protein, contributes to *Botrytis cinerea* virulence and elicits the hypersensitive response in the host. New Phytol 192:483–495

Friesen TL, Faris JD, Solomon PS, Oliver RP (2008a) Host-specific toxins: effectors of necrotrophic pathogenicity. Cell Microbiol 10:1421–1428

Friesen TL, Zhang ZC, Solomon PS, Oliver RP, Faris JD (2008b) Characterization of the interaction of a novel *Stagonospora nodorum* host-selective toxin with a wheat susceptibility gene. Plant Physiol 146:682–693

Fujikawa T, Kuga Y, Yano S, Yoshimi A, Tachiki T, Abe K, Nishimura M (2009) Dynamics of cell wall components of *Magnaporthe grisea* during infectious structure development. Mol Microbiol 73:553–570

Fujikawa T, Sakaguchi A, NishizawaY KY, Minami E, Yano S, Koga H, Meshi T, Nishimura M (2012) Surface α-1,3-glucan facilitates fungal stealth infection by interfering with innate immunity in plants. PLoS Pathog 8:e1002882

Fujisaki K, Abe Y, Ito A, Saitoh H, Yoshida K, Kanzaki H, Utsushi H, Yamashita T, Kamoun S, Terauchi R (2015) Rice Exo70 interacts with a fungal effector, AVR-Pii, and is required for AVR-Pii-triggered immunity. Plant J 83:875–887

Gaderer R, Lamdan NL, Frischmann A, Sulyok M, Krska R, Horwitz BA, Seidl-Seiboth V (2015) Sm2, a paralog of the *Trichoderma* cerato-platanin elicitor Sm1, is also highly important for plant protection conferred by the fungal-root interaction of *Trichoderma* with maize. BMC Microbiol 15:1

Galletti R, De Lorenzo G, Ferrari S (2009) Host-derived signals activate plant innate immunity. Plant Signal Behav 4:33–34

Gao QM, Zhu S, Kachroo P, Kachroo A (2015) Signal regulators of systemic acquired resistance. Front Plant Sci 6:228

Glazebrook J (2005) Contrasting mechanisms of defence against biotrophic and necrotrophic pathogens. Annu Rev Phytopathol 43:205–227

Gohre V, Robatzek S (2008) Breaking the barriers: microbial effector molecules subvert plant immunity. Annu Rev Phytopathol 46:189–215

Gómez-Gómez L, Boller T (2000) FLS2: an LRR receptor-like kinase involved in the perception of the bacterial elicitor flagellin in Arabidopsis. Mol Cell 5:1003–1011

Granado J, Felix G, Boller T (1995) Perception of fungal sterols in plants (subnanomolar concentrations of ergosterol elicit extracellular alkalinization in tomato cells). Plant Physiol 107:485–490

Halim VA, Hunger A, Macioszek V, Landgraf P, NürnbergerT SD, Rosahl S (2004) The oligopeptide elicitor Pep-13 induces salicylic acid-dependent and-independent defence reactions in potato. Physiol Mol Plant Pathol 64:311–318

Hamada H, Kurusu T, Okuma E, Nokajima H, Kiyoduka M, Koyano T, Miyao A (2012) Regulation of a proteinaceous elicitor-induced Ca2+ influx and production of phytoalexins by a putative voltage-gated cation channel, OsTPC1, in cultured rice cells. J Biol Chem 287:9931–9939

Han QM, Liu W, Wei GR, Huang LL, Kang ZS (2012) Determination of BTH-induced wheat resistance to *Puccinia striiformis*. Chin J Eco-Agric 9:20

Heil M, Land WG (2014) Danger signals-damaged-self recognition across the tree of life. Front Plant Sci 5:578

Hemetsberger C, Herrberger C, Zechmann B, Hillmer M, Doehlemann G (2012) The *Ustilago maydis* effector Pep1 suppresses plant immunity by inhibition of host peroxidase activity. PLoS Pathog 8:e1002684

Henriquez MA, Wolski EA, Molina OI, Adam LR, Andreu AB, Daayf F (2012) Effects of glucans and eicosapentaenoic acid on differential regulation of phenylpropanoid and mevalonic pathways during potato response to *Phytophthora infestans*. Plant Physiol Biochem 60:119–128

Henry G, Thonart P, Ongena M (2012) PAMPs, MAMPs, DAMPs and others: an update on the diversity of plant immunity elicitors. Biotechnol Agron Soc 16:257

Herron SR, Benen JA, Scavetta RD, Visser J, Jurnak F (2000) Structure and function of pectic enzymes: virulence factors of plant pathogens. Proc Natl Acad Sci 97:8762–8769

Houterman PM, Speijer D, Dekker HL, de Koster CG, Cornelissen BJ, Rep M (2007) The mixed xylem sap proteome of *Fusarium oxysporum*-infected tomato plants. Mol Plant Pathol 8:215–221

Huang YJ, Li ZQ, Evans N, Rouxel T, Fitt BD, Balesdent MH (2006) Fitness cost associated with loss of the AvrLm4 avirulence function in *Leptosphaeria maculans* (phoma stem canker of oilseed rape). Eur J Plant Pathol 114:77–89

Huang YJ, Balesdent MH, Li ZQ, EvansN RT, Fitt BD (2010) Fitness cost of virulence differs between the AvrLm1 and AvrLm4 loci in *Leptosphaeria maculans* (phoma stem canker of oilseed rape). Eur J Plant Pathol 126:279–291

Huang CJ, Tsay JF, Chang SY, Yang HP, Wu WS, Chen CY (2012) Dimethyl disulfide is an induced systemic resistance elicitor produced by *Bacillus cereus* C1L. Pest Manag Sci 68:1306–1310

Iriti M, Faoro F (2003) Benzothiadiazole (BTH) Induces cell-death independent resistance in *Phaseolus vulgaris* against *Uromyces appendiculatus*. J Phytopathol 151:171–180

Jehle AK, Lipschis M, Albert M, Fallahzadeh-Mamaghani V, Fürst U, Mueller K, Felix G (2013) The receptor-like protein ReMAX of Arabidopsis detects the microbe-associated molecular pattern eMax from *Xanthomonas*. Plant Cell 25:2330–2340

Jones JD, Dangl JL (2006) The plant immune system. Nature 444:323–329

Jones DA, Thomas CM, Hammond-Kosack KE, Balint-Kurti PJ, Jones JD (1994) Isolation of the tomato Cf-9 gene for resistance to *Cladosporium fulvum* by transposon tagging. Science 666:789–789

de Jonge R, van Esse HP, Kombrink A, Shinya T, Desaki Y, Bours R, van der Krol S, Shibuya N, Joosten MH, Thomma BP (2010) Conserved fungal LysM effector Ecp6 prevents chitin-triggered immunity in plants. Science 329:953–955

de Jonge R, van Esse HP, Maruthachalam K, Bolton MD, Santhanam P, Saber MK, Thomma BP (2012) Tomato immune receptor Ve1 recognizes effector of multiple fungal pathogens uncovered by genome and RNA sequencing. Proc Natl Acad Sci 109:5110–5115

Kaku H, Nishizawa Y, Ishii-Minami N, Akimoto-Tomiyama C, Dohmae N, Takio K, Minami E, Shibuya N (2006) Plant cells recognize chitin fragments for defence signaling through a plasma membrane receptor. Proc Natl Acad Sci 103:11086–11091

Kaschani F, Shabab M, Bozkurt T, Shindo T, Schornack S, Gu C, van der Hoorn RA (2010) An effector-targeted protease contributes to defence against *Phytophthora infestans* and is under diversifying selection in natural hosts. Plant Physiol 154:1794–1804

Kasparovsky T, Milat ML, Humbert C, Blein JP, Havel L, Mikes V (2003) Elicitation of tobacco cells with ergosterol activates a signal pathway including mobilization of internal calcium. Plant Physiol Biochem 41:495–501

Kasparovsky T, Blein JP, Mikes V (2004) Ergosterol elicits oxidative burst in tobacco cells via phospholipase A 2 and protein kinase C signal pathway. Plant Physiol Biochem 42:429–435

Kawamura Y, Hase S, Takenaka S, Kanayama Y, Yoshioka H, Kamoun S, Takahashi H (2009) INF1 elicitin activates jasmonic acid-and ethylene-mediated signalling pathways and induces resistance to bacterial wilt disease in tomato. J Phytopathol 157:287–297

Keen NT, Partridg JE, Zaki AI (1972) Pathogen-produced elicitor of a chemical defence mechanism in soybeans monogenically resistant to *Phytophthora-megasperma* var *sojae*. In: Phytopathology. American Phytopathological Society, p 768

Kim YH, Yeo WH, Kim YS, Chae SY, Kim KS (2000) Antiviral activity of antibiotic peptaibols, chrysospemins B and D, produced by *Apiocrea* sp. 14T against TMV infection. J Microbiol Biotechnol 10:522–528

Kim PD, Sasek V, Burketova L, Copikova J, Synytsya A, Jindrichova B, Valentova O (2013) Cell wall components of *Leptosphaeria maculans* enhance resistance of *Brassica napus*. J Agric Food Chem 61:5207–5214

Kishimoto K, Kouzai Y, Kaku H, Shibuya N, Minami E, Nishizawa Y (2010) Perception of the chitin oligosaccharides contributes to disease resistance to blast fungus *Magnaporthe oryzae* in rice. Plant J 64:343–354

Klemptner RL, Sherwood JS, Tugizimana F, Dubery IA, Piater LA (2014) Ergosterol, an orphan fungal microbe-associated molecular pattern (MAMP). Mol Plant Pathol 15:747–761

Kromina KA, Dzhavakhiia VG (2006) Expression of the bacterial gene CspD in tobacco plants increases their resistance to fungal and viral pathogens. Mol Gen Mikrobiol Virusol 1:31–34

Kulye M, Liu HUA, Zhang Y, Zeng H, Yang X, Qiu D (2012) Hrip1, a novel protein elicitor from necrotrophic fungus, *Alternaria tenuissima*, elicits cell death, expression of defence-related genes and systemic acquired resistance in tobacco. Plant Cell Environ 35:2104–2120

Kunze G, Zipfel C, Robatzek S, Niehaus K, Boller T, Felix G (2004) The N terminus of bacterial elongation factor Tu elicits innate immunity in Arabidopsis plants. Plant Cell 16:3496–3507

Kurusu T, Yagala T, Miyao A, Hirochika H, Kuchitsu K (2005) Identification of a putative voltage-gated Ca^{2+} channel as a key regulator of elicitor-induced cell death and mitogen-activated protein kinase activation in rice. Plant J 42:798

Kurusu T, Hamada J, Nokajima H, Kitagawa Y, Kiyoduka M, Takahashi A, Koga J (2010) Regulation of MAMP-induced cell death, phytoalexin production & defence gene expression by Calcineurin B-like protein-interacting protein kinases, OsCIPK14/15, in rice cultured cells. Plant Physiol 153:678–692

Larroque M, Belmas E, Martinez T, Vergnes S, Ladouce N, Lafitte C, Dumas B (2013) Pathogen-associated molecular pattern-triggered immunity and resistance to the root pathogen *Phytophthora parasitica* in Arabidopsis. J Exp Bot 64:3615–3625

Lee SW, Jeong KS, Han SW, Lee SE, Phee BK, Hahn TR, Ronald P (2008) The *Xanthomonas oryzae* pv. *oryzae* PhoPQ two-component system is required for AvrXA21 activity, hrpG expression, and virulence. J Bacteriol 190:2183–2197

Lee SW, Han SW, Sririyanum M, Park CJ, Seo YS, Ronald PC (2009) A type I-secreted, sulfated peptide triggers XA21-mediated innate immunity. Science 326:850–853

Lemes E, Mackowiak CL, Blount A, Marois JJ, Wright DL, Coelho L, Datnoff LE (2011) Effects of silicon applications on soybean rust development under greenhouse and field conditions. Plant Dis 95:317–324

Liang Y, Tóth K, Cao Y, Tanaka K, Espinoza C, Stacey G (2014) Lipochitooligosaccharide recognition: an ancient story. New Phytol 204:289–296

Liebrand TW, van den Berg GC, Zhang Z, Smit P, Cordewener JH, America AH, Thomma BP (2013) Receptor-like kinase SOBIR1/EVR interacts with receptor-like proteins in plant immunity against fungal infection. Proc Natl Acad Sci U S A 110:10010–10015

Liebrand TW, van den Burg HA, Joosten MH (2014) Two for all: receptor-associated kinases SOBIR1 and BAK1. Trends Plant Sci 19:123–132

Liu W, Wang GL (2016) Plant innate immunity in rice: a defence against pathogen infection. Natl Sci Rev 3:295–308

Liu B, Li JF, Ao Y, Li Z, Liu J, Feng D, Qi K, He Y, Zeng L, Wang J, Wang HB (2013a) OsLYP4 and OsLYP6 play critical roles in rice defence signal transduction. Plant Signal Behav 8:e22980

Liu W, Liu J, Ning Y, Ding B, Wang X, Wang Z, Wang GL (2013b) Recent progress in understanding PAMP-and effector-triggered immunity against the rice blast fungus *Magnaporthe oryzae*. Mol Plant 6:605–620

Lombardi L, Faoro F, Luti S, Baccelli I, Martellini F, Bernardi R, Pazzagli L (2013) Differential timing of defence-related responses induced by cerato-platanin and cerato-populin, two non-catalytic fungal elicitors. Physiol Plant 149:408–421

Ma Z, Song T, Zhu L, Ye W, Wang Y, Shao Y, Tyler BM (2015) A *Phytophthora sojae* glycoside hydrolase 12 protein is a major virulence factor during soybean infection and is recognized as a PAMP. Plant Cell 27:2057–2072

Maffei ME, Arimura GI, Mithöfer A (2012) Natural elicitors, effectors and modulators of plant responses. Nat Prod Rep 29:1288–1303

Maffi D, Iriti M, Pigni M, Vannini C, Faoro F (2011) *Uromyces appendiculatus* infection in BTH-treated bean plants: ultrastructural details of a lost fight. Mycopathologia 171:209–221

Marshall R, Kombrink A, Motteram J, Loza-Reyes E, Lucas J, Hammond-Kosack KE, Ruddv JJ (2011) Analysis of two in planta expressed LysM effector homologs from the fungus *Mycosphaerella graminicola* reveals novel functional properties and varying contributions to virulence on wheat. Plant Physiol 156:756–769

Mazzotta S, Kemmerling B (2011) Pattern recognition in plant innate immunity. J Plant Pathol 93:7–17

Ménard R, Alban S, de Ruffray P, Jamois F, Franz G, Fritig B, Kauffmann S (2004) β-1,3 Glucan sulfate, but not β-1,3 glucan, induces the salicylic acid signaling pathway in tobacco and Arabidopsis. Plant Cell 16:3020–3032

Mentlak TA, Kombrink A, Shinya T, Ryder LS, Otomo I, Saitoh H, Talbot NJ (2012) Effector-mediated suppression of chitin-triggered immunity by *Magnaporthe oryzae* is necessary for rice blast disease. Plant Cell 24:322–335

Miao W, Wang X, Li M, Song C, Wang Y, Hu D, Wang J (2010) Genetic transformation of cotton with a harpin-encoding gene hpa Xoo confers an enhanced defence response against different pathogens through a priming mechanism. BMC Plant Biol 10:1

Mishra AK, Sharma K, Misra RS (2009) Purification and characterization of elicitor protein from *Phytophthora colocasiae* and basic resistance in *Colocasia esculenta*. Microbiol Res 164:688–693

Mishra AK, Sharma K, Misra RS (2012) Elicitor recognition, signal transduction and induced resistance in plants. J Plant Interact 7:95–120

Miya A, Albert P, Shinya T, Desaki Y, Ichimura K, Shirasu K, Shibuya N (2007) CERK1, a LysM receptor kinase, is essential for chitin elicitor signaling in Arabidopsis. Proc Natl Acad Sci U S A 104:19613–19618

Monaghan J, Zipfel C (2012) Plant pattern recognition receptor complexes at the plasma membrane. Curr Opin Plant Biol 15:349–357

Montesano M, Brader G, Palva ET (2003) Pathogen derived elicitors: searching for receptors in plants. Mol Plant Pathol 4:73–79

Mueller AN, Ziemann S, Treitschke S, Abmann D, Doehlemann G (2013) Compatibility in the *Ustilago maydis*–maize interaction requires inhibition of host cysteine proteases by the fungal effector Pit2. PLoS Pathog 9:e1003177

Najihah NI, Hanafi MM, Idris AS, Hakim MA (2015) Silicon treatment in oil palms confers resistance to basal stem rot disease caused by *Ganoderma boninense*. Crop Prot 67:151–159

Nars A, Lafitte C, Chabaud M, Drouillard S, Mélida H, Danoun S, Barker DG (2013) *Aphanomyces euteiches* cell wall fractions containing novel glucan-chitosaccharides induce defence genes and nuclear calcium oscillations in the plant host *Medicago truncatula*. PLoS One 8:e75039

Naveen J, Hariprasad P, Nayaka SC, Niranjana SR (2013) Cerebroside mediated elicitation of defence response in chilli (*Capsicum annuum* L.) against *Colletotrichum capsici* infection. J Plant Interact 8:65–73

Newman MA, Roepenack-Lahaye V, Parr A, Daniels MJ, Dow JM (2002) Prior exposure to lipopolysaccharide potentiates expression of plant defences in response to bacteria. Plant J 29:487–495
Newman MA, Sundelin T, Nielsen JT, Erbs G (2013) MAMP (microbe-associated molecular pattern) triggered immunity in plants. Front Plant Sci 4:139
Nguyen HP, Chakravarthy S, Velásquez AC, McLane HL, Zeng L, Nakayashiki H, Martin GB (2010) Methods to study PAMP-triggered immunity using tomato and *Nicotiana benthamiana*. Mol Plant Microbe Interact 23:991–999
Noda J, Brito N, González C (2010) The *Botrytis cinerea* xylanase Xyn11A contributes to virulence with its necrotizing activity, not with its catalytic activity. BMC Plant Biol 10:38
O'Leary BM, Neale HC, Geilfus CM, Jackson RW, Arnold DL, Preston GM (2016) Early changes in apoplast composition associated with defence and disease in interactions between *Phaseolus vulgaris* and the halo blight pathogen *Pseudomonas syringae* pv. *phaseolicola*. Plant Cell Environ 39:2172–2184
de Oliveira AL, Gallo M, Pazzagli L, Benedetti CE, Cappugi G, Scala A, Cicero DO (2011) The structure of the elicitor cerato-platanin, first member of the CP fungal protein family, reveals a double ψβ-barrel fold and carbohydrate binding. J Biol Chem 286:17560–17568
Oliveira-Garcia E, Deising HB (2013) Infection structure-specific expression of β-1,3-glucan synthase is essential for pathogenicity of *Colletotrichum graminicola* and evasion of β-glucan–triggered immunity in maize. Plant Cell 25:2356–2378
Oliveira-Garcia E, Deising HB (2016) Attenuation of PAMP-triggered immunity in maize requires down-regulation of the key β-1,6-glucan synthesis genes KRE5 and KRE6 in biotrophic hyphae of *Colletotrichum graminicola*. Plant J 87:355–375
Oliver RP, Solomon PS (2010) New developments in pathogenicity and virulence of necrotrophs. Curr Opin Plant Biol 13:415–419
Oome S, Raaymakers TM, Cabral A, Samwel S, Böhm H, Albert I, Van den Ackerveken G (2014) Nep1-like proteins from three kingdoms of life act as a microbe-associated molecular pattern in Arabidopsis. Proc Natl Acad Sci U S A 111:16955–16960
Osman H, Vauthrin S, Mikes V, Milat ML, Panabières F, Marais A, Brunie S, Maume B, Ponchet M, Blein JP (2001) Mediation of elicitin activity on tobacco is assumed by elicitin-sterol complexes. Mol Biol Cell 12:2825–2834
Ouyang Z, Li X, Huang L, Hong Y, Zhang Y, Zhang H, Song F (2015) Elicitin-like proteins Oli-D1 and Oli-D2 from *Pythium oligandrum* trigger hypersensitive response in *Nicotiana benthamiana* and induce resistance against *Botrytis cinerea* in tomato. Mol Plant Pathol 16:238–250
Ozeretskovskaya OL, Vasyukova NI, Perekhod EA, Chalenko GI, Il'inskaya LI, Gerasimova NG (2001) Plant resistance suppressors in the pathosystem formed by potato and the causal agent of late blight. Appl Biochem Microbiol 37:506–511
Park K, Kloepper JW, Ryu CM (2008) Rhizobacterial exopolysaccharides elicit induced resistance on cucumber. J Microbiol Biotechnol 18:1095–1100
Pawlowski ML, Bowen CR, Hill CB, Hartman GL (2016) Responses of soybean genotypes to pathogen infection after the application of elicitors. Crop Prot 87:78–84
Pazzagli L, Cappugi G, Manao G, Camici G, Santini A, Scala A (1999) Purification, characterization, and amino acid sequence of cerato-platanin, a new phytotoxic protein from *Ceratocystis fimbriata f.* sp. *platani*. J Biol Chem 274:24959–24964
Peng DH, Qiu DW, Ruan LF, Zhou CF, Sun M (2011) Protein elicitor PemG1 from *Magnaporthe grisea* induces systemic acquired resistance (SAR) in plants. Mol Plant Microbe Interact 24:1239–1246
Petutschnig EK, Jones AM, Serazetdinova L, Lipka U, Lipka V (2010) The lysin motif receptor-like kinase (LysM-RLK) CERK1 is a major chitin-binding protein in *Arabidopsis thaliana* and subject to chitin-induced phosphorylation. J Biol Chem 285:28902–28911
Possa K, Tenente R, Chaves I, Pinheiro C, Renaut J, Cândido R, Guerra-Guimarães L (2014) An overview of protein changes in Coffea arabica leaves upon treatment with inducers of resistance to coffee rust. Paper presented at 25th International Conference on Coffee Science (ASIC), Armenia, Quindío, Colombia, 8–13 September, 2014, pp 42–46

Postma J, Liebrand TW, BiG EA, Bye RR, Mbengue M, Robatzek S (2016) Avr4 promotes Cf-4 receptor-like protein association with the BAK1/SERK3 receptor-like kinase to initiate receptor endocytosis and plant immunity. New Phytol 210:627–642

Pozza EA, Pozza AAA, Botelho DMDS (2015) Silicon in plant disease control. Rev Ceres 62:323–331

Rodrigues FA, Duarte HSS, Domiciano GP, Souza CA, Korndörfer GH, Zamnolim L (2009) Foliar application of potassium silicate reduces the intensity of soybean rust. Australas Plant Pathol 38:366–372

Rodrigues FA, Resende RS, Dallagnol LJ, Datnoff LE (2015) Silicon potentiates host defence mechanisms against infection by plant pathogens. In: Silicon and plant diseases. Springer International Publishing, Cham, pp 109–138

Ron M, Avni A (2004) The receptor for the fungal elicitor ethylene-inducing xylanase is a member of a resistance-like gene family in tomato. Plant Cell 16:1604–1615

Rooney HC, van't Klooster JW, van der Hoorn RA, Joosten MH, Jones JD, de Wit PJ (2005) *Cladosporium* Avr2 inhibits tomato Rcr3 protease required for Cf-2-dependent disease resistance. Science 308:1783–1786

Rossard S, Luini E, Pérault JM, Bonmort J, Roblin G (2006) Early changes in membrane permeability, production of oxidative burst and modification of PAL activity induced by ergosterol in cotyledons of *Mimosa pudica*. J Exp Bot 57:1245–1252

Rossard S, Roblin G, Atanassova R (2010) Ergosterol triggers characteristic elicitation steps in *Beta vulgaris* leaf tissues. J Exp Bot 61:1807–1816

Rotblat B, Enshell-Seijffers D, Gershoni JM, Schuster S, Avni A (2002) Identification of an essential component of the elicitation active site of the EIX protein elicitor. Plant J 32:1049–1055

Ruocco M, Lanzuise S, Lombardi N, Woo SL, Vinale F, Marra R, Turrà D (2015) Multiple roles and effects of a novel *Trichoderma* hydrophobin. Mol Plant Microbe Interact 28:167–179

Salas-Marina MA, Isordia-Jasso MI, Islas-Osuna MA, Delgado-Sánchez P, Jiménez-Bremont JF, Rodríguez-Kessler M, Casas-Flores S (2015) The Epl1 and Sm1 proteins from *Trichoderma atroviride* and *Trichoderma virens* differentially modulate systemic disease resistance against different life style pathogens in *Solanum lycopersicum*. Front Plant Sci 6:1–13

Sánchez-Vallet A, Saleem-Batcha R, Kombrink A, Hansen G, Valkenburg DJ, Thomma BP, Mesters JR (2013) Fungal effector Ecp6 outcompetes host immune receptor for chitin binding through intrachain LysM dimerization. Elife 2:e00790

Schurt DA, Cruz MF, Nascimento KJ, Filippi MC, Rodrigues FA (2014) Silicon potentiates the activities of defence enzymes in the leaf sheaths of rice plants infected by *Rhizoctonia solani*. Trop Plant Pathol 39:457–463

Schweizer P, Pokorny J, Schulze-Lefert P, Dudler R (2000) Double-stranded RNA interferes with gene function at the single-cell level in cereals. Plant J 24:895–903

Shcherbakova LA, Odintsova TI, Stakheev AA, Fravel DR, Zavriev SK (2015) Identification of a novel small cysteine-rich protein in the fraction from the biocontrol *Fusarium oxysporum* strain cs-20 that mitigates fusarium wilt symptoms and triggers defence responses in tomato. Front Plant Sci 6:1207

Shibuya N, Minami E (2001) Oligosaccharide signalling for defence responses in plant. Physiol Mol Plant Pathol 59:223–233

Shimizu T, Nakano T, Takamizawa D, Desaki Y, Ishii-Minami N, Nishizawa Y, Shibuya N (2010) Two LysM receptor molecules, CEBiP and OsCERK1, cooperatively regulate chitin elicitor signaling in rice. Plant J 64:204–214

da Silva FG, Shen Y, Dardick C, Burdman S, Yadav RC, de Leon AL, Ronald PC (2004) Bacterial genes involved in type I secretion and sulfation are required to elicit the rice Xa21-mediated innate immune response. Mol Plant Microbe Interact 17:593–601

Srivastava P, George S, Marois JJ, Wright DL, Walker DR (2011) Saccharin-induced systemic acquired resistance against rust (*Phakopsora pachyrhizi*) infection in soybean: Effects on growth and development. Crop Prot 30:726–732

Stassen JH, Van den Ackerveken G (2011) How do oomycete effectors interfere with plant life? Curr Opin Plant Biol 14:407–414

Stergiopoulos I, van den Burg HA, Ökmen B, Beenen HG, van Liere S, Kema GH, de Wit PJ (2010) Tomato Cf resistance proteins mediate recognition of cognate homologous effectors from fungi pathogenic on dicots and monocots. Proc Natl Acad Sci U S A 107:7610–7615

Tian M, Win J, Song J, van der Hoorn R, van der Knaap E, Kamoun S (2007) A *Phytophthora infestans* cystatin-like protein targets a novel tomato papain-like apoplastic protease. Plant Physiol 143:364–377

Tomlin CDS (2001) The pesticide manual. British Crop Protection Council, Berkshire

Trdá L, Boutrot F, Claverie J, Brulé D, Dorey S, Poinssot B (2015) Perception of pathogenic or beneficial bacteria and their evasion of host immunity: pattern recognition receptors in the frontline. Front Plant Sci 6:219

Trouvelot S, Héloir MC, Poinssot B, Gauthier A, Paris F, Guillier C, Daire X (2014) Carbohydrates in plant immunity and plant protection: roles and potential application as foliar sprays. Front Plant Sci 5:592

Varnier AL, Sanchez L, Vatsa P, Boudesocque L, Garcia-Brugger ANGELA, Rabenoelina F, Clement C (2009) Bacterial rhamnolipids are novel MAMPs conferring resistance to *Botrytis cinerea* in grapevine. Plant Cell Environ 32:178–193

Vatsa P, Chiltz A, Luini E, Vandelle E, Pugin A, Roblin G (2011) Cytosolic calcium rises and related events in ergosterol-treated *Nicotiana* cells. Plant Physiol Biochem 49:764–773

Viterbo ADA, Wiest ARIC, Brotman Y, Chet ILAN, Kenerley C (2007) The 18mer peptaibols from *Trichoderma virens* elicit plant defence responses. Mol Plant Pathol 8:737–746

Vorhölter FJ, Wiggerich HG, Scheidle H, Sidhu VK, Mrozek K, Küster H, Niehaus K (2012) Involvement of bacterial TonB-dependent signaling in the generation of an oligogalacturonide damage-associated molecular pattern from plant cell walls exposed to *Xanthomonas campestris* pv. *campestris* pectate lyases. BMC Microbiol 12:239

Wahlroos T, Susi P (2015) Differential effects of viral coat protein on induction of hypersensitive response and systemic movement of Tobacco mosaic virus and Tobacco mild green mosaic virus in *Nicotiana megalosiphon*. J Plant Pathol Microbiol 6:1

Walters DR, Ratsep J, Havis ND (2013) Controlling crop diseases using induced resistance: challenges for the future. J Exp Bot 64:1263–1280

Wan J, Zhang XC, Neece D, Ramonell KM, Clough S, Kim SY, Stacey MG, Stacey G (2008) A LysM receptor-like kinase plays a critical role in chitin signaling and fungal resistance in Arabidopsis. Plant Cell 20:471–481

Wan J, Tanaka K, Zhang XC, Son GH, Brechenmacher L, Nguyen THN, Stacey G (2012) LYK4, a lysin motif receptor-like kinase, is important for chitin signaling and plant innate immunity in Arabidopsis. Plant Physiol 160:396–406

Wang B, Yang X, Zeng H, Liu H, Zhou T, Tan B, Qiu D (2012) The purification and characterization of a novel hypersensitive-like response-inducing elicitor from *Verticillium dahliae* that induces resistance responses in tobacco. Appl Microbiol Biotechnol 93:191–201

Wang Y, Song J, Wu Y, Odeph M, Liu Z, Howlett BJ, Yang Q (2013) Eplt4 proteinaceous elicitor produced in *Pichia pastoris* has a protective effect against *Cercosporidium sofinum* infections of soybean leaves. Appl Biochem Biotechnol 169:722–737

Wang Y, Wu J, Kim SG, Tsuda K, Gupta R, Park SY, Kang KY (2016) *Magnaporthe oryzae*-secreted protein MSP1 induces cell death and elicits defence responses in rice. Mol Plant Microbe Interact 29:299–312

Wawra S, Fesel P, Widmer H, Timm M, Seibel J, Leson L, Kesseler L, Nostadt R, Hilbert M, Langen G, Zuccaro A (2016) The fungal-specific β-glucan-binding lectin FGB1 alters cell-wall composition and suppresses glucan-triggered immunity in plants. Nat Commun 7:13188

Wiesel L, Newton AC, Elliott I, Booty D, Gilroy EM, Birch PR, Hein I (2014) Molecular effects of resistance elicitors from biological origin and their potential for crop protection. Front Plant Sci 5:655

Willmann R, Lajunen HM, Erbs G, Newman MA, Kolb D, Tsuda K, Katagiri F, Fliegmann J, Bono JJ, Cullimore JV, Jehle AK (2011) Arabidopsis lysin-motif proteins LYM1 LYM3 CERK1 mediate bacterial peptidoglycan sensing and immunity to bacterial infection. Proc Natl Acad Sci U S A 108:19824–19829

Wong MY, Sariah M (2007) Evaluation on the usage of a plant activator (acibenzolar-s-methyl) for the control of basal stem rot (BSR) of oil palm (*Elaeis guineensis*). ISSAAS J 13:126–149
Wulff BB, Thomas CM, Smoker M, Grant M, Jones JD (2001) Domain swapping and gene shuffling identify sequences required for induction of an Avr-dependent hypersensitive response by the tomato Cf-4 and Cf-9 proteins. Plant Cell 13:255–272
Yamaguchi T, Maehara Y, Kodama O, Okada M, Matsumura M, Shibuya N (2002) Two purified oligosaccharide elicitors, N-acetylchitohepatose and tetraglucosyl glucitol, derived from *Magnaporthe grisea* cell walls, synergistically activate biosynthesis of phytoalexin in suspension-cultured rice cells. JPlant Physiol 159:1147–1149
Yamaguchi Y, Pearce G, Ryan CA (2006) The cell surface leucine-rich repeat receptor for AtPep1, an endogenous peptide elicitor in Arabidopsis, is functional in transgenic tobacco cells. Proc Natl Acad Sci U S A 103:10104–10109
Yamaguchi Y, Huffaker A, Bryan AC, Tax FE, Ryan CA (2010) PEPR2 is a second receptor for the Pep1 and Pep2 peptides and contributes to defence responses in Arabidopsis. Plant Cell 22:508–522
Yang Y, Zhang H, Li G, Li W, Wang XE, Song F (2009) Ectopic expression of MgSM1, a cerato-platanin protein from *Magnaporthe grisea*, confers broad-spectrum disease resistance in Arabidopsis. Plant Biotechnol 7:763–777
Yin H, Du Y, Dong Z (2016) Chitin oligosaccharide and chitosan oligosaccharide: two similar but different plant elicitors. Front Plant Sci 7:522
Zeilinger S, Gupta VK, Dahms TE, Silva RN, Singh HB, Upadhyay RS, Tsui CKM (2015) Friends or foes? Emerging insights from fungal interactions with plants. FEMS Microbiol Rev 40:182–207
Zhang Y, Yang X, Liu Q, Qiu D, Zhang Y, Zeng H, Mao J (2010) Purification of novel protein elicitor from *Botrytis cinerea* that induces disease resistance and drought tolerance in plants. Microbiol Res 65:142–151
Zhang W, Yang X, Qiu D, Guo L, Zeng H, Mao J, Gao Q (2011) PeaT1-induced systemic acquired resistance in tobacco follows salicylic acid-dependent pathway. Mol Biol Rep 38:2549–2556
Zhang G, Cui Y, Ding X, Dai Q (2013a) Stimulation of phenolic metabolism by silicon contributes to rice resistance to sheath blight. J Plant Nutr Soil Sci 176:118–124
Zhang W, Fraiture M, Kolb D, Löffelhardt B, Desaki Y, Boutrot FF, Brunner F (2013b) Arabidopsis receptor-like protein30 and receptor-like kinase suppressor of BIR1-1/EVERSHED mediate innate immunity to necrotrophic fungi. Plant Cell 25:4227–4241
Zhang L, Kars I, Essenstam B, Liebrand TW, Wagemakers L, Elberse J, van KanJ A (2014) Fungal endopolygalacturonases are recognized as microbe-associated molecular patterns by the Arabidopsis receptor-like protein responsiveness to botrytis polygalacturonases1. Plant Physiol 164:352–364
Zhang Y, Zhang Y, Qiu D, Zeng H, Guo L, Yang X (2015) BcGs1, a glycoprotein from *Botrytis cinerea*, elicits defence response and improves disease resistance in host plants. Biochem Biophys Res Commun 457:627–634
Zipfel C (2014) Plant pattern-recognition receptors. Trends Immunol 35:345
Zipfel C, Kunze G, Chinchilla D, Caniard A, Jones JD, Boller T, Felix G (2006) Perception of the bacterial PAMP EF-Tu by the receptor EFR restricts Agrobacterium-mediated transformation. Cell 125:749–760

Chapter 15
Induction of Systemic Resistance for Disease Suppression

Kalaivani K. Nadarajah

15.1 Introduction

Plants are dependent on nutrient acquisition from soil. Plant roots secrete a considerable measure of chemicals into the rhizosphere which influences growth, development, and acclimatization to environmental stresses (Vallad and Goodman 2004; Van Loon and Glick 2004). The microbial population within the rhizospheric region will similarly contribute chemical constituents that affect the microbial population as well as the plant. The dynamic nature of the rhizospheric microflora allows for an interplay between pathogenic and beneficial microorganisms. This therefore results in the organisms interacting via synergistic or antagonistic interactions (Beardon et al. 2014) where signals are being exchanged between the microorganisms and the root systems that effectively form an active belowground association (Weller et al. 2002; Van Loon and Bakker 2005). These belowground interactions are functional as long as the microbial–plant systems are kept alive to buffer the activity in the rhizospheric environment. These root microbe interactions can result in variation in effect against soil pathogens, microbial propagation, and colonization of the roots (Somers et al. 2004; De Vleesschauwer et al. 2009).

Beneficial organisms such as PGPR and plant growth promoting fungi (PGPF) control plant diseases through suppression of pathogenic soil organisms and induction of host systemic resistance. The presence of these organisms consistently induces resistance in the host beyond basal levels which acts to protect against a host of non-beneficial organisms in its surrounding. *Acinetobacter*, *Azospirilium*, *Rhizobium*, *Pseudomonas*, and *Bacillus* have been reported as efficient inducers of systemic resistance in leguminous and nonleguminous plants. In addition, *Trichoderma* spp., *Penicillium simplicissimum*, *Piriformospora indica*, *Phoma* sp.,

K.K. Nadarajah (✉)
Faculty of Science and Technology, School of Environmental and Natural Resource Sciences, National University of Malaysia, 43600 UKM, Bangi, Selangor, Malaysia
e-mail: vani@ukm.edu.my

S.N.A. Abdullah et al. (eds.), *Crop Improvement*,
DOI 10.1007/978-3-319-65079-1_15

non-pathogenic *Fusarium oxysporum*, and arbuscular mycorrhizal fungi have been listed as PGPF that have successfully suppressed diseases in several plant systems (Bakker et al. 2013).

While a large number of these strains interact and produce beneficial outcomes on varying hosts, certain strains have shown specificity indicating that the plant–microbial interaction are regulated by host variety, soil conditions, and microbial populations. Certain rhizospheric organisms produce determinants such as lipopolysaccharides, siderophores, lytic enzymes, exopolysaccharides, lipopeptides, and others (Nadarajah 2016a). These determinants trigger pathways that result in the activation of defense-related genes and responses downstream. Jasmonic acid (JA) and ethylene (ET) regulate rhizobacterial-induced systemic resistance (ISR), while systemic acquired resistance (SAR) is controlled by SA (Van Loon and Glick 2004; Haas and Défago 2005). Though both mechanisms induce host systemic resistance, they remain distinct.

While an array of microorganisms have been identified as potential biocontrol agents, only a handful have had their mechanisms elucidated (Heil and Bostock 2002; Choudhary et al. 2007). Mutants have proven to be a wonderful tool in studying the role of determinants in the mechanism of disease suppression as seen in the repression of *F. oxysporum* f. sp. *raphani* by *P. putida* WCS358 (Raaijmakers et al. 1995). Similarly cucumber plants challenged with *Colletotrichum orbiculare* showed inhibition of anthracnose symptoms post inoculation with several strains of PGPR (Wei et al. 1991). These experiments imply that the area of antagonistic influence of PGPR is not confined to the rhizosphere, but develops from below ground into above ground defense elicitations. Hence various studies and experimentations have concluded that the heightened level of resistance *in planta* was mediated through an immune response that was activated in response to rhizobacteria-ISR. As extensive reviews of these organisms have been presented elsewhere (Van Loon et al. 1998; Pozo and Azcon-Aguilar 2007), this chapter ventures to present the mechanisms, signaling pathways, comparisons between ISR and SAR, in addition to differences between these defense mechanisms that collectively work to defend plants against their hostile environment (Kloepper et al. 2004; Van Loon and Bakker 2006; De Vleesschauwer et al. 2008, 2009; Walters et al. 2013).

15.2 Induced Systemic Resistance (ISR): The Mechanism

ISR and SAR are two major players in induced plant resistance. While both contribute towards resistance, one major difference between these systems lies in the inducers, where contrary to ISR, SAR is induced in response to pathogens which results in subsequent protection from infections against a broad host of attackers (Walters et al. 2013; Pieterse et al. 2014). Further ISR and SAR are not just expressed within the locality of induction but are transmitted to distant tissues through systemic spread of signal molecules (Van Loon et al. 2008). ISR like SAR is regulated by signaling pathways that are interlinked and regulated by signal molecules/hormones (Pieterse et al. 2012). In the following segments, ISR and the contribution of ISR in agricultural practices, specifically in disease suppression, will be discussed.

15.3 Pathogen-Induced SAR

SAR has been studied locally and systemically in various plant systems. The local response includes the production of physical and chemical responses such as structural modification to the cell walls, production of phytoalexins and pathogenesis-related (PR) proteins, and hypersensitive response (HR) (Hunt and Ryals 1996; Lamb and Dixon 1997; Van Loon 1997; Van Loon and Van Strien 1999; Métraux 2002; Durrant and Dong 2004; Conrath et al. 2006). Although HR is produced in both compatible and incompatible gene-for-gene interactions (Hammond-Kosack and Jones 1997; Ellis et al. 2002), at the molecular and cellular level, HR is dispersed through uninfected tissues to trigger systemic resistance in the whole plant (Stone et al. 2000). While changes such as lignification and callous deposition are brought about post infection, the systemic transmission results in PR protein production prior to infection (Sticher et al. 1997; Dong 2004). This rapid response of distant tissue is referred to as conditioning which involves systemic signal molecule(s) such as SA. SA and other related inducers (2,6-dichloroisonicotinic acid [INA] or benzothiadiazole [BTH]) are able to promote superoxide production in the cell resulting in the production of reactive oxygen species (ROS), which ultimately activates downstream host defense enzymes such as phenylalanine ammonia-lyase (PAL) and lipoxygenase (LOX) (Katz et al. 1998; Thulke and Conrath 1998; Kauss et al. 1999; Conrath et al. 2002). Another player in the induction of pathogen-derived resistance, β-aminobutyric acid (BABA), retains effective induction even in plants with impaired SA, JA, and ET pathways (Zimmerli et al. 2000). However BABA is only able to protect mutants insensitive to JA and ET but remains ineffective in rescuing mutants defective in SAR signaling.

15.3.1 SAR Signaling

Endogenous SA has been experimentally proven to induce SAR (Van Loon and Antoniw 1982; Van Loon et al. 2008) resulting in increase of SA post induction in local and distant tissues through phloem transport (Malamy et al. 1990; Métraux et al. 1990; Verberne et al. 2003; Durrant and Dong 2004; Van Loon et al. 2008). The salicylate hydrolase defective mutant, *NahG*, that reduces SA to catechol leaving it incapable of inducing SAR was used to study the role of SA in SAR. The lack of SAR in these plants may be “rescued” through the application of exogenous INA and BTH (Ryals et al. 1996; Sticher et al. 1997; Conrath et al. 2002). Subsequently in establishing the mobile signal(s) involved in SAR, there are two possibilities: (1) SA is not the mobile signal as the rootstock-scion experiment showed induction of SAR despite no accumulation of SA in the NahG rootstock; and (2) SA as a versatile signal that is transported to distal tissues ensuing SA generation in distant tissues. Further, the presence of SA in the phloem of plants has been linked to the transport of this signal molecule within the plant to distal organs thus lending towards SAR. The overexpression of salicylate hydroxylase focused in phloem tissue of tobacco resulted in diminished SAR thus supporting a role for SA in systemic signaling (Mur et al. 2000). Another compound, methyl salicylate (MeSA) was observed in tobacco to elicit defense response. As such, MeSA was proposed as a component that acts with SA in *in planta* communication and signaling. It is therefore likely that SA as well as other systemic signals (azelaic acid, diterpenoid dehydroabietinal, glycerol-3-phosphate-dependent factor, pipecolic acid) could be involved in SAR (Shulaev et al. 1995; Seskar et al. 1998; Pieterse et al. 2014). The effective function of SA is dependent on the presence of an ankyrin protein that changes the oligomeric state of NPR1 to monomers (Cao et al. 1997). Pathogenesis-related (PR) proteins are produced from the interaction between NPR1 and transcription factors (Dong 2004). PR proteins are affected by SAR and therefore are suitable markers to study induced resistance (Kessmann et al. 1994) and remain the hallmark of SAR induction.

15.3.2 SA Mode in SAR

Catalase and ascorbate peroxidase act as SA-binding proteins that result in the formation of phenolic radicals involved in lipid peroxidation. Lipid peroxidation remains a crucial process in ensuring defense gene activation (Farmer and Mueller 2013), hence requiring the proper execution of its production at the right location and time. Other SA-binding proteins (SAPs) that demonstrate a higher affinity for SA and its analogs were identified (Bakker et al. 2014). While the biological significance of these SAPs remains unresolved, they provide an interesting view in comprehending the method of SA activity. SA- and pathogen-inducible protein kinase (SIPK), a MAP kinase member was isolated and studied in tobacco (Zhang and Klessig 1997; Zhang et al. 2002). Various studies have focused on the upstream regulatory sequences (URS) of *PR-1* promoter, one of the terminating reactions in SAR. The TGACG sequence in the URS of *PR-1* is perceived by a bZIP family TGA transcription factor (Lebel et al. 1998). TGAs were likewise found to interact with the NPR1 protein, providing a connection between *NPR1* and SA-induced *PR-1* expression (Lebel et al. 1998; Zhang et al. 1999; Després et al. 2000; Zhou et al. 2000). PR expression is suppressed when *SNI1* (negative regulator) binds to the DNA or transcription factors (Li et al. 1999). Other research groups have looked into a SA- and pathogen-inducible WRKY DNA-binding elements that recognize specific sequences on the promoter sequence of chitinase gene in tobacco. This study discovered that protein phosphorylation is essential for the function of WRKY DNA-binding components, thus underscoring the function of kinases in SA signaling (Yang et al. 1999).

15.3.3 SAR-Signaling Network

Ding et al. (2015) conducted a genetic screen of SA mutants via biosensor technique (Marek et al. 2010) which identified upstream (*EDS1, PAD4, NDR1*) and downstream components (e.g., *NPR1*), transcription factors (*CBP60g, SARD1*), and metabolic enzymes (*EPS1, PBS3*) that are crucial for SA signaling (Cao et al. 1997; Ryals et al. 1997; Jirage et al. 1999; Zhang et al. 2010; Wang et al. 2011; Yezhang et al. 2015). These screens identified two leucine rich repeats (LRR—NBS; LRR and TRI:NBS:LRR) as effectors of signal transduction post infection by avirulent pathogens (Glazebrook 1999). These pathways in the end merge at the DND1 protein which controls the development of HR cell death (Clough et al. 2000) (Fig. 15.1). In the event of being induced by a virulent pathogen, *PAD4* is activated resulting in phytoalexin production (Jirage et al. 1999). This activation further results in the downstream activation of *SID1* and *SID2* that controls SA generation (Nawrath and Metraux 1999). Studies on the *SID1* gene has shown that it is associated with the SOS response of the cell. The SOS response is elicited upon introduction of stress into the cell system. While *EDS5/SID1* expression is independent of SA, *EDS5/SID2* genes encode a MATE transporter and an ICS1 enzyme which are crucial in the SA synthesis (Ding et al. 2015) (Fig. 15.1). The *sln* mutants together with *eds5*, *sid2*, and *pad4* are involved in SA accumulation (Ryals et al. 1997; Jirage et al. 1999; Nawrath and Métraux, 1999; Nawrath et al. 2002; Ding et al. 2015). NPR1 acts as a feedback inhibitor of SA biosynthesis following accumulation of SA in response to infection and infestation in *NPR1* gene mutants. In addition to components upstream of SA biosynthesis (Clarke et al. 2000; Zhang et al. 2010), NPR1-independent defense responses is triggered by *EDS5*, *PAD4*, *SID2*, and *SLN* genes (Glazebrook 1999, 2001). The *sln1* mutant influences the PR protein expression; hence this goes against the proposed involvement of SA-independent pathway in the regulation of *PR* gene expression.

15.4 Rhizobacteria-ISR

The microorganisms within the rhizosphere of the soil have an important role in the general well-being and health of plants. The bacteria and fungi within the rhizosphere can either be in a beneficial relationship or negatively affect the plant or microbial population (Nadarajah 2016a, b). Among the functions attributed to these organisms are the ability to participate in the nutrient cycles, nutrient acquisition, and management of biotic (Bakker et al. 2007; Van der Heijden et al. 2008; Khan et al. 2009; Kraiser et al. 2011; Berendsen et al. 2012) and abiotic stresses (Yang et al. 2004). These functions are associated with a staggeringly dynamic and complex microbiome within the rhizosphere (Berendsen et al. 2012; Hartmann et al. 2009; Raaijmakers et al. 2009). Through the utilization of molecular techniques, it is expected that the repertoire of microbiome identified that are in

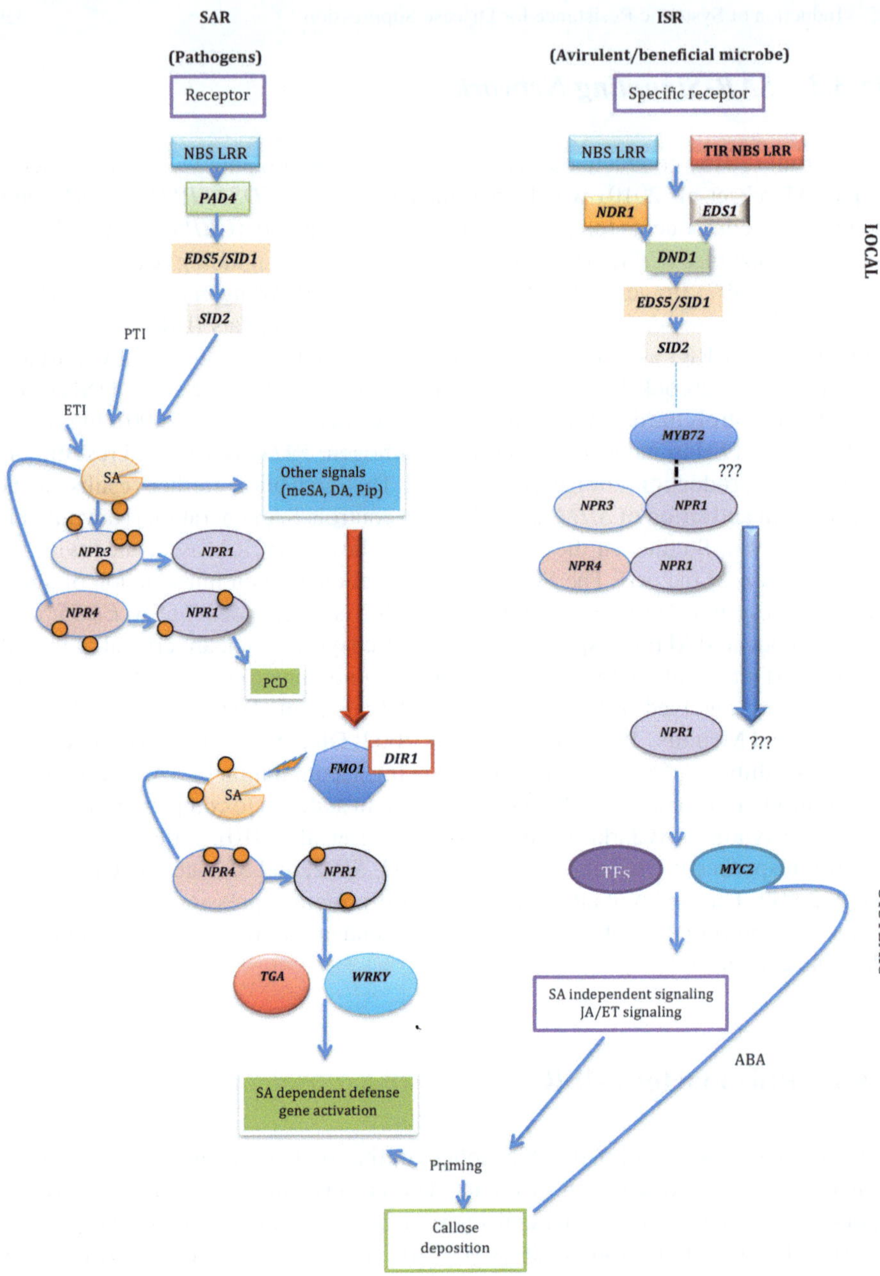

Fig. 15.1 Schematic representation of the components involved in the activation of SAR and ISR. These protein perceive and transmit the signal in SAR and ISR. *Solid lines* and *arrows* are confirmed connections. *Dashes* imply further study required. *Solid arrows* indicate the transmission from local site of elicitation to distant tissue transmission. *Abbreviations*: *NB-LRR* Nucleotide-Binding-Leucine-Rich Repeat, *PCD* Programmed Cell Death, *PRR* Pattern-Recognition Receptor, *PTI* PAMP-Triggered Immunity, *SA* Salicylic Acid, *TF* Transcription Factor, *PAD* Phytoalexin Deficient, *EDS* Enhanced Disease Susceptibility, *SID* SA Induction-Deficient, *DND* Defense, No Death, *NDR* Non-Race Specific Disease Resistance, *DIR* Defective In Induced Resistance, *FMO* Flavin-Dependent Monooxygenase, *NPR* Non-Expressor PR, *TGA* TGACG-binding protein (TF), *MYB* Myeloblastosis (TF), and *MYC* TF regulator

interaction with plants and associated with ISR against biotic and abiotic stresses will increase. Studies will not be limited to only identification of new rhizospheric microorganisms but to the mechanisms, key players, and pathways involved in these processes.

15.4.1 Beneficial Microbes and the SAR Pathway

While SAR is a complex mechanism of systemic resistance in plants, ISR prompts a more complex defense in response to non-pathogenic rhizobacteria. Due to the systemic response elicited by ISR, it was at one point assumed to be SAR. This misconception was debunked by Hoffland et al. (1995) who established that ISR against *F. oxysporum* was induced by *P. fluorescens* WCS417r in radish without the trademark PR production as seen in SAR. Similar findings were described by Pieterse et al. (1996) in Arabidopsis. This was further corroborated when *NahG* resulted in an ISR response post treatment with WCS417r-ISR, indicating the involvement of an SA-independent pathway and separating this process from SAR (Pieterse et al. 1996). The same is seen in response to *P. putida* WCS358r (Raaijmakers et al. 1995; Van Wees et al. 1999) and *Serratia marcescens* 90-166 where the loss of SA production induced resistance against *Colletotrichum orbiculare* and *P. syringae* pv. tomato (Press et al. 1997). However when strain 7NKS2 was used in treatment of *NahG* mutants in Arabidopsis and tomato, ISR was abolished against TMV and *Botrytis cinerea* (De Meyer et al. 1999a; Audenaert et al. 2002). Ryu et al. (2003) also encountered similar findings in Arabidopsis, with *B. pumilus* SE34 against *P. syringae pv maculicola*. Further, Maurhofer et al. (1998) observed SA-dependent SAR elicitation post treatment with *P. fluorescens* P3 overexpressing SA biosynthesis gene cluster of *P. aeruginosa* PAO1. Additionally *P. fluorescens* SS101, *Paenibacillus alvei* K165 (Tjamos et al. 2005; Van de Mortel et al. 2012), and *Trichoderma* PGPF (Mathys et al. 2012; Martínez-Medina et al. 2013) were also reported to induce SA-dependent SAR. The requirement for SA in 7NKS2 was substantiated through utilization of bacterial mutants defective in SA production (De Meyer and Höfte 1997; De Meyer et al. 1999b; Audenaert et al. 2002). However following this finding, further experiments were conducted by Van Loon and Bakker (2005) who inferred that SA-independent pathways is the main regulatory pathway in rhizobacteria-mediated ISR (De Meyer et al. 1999a). Although some PGPR produce SA but it is still not the main signal involved in elicitation of ISR (Ran et al. 2005; Djavaheri et al. 2012). The SA produced however binds with siderophores in iron-limiting condition and thus is not directly involved in SAR. Meanwhile, in cases where SA is produced by PGPR, ROS is an elicitor required to activate SAR. Further these responses are not dependent on accumulation of SA but rather on increasing the sensitivity of tissue to SA (Van Loon and Bakker 2003).

15.4.2 *NPR1* as a Common Component of ISR and SAR

Transmission of SAR to distal organs requires mediators. The transmitted SAR signal is chaperoned by one such mediator, Defective In Induced Resistance1 (*DIR1*) (Champigny et al. 2013) which assist Flavin-Dependent Monooxygenase 1 (*FMO1*) in receiving and amplifying signals for long distance SAR signaling (Mishina and Zeier 2007) (Fig. 15.1). The well-characterized transcriptional co-regulator, *NPR1*, plays a role in SA accumulation and in the SAR-signaling pathway (Dong 2004; Vlot et al. 2009; Pajerowska-Mukhtar et al. 2013). Pieterse et al. (1998) had implicated *NPR1* in ISR through studies conducted on Arabidopsis. In studying the activation of ISR post treatment with *P. fluorescens* WCS417r and numerous other PGPR and PGPF, Pieterse et al. (1998) found a connection between *NPR1* and the JA/ET-signaling pathways (Lavicoli et al. 2003; Ryu et al. 2003; Stein et al. 2008; Weller et al. 2012). This therefore demonstrates that SA signaling in response to either an avirulent pathogen or rhizobacteria can activate *NPR1*. While the role of *NPR1* in SA signaling has been connected to nuclear function, recent studies have provided information that the *NPR1* component of the JA/ET signaling is within the cytosol (Spoel et al. 2003; Stein et al. 2008; Ramirez et al. 2010). Both ISR and SAR defense mechanisms have additive effect within the host. At this juncture, it is difficult to ascertain the specific molecular mechanism involved in the *NPR1* mediated JA/ET based ISR induction in host (Van Wees et al. 2000). Pieterse et al. (2014) reported that plant roots expressed high levels of *NPR1, NPR3,* and *NPR4* suggesting a crucial role for these genes in belowground interactions. Both NPR3 and NPR4 together with Cullin 3 (CUL3) ubiquitin E3 ligase are involved in the degradation of NPR1. NPR3 degrades NPR1 when the levels of SA are high causing localized cell death during effector triggered immunity (ETI), while at lower SA levels, NPR4 maintains NPR1 during pathogen-associated molecular patterns triggered immunity (PTI) and thus results in PR expression (Fig. 15.1). In ISR, NPR1 itself acts to mediate the systemic response together with MYC and TFs and the JA/ET pathways. Though *NPR1* is a shared component of ISR and SAR, the mechanism downstream of *NPR1* perception is different as SAR results in PR accumulation while ISR does not. This could perhaps be due to lower levels of SA-induced ISR that perhaps was insufficient to result in PR production. Studies with the *npr1* mutant plants that did not express ISR post cultivation with WCS417 indicated that the expression was dependent on regulation and sensitivity and not towards the SA levels within the host. However, further research is required to understand the role of *NPR1* and the possible involvement of other regulatory factors in the SA-NPR1 interaction in ISR (Pieterse et al. 2012).

15.4.3 Other Pathways That Control ISR

As mentioned in the above sections, the JA/ET-signaling pathway is crucial in the induction of ISR in plants. Arabidopsis JA (*jar1, coi1, jin1*) and ET (*ein2, etr1, eir1, ein3*) mutants were utilized to establish the function of JA/ET in the plant immune system (Thomma et al. 2001). When these mutants were treated with PGPR (*Pseudomonas* CHA0, *P. fluorescens* WCS417r–ISR, *P. syringae* pv. maculicola, *P. fluorescens* Q2-87, *S. marcescens* 90–166) (Pieterse et al. 1998; Ryu et al. 2003; Pozo et al. 2008) and PGPF (*P. indica, Penicillium* sp. GP16-2, *Trichoderma harzianum* T39) they failed to induce ISR confirming the role of JA and ET in ISR (Ryu et al. 2004, Stein et al. 2008, Weller et al. 2012; Pieterse et al. 2014). Similar observations were also made in other plant systems, thus supporting the notion that in SA-independent ISR, JA/ET are the main regulators of plant immunity (Yan et al. 2002; De Vleesschauwer et al. 2008; Van der Ent et al. 2009). These pathways are also effective against necrotrophs and herbivores (Van Loon et al. 2008; Van Wees et al. 2008; De Vleesschauwer et al. 2009; Ding et al. 2015; Yezhang et al. 2015).

The *Jar1* gene encodes JA-amino acid synthetase which activates the JA signaling. Treatment of wild-type plants with meJA and the ET precursor1-aminocyclopropane-1-carboxylate (ACC) did elicit a response similar to rhizobacterial colonization in plants. However, when treated with these beneficial organisms, endogenous JA levels did not increase which led to the conclusion that the signaling was dependent on JA responsiveness (Pieterse et al. 2000; Staswick and Tiryaki 2004; Van Loon and Bakker 2005). Further, Knoester et al. (1999) using ethylene insensitive mutants demonstrated impaired ISR implicating the requirement of complete and functional ET pathway for proper ISR function.

15.5 Elicitor Molecules in Rhizobacteria-ISR

The organisms that result in ISR do not cause any damage to host. Hence this has resulted in the early conclusion that the chemical compounds resulting in ISR and those resulting in SAR/HR are different. Unlike SAR, ISR is not dependent on localized cell death but rather on the production of elicitors/determinants that trigger the mechanism (Ebel and Mithöfer 1998). A host of chemical determinants have been identified as capable of inducing resistance either individually or in combination (Bakker et al. 2003). These determinants however seem to share similarities in the defense reaction elicited within the host (Gómez-Gómez 2004; Nürnberger et al. 2004). For instance, crude cell wall extracts and lipopolysaccharides (LPS) of *P. fluorescens* WCS358 resulted in the activation of defense-associated reactions in Arabidopsis (Raaijmakers et al. 1995; Van Wees et al. 1999; Meziane et al. 2005; Nadarajah 2016a) and reduced disease symptoms in pathogen challenged plants. Mutant analysis displayed a redundancy in elicitors as the lack of either O antigenic side chains in Lipopolysaccharide (LPS) or flagellin in these mutants still resulted in induced resistance, as the presence of either one elicitor compounds was sufficient to elicit a response. However, not all strains of *P. fluorescence* can elicit resistance in Arabidopsis or other plants (Van Wees et al. 1997). This variation may be due to differences in the chemical composition or structure of their determinants. For instance, it was reported that the O-antigenic side-chain of LPS differs from strain to strain probably resulting in perception specificity towards different plant species. Examples of specificity can be seen in application of LPS from *Burkholderia cepacia* against *Phytophthora nicotianae* strain ASP B 2D in tobacco (Coventry and Dubery 2001) and the efficient control of the nematode, *Globodera pallida* with LPS from *Rhizobium etli* strain G12 in potato (Reitz et al. 2002). Different species or strains of these rhizobacteria resulted in either induction or no response in the host.

Siderophore is another determinant that is involved in the induction of ISR. As there is a redundancy of determinants in rhizobacteria, ISR may be induced by different components in different plant species as exhibited by 7NSK2 in bean and tomato where SA and siderophores were implicated in the response (Audenaert et al. 2002). As siderophores are produced under iron-limiting conditions, it not only inhibits the pathogens within the soil but also induces systemic reaction within the host. However, while all siderophores are able to utilize iron, not every siderophore elicits ISR due to the differences in their chemical structure. Some siderophores produced by the rhizobacteria are pseudomonine, pyochelin, and pseudobactin (Nadarajah 2016a). Some examples of siderophore utilizing rhizobacteria are WCS374 (Leeman et al. 1996; Djavaheri et al. 2012), *P. aeruginosa* 7NSK2 (Audenaert et al. 2002), *Serratia marcescens* 90-166 (Press et al. 1997), and *P. fluorescens* CHA0 (Maurhofer et al. 1994; Weller et al. 2004).

Antibiotics play dual function in the rhizosphere as a microbial antagonist and a defense activator (Fernando et al. 2005). PGPR have been associated with producing more than one antibiotic which relates to its usefulness against phytopathogens (Glick et al. 2007). Diffusible (e.g., phenazines, pyoluteorin, pyrrolnitrin, cyclic lipopeptides (CLP)) and volatile (HCN) antimicrobial products are classified into six groups and interact effectively against microorganisms, nematodes, and plants (Haas and Défago 2005; Raaijmakers et al. 2010). The pyocyanin and pyochelin siderophore from 7NSK2 elicit ISR in conjunction with the 2,4-diacetylphloroglucinol (DAPG) antibiotic (Lavicoli et al. 2003) in tomato. DAPG likewise acts as an inducer in Q2-87 and CHA0 inducing resistance in tomato against *Meloidogyne javanica* (Siddiqui and Shaukat, 2003; Weller et al. 2004). These reports on DAPG suggest that there may be other rhizobacteria and antibiotics capable of eliciting ISR in plants. Pyrrolnitrin produced by the *P. fluorescens* (BL915) prevents damping-off by *Rhizoctonia solani* in cotton while phenazine producing pseudomonads possess redox potential with the ability to suppress various pathogens (Chin-A-Woeng et al. 2003). Phenazine-1-carboxamide that was isolated and studied from roots of tomato was able to mobilize iron from soil (*P. chlororaphis* PCL1391) (Haas and Défago 2005). A large number of Pseudomonads and *Bacillus* spp. have been reported to produce various antimicrobial compounds that have selective effect against various host and environments (Beneduzi et al. 2013).

Studies have shown that the interaction between these rhizobacteria and plant roots are dependent on plant variety, environmental conditions, and soil community (Ton et al. 1999; Nadarajah 2016b). While certain strains are perfect inducers of resistance in various plant species, most show tight specificity to root cell surface receptors (Van Loon et al. 1998). For example, WCS358 stimulates resistance in tomato, Arabidopsis, and bean (Raaijmakers et al. 1995; Meziane et al. 2005), and fails to do so in carnation or radish (Leeman et al. 1995). Other strains such as WCS374 induced a powerful response in radish (Leeman et al. 1995) while another, WCS417, could successfully elicit a response in all the above five species of plants (Leeman et al. 1995; Bakker et al. 2013, 2014). Over the course of the last two

Table 15.1 Example of beneficial microbes and their determinants involved in disease suppression

Host	Pathogen	Beneficial microbe	Determinant	Reference
Arabidopsis	*Erwinia caratovora*	*B. amyloliquefaciens* IN937a	2,3-Butanediol	Ryu et al. (2003, 2004)
	P. syringae pv maculicola	*B. substilis* GB03	2,3-Butanediol	Lavicoli et al. (2003)
		B. pumilus SE34;T4	SA	Weller et al. (2004)
		P. fluorescens CHA0	2,4-DAPG	Meziane et al. (2005)
	Peronospora parasitica	*P. fluorescens* Q2-87	2,4-DAPG	
	P. syringae pv tomato	*P. putida* WCS358	LPS, siderophore	
Tobacco	TNV	*P. fluorescens* CHA0	Siderophore	Maurhofer et al. (1994)
	Peronospora tabacina	*B. pumilus* SE34	SA	De Meyer et al. (1999a)
	TMV	*P. aeruginosa* 7NKS2	SA	De Meyer et al. (1999b)
Tomato	*Botrytis cinerea*	*P. aeruginosa* 7NKS2	Pyochelin, pyocyanin	Audenaert et al. (2002)
	Meloidogyne javanica	*P. fluorescens* CHA0, *P. putida* WCS358	2,4-DAPG, LPS, siderophore	Siddiqui and Shaukat (2003)
	P. syringae			Meziane et al. (2005)
Bean	*P. syringae*	*P. aeruginosa* 7NKS2	SA	De Meyer et al. (1999a)
		P. putida WCS358	LPS, siderophore	Meziane et al. (2005)
		P. putida BTP1	Iron regulated metabolite, hexanal	
Rice	*Magnaporthe oryzae*	*P. aeruginosa* 7NSK2	SA, LPS, siderophores	De Vleesschauwer et al. (2006, 2008, 2009)
	Rhizoctonia solani	*P. fluorescens* WCS374r	LPS, siderophores	
	Magnaporthe oryzae	*Serratia plymuthica* IC1270	SA, LPS, siderophores	
	Cochliobolus myiabeanus			

decades, many ISR determinants have been identified in certain rhizobacterial species. Some examples are provided in Table 15.1.

15.5.1 Key Early Root-Specific Regulator in ISR

Although signaling for ISR starts at the root–microbe interface, not much research has been done to investigate the signaling components involved at eliciting ISR at the root level. Knoester et al. (1999) in studying the root interaction in ISR used a root ET insensitive mutant (*eir1*) which exhibited the involvement of ET in the transmission of ISR, which was aided perhaps by some other regulatory elements. Further *MYB72* was identified as a transcription factor that is involved in the signal transduction from the root as observed in response to *P. fluorescens* WCS417r in Arabidopsis (Verhagen et al. 2004; Pieterse et al. 2014). *MYB72* shows high levels of expression in PGPR-induced roots and no expression was detected in the phloem of uninduced plants. PGPR (*P. putida* WCS358, *P. fluorescens* WCS417r) and PGPF (*Trichoderma* spp.) treated mutant *MYB72* plants showed no ISR response hence indicating a significant role for this factor in ISR (Segarra et al. 2009). However, these studies showed that an overexpression of *MYB72* did not result in enhanced ISR but rather is dependent on iron-limiting conditions making a connection between iron equilibrium and ISR induction (Van der Ent et al. 2008; Palmer et al. 2013). Treatment with ISR-inducing *Pseudomonas* strains resulted in the co-regulation of iron deficient marker genes (*FRO2, IRT1*) and *MYB72* in Arabidopsis (Zamioudis et al. 2013). Transcriptome profiling of mutant *myb72* and wild-type Arabidopsis provided evidence that iron deficiency response genes were the most dominant species found in roots. PGPR and PGPF however are known to produce siderophores which result in iron uptake from environment therefore resulting in the iron deficient environment. In order to establish if the siderophores are responsible for the deficiency, a siderophore mutant was utilized which exhibited normal *MYB72, FRO2,* and *IRT1* gene activity confirming the role of these microbes in iron deficiency. This interaction requires further study for a better understanding of the connection between iron limitation and siderophore function in ISR (Zhang et al. 2007).

15.6 Expression of ISR

The consequence of ISR expression leads to reduced disease incidence as well as severity post treatment. While ISR executes its defense from belowground, SAR spreads its defense to distal organs from site of pathogenic infection. While both share some overlap in the mechanism of defense moderation, their signals differ. Studies have also shown that due to these differences in signals and moderation, SAR is more effective against biotrophic pathogens while ISR are active against necrotrophs (Bakker et al. 2013). Therefore through the activation of JA and SA-dependent pathways, plants defends themselves against different pathogens in different plant species. This preparatory state of the plant to defend against invasion is called "priming" where there is enhanced level of cellular defense resulting in improved resistance (Van Wees et al. 2008). Various studies conducted on PGPR and PGPF have shown a role for priming in ISR defense (Van Loon and Bakker 2005; Wang et al. 2005). Priming is an important biological and chemical process that is fit and cost effective in adapting plants to its hostile environment (Pozo et al. 2008; Conrath 2011). In addition to the chemical changes observed within the plants in defense, there are structural changes such as callose deposition observed at the site of pathogen entry as seen in *P. fluorescens* WCS417r treated Arabidopsis (Van der Ent et al. 2009). Abscisic acid (ABA) has been indicated as essential in primed response against insects and pathogens in ISR (Corné et al. 2013; Vos et al. 2013). Besides callose deposition, *Bacillus subtilis* FB17 was observed to aggregate around the roots of plants infected by *P. syringae.* The presence of this organism induced stomatal closure and thence reduced the potential of invasion by foliar pathogens through the stomata (Walters et al. 2008). Transcription factors have a responsibility in signaling and regulating the primed state. These factors remain inactive in a non-induced stage but are rapidly activated when the host is affected by pathogens or insects. One transcriptional factor that has been linked to regulation and signaling of ISR is a member of the AP2/ERF family (Van der Ent et al. 2009). These factors have been linked with JA/ET regulated genes and are directly linked to ISR expression (Verhagen et al. 2004). Promoter region analysis of these ISR-related genes revealed the presence of cis-acting G box motif. These motifs are linked to a regulator of JA dependent response, MYC2 (Pozo et al. 2008; Stein et al. 2008) (Fig. 15.1) which is required for proper execution of this pathway. Out of the genes expressed in Arabidopsis, only ~1% of these genes are expressed at the root level and there is no constancy in the expression level observed in the distal leaves (Verhagen et al. 2004).

15.7 Disease-Suppressed Soil

Disease suppressive soil has been described as soil that shows suppression of pathogen through competition for nutrient, antagonism, lytic enzymes, quorum sensing, and various other means by which a non-beneficial microbial population is kept at check (Weller et al. 2002; Loper et al. 2012; Philippot et al. 2013). Through these belowground activities, damage is reduced significantly to the host or the establishment of disease becomes less important over time in a particular soil (Mazzola 2002). Beneficial microbes have been used to control agriculturally significant organisms such as *Gaeumannomyces graminis* var. *tritici* through the production of DAPG on Take All Disease. Over successive events of take all in a particular location, the soil eventually became suppressive towards the pathogen. This has been seen in events where monocultures were grown over a period of time resulting in inhibition of the pathogen due to eventual multiplication and high dosage levels of the beneficial microbes (*Pseudomonas fluorescens*) within the soil (Weller et al. 2002). This disease suppressive soil can also be used in amending condusive soil to reduce disease incidences (Raaijmakers and Weller 1998). Another example in disease suppresiveness is against Fusarium wilt (Alabouvette 1999), and *Rhizoctonia solani* infestations (Mendes et al. 2011). The competition for nutrient and the production of phenazines appear to reduce the wilt symptom in infected soil (Mazurier et al. 2009). In each incident, it has been reported that while there may be a dominant microbe facilitating this suppression, in most cases it will involve a consortium. This consortium may be made up of microbes from the groups: *Azospirillum, Bulkholderia, Comamonas, Gluconacetobacter, Pseudomonas*, and *Sphingomonadaceae* genus (Kyselkova et al. 2009). Through Chip technology, 17 taxa of β and γ-proteobacteria and firmicutes were linked to disease suppressiveness (Mendes et al. 2011). Most often, disease suppressiveness has been linked to antibiosis (Raaijmakers et al. 2002), siderophore producing Pseudomonaceae (Duijff et al. 1998; Zhang et al. 2007) and ISR (Bakker et al. 2007) subsequently resulting in reduced disease incidence and severity (Pieterse et al. 2013).

15.8 Concluding Remarks and Future Prospects

Much research has been devoted towards understanding the role of beneficial microbes in the elicitation of plant immunity and its specific role in ISR since its discovery more than two decades ago. The plant immune system is unique in a way that it is activated to fend off enemies while it remains suppressed to support beneficial interactions. Both these interactions of the plant immune system are in play in ISR to benefit the host. It remains to be deciphered how a phenomenon that enhances plant immunity towards both biotic and abiotic stresses can at the same time contribute towards improved growth and development in the host. One would expect that the initial approach would be to try and determine or understand the "message" transmitted at the point of contact and how this message is then amplified and transmitted to the rest of the plant. We should also look at how both the ISR and SAR mechanisms overlap and what are the shared or different points of regulation between these mechanisms in incurring an effective immune response in host. The perception of the signal by the receptors as well as the regulators involved in the long distance signaling and perception of the signal in both ISR and SAR is still not completely understood. We believe through the use of the "omics" platforms, a wealth of information will surface not only to enable us to better understand the key players in these processes but to add on and gap fill on issues such as regulators, receptors, signal molecules, pathways, and other participants of the complex system of plant–microbe interactions in ISR and SAR. While we may have acquired sufficient knowledge on how the soil microbiomes improve plant heath and development, we are still vague on how the host is able to shape the microbial community surrounding the roots to best benefit it. Likewise, while we may know the key processes involved in the perception and signaling of SAR in plant, there are still gaps in our knowledge with regards to the signals, the regulation, the perception and the mode of transmission of signals long distance.

As a major societal challenge is producing sufficient agricultural produce to meet the market and population demands, any development in science that lends towards this is a positive contribution. In this chapter, we see how both SAR and ISR are two main contributors of the plant defense mechanism and how a better knowledge and understanding of these can assist us with the challenge. ISR has been used in the past decades as biocontrols and in soil amendments all with the hope of reducing disease incidence and severity and at the same time contributing towards better yield and development. A better understanding of SAR on the other hand will likewise contribute towards the knowledge to enhance plant immunity through external stimuli, breeding and the utilization of transgenics towards generating crops with heightened defense mechanisms.

Some answers that may assist with obtaining a clearer and more well-defined picture of SAR and ISR in plants may arise from addressing the following questions:

1. How does the plant facilitate the colonization of a suitable community to enhance growth and immunity? How do these communities play a dual role in growth and immunity?
2. What are the long distance signal(s) involved in both ISR and SAR? How are they transmitted and how do they trigger ISR/SAR?
3. Are there any other regulators than *MYB72* in the root for ISR?
4. Are there any other transcription factors or regulators that are involved in ISR and SAR? What are their function and contribution in these processes?
5. Is there a role for autoregulation of mutualism in ISR?
6. How exactly does *NPR1* regulate ISR?
7. What are the differences and similarities in genes triggered by ISR and SAR in plants?

While the above questions are not the only areas left with gaps to fill, the constant inquisition on the above mechanisms will only increase our knowledge. However there is always a possibility that with new knowledge comes new questions and new issues to address.

References

Alabouvette C (1999) Fusarium wilt suppressive soils: an example of disease-suppressive soils. Australasian Plant Pathol 28:57–64

Audenaert K, Pattery T, Cornelis P, Hofte M (2002) Induction of systemic resistance to *Botrytis cinerea* in tomato by *Pseudomonas aeruginosa* 7NSK2: role of salicylic acid, pyochelin, and pyocyanin. Mol Plant-Microbe Interact 15:1147–1156

Bakker PAHM, Ran LX, Pieterse CMJ, Van Loon LC (2003) Understanding the involvement of rhizobacteria-mediated induction of systemic resistance in biocontrol of plant diseases. Can J Plant Pathol 25:5–9

Bakker PAHM, Pieterse CMJ, Van Loon LC (2007) Induced systemic resistance by fluorescent *Pseudomonas* spp. Phytopathology 97:239–243

Bakker PAHM, Berendsen RL, Doornbos RF, Wintermans PCA, Pieterse CMJ (2013) The rhizospere revisited: root microbiomics. Front Plant Sci 4:165

Bakker PAHM, Ran LX, Mercado-Blanco J (2014) Rhizobacterial salicylate production provokes headaches. Plant Soil 382:1–16

Beardon E, Scholes J, Ton J (2014) How do beneficial microbes induce systemic resistance? In: Walters RD, Newton AC, Gary DL (eds) Induced resistance for plant defense: a sustainable approach to crop protection. Wiley, Chichester, pp 232–248

Beneduzi A, Moreira F, Costa PB, Vargas LK, Lisboa BB, Favreto R et al (2013) Diversity and plant growth promoting evaluation abilities of bacteria isolated from sugarcane cultivated in the South of Brazil. Appl Soil Ecol 63:94–104

Berendsen RL, Pieterse CMJ, Bakker PAHM (2012) The rhizosphere microbiome and plant health. Trends Plant Sci 17:478–486

Cao H, Glazebrook J, Clarke JD, Volko S, Dong X (1997) The Arabidopsis NPR1 gene that controls systemic acquired resistance encodes a novel protein containing ankyrin repeats. Cell 88:57–63

Champigny MJ, Isaacs M, Carella P, Faubert J, Fobert PR, Cameron RK (2013) Long distance movement of DIR1 and investigation of the role of DIR1-like during systemic acquired resistance in Arabidopsis. Front Plant Sci 4:230

Chin-A-Woeng TFC, Bloemberg GV, Lugtenberg BJJ (2003) Phenazines and their role in biocontrol by *Pseudomonas* bacteria. New Phytol 157:503–523

Choudhary DK, Prakash A, Johri BN (2007) Induced systemic resistance (ISR) in plants: mechanism of action. Indian J Microbiol 47:289–297

Clarke JD, Volko SM, Ledford H, Ausubel FM, Dong X (2000) Roles of salicylic acid, jasmonic acid, and ethylene in *cpr*-induced resistance in Arabidopsis. Plant Cell 12:2175–2190

Clough SJ, Fengler KA, Yu IC, Lippok B, Smith RK Jr, Bent AF (2000) The Arabidopsis *dnd1* "defense, no death" gene encodes a mutated cyclic nucleotide-gated ion channel. Proc Natl Acad Sci USA 97:9323–9328

Conrath U (2011) Molecular aspects of defence priming. Trends Plant Sci 16:524–531

Conrath U, Pieterse CMJ, Mauch-Mani B (2002) Priming in plant-pathogen interactions. Trends Plant Sci 7:210–216

Conrath U, Beckers GJM, Flors V, García-Agustín P, Jakab G, Mauch F, Newman MA, Pieterse CM, Poinssot B, Pozo MJ, Pugin A, Schaffrath U, Ton J, Wendehenne D, Zimmerly L, Mouch-Mani B (2006) Priming: getting ready for battle. Mol Plant Microbe Interact 19:1062–1071

Corné MJP, Poelman EH, Van Wees SCM, Marcel D (2013) Induced plant responses to microbes and insects. Front Plant Sci 4:475

Coventry HS, Dubery IA (2001) Lipopolysaccharides from *Burkholderia cepacia* contribute to an enhanced defensive capacity and the induction of pathogenesis-related proteins in *Nicotiana tabacum*. Physiol Mol Plant Pathol 58:149–158

De Meyer G, Höfte M (1997) Salicylic acid produced by the rhizobacterium *Pseudomonas aeruginosa* 7NSK2 induces resistance to leaf infection by *Botrytis cinerea* on bean. Phytopathology 87:588–593

De Meyer G, Audenaert K, Höfte M (1999a) *Pseudomonas aeruginosa* 7NSK2-induced systemic resistance in tobacco depends on *in planta* salicylic acid accumulation but is not associated with *PR1a* expression. Eur J Plant Pathol 105:513–517

De Meyer G, Capieau K, Audenaert K, Buchala A, Métraux JP, Höfte M (1999b) Nanogram amounts of salicylic acid produced by the rhizobacterium *Pseudomonas aeruginosa* 7NSK2 activate the systemic acquired resistance pathway in bean. Mol Plant Microbe Interact 12:450–458

De Vleesschauwer D, Cornelis P, Höfte M (2006) Redox-active pyocyanin secreted by *Pseudomonas aeruginosa* 7NSK2 triggers systemic resistance to *Magnaporte grisea* but enhances *Rhizoctonia solani* susceptibility in rice. Mol Plant Microbe Interact 19:1406–1419

De Vleesschauwer D, Djavaheri M, Bakker PAHM, Hofte M (2008) *Pseudomonas fluorescens* WCS374r-induced systemic resistance in rice against *Magnaporthe oryzae* is based on pseudobactin-mediated priming for a salicylic acid-repressible multifaceted defense response. Plant Physiol 148:1996–2012

De Vleesschauwer D, Chernin L, Höfte MM (2009) Differential effectiveness of *Serratia plymuthica* IC1270-induced systemic resistance against hemibiotrophic and necrotrophic leaf pathogens in rice. BMC Plant Biol 9:9

Despres C, Delong C, Glaze S, Liu E, Fobert PR (2000) The Arabidopsis NPR1/Nim 1 protein enhances the DNA binding activity of a subgroup of the TGA family of bZIP transcription factors. Plant Cell 12:279–290

Ding Y, Shaholli D, Mou Z (2015) A large-scale genetic screen for mutants with altered salicylic acid accumulation in Arabidopsis. Front Plant Sci 5:763

Djavaheri M, Mercado-Blanco J, Versluis C, Meyer J-M, Van Loon LC, Bakker PAHM (2012) Iron regulated metabolites produced by *Pseudomonas fluorescens WCS374r are* not required for eliciting induced systemic resistance (ISR) against *Pseudomonas syringae* pv. tomato in Arabidopsis. Microbiol Open 1:311–325

Dong X (2004) NPR1, all things considered. Curr Opin Plant Biol 7:547–552

Duijff BJ, Pouhair D, Olivain C, Alabouvette C, Lemanceau P (1998) Implication of systemic induced resistance in the suppression of fusarium wilt of tomato by *Pseudomonas fluorescens* WCS417r and by nonpathogenic *Fusarium oxysporum* Fo47. Eur J Plant Pathol 104:903–910

Durrant WE, Dong X (2004) Systemic acquired resistance. Annu Rev Phytopathol 42:185–209

Ebel J, Mithöfer A (1998) Early events in the elicitation of plant defence. Planta 206:335–348

Ellis C, Karafyllidis I, Turner JG (2002) Constitutive activation of jasmonate signaling in an Arabidopsis mutant correlates with enhanced resistance to *Erysiphe cichoracearum, Pseudomonas syringae*, and *Myzus persicae*. Mol Plant Microbe Interact 15:1025–1030

Farmer EE, Mueller MJ (2013) ROS-mediated lipid peroxidation and RES-activated signaling. Annu Rev Plant Biol 64:1–22

Fernando WG, Ramarathnam R, Krishnamoorthy AS, Savchuk SC (2005) Identification and use of potential bacterial organic antifungal volatiles in biocontol. Soil Biol Biochem 37:955–964

Glazebrook J (1999) Genes controlling expression of defense responses in Arabidopsis. Curr Opin Plant Biol 2:280–286

Glazebrook J (2001) Genes controlling expression of defense responses in Arabidopsis. Curr Opin Plant Biol 4:301–308

Glick BR, Cheng Z, Czarny J, Duan J (2007) Promotion of plant growth by ACC deaminase-containing soil bacteria. Eur J Plant Pathol 119:329–339

Gómez-Gómez L (2004) Plant perception systems for pathogen recognition and defence. Mol Immunol 41:1055–1062

Haas D, Défago G (2005) Biological control of soilborne pathogens by fluorescent pseudomonads. Nat Rev Microbiol 4:1–13

Hammond-Kosack KE, Jones JD (1997) Plant disease resistance genes. Annu Rev Plant Physiol Plant Mol Biol 48:575–607

Hartmann A, Schmid M, van Tuinen D, Berg G (2009) Plant-driven selection of microbes. Plant and Soil 321:235–257

Heil M, Bostock RM (2002) Induced systemic resistance (ISR) against pathogens in the context of induced plant defences. Ann Bot 89:503–512

Hoffland E, Pieterse CMJ, Bik L, Van Pelt JA (1995) Induced systemic resistance in radish is not associated with accumulation of pathogenesis-related proteins. Physiol Mol Plant Pathol 46:309–320

Hunt M, Ryals J (1996) Systemic acquired resistance signal transduction. Crit Rev Plant Sci 15:583–606

Jirage D, Tootle TL, Reuber TL, Frost LN, Feys BJ, Parker JE, Ausubel FM, Glazebrook J (1999) *Arabidopsis thaliana* PAD4 encodes a lipase-like gene that is important for salicylic acid signaling. Proc Natl Acad Sci U S A 96:13583–13588

Katz VA, Oliver U, Thulke CU (1998) Benzothiadiazole primes parsley cells for augmented elicitation of defense responses. Plant Physiol 117:1333–1339

Kauss H, Fauth M, Merten A, Jeblick W (1999) Cucumber hypocotyls respond to cutin monomers via both an inducible and a constitutive H_2O_2-generating system. Plant Physiol 120:1175–1182

Kessmann H, Staub T, Ligon J, Oostendorp M, Ryals J (1994) Activation of systemic acquired disease resistance in plants. Eur J Plant Pathol 100:359–369

Khan MS, Zaidi A, Wani PA, Oves M (2009) Role of plant growth promoting rhizobacteria in the remediation of metal contaminated soils. Environ Chem Lett 7:1–19

Kloepper JW, Ryu C-M, Zhang SA (2004) Induced systemic resistance and promotion of plant growth by *Bacillus* spp. Phytopathology 94:1259–1266

Knoester M, Pieterse CMJ, Bol JF, Van Loon LC (1999) Systemic resistance in Arabidopsis induced by rhizobacteria requires ethylene-dependent signaling at the site of application. Mol Plant Microbe Interact 12:720–727

Kraiser T, Gras DE, Gutiérrez AG, González B, Gutiérrez RA (2011) A holistic view of nitrogen acquisition in plants. J Exp Bot 62:1455–1466

Kyselkova M, Kopecky J, Frapolli M, Defago G, Sagova-Mareckova M, Grundmann GL, Moënne-Loccoz Y (2009) Comparison of rhizobacterial community composition in soil suppressive or conducive to tobacco black root rot disease. ISME J 3:1127–1138

Lamb C, Dixon RA (1997) The oxidative burst in plant disease resistance. Annu Rev Plant Physiol Plant Mol Biol 48:251–275

Lavicoli A, Boutet E, Buchala A, Métraux J-P (2003) Induced systemic resistance in *Arabidopsis thaliana* in response to root inoculation with *Pseudomonas fluorescens* CHA0. Mol Plant Microbe Interact 16:851–858

Lebel E, Heifetz P, Thorne L, Uknes S, Ryals J, Ward E (1998) Functional analysis of regulatory sequences controlling *PR-1* gene expression in Arabidopsis. Plant J 16:223–233

Leeman M, Van Pelt JA, Den Ouden FM, Heinsbroek M, Bakker PAHM, Schippers B (1995) Induction of systemic resistance by *Pseudomonas fluorescens* in radish cultivars differing in susceptibility to fusarium wilt, using a novel bioassay. Eur J Plant Pathol 101:655–664

Leeman M, Den OFM, Van Pelt JA, Dirkx FPM, Steijl H, Bakker PAHM, Schippers B (1996) Iron availability affects induction of systemic resistance against fusarium wilt of radish by *Pseudomonas fluorescens*. Phytopathology 86:149–155

Li X, Zhang Y, Clarke JD, Li Y, Dong X (1999) Identification and cloning of a negative regulator of systemic acquired resistance, *SNI1*, through a screen for suppressors of *npr1-1*. Cell 98:329–339

Loper JE, Hassan KA, Mavrodi DV, Davis EW, Lim CK, Shaffer BT, Elbourne LD, Stockwell VO, Hartney SL, Breakwell K, Henkels MD, Tetu SG, Rengel LI, Kidarsa TA, Wilson NL, van de Mortel JE, Song C, Blumhagen R, Radune D, Hostetler JB, Brinkac LM, Durkin AS, Kluepfel DA, Wechter WP, Anderson AJ, Kim YC, Pierson LS 3rd, Pierson EA, Lindow SE, Kobayashi D, Raajimakers JM, Weller DM, Thomashow LS, Allen AE, Paulsen IT (2012) Comparative genomics of plant-associated *Pseudomonas* spp.: insights into diversity and inheritance of traits involved in multitrophic interactions. PLoS Genet 8:e1002784

Malamy J, Carr JP, Klessig DF, Raskin I (1990) Salicylic acid: a likely endogenous signal in the resistance response of tobacco to viral infection. Science 250:1002–1004

Marek G, Carver R, Ding Y, Sathyanarayan D, Zhang X, Mou Z (2010) A high-throughput method for isolation of salicylic acid metabolic mutants. Plant Methods 6:21

Martínez-Medina A, Fernández I, Sánchez-Guzmán MJ, Jung SC, Pascual JA, Pozo MJ (2013) Deciphering the hormonal signalling network behind the systemic resistance induced by *Trichoderma harzianum* in tomato. Front Plant Sci 4:206

Mathys J, De Cremer K, Timmermans P, Van KS, Lievens B, Vanhaecke M, Cammue BP, De Connick B (2012) Genome-wide characterization of ISR induced in *Arabidopsis thaliana* by *Trichoderma hamatum* T382 against *Botrytis cinerea* infection. Front Plant Sci 3:108

Maurhofer M, Hase C, Meuwly P, Métraux JP, Défago G (1994) Induction of systemic resistance of tobacco to tobacco necrosis virus by the root-colonizing *Pseudomonas fluorescens* strain CHA0: influence of the *gacA* gene and of pyoverdine production. Phytopathology 84:139–146

Maurhofer M, Reimmann C, Schmidli-Sacherer P, Heeb SD, Défago G (1998) Salicylic acid biosynthesis genes expressed in *Pseudomonas fluorescens* strain P3 improve the induction of systemic resistance in tobacco against tobacco necrosis virus. Phytopathology 88:678–684

Mazurier S, Corberand T, Lemanceau P, Raaijmakers JM (2009) Phenazine antibiotics produced by fluorescent pseudomonads contribute to natural soil suppressiveness to Fusarium wilt. ISME J 3:977–991

Mazzola M (2002) Mechanisms of natural soil suppressiveness to soilborne diseases. Antonie Van Leeuwenhoek 81:557–564

Mendes R, Kruijt M, De Bruijn I, Dekkers E, Van der Voort M, Schneider JH, Piceno YM, DeSantis TZ, Andersen GL, Bakker PA, Raaijmakers JM (2011) Deciphering the rhizosphere microbiome for disease-suppressive bacteria. Science 332:1097–1100

Métraux JP (2002) Recent breakthroughs in the study of salicylic acid biosynthesis. Trends Plant Sci 7:332–334

Métraux JP, Signer H, Ryals J, Ward E, Wyss-Benz M, Gaudin J, Raschdorf K, Schmid E, Blum W, Inverardi B (1990) Increase in salicylic acid at the onset of systemic acquired resistance in cucumber. Science 250:1004–1006

Meziane H, Van der SI, Van Loon LC, Hofte M, Bakker PAHM (2005) Determinants of *Pseudomonas putida* WCS358 involved in inducing systemic resistance in plants. Mol Plant Pathol 6:177–185

Mishina TE, Zeier J (2007) Pathogen-associated molecular pattern recognition rather than development of tissue necrosis contributes to bacterial induction of systemic acquired resistance in Arabidopsis. Plant J 50:500–513

Mur LA, Brown IR, Darby RM, Bestwick CS, Bi YM, Mansfield JW, Draper J (2000) A loss of resistance to avirulent bacterial pathogens in tobacco is associated with the attenuation of a salicylic acid-potentiated oxidative burst. Plant J 23:609–621

Nadarajah K (2016a) Induced systemic resistance in rice. In: Choudhary KD, Varma A (eds) Microbial-mediated induced systemic resistance in plants. Springer Singapore, Singapore, pp 103–124. doi:10.1007/978-981-10-0388-2_7

Nadarajah K (2016b) Rhizosphere interactions: life below ground. In: Choudhary KD, Varma S, Tuteja N (eds) Plant microbe interactions: an approach to sustainable agriculture. Springer Singapore, Singapore

Nawrath C, Metraux JP (1999) Salicylic acid induction-deficient mutants of Arabidopsis express *PR-2* and *PR-5* and accumulate high levels of camalexin after pathogen inoculation. Plant Cell 11:1393–1404

Nawrath C, Heck S, Parinthawong N, Metraux JP (2002) EDS5, an essential component of salicylic acid-dependent signaling for disease resistance in Arabidopsis, is a member of the MATE transporter family. Plant Cell 14:275–286

Nürnberger T, Brunner F, Kemmerling B, Piater L (2004) Innate immunity in plants and animals: striking similarities and obvious differences. Immunol Rev 198:249–266

Pajerowska-Mukhtar KM, Emerine DK, Mukhtar MS (2013) Tell me more: roles of NPRs in plant immunity. Trends Plant Sci 18:402–411

Palmer CM, Hindt MN, Schmidt H, Clemens S, Guerinot ML (2013) MYB10 and MYB72 are required for growth under iron-limiting conditions. PLoS Genet 9:e1003953

Philippot L, Raaijmakers JM, Lemanceau P, van der Putten WH (2013) Going back to the roots: the microbial ecology of the rhizosphere. Nat Rev Microbiol 11:789–799

Pieterse CMJ, Van Wees SCM, Hoffland E, Van Pelt JA, Van Loon LC (1996) Systemic resistance in Arabidopsis induced by biocontrol bacteria is independent of salicylic acid accumulation and pathogenesis related gene expression. Plant Cell 8:1225–1237

Pieterse CMJ, Van Wees SCM, Van Pelt JA, Knoester M, Laan R, Gerrits H, Weisbeek PJ, Van Loon LC (1998) A novel signaling pathway controlling induced systemic resistance in Arabidopsis. Plant Cell 10:1571–1580

Pieterse CMJ, Van Pelt JA, Ton J, Parchmann S, Mueller MJ, Buchala AJ, Métraux JP, Van Loon LC (2000) Rhizobacteria-mediated induced systemic resistance (ISR) in Arabidopsis requires sensitivity to jasmonate and ethylene but is not accompanied by an increase in their production. Physiol Mol Plant Pathol 57:123–134

Pieterse CMJ, Van der Does D, Zamioudis C, Leon-Reyes A, Van Wees SCM (2012) Hormonal modulation of plant immunity. Annu Rev Cell Dev Biol 28:489–521

Pieterse CMJ, Poelman EH, Van Wees SC, Dicke M (2013) Induced plant responses to microbes and insects. Front Plant Sci 4:475

Pieterse CMJ, Zamioudis C, Berendsen RL, Weller DM, Van Wees SCM, Bakker PAHM (2014) Induced systemic resistance by beneficial microbes. Annu Rev Phytopathol 52:347–375

Pozo MJ, Azcon-Aguilar C (2007) Unraveling mycorrhiza-induced resistance. Curr Opion Plant Biol 10:393–398

Pozo MJ, Van der Ent S, Van Loon LC, Pieterse CMJ (2008) Transcription factor MYC2 is involved in priming for enhanced defense during rhizobacteria-induced systemic resistance in *Arabidopsis thaliana*. New Phytol 180:511–523

Press CM, Wilson M, Tuzun S, Kloepper JW (1997) Salicylic acid produced by *Serratia marcescens* 91-166 is not the primary determinant of induced systemic resistance in cucumber or tobacco. Mol Plant Microbe Interact 10:761–768

Raaijmakers JM, Weller DM (1998) Natural plant protection by 2,4-diacetylphloroglucinol-producing *Pseudomonas* spp. in take-all decline soils. Mol Plant Microbe Interact 11:144–152

Raaijmakers JM, Leeman M, Van Oorschot MMP, Van der Sluis I, Schippers B, Bakker PAHM (1995) Dose-response relationships in biological control of Fusarium wilt of radish by *Pseudomonas* spp. Phytopathology 85:1075–1081

Raaijmakers JM, Vlami M, de Souza JT (2002) Antibiotic production by bacterial biocontrol agents. Antonie Van Leeuwenhoek 81:537–547

Raaijmakers JM, Paulitz TC, Steinberg C, Alabouvette C, Moënne-Loccoz Y (2009) The rhizosphere: a playground and battlefield for soilborne pathogens and beneficial microorganisms. Plant and Soil 321:341–361

Raaijmakers JM, de Bruijn I, Nybroe O, Ongena M (2010) Natural functions of lipopeptides from bacillus and pseudomonas: more than surfactants and antibiotics. FEMS Microbiol Rev 34:1037–1062

Ramirez V, Van der Ent S, Garcia-Andrade J, Coego A, Pieterse CMJ, Vera P (2010) OCP3 is an important modulator of NPR1-mediated jasmonic acid-dependent induced defenses in Arabidopsis. BMC Plant Biol 10:199

Ran LX, Van Loon LC, Bakker PAHM (2005) No role for bacterially produced salicylic acid in rhizobacterial induction of systemic resistance in Arabidopsis. Phytopathology 95:1349–1355

Reitz M, Oger P, Meyer A, Niehaus K, Farrand SK, Hallmann J, Sikora RA (2002) Importance of the O-antigen, core-region and lipid a of rhizobial lipopolysaccharides for the induction of systemic resistance in potato to *Globodera pallida*. Nematol 4:73–79

Ryals JA, Neuenschwander UH, Willits MG, Molina A, Steiner H-Y, Hunt MD (1996) Systemic acquired resistance. Plant Cell 8:1808–1819

Ryals JA, Weymann K, Lawton K, Friedrich L, Ellis D, Steiner H-Y, Johnson J, Delaney TP, Jesse T, Vos P, Uknes S (1997) The Arabidopsis NIM1 protein shows homology to the mammalian transcription factor inhibitor IκB. Plant Cell 9:425–439

Ryu C-M, Hu C-H, Reddy MS, Kloepper JW (2003) Different signaling pathways of induced resistance by rhizobacteria in *Arabidopsis thaliana* against two pathovars of *Pseudomonas syringae*. New Phytol 160:413–420

Ryu C-M, Farag MA, Hu CH, Reddy MS, Kloepper JW, Paré PW (2004) Bacterial volatiles induce systemic resistance in Arabidopsis. Plant Physiol 134:1017–1026

Segarra G, Van der Ent S, Trillas I, Pieterse CMJ (2009) MYB72, a node of convergence in induced systemic resistance triggered by a fungal and a bacterial beneficial microbe. Plant Biol 11:90–96

Seskar M, Shulaev V, Raskin I (1998) Endogenous methyl salicylate in pathogen-inoculated tobacco plants. Plant Physiol 116:387–392

Shulaev V, Leon J, Raskin I (1995) Is salicylic acid a transported signal of systemic acquired resistance in tobacco? Plant Cell 7:1691–1701

Siddiqui IA, Shaukat SS (2003) Suppression of root-knot disease by *Pseudomonas fluorescens* CHA0 in tomato: importance of bacterial secondary metabolite, 2,4-diacetylpholoroglucinol. Soil Biol Biochem 35:1615–1623

Somers E, Vanderleyden J, Srinivasan M (2004) Rhizosphere bacterial signalling, a love parade beneath our feet. Crit Rev Microbiol 30:205–240

Spoel SH, Koornneef A, Claessens SMC, Korzelius JP, Van Pelt MJ, Mueller AJ, Buchala J-P, Métraux R, Brown K, Kazan K (2003) NPR1 modulates crosstalk between salicylate- and jasmonate-dependent defense pathways through a novel function in the cytosol. Plant Cell 15:760–770

Staswick PE, Tiryaki I (2004) The oxylipin signal jasmonic acid is activated by an enzyme that conjugates it to isoleucine in Arabidopsis. Plant Cell 16:2117–2127

Stein E, Molitor A, Kogel KH, Waller F (2008) Systemic resistance in Arabidopsis conferred by the mycorrhizal fungus *Piriformospora indica* requires jasmonic acid signaling and the cytoplasmic function of *NPR1*. Plant Cell Physiol 49:1747–1751

Sticher L, Mauch-Mani B, Métraux JP (1997) Sistemathic acquired resistance. Annu Rev Phytopathol 35:235–270

Stone JK, Bacon CW, White JR (2000) An overview of endophytic microbe: endophytism defined. In: Bacon CW, White JF Jr (eds) Microbial endophytes. Marcel Dekker, New York, pp 3–30

Thomma BPHJ, Penninckx IAMA, Broekaert WF, Cammue BPA (2001) The complexity of disease signaling in Arabidopsis. Curr Opin Immunol 13:63–68

Thulke OU, Conrath U (1998) Salicylic acid has a dual role in the activation of defense-related genes in parsley. Plant J 14:35–42

Tjamos SE, Flemetakis E, Paplomatas EJ, Katinakis P (2005) Induction of resistance to Verticillium dahlia in *Arabidopsis thaliana* by the biocontrol agent K-165 and pathogenesis-related proteins gene expression. Mol Plant Microbe Interact 18:555–561

Ton J, Pieterse CMJ, Van Loon LC (1999) Identification of a locus in Arabidopsis controlling both the expression of rhizobacteria-mediated induced systemic resistance (ISR) and basal resistance against *Pseudomonas syringae* pv. Tomato. Mol Plant Microbe Interact 12:911–918

Vallad GE, Goodman RM (2004) Systematic acquired resistance and induced systematic resistance in conventional agriculture. Crop Sci 44:1920–1934

Van de Mortel JE, De Vos RCH, Dekkers E, Pineda A, Guillod L, Bouwmeester K, van Loon JJ, Dicke M, Raaijmakers JM (2012) Metabolic and transcriptomic changes induced in Arabidopsis by the rhizobacterium *Pseudomonas fluorescens* SS101. Plant Physiol 160:2173–2188

Van der Ent S, Verhagen BW, Van Doorn R, Bakker D, Verlaan MG, Pel MJ, Joosten RG, Proveniers MC, Van Loon LC, Ton J, Pieterse CM (2008) MYB72 is required in early signaling steps of rhizobacteria-induced systemic resistance in Arabidopsis. Plant Physiol 146:1293–1304

Van der Ent S, Van Hulten MHA, Pozo MJ, Czechowski T, Udvardi MK, Pieterse CM, Ton J (2009) Priming of plant innate immunity by rhizobacteria and β-aminobutyric acid: differences and similarities in regulation. New Phytol 183:419–431

Van der Heijden MGA, Bardgett RD, Van Straalen NM (2008) The unseen majority: soil microbes as drivers of plant diversity and productivity in terrestrial ecosystems. Ecol Lett 11:296–310

Van Loon LC (1997) Induced resistance in plants and the role of pathogenesis-related proteins. Eur J Plant Pathol 103:753–765

Van Loon LC, Antoniw JF (1982) Comparison of the effects of salicylic acid and ethephon with virus-induced hypersensitivity and acquired resistance in tobacco. Neth J Plant Pathol 88:237–256

Van Loon LC, Bakker PAHM (2003) Signalling in rhizobacteria-plant interactions. In: De Kroon H, Visser EJW (eds) Root ecology. Springer, Berlin, pp 297–330
Van Loon LC, Bakker PAHM (2005) Induced systemic resistance as a mechanism of disease suppression by rhizobacteria. In: Siddiqui ZA (ed) PGPR: biocontrol and biofertilization. Springer, Dordrecht, pp 39–66
Van Loon LC, Bakker PAHM (2006) Root-associated bacteria inducing systemic resistance. In: Gnanamanickam SS (ed) Plant-associated bacteria. Springer, Dordrecht, pp 269–316
Van Loon LC, Glick BR (2004) Increased plant fitness by rhizobacteria. In: Sandermann H (ed) Molecular ecotoxicology of plants. Springer, Berlin, pp 177–205
Van Loon LC, Van Strien EA (1999) The families of pathogenesis-related proteins, their activities, and comparative analysis of PR-1 type proteins. Physiol Mol Plant Pathol 55:85–97
Van Loon LC, Bakker PAHM, Pieterse CMJ (1998) Systemic resistance induced by rhizosphere bacteria. Annu Rev Phytopathol 36:453–483
Van Loon LC, Bakker PA, van der Heijdt WH, Wendehenne D, Pugin A (2008) Early responses of tobacco suspension cells to rhizobacterial elicitors of induced systemic resistance. Mol Plant Microbe Interact 21:1609–1621
Van Wees SCM, Pieterse CMJ, Trijssenaar A, Van't Westende Y, Hartog F, Van Loon LC (1997) Differential induction of systemic resistance in Arabidopsis by biocontrol bacteria. Mol Plant Microbe Interact 10:716–724
Van Wees SCM, Luijendijk M, Smoorenburg I, Van Loon LC, Pieterse CMJ (1999) Rhizobacteria mediated induced systemic resistance (ISR) in Arabidopsis is not associated with a direct effect on expression of known defense-related genes but stimulates the expression of the jasmonate-inducible gene Atvsp upon challenge. Plant Mol Biol 41:537–549
Van Wees SCM, De Swart EAM, Van Pelt JA, Van Loon LC, Pieterse CMJ (2000) Enhancement of induced disease resistance by simultaneous activation of salicylate- and jasmonate-dependent defense pathways in Arabidopsis thaliana. Proc Natl Acad Sci U S A 97:8711–8716
Van Wees SCM, Van der Ent S, Pieterse CMJ (2008) Plant immune responses triggered by beneficial microbes. Curr Opin Plant Biol 11:443–448
Verberne MC, Hoekstra J, Bol JF, Linthorst HJM (2003) Signaling of systemic acquired resistance in tobacco depends on ethylene perception. Plant J 35:27–32
Verhagen BWM, Glazebrook J, Zhu T, Chang H-S, Van Loon LC, Pieterse CMJ (2004) The transcriptome of rhizobacteria-induced systemic resistance in Arabidopsis. Mol Plant Microbe Interact 17:895–908
Vlot AC, Dempsey DA, Klessig DF (2009) Salicylic acid, a multifaceted hormone to combat disease. Annu Rev Phytopathol 47:177–206
Vos IA, Pieterse CMJ, Van Wees SCM (2013) Costs and benefits of hormone-regulated plant defences. Plant Pathol 62:43–55
Walters DR, Paterson L, Walsh DJ, Havis ND (2008) Priming for plant defense in barley provides benefits only under high disease pressure. Physiol Mol Plant Pathol 73:95–100
Walters DR, Ratsep J, Havis ND (2013) Controlling crop diseases using induced resistance: challenges for the future. J Exp Bot 64:1263–1280
Wang YQ, Ohara Y, Nakayashiki H, Tosa Y, Mayama S (2005) Microarray analysis of the gene expression profile induced by the endophytic plant growth–promoting rhizobacteria, *Pseudomonas fluorescens* FPT9601-T5 in Arabidopsis. Mol Plant Microbe Interact 18:385–396
Wang L, Tsuda K, Truman W, Sato M, Nguyen le V, Katagiri F, Glazebrook J (2011) CBP60g and SARD1 play partially redundant critical roles in salicylic acid signaling. Plant J 67:1029–1041
Wei G, Kloepper JW, Tuzun S (1991) Induction of systemic resistance of cucumber to *Colletotrichum orbiculare* by select strains of plant-growth promoting rhizobacteria. Phytopathology 81:1508–1512
Weller DM, Raaijmakers JM, McSpadden Gardener BB, Thomashow LS (2002) Microbial populations responsible for specific soil suppressiveness to plant pathogens. Annu Rev Phytopathol 40:309–348
Weller DM, Van Pelt JA, Mavrodi DV, Pieterse CMJ, Bakker PAHM, Van Loon LC (2004) Induced systemic resistance (ISR) in Arabidopsis against *Pseudomonas syringae* pv. tomato

by 2,4-diacetylphloroglucinol (DAPG)-producing *Pseudomonas fluorescens*. Phytopathology 94:S108

Weller DM, Mavrodi DV, Van Pelt JA, Pieterse CMJ, Van Loon LC, Bakker PAHM (2012) Induced systemic resistance (ISR) in *Arabidopsis thaliana* against *Pseudomonas syringae* pv. tomato by 2,4-diacetylphloroglucinol-producing *Pseudomonas fluorescens*. Phytopathology 102:403–412

Yan Z, Reddy MS, Ryu C-M, McInroy JA, Wilson M, Kloepper JW (2002) Induced systemic protection against tomato late blight elicited by plant growth–promoting rhizobacteria. Phytopathology 92:1329–1333

Yang P, Chen C, Wang Z, Fan B, Chen Z (1999) A pathogen- and salicylic acid-induced WRKY DNA-binding activity recognizes the elicitor response element of the tobacco class I chitinase gene promoter. Plant J 18:141–149

Yang YN, Qi M, Mei CS (2004) Endogenous salicylic acid protects rice plants from oxidative damage caused by aging as well as biotic and abiotic stress. Plant J 40:909–919

Yezhang D, Danjela S, Zhonglin M (2015) A large-scale genetic screen for mutants with altered salicylic acid accumulation in Arabidopsis. Front Plant Sci 5:763. doi:10.3389/fpls.2014.00763

Zamioudis C, Mastranesti P, Dhonukshe P, Blilou I, Pieterse CMJ (2013) Unraveling root developmental programs initiated by beneficial *Pseudomonas* spp. bacteria. Plant Physiol 162:304–318

Zhang S, Klessig DF (1997) Salicylic acid activates a 48-kD MAP kinase in tobacco. Plant Cell 9:809–824

Zhang Y, Fan W, Kinkema M, Li X, Dong X (1999) Interaction of NPR1 with basic leucine zipper protein transcription factors that bind sequences required for salicylic acid induction of the *PR-1* gene. Proc Natl Acad Sci U S A 96:6523–6528

Zhang S, Moyne AL, Reddy MS, Kloepper JW (2002) The role of salicylic acid in induced systemic resistance elicited by plant growth-promoting rhizobacteria against blue mold of tobacco. Biol Control 25:288–296

Zhang Z, Li Q, Li Z, Staswick PE, Wang M, Zhu Y, He Z (2007) Dual regulation role of GH3.5 in salicylic acid and auxin signaling during *Arabidopsis-Pseudomonas syringae* interaction. Plant Physiol 145:450–464

Zhang Y, Xu S, Ding P, Wang D, Cheng YT, He J, Gao M, Xu F, Li Y, Zhu Z, Li X, Zhang Y (2010) Control of salicylic acid synthesis and systemic acquired resistance by two members of a plant-specific family of transcription factors. Proc Natl Acad Sci U S A 107:18220–18225. doi:10.1073/pnas.1005225107

Zhou JM, Trifa Y, Silva H, Pontier D, Lam E, Shah J, Klessig DF (2000) *NPR1* differentially interacts with members of the TGA/OBF family of transcription factors that bind an element of the *PR-1* gene required for induction by salicylic acid. Mol Plant Microbe Interact 13:191–202. doi:10.1094/MPMI.2000.13.2.191

Zimmerli L, Jakab G, Métraux J-P, Mauch-Mani B (2000) Potentiation of pathogen-specific defense mechanisms in Arabidopsis by β-aminobutyric acid. Proc Natl Acad Sci U S A 97: 12920–12925

Chapter 16
Tools and Targeted Genes for Plant Disease Detection

Umaiyal Munusamy and Siti Nor Akmar Abdullah

16.1 Introduction

Plants are believed to be among the first living organisms to be present on the earth (Biello 2009). Therefore, due to their existence the earth is called a green planet. Their presence is inevitable for clean air, food, and water. Without plants, the ecosystem and human life on this earth cannot exist (Laurent et al. 2017). Therefore, we grow plants in our homes, buildings, gardens, farms, and by the roadsides. People from many countries rely on plant products for their livelihood and income (Dash et al. 2016) which indirectly contribute to the economy of the local community. Plant contributes significantly to food and nutritional security as well as poverty alleviation in the world (Passos et al. 2013). Hence, the relationship between plants, people, and the environment is so much interlinked that we cannot separate them (Laurent et al. 2017). However, a major constraint for plants is the infection by plant pathogens (Dean et al. 2012; Yim et al. 2014). There are various types of plant diseases like blights (Kheiri et al. 2017), rots (Nicosia et al. 2016), cankers (Cimmino et al. 2017), wilts (Gayatridevi et al. 2012), and rusts (Shabana et al. 2017) that have been reported. All of these diseases are fatal to the plant if they are not detected early. Usually at the earlier stage, plants react to the pathogen attack through activation of defense response genes (Table 16.1).

Plants employ two channels of defense mechanism: local immunity which is nearby the pathogen infected spot (Gayatridevi et al. 2012) and systemic immunity which is further away from the pathogen infected site (Dempsey and Klessig 2012). The best way to save a plant is either to prevent the infection (Pscheidt and Ocamb

U. Munusamy • S.N.A. Abdullah (✉)
Institute of Plantation Studies, Universiti Putra Malaysia,
43400 UPM, Serdang, Selangor, Malaysia
e-mail: snaa@upm.edu.my

S.N.A. Abdullah et al. (eds.), *Crop Improvement*,
DOI 10.1007/978-3-319-65079-1_16

Table 16.1 Candidate defense response genes during microbial pathogen attack in plant

Protein	Gene abbreviation	Pathway involved	References
Catalase	CAT	Metabolic	Gayatridevi et al. (2012), Amoako et al. (2015)
Thaumatin	THAU	Plant resistance Plant defense	Singh et al. (2013)
Pectin acetylesterase	PAE	Metabolic	Vercauteren et al. (2002)
Chitinase	CHI	Plant defense	Punja and Zhang (1993), Sharma et al. (2011), Gupta et al. (2013)
Relative glucose content	RGC	Primary metabolism	Lecompte et al. (2017)
Elongation factor	EF	Protein elongation	Hwang et al. (2015)
WRKY factors	WRKY	Plant defense	Eulgem (2006), Pandey and Somssich (2017)
Phenylalanine amino lyase	PAL	Phenylpropanoid biosynthesis	Huang et al. (2010)
Glucanase	Glu	Plant defense	Gupta et al. (2013)
Abscisic acid	ABA	JA and SA signalling	Lecompte et al. (2017)

2002) or to detect the infection at the local immunity stage (earlier stage) by utilizing various detection tools (Feau et al. 2009). Tools such as scoring analysis (Mak et al. 2001), quantitative real-time PCR analysis (Hao et al. 2013), DNA macroarray, enzymatic and antibody assays (Madadkhah et al. 2012) and sophisticated tools such as RNA sequencing (Yang et al. 2017) can be used to detect plant infection at the earlier stage.

16.2 Types of Plant Disease

16.2.1 Blights

Blight is a symptom of an infection by pathogenic organisms on the infected plant (Basim et al. 2017). It involves chlorosis and browning of the tissues. The early appearance of lesions on leaf tissues of plants suffered from blight rapidly spread to the surrounding tissues (Schumann 2017). The lesion will expand to kill the entire area and reach the plant organs such as the leaves, branches, twigs, or floral organs followed by the death of the whole plant (da Silva et al. 2017). There are a few types of blights:

- Late blight (Whisson et al. 2016)
- Fire blight (Sharifazizi et al. 2017)
- Early blight (Singh et al. 2017)
- Leaf blight (Santha Lakshmi Prasad et al. 2017
- Fusarium head blight (Dweba et al. 2017)

Usually, the naming of blights referred to their causative agent, such as Colletotrichum blight is based on the fungi *Colletotrichum capsici* (Shenoy et al. 2007) while Phytophthora blight refers to the water mold *Phytophthora parasitica* (Babadoost and Islam 2002).

16.2.2 Cankers

Canker appears as a small area and grows slowly around the dead tissues (Schumann 2017). The majority of canker-causing organisms such as fungi (Cortinas et al. 2006) and bacteria (Zhao et al. 2015) are bound to unique host species or genus, however a few attack a wide range of plants. There are different types of cankers, for examples, *Colletogloeopsis zuluensis* (Wingfield et al. 1997) causes a serious canker disease on Eucalyptus stem (Cortinas et al. 2006), *Pseudomonas syringae* pv. Actinidia causes canker on kiwifruit (Prencipe et al. 2016). Canker also tends to attack fruit trees (Wicaksono et al. 2017) and vegetables (Kasselaki et al. 2011). Although fungicides or bactericides can treat some cankers, often the only available treatment is to destroy the canker infected plants (Sosnowski et al. 2009).

16.2.3 Rots

Rot or decay is caused by soil-borne bacteria (Lei et al. 2017) and fungi (Malcolm et al. 2013) such as *Penicillium digitatum, Penicillium italicum, Geotrichum citri-aurantii, Xanthomonas campestris* (Liu et al. 2015), *Sclerotinia sclerotiorum* (Elsheshtawi et al. 2017), and *Fusarium* (Zhou et al. 2017). They are characterized as plant decomposer and appears as hard, dry, spongy, watery, mushy, or slimy characters in any plants or plant parts (Agrios 2005). Plants such as cucumber (Zhou et al. 2017), peanut (Thiessen and Woodward 2012), and blueberry (Haralson et al. 2013) are some of the examples that was reported on rot disease.

16.2.4 Rusts

Rusts are mainly caused by the fungal pathogen of the order Pucciniales (Shabana et al. 2017). A single species of these fungi is able to produce several spore-producing structures such as spermagonia, aecia, uredinia, telia, and basidia via the successive reproductive stages (Callan and Carris 2004). Rust typically affects healthy plants, and the infection is confined to certain plant parts such as leaves and stem (Pretorius et al. 2012). The systemic mechanism of infection that causes rusts may lead to deformities such as growth retardation, development of galls, or hypertrophy of the affected tissues (Robert-Seilaniantz et al. 2011). Severely infected plants may appear stunted, chlorotic (yellowed), or discolored (Martinez and Woodward 2015). The fungus that causes rust sporulates on the affected plant parts which are often observed as colored powder, consisting of tiny aeciospores while on vegetation they produces pustules or uredia (Mueller et al. 2004). During late spring or early summer, yellow, orange, or brown hairlike or ligulate structures called telia grow on the leaves or appear from woody hosts barks. These telia produce teliospores which will germinate into aerial basidiospores for dispersal and causing further infection (Crawford et al. 2014). Some examples of plant infected with rusts disease are rye, wheat, and cabbage (Pretorius et al. 2012; Liu et al. 2015).

16.2.5 Wilts

Wilts disease is mainly caused by *Fusarium* (Sutherland et al. 2013). Its mycelium invades the vascular tissue through the roots (Magliano and Kikot 2010) and block the xylem vessels by its spores (Di Pietro et al. 2003). This infection may occur at any age of the plant by causing discoloration and wilting symptom finally resulting in the collapse of the infected plant (Nel et al. 2007; Wu et al. 2010). *Fusarium* is a soil-borne pathogen and its complex pathogenic structure enables it to infect a wide range of plant hosts (DeVay et al. 1997). It usually attacks many crops including cotton (DeVay et al. 1997), watermelon (Tziros Georgios et al. 2007), tomato (Singh et al. 2010), lettuce (Cabral and Reis 2013), and guava (Amadi et al. 2014).

16.3 Tools for Plant Disease Detection

Disease development in plants is influenced by many factors including genotypes of the plant and pathogen, the plant age during infection, environment (e.g., soil and water source), weather (e.g., temperature, wind, and rain), or types of infection which is either single or mixed infections. Due to variation inherent in these factors, the diagnosis process becomes difficult. Detection at the early stage of pathogen infection in the plants as well as at the beginning of an epidemic will at least prevent the infected plant from rotting. Therefore, tools are developed to assist in rapid assessment at the early stage of plant pathogen infection.

16.3.1 Scoring Tools

Visual scoring of disease symptoms is used to identify tolerant plant lines. These scoring serve as an indirect measurement of the severity of infection by the fungus (Ali and McNear 2014). There are many papers reported on the different types of developed scoring tools from various laboratories for early detection of the disease infection in plants by *Fusarium* (Hall 1989; Burkholder et al. 2001; Churchill 2011). Therefore, it is important to standardize the scoring tools based on several characteristics as discussed below and as summarized in Fig. 16.1 (Hall 1989; Churchill 2011).

In the laboratory, during the development of the disease scoring tools, one factor that needs to be standardized is the size of the inoculum which should be produced under constant condition. For example, 10^5 to 10^6 are the economical spores concentration in the field, therefore, this range should be used for the development of the disease scoring tools (Burkholder et al. 2001). In addition, the mode of application should not be changed between experiments and it is best to avoid usage of external solvents such as wetting agent as it may affect the real event found in the field. An appropriate time period should be set since different duration of spore application onto plant material will have different effect on the plant defense mechanism (Chen et al. 2015).

In every type of infection, appropriate plant materials should be chosen, for example, root should be used for root infection studies while leaf for leaf infection studies. This is because different genes are regulated differently in different parts of the plant during infection. Furthermore, plants of different ages and developmental stages will respond differently towards the infection. For example, disease symptoms on the suckers may be different from symptoms on the matured plant while tissue culture plantlet may respond differently compared to plant grown in control growth room. Clearly, the physiological state and the development stage of the plant have a big influence on the phenotypic symptoms (Churchill 2011).

The plants should be grown under conditions with constant temperature, humidity, and light in order to achieve maximum homogeneity. The growth room needs to mimic the outdoor conditions and the quality and quantity of applied nutrients such as the fertilizer should not be changed. Each plant species should have its own standardized methods while closely related plant varieties can use a single standard protocol (Chen et al. 2015).

Cross confirmation for quality control and assurance in a standardized bioassay is critical. Cross confirmation of the data can be carried out by independent laboratories as well as among experts in the field. Confirmation of the results by proven specialist in the field is the hallmark of good science. The integrity of the findings should be strengthened against false positive as well as false negative (Churchill 2011). The standardized bioassay for detecting the *Fusarium* infection should be considered as a rigidly defined technique so that the standardized steps can accommodate samples from a variety of environments (Brunner et al. 2012).

16.3.2 Quantitative Real-Time PCR Analysis

Quantitative real-time PCR (qPCR) produces specific fluorescent labelled PCR product, measures the kinetics reaction at the early PCR stage, and quantifies specific targets in the test samples. New chemistries and recipes for qPCR are continuously being produced by many scientific companies. These available kits will ease

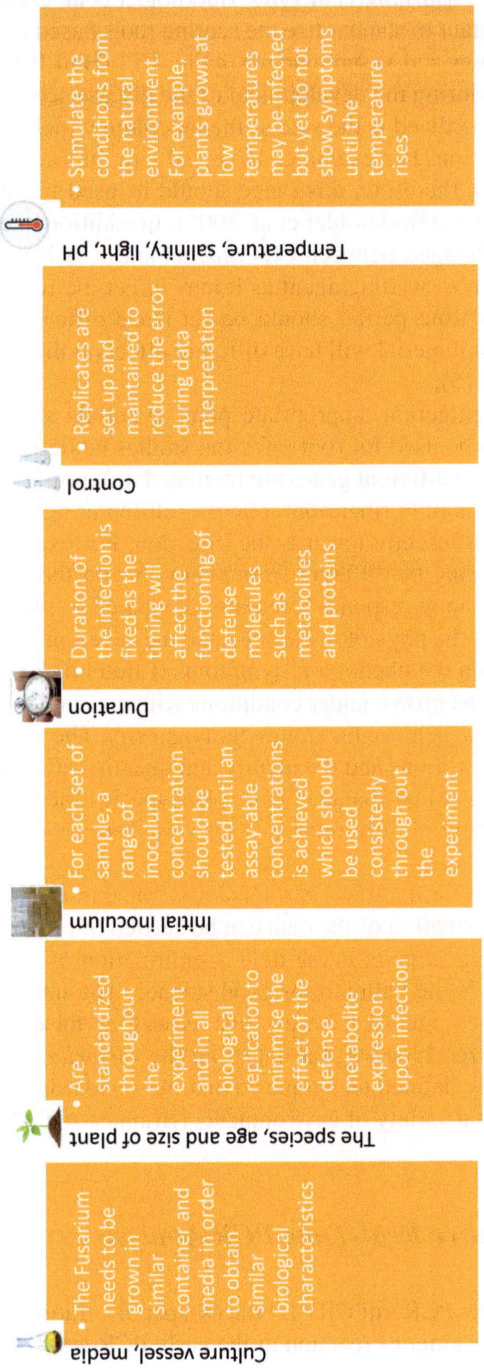

Fig. 16.1 Standardization of scoring method for pathogen detection in infected plant

the detection of PCR products separately from nonspecific products (Lifetechnologies 2012). Many new real-time PCR-based pathogen identification methods were developed for detecting the presence of particular pathogens (Vandesompele et al. 2002; Hao et al. 2013; Yim et al. 2014). Quantitative real-time nucleic acid amplifications have two major advantages during the detection of plant disease, (a) the ability to detect products as they are being amplified (Smith and Osborn 2009) and (b) the potential to design quantitative assays (Valasek and Repa 2005).

Real-time detection is becoming more important than endpoint PCR as detecting a signal while a particular nucleic acid is being amplified may speed up the detection assay whereby in the endpoint PCR, the specific products are detected only after the termination of the amplification (Brookman-Amissah et al. 2012). Many of the amplicons of qPCR are quite short, ranging from 50 to 200 bases, enabling RNA that is significantly degraded to be used as samples for the analysis (Romero et al. 2014). However, degradation of the target environmental RNA samples caused by nucleases, low number of targets, or reagent problems may in certain circumstances produce false positive/negative results (Schrader et al. 2012). For assessment of RNA degradation level, the RNA integrity number (RIN) value is required. The RNA integrity of 25S, 18S, and 16S rRNAs acts as indicators of RNA quality (Fleige and Pfaffl 2006).

Besides, as mentioned above, the problem with mutation at the annealing site of the primer generating false negative results is usually encountered when environmental samples are assessed (Broeders et al. 2014). Therefore, an internal positive control is necessary. This is achieved by including amplification of a housekeeping gene in the assays (Schena et al. 2013). With Normfinder, the selection of the housekeeping genes can be validated through optimal reference gene algorithms (Vandesompele et al. 2002). The stability of a reference gene is measured based on the M values. The lower the M_{value}, the better it is. Therefore, the best reference gene to normalize the targeted gene shall have the lowest M_{value}. In addition, primers are selected based on their ability in producing single peaks in the melting curve analysis conducted at 85 °C (the standard temperature) (Hao et al. 2013). A single species of DNA molecule in the reaction is represented by a single peak in melting curve analysis while primer dimer or nonspecific amplification will produce double or multiple peaks (Balcells et al. 2011). Primers chosen from this method offer results that are highly specific and reproducible in the qPCR assay (Lin et al. 2014).

The contamination of nucleic acids from external sources, such as exogenous DNA from cultures or from previous experiments, is a potential risk as the unwanted PCR products will accumulate as well (Okubara et al. 2010). This contamination risk can be minimized by using negative control, positive control, and reagent control in each PCR run. This problem can be overcome through melting curve analysis as mentioned above. This technology enables rapid, specific, and sensitive detection of nucleic acid with potential application in identifying pathogenic microorganisms infection in the environments that are normally difficult and/or laborious to culture (Ahmed and Huygens 2008). The application of real-time PCR to identify and quantify fungal and oomycete pathogens in biological samples such as tomato was described by Lievens et al. (2003).

16.3.3 DNA Macroarray

Macroarray is also known as reverse dot blot hybridization (RDBH) (Xiong et al. 2006). The term macroarray is used to refer to a larger array with fewer spots (Narayanasamy 2011) in contrast to microarray which contains a few thousands spots. This technique was originally developed as a method for clinical studies (Trad et al. 2004) but now it has also been successfully applied in plant pathology studies (Sholberg et al. 2005). DNA macroarrays have been successfully used to discriminate and identify DNA samples from specific oomycete, nematode, fungi, virus, and bacterial cultures from complex environmental samples without the need for isolation in culture (Tsui et al. 2011). In DNA macroarray, the non-labelled pathogen-specific oligonucleotide probes are fixed on a filter or membrane such as a nitrocellulose/nylon membrane (Goel et al. 2004). Whereas the target labelled DNA or PCR product is provided in a buffer that will enhance binding with the complementary non-labelled oligonucleotide on the membrane through hybridization under different stringency conditions (Lievens et al. 2003).

Several reports on the applications of this technique for screening and identification of plant pathogens have been reported. He et al. (2012) uses species-specific DNA macroarray based on rDNA-ITS to study pathogenic and nonpathogenic fungi and bacteria in an apple orchard. They reported successful detection of major microbial species simultaneously and the technique appeared to be a robust tool to monitor apple phyllosphere microbial community under specific disease management or certain environmental condition. The data generated provided useful information on antagonistic and cooperative interactions between specific pathogens in the orchard. Zhang et al. (2007) reported detection of Solanaceous pathogens in the *Fusarium solani* species using DNA macroarray. They were able to perform a highly specific detection of both inoculated plant materials as well as infected plant samples from the field through optimizing the length of the DNA probe and hybridization temperature.

The sensitivity offered by the macroarray technology is high enough to detect pathogens at densities in which they are likely to occur in the field (Zhang et al. 2007). In comparison to PCR-based method, it produces more rapid result as it does not require separation of DNA fragments through electrophoresis. Besides that, DNA macroarray also allows rapid detection in a single assay with high accuracy that even one single-nucleotide polymorphism (SNP) difference in the target DNA can be discriminated (Lievens et al. 2003).

16.3.4 RNA Sequencing

Early detection of pathogen infection on plants is possible through RNA sequencing analysis (Petre et al. 2012). The key strength in RNA sequencing is that it enables efficient and comprehensive analysis of the transcriptome to discover alternate transcripts, gene fusions, and allele-specific expression even from partially degraded sample. It delivers highly accurate quantification of gene and transcript abundance, captures both known and novel features, and provides an informative biomarkers for a range of phenotypes (Kimbrel et al. 2011; Illumina 2014; Conesa et al. 2016).

The exploration of eukaryotic gene expression has been revolutionized by the advent of RNA sequencing as it provides novel insights of expression levels that occur during development, disease, and stress conditions. It allows the elucidation of gene and protein function and interactions. It also enables identification of tissue and developmentally specific sets of RNA transcripts produced by a plant or microbial genome (Martin et al. 2013). RNA-seq based transcriptome analysis of Musa and its bacterial pathogen during infection produces a robust and sensitive detection that allows access to identification of specific cues in host defense mechanism (Guo et al. 2014). However, sometimes low percentage of mapping against the plant reference genome makes this important analytical step becomes difficult (BGI 2015). This happens due to the increase in the level of infection by the pathogen; hence a greater proportion of pathogen RNA molecule than plant RNA is present in the sample. On the other hand, low level of expression may suggest that only a few cells are expressing the genes or the genes somehow are subjected to tight regulation or perhaps they are involved in more specialized role in metabolism or development upon infection (Zhao et al. 2014).

RNA-seq provides several advantages over hybridisation-based approaches because it offers higher sensitivity and reproducibility, has lower technical variation, and is not limited by prior knowledge of the organism genome (Wang et al. 2009) making it particularly attractive for non-model organisms (Unamba et al. 2015) and RNA-seq data obtained is considered unbiased (Zheng et al. 2011). Until today, RNA sequencing has provided new insights into a broad range of studies from complex disease (Costa et al. 2013) to the ripening process of grapes (Guo et al. 2014). It can be applied to a wide scope of scientific questions through comparing gene expression profile between samples (Yang et al. 2017) such as pathogen infected and non-infected tissues (Sun et al. 2013). RNA-sequencing is well-established as a valuable platform with expanded applications in plant biology research (Illumina 2014). The recent development in sequencing technologies, which include increase in read lengths and numbers of reads per run, as well as advancement in computational tools for sequence assembly, analysis, and integration with orthogonal data sets further enhanced RNA-seq applications (Conesa et al. 2016).

16.3.5 Enzyme and Antibody Assays

Enzymatic activities have been utilized in disease resistance and in plant defense mechanism (Anusuya and Sathiyabama 2014). Different types of enzymes from the plant defense response pathway are targeted for monitoring plant disease development. Plant lignin biosynthetic enzymes (Moerschbachkr et al. 1988), chitinase (CHI), protease, polyphenol oxidase (PAL), and peroxidase (Madadkhah et al. 2012) are some of the examples. Traditional enzyme assays are time consuming (Hammerschmidt et al. 1982). But enzyme assay kits such as the peroxidase assay (Nandi et al. 2013), PAL activity assay, and catalase assay (Iwase et al. 2013) offer the opportunity for plant pathology studies to be carried out more conveniently and faster.

However, in general enzyme assay is still relatively laborious (Fernando and Soysa 2015), slow (Kumar Panda 2012), needs a relatively large amount of samples, and difficult for high throughput analysis (Fernando and Soysa 2015). Therefore, next generation enzyme assay using specific antibodies against the protein (enzyme) for plant disease monitoring is preferred. This is carried out using the well-known method known as the enzyme linked immunosorbent assay (ELISA) (Lacroix et al. 2016). The advantages of ELISA include the ability to screen a large number of samples, high sensitivity, and fast speed of reaction. ELISA was successfully used to detect infection on tomato (Sevik and Kose-Tohumcu 2011), potato (Boonham et al. 2003), kiwifruit (Cimmino et al. 2017), and on field trees (Martin et al. 2000).

Antibody-based detection of pathogens has undergone rapid development and various innovative ideas for field applications have been reported. For example, one-step lateral-flow tests were developed for detection and identification of plant viruses in the field. In this assay, specific monoclonal and polyclonal antibodies were used in an immunochromatographic format, incorporating antibody-coated latex particles. Disease diagnosis within 3 min with 100% accuracy when tested on Potato Y potyvirus and Potato X potexvirus was achieved through this method (Danks and Barker 2000; Sastry 2013; Shibaei et al. 2017).

16.4 Overlapping Cellular Responses upon Plant Pathogen Attack

16.4.1 Differential Gene Expression

In response to pathogen infection, plants have evolved complex signalling network and various defense pathways. The first line of defense includes the physical barriers such as cuticles and cell wall in the plant. Lignification is one of the primary channels in restricting pathogen development, and it occurs early in the host pathogen interaction and should be localized near the infection site of the pathogen (Ebrahim et al. 2011). While, chemical barriers such as phytoanticipins, phenols, defensins, and peptides represent the second line of defense (Kunkel and Brooks 2002).

The expressed genes that were annotated basically fall in three different categories which are cellular component, molecular function, and biological process. Each gene demonstrated a specific pattern of expression during infection. Published findings in many plants showed that pathogen responsive 10 (PR10) are induced as early response just upon microbial attack. This is to enhance the defense response in the plant (Hoffmann-Sommeergruber 2002; Dolezal et al. 2014). The hypersensitive response functions to curb the spread of the infection by the plant pathogen. This is induced by PR proteins and rapid cell death at the site of pathogen invasion (Yadav et al. 2016). The PR proteins exhibit strong antifungal activities and are essential during pathogen attack (Yadav et al. 2016). In plants there are many PR proteins such as PR1, PR3, PR4, and PR10. PR1 shows antifungal properties at micromolar levels against various pathogenic fungi. While, PR3 proteins destroy the chitin polymer in fungal cell wall to inhibit further fungal growth. PR4 inhibits the fungal spread by binding to the β-chitin in the cell wall of attacking fungi (Yadav et al. 2016).

Van Den Berg et al. (2007) reported that catalase (CAT) gene expression was enhanced in infected tolerant and susceptible plantlets. Increase in the production of PAL is also a way to inhibit the penetration of fungal through the cell wall of banana roots (Mohd Fishal et al. 2010). Furthermore, release of CHI causes degradation of the fungal cell wall. This is because once attacked by pathogen, plant will release CHI to degrade fungal cell wall that mostly consists of chitin (Chen et al. 2015). In certain reported data PAL and CAT are downregulated even though they are vital enzymes in phenolic biosynthesis which links primary metabolism (shikimic acid pathway) to secondary metabolism (Sun et al. 2013; EL-Khallal 2007). Therefore, early detection at the level of pathogen infection measured by the up/downregulation of these genes can provide real-time data. This will allow us to determine the seriousness of the infection for further remedial actions.

16.4.2 Ethylene Activated Signalling Pathway

Plants will respond to pathogen infection by inducing broad spectrum resistance system called systematic acquired resistance (SAR). Induction of systemic acquired resistance includes the production of ethylene (Ecker 1995; Ecker and Guo 2004). In plant, ethylene is involved in physiological and development processes (Ouaked et al. 2003). Systemic acquired resistance is triggered by normal wounding and through pathogen infection and these correlate with the induction of an array of PR proteins (Huot et al. 2014). During early infection, ethylene plays a role in responding towards pathogen attack while at the later stage of infection ethylene stimulates synthesis of Acetyl-CoA carboxylase (ACC) synthase enzyme which thereby provide a means for autoregulation of the defense mechanism (Ouaked et al. 2003). By quantifying ethylene gas produced by the infected plant sample we can identify the stage of infection, for example, a decrease in ethylene emission from an infected tomato after attaining maximal expression correlated with an advanced fungal infection stage (Cristescu et al. 2002).

16.4.3 Transcription Factor

Transcription factors have been used as an indicator for the activation of plant defense mechanism. Hong et al. (2013) assayed the expression of transcription factors during *Colletotrichum gloeosporioides* infection for 9 days. It was found that the ethylene response factor (ERF) functions as a transcription factor that integrates signals from the ethylene and jasmonic acid pathways. The ERF1 and EF belongs to a large family of APETALA2-domain-containing transcription factors that bind to a GCC-box present in the promoters of many ethylene inducible and defense related genes (Guo and Ecker, 2004). While, WRKY transcription factor that was identified in the transcriptome data set (Phukan et al. 2016) belongs to a large family of transcription factors. It regulates the expression of the defense genes that contain sequence specific DNA binding site. The transcripts of many WRKYs are strongly induced by abiotic and biotic stresses coupled with salicyclic acid or hydrogen peroxidase (Ren et al. 2010). Jiang et al. (2016) reported that WRKY45 was responsible for inducing SA responsive gene. However, WRKY transcription factors functioned within a short period of time upon infection in order to provide a rapid response towards the infection by the pathogen (Jiang et al. 2016).

16.4.4 Jasmonic Acid and Abscisic Acid

Jasmonic acid (JA) responses are initiated by necrotrophs organism such as *Fusarium* (Thakur and Sohal 2013; Yi et al. 2014). Jasmonic acid was proposed as the systemic signal for SAR (Dampsey and Klessig 2012). Studies have been reported that the JA level will increase upon infection (Fu and Dong 2013) and its deficiency will increase the susceptibility of the plant to pathogens (Thaler et al. 2004). *PLANT DEFENSIN1.2, THIONIN2.1, HEVEIN-LIKE PROTEIN CHI, CAT and PAL* genes expression serves as markers for JA level (Kunkel and Brooks 2002; Sun et al. 2013). Jasmonic acid also will trigger many PR proteins (El-Khallal 2007) and moderate many differential gene expression either negatively or positively with the availability of other plant hormones. On the other hand, abscisic acid is able to enhance resistance to fungal infection by stimulating callose deposition in the apoplast that inhibits pathogen entry. However, interpretation of abscisic acid is always complicated as it works together with JA, salicylic acid, and ethylene (Finkelstein 2013). Abscisic acid promotes early defense responses but is likely to inhibit late defense responses by suppressing salicyclic acid dependent responses and modulating JA and ethylene dependent defense (Fujii 2014). This demonstrates that plants hormonal response to one stress influences its responses to additional stresses (Thaler et al. 2004). Interaction between pathogen and host can be identified through the detection of abscisic acid as it is the best moderator during any plant pathogen interaction. This result will provide some insights on the different stages of the infection process (Ulferts et al. 2015).

16.4.5 Wounding

Wounding response is usually identified in the infected wounded plant root. Minimal expression of glutathione S-transferase (GST) can be detected in the wounded plant (Mayer et al. 2017). The mechanical wounding itself will trigger healing process of the damaged tissues to prevent further deterioration by activating other defense mechanisms (León et al. 2001). In addition, GSTs is important in the suppression of necrosis due to pathogen attack by detoxifying organic hydroperoxides of fatty acids generated from peroxidation of membranes (Mayer et al. 2017). Even though wounding is another type of environmental stresses, during this event, pathogen will easily penetrate and infect into plant cell (Ouaked et al. 2003). According to Mayer et al. (2017), the highest level of GST gene expression was measured in mycorrhizal infected plants during high temperature stress. These earlier triggered responses will block the pathogen to penetrate further.

In addition, there are other structurally different molecules that play regulatory roles in wound signalling such as oligopeptide system and plant hormones such as JA, ethylene, and abscisic acid (León et al. 2001). Simultaneous action of different signals and regulators is essential for the wound responses, however the way the integration is achieved may vary among plant species (Nguyen et al. 2016). Therefore, transcriptome or any sort of experimental data for each and every infected plant species need to be generated in order to determine the pattern and stage of infection which can be utilized in further analysis.

16.4.6 Oxidative Stress

Oxidative stress is another outcome of the cellular responses following pathogen infection. This involves the production of reactive oxygen species. The hydrogen peroxidase (H_2O_2) and oxygen (O_2) are the important elements to measure the oxidative stress level (Sun et al. 2013). Both of these compounds will accumulate in order to give the plant enough time to mobilize various other defense response mechanisms (EL-Khallal 2007). However, high accumulation of this compound will damage the plant cellular membrane until other defense mechanism takes place (Sun et al. 2013). Oxidative stress is measured based on GST6 and GST3 expression and both of these genes are overexpressed during the early infection stage. GSTs appear to play a wider role in plant–microbe interactions, as expression of at least one GST is enhanced by infection with various types of fungi, including mycorrhizal species (Dalton et al. 2009).

16.5 Conclusion and Future Prospects

In this article, we reviewed the different types of plant disease and the molecular tools that can be used to detect the infection at an early stage. The pathogen or the selected plant defense response gene expressed during the early stage of infection chosen as a marker can be detected with the available molecular tools. This early detection could serve as a guide for plant disease management before phenotypic symptoms can be observed, as there is no single solution to overcome plant infection apart from performing continuous monitoring. Only through early detection, infected plant could be saved from deleterious scenario. New technology and innovations on sensors, chip or robotic may replace current tools for efficient early detection. This sophisticated tool will enable aerial monitoring and effective early warning on potential outbreak.

References

Agrios GN (2005) Plant pathology, 5th edn. Academic, New York. ISBN: 0120445654, 952p

Ahmed W, Huygens F, Goontilleke A, Gardner T (2008) Real-time PCR detection of pathogenic microorganisms in roof-harvested rainwater in Southeast Queensland Australia. Appl Environ Microbiol 74:5490–5496

Ali MB, McNear DH Jr (2014) Induced transcriptional profiling of phenylpropanoid pathway genes increased flavonoid and lignin content in Arabidopsis leaves in response to microbial products. BMC Plant Biol 14:1–14

Amadi JE, Nwaokike P, Olahan GS, Garuba T (2014) Isolation and identification of fungi involved in the post-harvest spoilage of guava (*Psidium guajava*) in Awka metropolis. Int J Eng Appl Sci 4:7–12

Amoako S, Yahaya A, Sarfo JK (2015) Catalase activity of cassava (*Manihot esculenta*) plant under African cassava mosaic virus infection in cape coast, Ghana. Afr J Biotechnol 14:1201–1206

Anusuya S, Sathiyabama M (2014) Effect of chitosan on rhizome rot disease of turmeric caused by *Pythium aphanidermatum*. ISRN Biotechnol 2014., Article ID 305349, 5 p:1

Babadoost M, Islam SZ (2002) Phytophthora blight on pumpkin. Online Plant Health Prog. doi:10.1094/PHP-2002-1216-01-DG

Balcells I, Cirera S, Busk PK (2011) Specific and sensitive quantitative RT-PCR of miRNAs with DNA primers. BMC Biotechnol 11:70

Basım E, Basım H, Abdulai M, Baki D, Oztürk N (2017) Identification and characterization of *Alternaria alternata* causing leaf spot of olive tree (*Olea europaea*) in Turkey. Crop Prot 92:79e88

BGI (2015) Application Note: Transcriptome sequencing & RNA-Seq for quantification

Biello D (2009) Can carbon capture and storage save coal? Sci Am, pp 1–6

Boonham N, Walsh K, Smith P, Madagan K, Graham I, Barker I (2003) Detection of potato viruses using microarray technology: towards a generic method for plant viral disease diagnosis. J Virol Methods 108:181–187

Broeders S, Huber I, Grohmann L, Berben G, Taverniers I, Mazzara M, Roosensa N, Morisset D (2014) Guidelines for validation of qualitative real-time PCR methods. Trends Food Sci Technol 37:115–126

Brookman-Amissah N, Packer H, Prediger E, Sabel J (2012) qPCR application guide: experimental overview, protocol, troubleshooting, 3rd edn. Integrated DNA Technologies, Coralville

Brunner K, Farnleitner A, Mach RL (2012) Novel methods for the quantification of pathogenic fungi in crop plants: quantitative PCR and ELISA accurately determine Fusarium biomass. In: Dr. Cumagun CJ (ed) Plant pathology. InTech Open Access Publisher, Rijeka

Burkholder JM, Marshall HG, Glasgow HB, Seaborn DW (2001) The standardized fish bioassay procedure for detecting and culturing actively toxic *Pfiesteria* used by two reference laboratories for atlantic and gulf coast states. Environ Health Perspect 109:745–756

Cabral CS, Reis A (2013) Screening of lettuce accessions for resistance to *Fusarium oxysporum* f. sp. *lactucae* race 1. Trop Plant Pathol 38:275–281

Callan BE, Carris LM (2004) Fungi on living plant substrata, including fruits. In: Mueller GM, Bills GF, Foster MS (eds) Biodiversity of fungi—inventory and monitoring methods. Elsevier Academic Press, Amsterdam, pp 105–126

Chen Y, Xu H, Zhou M, Wang Y, Wang S, Zhang J (2015) Salecan enhances the activities of β-1,3-glucanase and decreases the biomass of soil borne fungi. PLoS One 10(8):e0134799

Churchill ACL (2011) *Mycosphaerella fijiensis*, the black leaf streak pathogen of banana: progress towards understanding pathogen biology and detection, disease development, and the challenges of control. Mol Plant Pathol 12:307–328

Cimmino A, Iannaccone M, Petriccione M, Masi M, Evidente M, Capparelli R, Scortichini M, Evidente A (2017) An ELISA method to identified the phytotoxic *Pseudomonas syringe* pv. *Actinidiae* of kiwifruit bacterial canker. Phytochem Lett 19:136–140

Conesa A, Madrigal P, Tarazona S, Gomez-Cabrero D, Cervera A, McPherson A, Szcześniak WM, Gaffney DJ, Elo LL, Zhang X, Mortazavi A (2016) A survey of best practices for RNA-seq data analysis. Genome Biol 17:13

Cortinas MN, Crous PW, Wingfield BD, Wingfield MJ (2006) Multi-gene phylogenies and phenotypic characters distinguish two species within the *Colletogloeopsis zuluensis* complex associated with eucalyptus stem cankers. Stud Mycol 55:133–146

Costa V, Aprile M, Esposito R, Ciccodicola A (2013) RNA-Seq and human complex disease: recent accomplishments and future perspectives. Eur J Hum Genet 21:134–142

Crawford MJ, Li X, Kapoor M, Williams DJ (2014) Patent application title: methods and compositions for plant pest control. Patent application number: 20140215656

Cristescu SM, De Martinis D, te Lintel HS, Parker DH, Harren FJM (2002) Ethylene production by Botrytis cinerea in vitro and in tomatoes. Appl Environ Microbiol 68:5342–5350

Dalton DA, Boniface C, Turner Z, Lindahl A, Kim HJ, Jelinek L, Govindarajulu M, Finger RE, Taylor CG (2009) Physiological roles of glutathione *S*-transferases in soybean root nodules. Plant Physiol 150:521–530

Danks C, Barker I (2000) On-site detection of plant pathogens using lateral-flow devices. EPPO Bull 30:421–426

Dash M, Behera B, Rahut DB (2016) Determinants of household collection of non-timber forest products (NTFPs) and alternative livelihood activities in Similipal Tiger Reserve, India. For Policy Econ 73:215–228

Dean R, Van Kan JAL, Pretorius ZA, Hammond-Kosack KE, Di Pietro A, Spanu PD, Rudd JJ, Dickamn M, Kahmann R, Ellis J, Foster GD (2012) The top 10 fungal pathogens in molecular plant pathology. Mol Plant Pathol 13:414–430

Dempsey DA, Klessig DF (2012) SOS-too many signal for systemic acquired resistance? Trends Plant Sci 17:538–545

DeVay JE, Gutierrez AP, Pullman GS, Wakeman RJ, Garber RH, Jeffers DP, Smith SN, Goodell PB, Roberst PA (1997) Inoculum densities of *Fusarium oxysporum* f. sp. *vasinfectum* and Meloidogyne incognita in relation to the development of fusarium wilt and the phenology of cotton plants (*Gossypium hirsutum*). Phytopathology 87:341–346

Di Pietro A, Madrid MP, Caracuel Z, Delgado-Jarana J, Roncero MIG (2003) *Fusarium oxysporum*: exploring the molecular arsenal of a vascular wilt fungus. Mol Plant Pathol 4:315–325

Dolezal AL, Xiaomei S, Brian GRO, Nielsen DM, Woloshuk CP, Boston RS, Payne GA (2014) A*spergillus flavus* infection induces transcription land physical changes in developing maize kernels. Front Microbiol 5:1–10

Dweba CC, Figlan S, Shimelis HA, Motaung TE, Sydenham S, Mwadzingeni TTJ (2017) Fusarium head blight of wheat: pathogenesis and control strategies. Crop Prot 91:114–122

Ebrahim SB, Usha K, Singh B (2011) Pathogenesis related (PR) proteins in plant defense mechanism. FORMATEX, Spain, pp 1043–1054

Ecker JR (1995) The ethylene signal transduction pathway in plants. Science 268:667–675

Ecker JR, Guo H (2004) The ethylene signalling pathway: new insights. Curr Opin Plant Biol 7:40–49

El-Khallal SM (2007) Induction and modulation of resistance in tomato plants against fusarium wilt disease by bioagent fungi (arbuscular mycorrhiza) and/or hormonal elicitors (Jasmonic Acid & Salicylic Acid): 1-changes in growth, some metabolic activities and endogenous hormones related to defence mechanism. Aust J Basic Appl Sci 1:691–705

Elsheshtawi M, Elkhaky MT, Sayed SR, Bahkali AH, Mohammed AA, Gambhir D, Mansourf AS, Elgorban AM (2017) Integrated control of white rot disease on beans caused by Sclerotinia sclerotiorum using Contans and reduced fungicides application. Saudi J Biol Sci 24:405–409

Eulgem T (2006) Dissecting the WRKY web of plant defense regulators. PLoS Pathog 2:e126
Feau N, Vialle A, Allaire M, Tanguay P, Joly DL, Frey P, Callan BE, Hamelin RC (2009) Fungal pathogen (MIS) identifications; a case study with DNA barcodes on Melampsora rusts of aspen and white poplar. Mycol Res 113:713–724
Fernando CK, Soysa P (2015) Optimized enzymatic colorimetric assay for determination of hydrogen peroxide (H_2O_2) scavenging activity of plant extracts. MethodsX 2:283–291
Finkelstein R (2013) Abscisic acid synthesis and response. Arabidopsis Book 11:e0166
Fleige S, Pfaffl MW (2006) RNA integrity and the effect on the realtime qRT-PCR performance. Mol Asp Med 27:126–139
Fu ZQ, Dong X (2013) Systemic acquired resistance: turning local infection into global defense. Annu Rev Plant Biol 64:839–863
Fujii H (2014) Abscisic acid implication in plant growth and stress responses. In: Tran LS, Pal S (eds) Phytohormones: a window to metabolism, signaling and biotechnological applications. Springer, New York, pp 37–54
Gayatridevi S, Jayalakshmi SK, Sreeramulu K (2012) Salicylic acid is a modulator of catalase isozymes in chickpea plants infected with *Fusarium oxysporum* f. sp. *ciceri* plant. Physiol Biochem 52:154–161
Goel R, Kumar P, Sinha R, Ambwani S (2004) Dot-blot assay using revere probe genomic hybridization of repetitive DNA for environment monitoring of fluorescent pseudomonads. Indian J Biotechnol 3:82–85
Guo H, Ecker JR (2004) The ethylene signalling pathway: new insights. Curr Opin Plant Biol 7:40–49
Guo L, Han L, Yang L, Zeng H, Fan D et al (2014) Genome and transcriptome analysis of the fungal pathogen *Fusarium oxysporum* f. sp. *cubense* causing banana vascular wilt disease. PLoS One 9(4):e95543. doi:10.1371/journal.pone.0095543
Gupta P, Ravi I, Sharma V (2013) Induction of β-1,3-glucanase and chitinase activity in the defense response of *Eruca sativa* plants against the fungal pathogen *Alternaria brassicicola.* J Plant Interact 8:155–161
Hall RA (1989) Pathogenicity of fungi and bioassay design. Mem Inst Oswaldo Cruz 84:47–50
Hammerschmidt LI, Nuckles EM, Kuc J (1982) Association of enhanced peroxidase activity with induced systemic resistance of cucumber to Colletotrichum I'agenarium. Physiol Plant Pathol 20:73–82
Hao GJ, Zhang K, Zhang JY, Wang XR, Qin Z, Wang XZ, Wang L, Meng JR, Yang ZQ, Li JX (2013) RT-qPCR analysis of dexB and gale gene expression of *Stretococcus alactolyticus* in Astragalus membranous fermentation. Appl Microbial Biotechnol 97:6009–6018
Haralson JC, Brannen PM, NeSmith DS, Schem H (2013) Chemical control of Cylindrocladium and Rhizoctonia root rots in blueberry propagation. Crop Prot 44:1–5
He YH, Isono S, Shibuya M, Tsuji M, Adkar Purushothama CR (2012) Oligo-DNA custom macroarray for monitoring major pathogenic and non-pathogenic fungi and bacteria in the phyllosphere of apple trees. PLoS One 7:e34249
Hoffmann-Sommeergruber K (2002) Pathogenesis-related (PR)-proteins identified as allergens. Biochem Soc Trans 30:930–935
Hong SY, Roze LV, Linz JE (2013) Oxidative stress related transcription factors in the regulation of secondary metabolism. Toxins (Basel) 5:683–702
Huang J, Gu M, Lai Z, Fan B, Shi K, Zhou YH, Yu JQ, Chen Z (2010) Functional analysis of the Arabidopsis PAL gene family in plant growth, development and response to environmental stress. Plant Physiol 153:1526–1538
Huot B, Yao J, Montgomery BL, He SY (2014) Growth-defense tradeoffs in plants: a balancing act to optimize fitness. Mol Plant 7:1267–1287
Hwang J, Lee S, Lee JH, Kang WH, Kang JH, Kang MY, Oh CS, Kang BC (2015) Plant translation elongation factor 1Bβ facilitates potato virus X (PVX) infection and interacts with PVX triple gene block protein 1. PLoS One 10:e0128014
Illumina (2014) Next generation sequencing: application guide, pp 3–110

Iwase T, Tajima A, Sugimoto OK, Hironaka I, Kamata Y, Takada K, Mizunoe Y (2013) A simple assay for measuring catalase activity: a visual approach. Sci Rep 3:3081

Jiang JJ, Ma S, Ye N, Jiang M, Cao J, Jianhua Z (2016) WRKY transcription factors in plant responses to stresses. J Integr Plant Biol 59:86–101

Kasselaki AM, Goumas D, Tamm L, Fuchs J, Cooper J, Leifert C (2011) Effect of alternative strategies for the disinfection of tomato seed infected with bacterial canker (*Clavibacter michiganensis* subsp. *michiganensis*). NJAS Wagen J Life Sci 58:145–147

Kheiri A, Moosawi Jorf SA, Malihipour A, Saremi H, Nikkhah M (2017) Synthesis and characterization of chitosan nanoparticles and their effect on fusarium head blight and oxidative activity in wheat. Int J Biol Macromol 102:526–538

Kimbrel JA, Di Y, Cumbie JS, Chang JF (2011) RNA-Seg for plant pathogenic bacteria. Genes (Basel) 2:689–705

Kumar Panda S (2012) Assay guided comparison for enzymatic and non-enzymatic antioxidant activities with special reference to medicinal plants. In: Prof. Mohammed Amr El-Missiry (ed) Antioxidant enzyme. InTech, Rijeka

Kunkel N, Brooks DM (2002) Cross talk between signalling pathways in pathogen defense. Curr Opin Plant Biol 5:325–331

Lacroix B, Ryan J, Dumont J, Maddox PS, Maddox AS (2016) Identification of microtubule growth deceleration and its regulation by conserved and novel proteins. Mol Biol Cell 27:1479–1487

Laurent L, Marell A, Korboulewsk N, Said S, Balandier P (2017) How does disturbance affect the intensity and importance of plant competition along resource gradients? For Ecol Manag 391:239–245

Lecompte F, Nicot PC, Ripoll J, Abro MA, Raimbault AK, Lopez-Lauri F, Bertin N (2017) Reduced susceptibility of tomato stem to the necrotrophic fungus *Botrytis cinerea* is associated with a specific adjustment of fructose content in the host sugar pool. Ann Bot 119:931–943

Lei F, Fu J, Zhou R, Wang D, Zhang A, Ma W, Zhang LX (2017) Chemotactic response of ginseng bacterial soft-rot to ginseng root exudates. Saudi J Biol Sci Article in Press

León J, Rojo E, Sánchez-Serrano JJ (2001) Wound signalling in plants. J Exp Bot 52:1–9

Lievens B, Brouwe M, Vanachter Alfons CRC, Andre Levesque C, Cammue Bruon PA, Thomma Bart PHJ (2003) Design and development of a DNA array for rapid detection and identification of multiple tomato vascular wilt pathogens. FEMS Microbiol Lett 223:113–122

LifeTechnologies (2012) Real time PCR handbook, pp 1–66

Lin Y, Zhang C, Lan H, Gao S, Liu H, Liu J, Cao M, Pan G, Rong T, Zhang S (2014) Validation of potential reference genes for qPCR in maize across abiotic stresses, hormone treatments, and tissue types. PLoS One 9:e95445

Liu YC, Wang SC, Yu YJ, Fung KM, Yang MT, Tseng YH, Tsai SF, Sun HS, Lyu PC, Chou S-H (2015) Complete genome sequence of Xanthomonas campestris pv. campestris strain 17 from Taiwan. Genome Announc 3:e01466–e01415

Madadkhah E, Lotfi M, Nabipour A, Rahmanpour S, Banihashemi Z, Shoorooei M (2012) Enzymatic activities in roots of melon genotypes infected with *Fusarium oxysporum* f. sp. *melonis* race 1. Sci Hortic 135:171–176

Magliano ATM, Kikot GE (2010) Microscopy techniques to analyse plant-pathogen interactions: *Fusarium graminearum* as pathogen of wheat. In: Mendez-Vilas A, Diaz J (eds) Microscopy: science, technology, applications and education. Formatex, Spain, pp 1108–1114

Mak C, Mohamed AA, Liew KW, Ho YW (2001) Early screening technique for *Fusarium* wilt resistance in banana micropropagated plants. Proceedings of a meeting held in Leuven, Belgium, 24–28 September 2001, pp 219–227

Malcolm GM, Kuldau GA, Gugino BK, Jiménez-Gasco Mdel M (2013) Hidden host plant associations of soilborne fungal pathogens: an ecological perspective. Phytopathology 103:538–544

Martin RR, James D, Andre Levesque C (2000) Impacts of molecular diagnostic technologies on plant disease management. Annu Rev Phytopathol 38:207–239

Martin LBB, Fei Z, Giovannoni JJ, JKC R (2013) Catalyzing plant science research with RNS-seq. Front Plant Sci 4:1–10

Martinez A and Woodward J (2015) Common Landscape Diseases in Bulletin 1238 Reviewed May 2015 UGA Extension Bulletin 1238

Mayer Z, Hong Duc N, Posta K (2017) Gene expression of *glutathione-S-transferase* in sunflower (*Helianthus annuus* L.) inoculated with arbuscular mycorrhizal fungi under temperature stresses. Columella 4:69–70

Moerschbachkr BM, Flott BI, Noll V, Rlusi-NKR J (1988) Lignin biosynthetic enzymes in stem rust inlcctcd resistant and susceptible near(ist) gemc heat lines. Physiol Mol Plant Pathol 33:33–46

Mohd Fishal EM, Meon S, Mui Yun W (2010) Induction of tolerance to fusarium wilt and defense related mechanisms in the plantlets of susceptible berangan banana pre-inoculated with *Pseudomonas* sp. (UPMP3) and *Burkholderia* sp. (UPMB3). Agric Sci China 9:1140–1149

Mueller DS, Jeffers SN, Buck JW (2004) Effect of timing of fungicide applications on development of rusts on daylily, geranium, and sunflower. Plant Dis 88:657–661

Nandi S, Dutta S, Mondal A, Adhikari A, Nath R, Chattopadhaya A, Chaudhuri S (2013) Biochemical responses during the pathogenesis of Sclerotium rolfsii on cowpea. Afr J Biotechnol 12:3968–3977

Narayanasamy P (2011). Microbial plant pathogens detection and disease diagnosis: viral and viroids pathogens, vol 3. Springer, London

Nel B, Steinberg G, Labuschagne N, Viljoen A (2007) Evaluation of fungicides and sterilises for potential application in the management of fusarium wilt of banana. Crop Prot 26:697–705

Nguyen D, Rieu I, Mariani C, van Dam NM (2016) How plants handle multiple stresses: hormonal interactions underlying responses to abiotic stress and insect herbivory. Plant Mol Biol 91:727–740

Nicosia MGLD, Pangallo S, Raphael G, Romeo FV, Strano MC (2016) Control of postharvest fungal rots on citrus fruit and sweet cherries using a pomegranate peel extract. Postharvest Biol Technol 114:54–61

Okubara PA, Li C, Schroeder KL, Schumacher RT, Lawrence N (2010) Improved extraction of Rhizoctonia and Pythium DNA from wheat roots and soil samples using pressure cycling technology. Can J Plant Pathol 29:304–310

Ouaked F, Rozhon W, Lecourieux D, Hirt HA (2003) MAPK pathway mediates ethylene signaling in plants. EMBO J 22:1282–1288

Pandey SP, Somssich IE (2017) The role of WRKY transcription factors in plant immunity. Plant Physiol 150:1648–1655

Passos MAN, De Cruz VO, Emediato FL, De Teixeira CC, Renno Azevedo VC, Brasileiro ACM, Amorim EP, Ferreira CF, Martins NF, Togawa RC, Pappas Junior GJ, Da Silva OB Jr, Miller RNG (2013) Analysis of the leaf transcriptome of *Musa acuminate* during interaction with *Mycocphaerella musicola*: gene assembly, annotation and marker development. BMC Genomics 14:78–95

Petre B, Morin E, Tisserant E, Hacquard S, Da Silva C et al (2012) RNA-Seq of early-infected poplar leaves by the rust pathogen *Melampsora laricipopulina* uncovers PtSultr a fungal-induced host sulfate transporter. PLoS One 7:e44408

Phukan UJ, Jeena GS, Shukla RK (2016) WRKY transcription factors: molecular regulation and stress responses in plants. Front Plant Sci 7:760–773

Prencipe S, Nari L, Vittone G, Gullino ML, Spadaro D (2016) Effect of bacterial canker caused by *Pseudomonas syringae* pv. *actinidiae* on postharvest quality and rots of kiwifruit 'Hayward'. Postharvest Biol Technol 113:119–124

Pretorius ZA, Szabo LJ, Boshoff WHP, Herselman L, Visser B (2012) First report of a new TTKSF race of wheat stem rust (*Puccinia graminis* f. sp. *tritici*) in South Africa and Zimbabwe. Disease Notes 96:590

Pscheidt JW, Ocamb CM (2002) Pacific northwest plant disease management handbook. Agricultural Experiment Station, Oregon State University, Corvallis, OR

Punja ZK, Zhang YY (1993) Plant chitinase and their roles in resistance to fungal diseases. J Nematol 25:526–540

Ren X, Chen Z, Liu Y, Zhang H, Zhang M, Liu Q, Hong X, Zhu JK, Gong Z (2010) AB03, a WRKY transcription factor, mediates plant responses to abscisic acid and drought tolerance in Arabidopsis. Plant J 63:417–429

Robert-Seilaniantz A, Grant M, Jones JD (2011) Hormone crosstalk in plant disease and defense: more than just jasmonate-salicylate antagonism. Annu Rev Phytopathol 49:317–343

Romero IG, Pai AA, Tung J, Gilad Y (2014) RNA-seq: impact of RNA degradation on transcript quantification. BMC Biol 12:42

Sastry KS (2013) Plant virus and viroid diseases in the tropic volume 1 Introduction of plant viruses and subviral agents, classification, assesssment of loss, transmission and diagnosis Springer Dordrecht Heidelberg New York London, pp 353

Schena L, Nicosia LDMG, Sanzani SM, Faedda R, Ippolito A, Cacciola SO (2013) Development of quantitative PCR detection methods for phytopathogenic fungi and oomycetes. J Plant Pathol 95:7–24

Schrader A, Schielke L, Ellerbroek L, Johne R (2012) PCR inhibitors-occurrence, properties and removal. J Appl Microbiol 113:1014–1026

Schumann GL (2017) The Irish potato famine and the birth of plant pathology in plant diseases: their biology and social impact. APS, College Park

Sevik MA, Kose-Tohumcu E (2011) The ELISA analysis results in tomato (Lycopersicon esculentum mill.) seed health testing for tobacco mosaic virus. Zemdirbyste Agric 98:301–306

Shabana YM, Abdalla ME, Shahin AA, El-Sawy MM, Draz IS, Youssif AW (2017) Efficacy of plant extracts in controlling wheat leaf rust disease caused by *Puccinia triticina*. Egypt J Basic Appl Sci 4:67–73

Sharifazizi M, Harighi B, Sadeghi A (2017) Evaluation of biological control of Erwinia amylovora, causal agent of fire blight disease of pear by antagonistic bacteria. Biol Control 104:28–34

Sharma N, Sharma KP, Gaur RK, Gupta VK (2011) Role of chitinase in plant defense. Asian J Biochem 6:29–37

Shenoy BD, Jeewon R, Hyde KD (2007) Impact of DNA sequence-data on the taxonomy of anamorphic fungi. Fungal Divers 26:1–54

Shibaei N, Majidi J, Razavi K, Karkhane AA, Sokhandan-Bashir N, Aghebati-Maleki L (2017) Production and characterization of monoclonal antibody against recombinant virus coat protein CP42. Iran J Allergy Asthma Immunol 16:60–71

Sholberg P, Gorman O, Bedford DK, Lévesque CA (2005) Development of a DNA macroarray for detection and monitoring of economically important apple diseases. Plant Dis 89:1143–1150

da Silva JC, Suassuna ND, Bettiol W (2017) Management of Ramularia leaf spot on cotton using integrated control with genotypes, a fungicide and Trichoderma asperellum. Crop Prot 94:28e32

Singh PK, Mishra M, Vyas D (2010) Interactions of vesicular arbuscular mycorrhizal fungi with fusarium wilt and growth of the tomato. Indian Phytopathol 63:30–34

Singh NK, Kumar KRR, Kumar D, Shukla P, Kirti PB (2013) Characterization of pathogen induced thaumatin-like protein gene *AdTLP* from *Arachis diogoi,* a wild peanut. PLoS One 8:e83963

Singh AK, Raib N, Singh RK, Saha S, Rai RK, Singh RP (2017) Genetics of resistance to early blight disease in crosses of wild derivatives of tomato. Sci Hortic 219:70–78

Smith C, Osborn M (2009) Advantages and limitations of quantitative PCR (qPCR)-based approaches in microbial ecology. FEMS Microbiol Ecol 67:6–20

Sosnowski MR, Fletcher JD, Daly AM, Rodoni BC, Viljanen-Rollinson SLH (2009) Techniques for the treatment, removal and disposal of host material during programmes for plant pathogen eradication. Plant Pathol 58:621–635

Sun L, Yang DL, Kong Y, Chen Y, Li XZ, Zeng LJ (2013) Sugar homeostasis mediated by cell wall invertase grain incomplete filling 1 (GIF1) plays a role in pre-existing and induced defence in rice. Mol Plant Pathol 15:161–173

Sutherland A, Viljoen A, Myburg AA, Van den Berg N (2013) Pathogenicity associated genes in *Fusarium oxysporum* f. sp. *cubense* race 4. S Afr J Sci 109:1–10

Thakur M, Sohal BS (2013) Role of elicitors in inducing resistance in plants against pathogen infection: review article. ISRN Biochem Article ID 762412: 10p

Thaler JS, Agrawal AA, Halitschke R (2004) Salicylate-mediated interactions between pathogens and herbivores. Ecology 91:1075–1082

Thiessen LD, Woodward JE (2012) Diseases of peanut caused by soilborne pathogens in the Southwestern United States. ISRN Agron Article ID 517905

Trad S, Allignet J, Frangeul L, Davi M, Vergassola M, Couve E, Morvan Z, Kechrid A, Buchrieser C, Glase P, El Solh N (2004) DNA macroarray for identification and typing of *Staphylococcus aureus* isolates. J Clin Microbiol 42:2054–2064

Tsui CKM, Woodhall J, Chen W, André Lévesque C, Lau A, Schoen CD, Baschien C, Najafzadeh MJ, Sybren de Hoog G (2011) Molecular techniques for pathogen identification and fungus detection in the environment. IMA Fungus 2:177–189

Tziros Georgios T, Lagopodi Anastasia L, Tzavella-Klonari K (2007) Reduction of fusarium wilt in watermelon by *Pseudomonas Chlororaphis* Pcl1391 and *P. Fl Uorescens* Wcs365. Phytopathol Mediterr 46:320–323

Ulferts S, Delventhal R, Splivallo R, Krlovsky P, Schaffrath U (2015) Abscisic acid negatively interferes with basal defence of barley against *Magnaporthe oryzae*. BMC Plant Biol 15:7

Unamba CIN, Nag A, Sharma RK (2015) Next generation sequencing technologies: the doorway to the unexplored genomics of non-model plants. Front Plant Sci 6:1074

Valasek MA, Repa JJ (2005) The power of real-time PCR. Adv Physiol Educ 29:151–159

Van Den Berg N, Berger DK, Hein I, Birch PRJ, Wingfield MJ, Viljoen A (2007) Tolerance in banana to fusarium wilt is associated with early up-regulation of cell wall-strengthening genes in the roots. Mol Plant Pathol 8:333–341

Vandesompele J, De Preter K, Pattyn F, Poppe B, Van Roy N, De Paepe A, Speleman F (2002) Accurate normalization of real time quantitative RT-PCR data by geometric averaging of multiple internal control genes. Genome Biol 3:RESEARCH0034:research0034.1

Vercauteren I, de Almeida EJ, de Groodt R, Gheysen G (2002) An *Arabidopsis thaliana* pectin acethylesterase gene is upregulated in nematode feeding sites induced by root-knot and cyst nematodes. Mol Plant-Microbe Interact 15:404–407

Wang Z, Gerstein M, Snyder M (2009) RNA-Seq: a revolutionary tool for transcriptomics. Nat Rev Genet 10:57–63

Whisson SC, Boevink PC, Wang S, Birch PRJ (2016) The cell biology of late blight disease. Curr Opin Microbiol 34:127–135

Wicaksono WA, Jones EE, Casonato S, Monk J, Ridgway HJ (2017) Biological control of pseudomonas syringae pv. actinidiae (Psa), the causal agent of bacterial canker of kiwifruit, using endophytic bacteria recovered from a medicinal plant. Biol Control Article in Press

Wingfield MJ, Crous PW, Couthinho TA (1997) A serious new canker disease of eucalyptus in South Africa caused by a new species of Coniothyrium. Mycopathologia 136:139–145

Wu YL, Yi GJ, Peng XX (2010) Rapid screening of Musa species for resistance to fusarium wilt in an in vitro bioassay. Eur J Plant Pathol 128:409–415

Xiong L, Kong F, Yang Y, Cheng J, Gilbert GL (2006) Use of PCR and reverse line blot hybridization microarray based on 16S-23SrRNA gene internal transcribed spacer sequences for rapid identification of 34 *Mycobacterium* species. J Clin Microbiol 44:3544–3550

Yadav IS, Sharma A, Kaur S, Nahar N, Bhardwaj SC, Sharma TR, Chhuneja P (2016) Comparative temporal transcriptome profiling of wheat near isogenic line carrying Lr57 under compatible and incompatible interactions. Front Plant Sci 7:1943

Yang Y, Yan P, Yi C, Li W, Chai Y, Fei L, Gao P, Zhao H, Wang Y, Timko MP, Wang B, Han S (2017) Transcriptome-wide analysis of jasmonate-treated BY-2 cells reveals new transcriptional regulators associated with alkaloid formation in tobacco. J Plant Physiol 215:1–10

Yi SY, Shirasu K, Moon JS, Lee S-G, Kwon S-Y (2014) The activated SA and JA signaling pathways have an influence on flg22-triggered oxidative burst and callose deposition. PLoS One 9(2):e88951

Yim WJ, Kim KY, Lee YW, Sundaram SP, Lee Y, Sa TM (2014) Real time expression of ACC oxidase and PR protein genes mediated by *Methylobacterium* spp. in tomato plants challenged with *Xanthomonas campestris* pv *Vesicatoria*. J Plant Physiol 171:1064–1075

Zhang N, Geiser DM, Smart CD (2007) Macroarray detection of solanaceous plant pathogens in the *Fusarium solani* species complex. Plant Dis 91:1612–1620

Zhao S, Fu Y, Mersha Z, Pernezny K (2014) Assessment of copper resistance in *Pseudomonas syringae* pv. *Phaseolicola*, the pathogen of halo blight on snap bean. Crop Prot 98:8–15

Zhao Z, Gao X, Yang D, Huang L, Qin H, Kang ZS, Wang N (2015) Field detection of canker-causing bacteria on kiwifruit trees: *Pseudomonas syringae* pv. *actinidiae* is the major causal agent. Crop Prot 75:55–62

Zheng W, Chung LM, Zhao H (2011) Bias detection and correction in RNA sequencing data. BMC Bioinf 12:290–304

Zhou J, Wang M, Sun Y, Gu Z, Wang R, Saydin A, Shen Q, Guo S (2017) Nitrate increased cucumber tolerance to fusarium wilt by regulating fungal toxin production and distribution. Toxins 9:100

Index

A
Abiotic stresses, 38, 40, 72, 73, 78–81, 85–87
affecting oil palm productivity, 73, 74
and fruit development
AP2-ERF superfamily, 78, 79
families of transcription factors, 79–81
and symptoms, 75
transcription factors in oil palm, 84
CBF, 87
DREB1, 86, 87
EABF and EABF1, 85
Abscisic acid (ABA), 75, 82
Acacia
A. auriculiformis, 117, 118, 121–123, 125–127
A. mangium, 117, 118, 121–123, 125–127
development, 126, 127
ESTs, 119, 120
linkage mapping, 128–130
QTL analysis, 130–133
Aculeiferum, 117
Agrobacterium, 192, 196–199, 201, 202, 204, 205, 208, 209
Ajmalicine, 139, 151
Alkaloids, 137, 144, 145
Amplified fragment length polymorphism (AFLP), 52, 53
Anthocyanins, 151, 156
Anthraquinones, 150
Anthriscus sylvestris, 150
Antioxidant dietary fibre (ADF), 280
APETALA2/ethylene response factor (AP2/ERF), 78, 79
Aptamer, 104
aptasensors, 105, 107
immobilization, 107
production, 105
Arabidopsis thaliana, 8, 30, 132, 149
Arbuscular mycorrhiza (AM), 2
Auxins, 84
Avirulence proteins, 321

B
Benzothiadiazole, 322, 323
Benzylisoquinoline, 158
Berberine, 158
Beta-aminobutyric acid (BABA), 322
Beta-D-xylosidase, 30
Biacore-based SPR system, 110
Bioreactors, metabolite production, 171, 172, 174–177, 182–185
astaxantin, 171
biological drugs, 171
cultivation, living organisms, 170
disposable bioreactors, 185–187
economic analysis
extraction and purification, 171
medical herbs, 174
plant cell culture, 171
production costs, 172, 174, 175
research activity, 177
rosmarinic acid, 175
substrate costs, 175, 176
ethanol, lactic acid and amino acids, 170
fed-batch operations, 182
fermentation, 169
food colors and flavors market, 170
food, clothes and transportation, 169
gas phase (*see* Gas phase bioreactors)

S.N.A. Abdullah et al. (eds.), *Crop Improvement*,
DOI 10.1007/978-3-319-65079-1

Bioreactors, metabolite production (*cont.*)
growth environment, 177
hairy roots, 181
liquid phase, 178–180
monitoring and control instruments, 169
novelties, 180, 181
secondary metabolites, 170
Biosensors, 103–105
Blast2GO, 8
BLASTX, 7
Blights, 360, 361
Boraginaceae, 144
Botrytis cinerea, 13
Bradyrhizobium, 117
Bremia lactucae, 13
Brown rice, 292–293
and digestibility, 293
heat and breakage, 294

C

C_3 species, 30, 33
C_4 photosynthesis, 33
Caffeic acid O-methyltransferase (COMT), 120
Caffeoyl-CoA O-methyltransferase (CCoAOMT), 120
Calvin cycle, 27
Camellia sinensis, 53
Cankers, 361
Carbon dioxide (CO_2), 27
Catharanthine, 139
Catharanthus roseus, 139, 149
cDNA fragments, 5
Cell wall material (CWM), 279, 280
Cellulose synthase (CesA), 120
Cerato-platanins (CPs), 320, 321
Chalcone synthase, 132
Charentais genetic, 191
Chenopodium quinoa, 48
Chimeras
identification, 207
occurrence, 206, 207
Chlorophyll a-b binding protein, 35
Chloroplasts, 29
Choanephora infundibulifera, 148
Cinnamate-4-hydroxylase (C4H), 120
Cinnamoyl-CoA reductase (CCR), 120
Cinnamyl-alcohol dehydrogenase (CAD), 120
Cis/intragenic plants generation techniques, 197–205
co-transformation, 199
pMF marker-free technology vectors, 201–204
site-specific recombination, 200, 201
transformation without selection, 198, 199
Cisgenesis
advantages and drawbacks, 205, 206
crop improvement, 191–209
Citrullus colocynthis, 61
Citrullus lanatus, 61
Citrus psorosis virus, 101
Cleaved amplified polymorphic sites (CAPSs), 53, 124
Climacteric fruits
development and ripening, 74
CO_2 concentration, 28
Cold stress, 87
Constitutive triple response1 (CTR1), 157
Conventional plant breeding, 191–193
Coptis japonica, 146
Crop diseases, sensing, 105, 107–111
antibody, 109
immunosensors, 108, 109
production, 108
aptamer, 104
aptasensors, 105, 107
immobilization, 107
production, 105
array-based techniques, 112
biosensors, 103–105
causative agents, 103
ELISA strategy, 110
future prospects, 111, 112
pathogens, 102
potential techniques, 113
SELEX process, 106
using DNA, 111
nucleic acid sensors, 110, 111
nucleic acids, synthesis, 109
Crop genetic improvement, 49–56, 191
Crop protection
bacterial and fungal resistance, 237
insects and pest resistance, 233–237
virus resistance, 233
Crop utilisation, food waste, 277–283
education and engagement, consumers, 276
financial pressures, 276
product development
enzymatic activity, 277
fibre and CWM, 279, 280
financial sustainability, 277
format NPD, 277–279
pigments, 281, 282
polyphenols and bioactive compounds, 280, 281

proteins, 282
substrates, 283
valorisation, 279
Cucurbitaceae, 61
CWM. *See* Cell wall material (CWM)
Cytokinines, 148

D
Danger-associated molecular patterns (DAMPs), 305
De novo assembly, 17
Defence-eliciting capability, 308–313
Dehydrin (DHN), 40
1-Deoxy-D-xylulose phosphate (DXP), 139
Deoxypodophyllotoxin, 150
Diameter at breast high (DBH), 123
Dickeya dianthicola, 108
Diffusion bioreactors, 181
Dioscorea balcanica, 138
Diosgenin, 151
Disposable bioreactors, 185–187
Diurnal cycle, 37, 38
Diversity Arrays Technology (DArT), 55, 56
DNA-based marker systems, 50, 58, 124–126
Draft genomes, 57
Drought stress response, 86, 87

E
Echinacea purpurea, 148
Ectomycorrhizal interaction, 12
Effector proteins, 317
Effector-triggered immunity (ETI), 306
Eleutherococcus senticosus, 148
Elicitors, 306–314
Endemic endophyte, 8
Enzyme-linked immunosorbent assay (ELISA), 108
Ethylene insensitive-like proteins (EILs), 82, 83, 157
E-values, 7
Expressed sequence tags (EST, , ,), 18, 48, 53, 60, 118–120

F
Fagaceae, 143
FASTX-Toolkit, 6
Ferredoxin NADP reductase (FNR), 35, 37, 40
Fluorescence-based detection system, 110
Food waste and supply chain sustainability
causes, 273–275
definition, 272
food security, 271
pressures, 275, 276
production, post-harvest and processing stages, 272
physical losses, 295–297
volumes, 272, 273
Fresh fruit bunch (FFB), 38
Funneliformis mosseae, 8
Fusarium, 40, 41

G
Ganoderma boninense, 40, 41
Gas phase bioreactors
advantages and disadvantages, 183
commercialization, 183
description, 182
mist/spray, 184, 185
modular approach, 183
oxygen supply, 183
plant tissues and organs, 183
trickle-bed, 185
Gene expression analysis, 86
Gene ontology (GO), 7
Gene-for-gene resistance, 3
Genetic improvement
AFLP, 52, 53
DArT, 55, 56
RAPD, 51, 52
RFLP, 51
SNP, 54, 55
SSR, 53, 54
Genetically modified crops (GM crops), 192, 201, 206
Genetically modified organisms (GMOs), 197
Genotype-by-sequencing (GbS), 55
Gibberellic acid (GA3), 148
Ginsenoside, 150
Global food supply, 1
Glucose 6-phosphate (G6P), 35
Glycyrrhizin inflata, 148
Good manufacturing practice (GMP), 145
Grading, 297–298
Grapevine fanleaf virus, 101
Greenforce CuCa, 322

H
Heliotropium spp., 144
Hemibiotrophs, 2
High-quality rice, 298–300
Homeobox (HB), 122

Hulling Machine (Huller), 291, 292
Hyoscyamine, 145
Hyoscyamus niger, 152
Hyperforin, 154, 155
Hypericins, 154
Hypericum perforatum, 154, 155

I
Illumina HiSeq sequencing platform, 5
Immunosensors, 108, 109
Indole acetic acid (IAA), 148, 322
Induced systemic resistance (ISR), 2, 3, 306
  antibiotics, 344
  bacteria and fungi, 339
  beneficial microbes and determinants, 345
  biocontrols and soil amendments, 349
  expression, 346, 347
  JA/ET-signaling pathway, 342
  Jar1 gene, 343
  LPS, 343
  mechanism, 336–337
  microbes and SAR pathway, 341
  microorganisms, 339
  MYB72, 346
  NPR1, 342
  plant immunity, 343, 348
  pyocyanin and pyochelin, 344
  siderophores, 344
InterPro, 7
Intragenesis and crop improvement, 191
Isopentenyl transferase, 198
ISR. *See* Induced systemic resistance (ISR)

J
JA ZIM domain (JAZ), 158
Jasmonate-inducible alkaloid transporter 1 (JAT1), 157, 159
Jasmonic acid (JA), 2, 83

K
KASPar-based SNP array, 60
Kruskal-Wallis (KW) test, 130
Kyoto Encyclopedia of Genes and Genomes (KEGG), 7

L
Label-free method, 105
Laccaria bicolor, 12
Late embryogenesis abundant (LEA), 40
Leguminosae, 60, 61, 144
Lettuce mosaic virus, 101
Lignin, 118
Liliaceae, 143
Lipopolysaccharides (LPS), 343, 345
Liquid phase bioreactors
  aeration and mixing, 178
  airlift bioreactor, 179
  bubble column bioreactors, 179
  and gas phases bioreactors, 180
  homogeneity, 178
  microorganisms, 178
  oxygen supply, 178
  plant cell cultures, 179
  rotating drum bioreactors, 179
  shear stress, 178
  stirred tank bioreactor, 179
  temporary immersion bioreactors, 180
Lithospermum erythrorhizon, 150
LPS. *See* Lipopolysaccharides (LPS)

M
MAP-kinase kinase kinase (MAPKKK), 157
Marker-assisted selection (MAS), 49, 118
Medicago sativa, 55
Medicago truncatula, 126
Medicinal plants
  active compounds, 141, 142
  bioactive compounds, 147
  biotic and abiotic stresses, 138
  culture medium/subcellular compartments, 158, 159
  enzyme activity, 156
  metabolic pathway, 151–156
  PGRs, 148
  physicochemical, 148
  whole plant and tissue cultured materials, 149–151
Mentha piperita, 148
Mesocarp, 35
Metabolomics, 33
Metallothionein type 2, 40
Methylerythritol phosphate (MEP), 143
Mevalonic acid (MVA), 139, 152
MicroRNA analysis, 123, 124
Mimosoideae, 117
Mirtaceae, 143
Mist/spray bioreactors, 184, 185
Molecular markers
  allelic variation, 48
  crop improvement, 49–56
  underutilised crops, 56, 57
Moniliophthora perniciosa, 13
Multidrug and toxic compound extrusion (MATE), 158
Multispecies transcriptomics

pathogenic relationship, 12–14
profiling, 12
root colonization, 8
transcriptional changes, 8
Mutation breeding, 192
Mutualism relationship, 8–12
Myeloblastosis related proteins (MYB), 79
Myelocytomatosis related proteins (MYC), 79

N

NAC recognition sequence (NACRS), 81
NADH dehydrogenase (metabolism), 30
Naringenin-chalcone synthase, 132
National Center for Biotechnology Information (NCBI), 7
Necrotrophs, 2
Nepeta species, 149
Next-generation sequencing (NGS), 4, 54
Nicotiana rustica, 152
Nicotiana tabacum, 159
Nitrate-peptide transporter (NRT), 158
Nutritional losses, 297

O

Oases, 6
Ochratoxin A, 107
Octadecanoid-Responsive Catharanthus AP2/ERF-domain (ORCAs), 157
Oil palm, 28, 73–76, 78–92
abiotic stress, 74
affecting productivity, 73, 74
fruit development, 76
in plants, 72, 73
and symptoms, 75
future prospects, 92, 93
phytohormones as signaling molecules
ABA, 82
auxin, 84
ethylene, 82, 83
JA, 83
productivity, 72
ripe and unripe fruit bunches, 77
structure of transcription factors
AP2-ERF superfamily, 78, 79
families, 79–81
transcription factors in abiotic stress, 84, 91
Omics platforms, 29–34
biological replicates, 29
diurnal cycle, 37, 38
human population, 27
in leaf photosynthesis, 28
light-saturated, 28
mesocarp, 28
photosynthesis, 27, 28 (*see* Photosynthesis)
source-limited plant, 28
yield, 34–36
11-Oxo-β-amyrin, 153
Oxygen evolving enhancer protein 1 (OEE1), 40

P

Pacific Biosciences (PacBio), 15
Paddy
cleaning, 290, 291
drying, 290
at harvest, 288
palm oil, 71
rice, 289
Panax ginseng, 138
Papaver somniferum, 144
Parallel-gene expression analyses, 17
Pathogen-associated molecular patterns (PAMPs), 305
Pattern recognition receptors (PRRs), 305, 315, 316
PCR amplification, 110
P-DNA borders, 196
Pectobacterium atrosepticum, 108
Pectolytic enzymes, 320
Peptaibols, 320
Perfusion bioreactors, 181
PGPF. *See* Plant growth promoting fungi (PGPF)
PGPR. *See* Plant growth promoting rhizobacteria (PGPR)
Phaseolus vulgaris, 55
Phenolics, 137, 143, 144
Phenylalanine ammonia-lyase (PAL), 120, 322
Phenylpropanoid pathway, 13
Phlorisobutyrophenone, 155
Phosphate starvation, 12
Photosynthesis, 139
cultivated crops, 29
genes and metabolites, 36
genomics, 29
metabolomics, 33
plant phenomics, 33
proteomics, 30, 33
transcriptomics, 29, 30
Photosystem I (PSI), 32
Photosystem II (PSII), 32
Phyllodineae, 117
Phytohormone synthesis, 81, 132
ABA, 82
auxin, 84
ethylene, 82, 83
JA, 83

Pilocarpus jaborandis, 138
Piriformospora, 3
PIs. *See* Protease inhibitors (PIs)
Plant defence elicitors, 306, 307, 318
Plant disease, 359, 360, 362–371
  blights, 360, 361
  cankers, 361
  defense mechanism, 359
  ethylene activated signalling pathway, 369
  food and nutritional security, 359
  green planet, 359
  molecular tools
    disease development, 362
    DNA macroarray, 366
    enzyme and antibody assays, 367–368
    qPCR analysis, 363–365
    RNA sequencing, 366, 367
    scoring tools, 362, 363
    targeted genes, 365
  pathogen infection
    abscisic acid, 370
    candidate defense response genes, 359, 360
    gene expression, 368, 369
    jasmonic acid (JA), 370
    oxidative stress, 371
    transcription factors, 369, 370
    wounding, 370–371
  rots, 361
  rusts, 361, 362
  wilts, 362
Plant growth promoting fungi (PGPF), 335, 336, 341–343, 346, 347
Plant growth promoting rhizobacteria (PGPR), 336–339
  Colletotrichum orbiculare, 336
  disease suppressive soil, 347, 348
  ISR (*see* Induced systemic resistance (ISR))
  microbial population, 335
  microorganisms, 336
  PGPF, 335, 336
  plant roots, 335
  SAR (*see* Systemic acquired resistance (SAR))
  strains, 336
Plant growth regulators (PGRs), 149
Plant immunity, 314–317
Plant parasitic nematodes (PPN), 237
Plant phenomics, 33
Plant–fungus interactions, 14–17
  abiotic stresses, 2
  bacterial and fungal root symbioses, 2
  beneficial fungi, 2
  challenges
    data analysis, 15–17
    experimental design and replication, 14
    RNA preparation, 14–15
    sampling, 14–15
  environmental factors, 2
  fungal colonization, 3
  mycoparasitism, 3
  neutral, 4
  pathogenic, 3
  transcriptome, 4
Plasmopara viticola, 12
Podophyllum hexandrum, 138
Polishing
  facial improvement, 295
  structure, 295
Polymerase chain reaction (PCR), 49, 105, 124
Post-transcriptional gene silencing (PTGS), 217–227, 229, 230
Principal component plot (PCA), 14
Protease inhibitors (PIs)
  agriculture, 253
  anti-pathogenic properties, 254, 262
  anti-trypsin activity, 256
  chymotrypsin, pepsin and trypsin, 256
  classification, 257, 258
  defence mechanism, 257–259
  enzymes, inhibition effect, 256
  food preservation industry, 264
  genetic engineering, phytoprotection, 262, 263
  modes, 261
  overproduction, 259–261
  pest insects and pathogens, 254
  phytoprotection, 254
  plant pathogenesis, 254–256
  resistance development, 264
  turmeric, 265, 266
Protein-coding genes, 29
Proteomics, 30, 33
Pseudo-cereal, 58–60
Psophocarpus tetragonolobus, 48
PTGS. *See* Post-transcriptional gene silencing (PTGS)
Purine permease (PUP), 158
Putrescine N-methyl transferase (PMT), 152

Q

Quantitative real-time PCR (qPCR), 363–365

R

Ralstonia solanacearum, 101
Raman spectroscopy, 110
Random amplified polymorphic DNA (RAPD), 51, 52
Reactive oxygen species (ROS), 2, 73, 157
Recombinant inbred line (RIL), 60
Resistance (R) genes, 3
Restriction fragment length polymorphisms (RFLPs), 50, 51
Rhizobium, 117
Ribulose-1,5-biphosphate carboxylase/ oxygenase (RuBisCO), 28, 37
Rice grain, 288–290
Rice milling
  double losses, 298
  food scarcity, 287
  quality, 288
  staple foods, 287
  whitening and polishing, 298
Rice specification, 298, 300
RNA blot analysis, 18
RNA interference (RNAi), 196, 217, 218
RNA silencing, dsRNA, 232–243
  Agrobacterium, 229–231
  bacteria, 231–232
  crop protection (*see* Crop protection)
  development, 243
  gene expression, 217
  gene silencing, 229
  genetic engineering, 218
  Petunia hybrida flowers, 218
  phenotypes, crops, 234, 235
  PTGS and TGS, 218, 219, 229, 230
  quality traits, crop improvement
    amylopectin and amylose polysaccharides, 239
    caffeine, 240
    cotton seeds, 241
    flowers and fruits, 241–243
    genetic modification, 238
    onions, allinase, 240
    phytonutrients, 240
    plant seed oils, 239
    rice, 239
    shelf life, 238
    tobacco, 241
    triacylglycerol (TAG), 239
  RNAse III enzymes, 219
  siRNAs and miRNAs, 218, 219
  stability, 243
  stable transformation, 220–225
  TEV infection, 218
  tobacco ringspot virus, 217
  transient transformation, 226–228
  VIGS, 231
RNA-sequencing (RNA-seq) analysis
  DNA microarray, 4
  illumina platform, 5
  mutualism relationship, 8–12
  NGS sequencing, 5
  plant–microbe interactions, 5
  principles, 5–8
  quality, 6
  reproducibility, 6
  verification, 17–18
Rosmarinic acid, 151
Rots, 361
Rusts, 361, 362

S

Saccharides, 317–319
Saccharin, 324
Salicylic acid (SA), 2
SAR. *See* Systemic acquired resistance (SAR)
Scopolamine, 145
Scopolia parviflora, 152
Seeds of discovery (SeeD), 55
Sequence tagged sites (STSs), 53
Serpentine, 139
Shikimic acid, 139
Signalling molecules, 320, 322
Silicon (Si), 323
Simple sequence repeats (SSRs), 48, 53, 54, 124
Single feature polymorphisms (SFPs), 63
Single nucleotide polymorphisms (SNPs), 35, 48, 54, 55, 124
SOAPdenovo-Trans, 6
Solanaceae, 144
Spent coffee grounds (SCG), 283
Stacking Genes or Alleles, 208
Sterols, 142, 319, 320
Stevia rebaudiana, 148
Stilbene synthase (STS), 144
Strictosidine, 139
Surface plasmon resonance (SPR), 110
SwissProt database, 7
Systemic acquired resistance (SAR), 306
  catalase and ascorbate peroxidase, 338
  lignification and callous deposition, 337
  microbes, 341
  NPR1, 342
  physical and chemical responses, 337
  plant immunity, 349
  signaling, 337–339

T

Tanshinone, 139, 152
Terpenes, 137, 142
Terpenoid biosynthesis, 13
Terpenoid indole alkaloids (TIAs), 139, 156
Tetrahydrocannabinoids, 143
TGS. *See* Transcriptional gene silencing (TGS)
Theobroma cacao, 150
Tissue culture, 145–147
Tobacco etch virus (TEV), 218, 233
Tomato bushy stunt virus, 101
TopHat, 6
TopHat2, 16
Trans-ABySS, 6
Transcription factors (TFs), 72, 78, 79, 82–92
  abiotic stress, 84
    AP2-ERF superfamily, 78, 79
    CBF, 87
    DREB1, 86, 87
    EABF and EABF1, 85
  AP2/ERF family, 78
  families of, 79–81
  oil palm fruits
    AP2, 89, 90
    CBF, 90–92
    MADS-box, 89
    overexpressing, 91
    regulation of fruit ripening, 88, 89
    tomato, 87, 88
  phytohormones, 81
    ABA, 82
    auxin, 84
    ethylene, 82, 83
    JA, 83
Transcriptional gene silencing (TGS), 219, 221, 222, 224, 229, 230
Transcriptome analysis, 2, 4, 31, 32
Transcriptome sequencing, 5, 121–123
Transcriptomics, 29, 30
Trickle-bed bioreactors, 185
Tryptophan decarboxylase (*Tdc*), 139
Turmeric, 265, 266

U

Underutilised crops, 56, 57
  application of knowledge, 63
  AVRDC, 61
  D. alata, 62
  Dioscorea species, 62
  DNA-based markers, 58, 59
  genomics resources, 63
  leguminosae, 60, 61
  MAS breeding programmes, 62
  pseudo-cereal, 58–60
  research, 48
  shortage and heat stress, 47
  yam (*Dioscorea* spp.), 62
Unhusked paddies, 292
Unigenes, 6
United Nations' Food and Agriculture Organization, 287

V

Verticillium dahlia, 14
Virus-induced gene silencing (VIGS), 226–228, 231
Vitaceae, 143
Vitis vinifera, 159

W

Western blot, 108
Whitening machine, 293, 294
Wilts, 362
Witches' broom disease (WBD), 13

X

Xylanases, 319

Y

Yield improvement, 34–36
Yield of finished rice, 301

MIX
Papier aus verantwortungsvollen Quellen
Paper from responsible sources
FSC® C105338

If you have any concerns about our products,
you can contact us on
ProductSafety@springernature.com

In case Publisher is established outside the EU,
the EU authorized representative is:
Springer Nature Customer Service Center GmbH
Europaplatz 3, 69115 Heidelberg, Germany

Printed by Libri Plureos GmbH
in Hamburg, Germany